Nanotechnology in Wound Healing

This book comprehensively reviews the role of engineered nanomaterials in wound healing. It presents green synthesized nanomaterials for wound healing and cell growth stimulation and explores the advanced application of nanocellulose. The book highlights innovative approaches like smart 3D printed hydrogels and drug-loaded nanomaterials, as well as natural bioactive compounds encapsulated within nanomaterials. It also covers advancements in nanomaterial-based drug delivery systems and the integration of artificial intelligence in wound healing. Further chapters discuss the role of the immune system and the extracellular matrix in wound healing, strategies to modulate immune responses, and recent advancements in tissue engineering. The book addresses the mechanisms, limitations, and future perspectives of nanomaterials in wound healing, and emphasizes the importance of toxicity studies.

Nanotechnology in Wound Healing

Innovations, Applications, and Future Perspectives

Edited by
Mohammed Gulzar Ahmed, Sumel Ashique,
Arshad Farid, and Gokhan Zengin

CRC Press
Taylor & Francis Group
Boca Raton London New York

CRC Press is an imprint of the
Taylor & Francis Group, an **informa** business

Designed cover image: Shutterstock

First edition published 2025
by CRC Press
2385 NW Executive Center Drive, Suite 320, Boca Raton FL 33431

and by CRC Press
4 Park Square, Milton Park, Abingdon, Oxon, OX14 4RN

CRC Press is an imprint of Taylor & Francis Group, LLC

ISBN: 978-1-032-99773-5 (hbk)
ISBN: 978-1-032-99774-2 (pbk)
ISBN: 978-1-003-60596-6 (ebk)

DOI: 10.1201/9781003605966

Typeset in Times
by Apex CoVantage, LLC

Contents

About the Editors

Dr. Mohammed Gulzar Ahmed serves as the principal and dean at the Yenepoya Pharmacy College & Research Centre, part of Yenepoya (deemed to be a university), located in Mangalore, Karnataka, India. With a career that spans over 22 years in teaching and an additional 3 years dedicated to research, he has made significant contributions to the field. His scholarly output includes more than 150 publications in internationally and nationally accredited journals. His expertise encompasses novel drug delivery systems, nanotechnology, targeted drug delivery, and stability studies. Dr. Ahmed has secured seven patents and has authored three book chapters and four books. He has also obtained a DST-SERB CRG grant, along with several seed grants from Yenepoya and RGUHS universities. He has been recognized by the university with awards for best researcher, best paper, and good teacher.

Dr. Sumel Ashique currently holds the position of assistant professor in Bharat Technology, West Bengal, India. He brings four years of teaching experience to his role. His academic contributions include over 180 publications in both international and national journals of repute. His areas of expertise are in drug delivery, nanotechnology, and targeted treatment strategies. He has been awarded four patents from IP Australia and has published six book chapters in international books. Additionally, he has published 20 book chapters and edited 20 books with international publishers.

Dr. Arshad Farid has been serving as the Director at the Gomal Center of Biochemistry and Biotechnology, Gomal University, D.I. Khan, Pakistan, since April 2014. With approximately 16 years of combined teaching and research experience, he has published more than 90 research articles in journals of international and national repute. His research is markedly interdisciplinary, covering areas such as medicinal plants, natural product chemistry, herbal drug development, ethnopharmacology, nanotechnology, drug delivery, and microbiology. He has contributed more than 20 book chapters and holds 15 international patents from countries including Germany, the UK, and South Africa. Additionally, he has presented over 15 abstracts at national and international conferences. Dr. Farid actively contributes to the scientific community through peer review of papers for more than 20 respected journals. He was a fellow of the British Council-Charles Wallace Fellowship program at the University of Aberdeen, Scotland, UK, in 2023, and previously received a fellowship in 2007 for training in environmental, agricultural, and analytical chemistry laboratory techniques at the Department of Chemistry, University of Glasgow, Scotland, UK. He is also a research fellow at INTI International University, Malaysia, from June 1, 2023, to December 31, 2025.

Dr. Gokhan Zengin is an alumnus of Selcuk University's Faculty of Science in Turkey, where he also earned his Ph.D. from the Natural and Applied Sciences Institute's Department of Biology. Presently, he holds a professorship at Selcuk University within the same faculty and department. His research focuses on biological activity, chromatographic techniques, and enzyme inhibition. Dr. Zengin has contributed 800 publications to his field of research, amassed over 18,000 citations, and delivered numerous presentations at scientific congresses.

1 Engineered Nanomaterials for Wound Healing

Popat Mohite,[1] *Abhijeet Puri,*[1] *Shubham Munde,*[1] *Nitin Ade,*[1] *Sonali Verma,*[1] *and Sudarshan Singh*[2,3*]

[1] AET's St. John Institute of Pharmacy and Research, Palghar, 401 404, India

[2] Faculty of Pharmacy, Chiang Mai University, Chiang Mai, 50200, Thailand

[3] Office of Research Administration, Chiang Mai University, Chiang Mai, 50200, Thailand

* Corresponding author: Sudarshan Singh

1.1 INTRODUCTION

Wound healing (WH) is a complex physiological cascade comprised of several synchronized phases including hemostasis, inflammation, proliferation, and remodeling. For the integrity and function of the skin to be restored, each stage is essential. Yet a variety of things may hinder this process including infection, long-term illness, and severe scarring, which leads to chronic wound conditions [1]. Chronic wound conditions such as diabetic foot ulcers, pressure ulcers, and venous leg ulcers provide healthcare difficulties due to their slow healing and susceptibility to infection. To successfully address these problems, traditional wound treatment methods such as bandages and antiseptics often prove inadequate, underscoring the urgent need for more sophisticated therapeutic approaches [2, 3]. Engineered nanomaterials have become a hot topic in medical science recently, with creative applications ranging from regenerative medicine to targeted drug delivery. These materials' nanoscale size enables greater interaction with biological systems at the molecular level, resulting in improved integration and functionality [4]. These materials can be customized to fit individual clinical needs, providing new opportunities for improving wound healing results. They are distinguished by their nanoscale size and distinct physicochemical features [5]. High surface area-to-volume ratios, improved bioactivity, and the capacity to be functionalized with therapeutic substances are just a few of the exceptional qualities of engineered nanomaterials, which also include nanoparticles, nanofibers, and nanotubes. Because of these characteristics, they are especially well-suited for use in wound care, where they can support tissue regeneration, offer antibacterial protection, and enable controlled drug release. For example, silver nanoparticles are well-known for their antibacterial characteristics, which make them suitable for wound infection prevention. Similarly, gold nanoparticles can stimulate cellular function including proliferation and differentiation, which are required for tissue repair [6]. The application of engineered nanomaterials in WH leverages multiple mechanisms to improve the healing process. These materials possess antimicrobial properties that prevent infections, a frequent barrier to effective healing. Furthermore, they aid in promoting cell growth and generation of new blood vessels and enable the controlled release of drug candidates directly at the wound site, thus enhancing the overall healing efficiency. The ability of nanomaterials to modulate the wound microenvironment by reducing inflammation and promoting a conducive environment for tissue regeneration is a significant advantage over traditional wound care methods [7]. The

DOI: 10.1201/9781003605966-1

review aims to explore the role of engineered nanomaterials in WH [8], examining the underlying mechanisms that contribute to their effectiveness, recent developments in the field, and their clinical applications. By delving into current research and development, we seek to provide a comprehensive overview of how these innovative materials are shaping the future of wound care.

1.2 PATHOGENESIS OF WOUND

The pathogenesis of wound healing involves the intricate and dynamic processes that occur from the moment of tissue injury due to external stimuli until complete healing is achieved. Majorly, the dermal wounds are colonized by aerobic and anaerobic pathogens. The role of microbes in wound infections is a serious concern since pathogens can impede the healing process, harm tissue, and result in serious consequences. Various microorganisms, such as fungi, viruses, and bacteria, are frequently linked to wound infections [9, 10] (**Table. 1.1**). The complex process of wound pathogenesis is broken down into a series of events: hemostasis, inflammation, proliferation, and remodeling. Each stage involves cellular and molecular activities aimed at restoring tissue integrity and function [11] (**Figure 1.1**).

TABLE 1.1
List of Potential Wound Pathogens Associated With Wound Progression

Pathogen	Species
Gram-positive bacteria	*Staphylococcus aureus*
	Methicillin-resistant
	Coagulase-negative Staphylococci
Gram-negative bacteria	*Escherichia coli*
	Pseudomonas aeruginosa
	Klebsiella pneumoniae
Fungi	*Candida spp.*
	Aspergillus spp.
	Mucor spp.
Viruses	*Varicella-zoster virus*
	Herpes simplex virus
	Cytomegalovirus

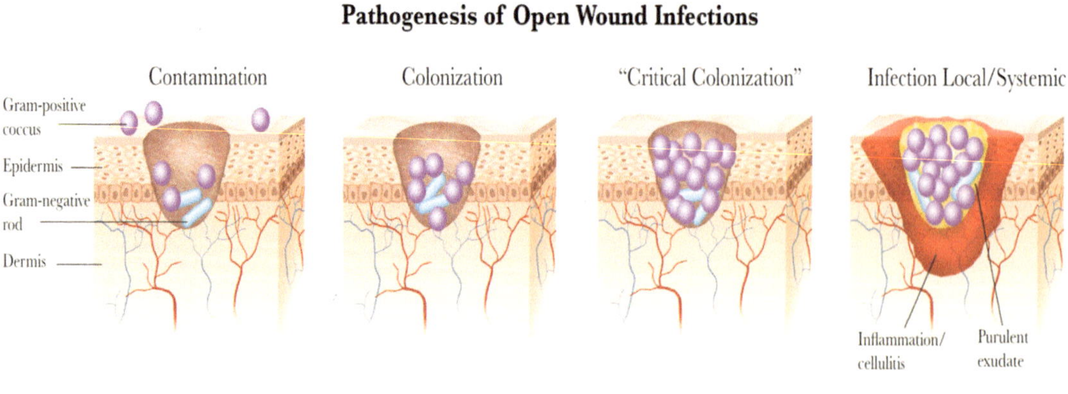

FIGURE 1.1 Pathogenesis of wound on skin surface. Reprinted from [18] under Creative Commons CC-BY-NC-ND.

1.2.1 Hemostasis

Hemostasis is the first stage of pathogenesis; it is the body's immediate response to injury to prevent excessive blood loss. Upon injury, blood vessels constrict, which results in reduced blood flow, also known as vasoconstriction. This event is referred to as the initiation phase. The second phase is platelet activation, where platelets attach to the exposed collagen in the injured endothelium. Platelets become activated and release granules containing cytokines, growth factors, and clotting factors. This aggregation of platelets creates a transient "platelet plug." The third phase is the coagulation cascade, in which intrinsic and extrinsic pathways are activated, leading to the conversion of prothrombin to thrombin, which converts fibrinogen to fibrin, forming the clot [12, 13].

1.2.2 Inflammation

Following hemostasis, blood vessels dilate, increasing blood flow and permeability to allow immune cells and nutrients, along with growth factors, to access the wound site. This phase is known as vasodilation. The second phase is immune cell infiltration. The first to arrive to phagocytose debris, germs, and foreign particles are neutrophils, which arrive in a matter of minutes. Proteolytic enzymes and reactive oxygen species are released by them, which can both further clean the wound and inflict more tissue damage. The third phase is macrophage activation, where macrophages replace neutrophils and play a critical role in wound healing, continuing phagocytosis and releasing cytokines that stimulate the next phase of healing [14].

1.2.3 Proliferation

Fibroblasts elevate and migrate into the wound, playing a crucial role in the production of extracellular matrix (ECM). They produce collagen and other ECM components that give growing tissue its structural support. Angiogenesis is the second phase of proliferation, where new blood vessels form through angiogenesis, driven by factors such as vascular endothelial growth factor. This neovascularization ensures an adequate supply of oxygen and nutrients, essential for tissue repair. The third phase is epithelialization, where keratinocytes at the wound edges proliferate and migrate across the wound bed, forming a new epithelial layer. This process re-establishes the skin's barrier function. The last phase is granulation tissue formation, where the combination of new capillaries, fibroblasts, and ECM forms granulation tissue, which fills the wound bed. Granulation tissue is typically rich in blood vessels, giving it a red and granular appearance [15].

1.2.4 Remodeling

Collagen remodeling occurs when the initially deposited type III collagen is replaced by the stronger type I collagen. This process involves the activity of matrix metalloproteinases and their inhibitors, which regulate the breakdown and deposition of ECM components. The second phase is wound contraction, where myofibroblasts, which have contractile properties, help in wound contraction by pulling the edges of the wound closer together, reducing the wound size. Scar formation is the last stage, where the vascularity of the scar decreases, and the ECM becomes more organized. The scar tissue formed is usually less elastic and less functional than the original tissue but provides necessary structural integrity [8, 16, 17].

1.3 PROGRESSION OF ACUTE TO CHRONIC WOUNDS

Progression from acute to chronic wound involves a complex interplay of factors that disrupt the normal healing process [19]. This condition is characterized by prolonged inflammation, impaired tissue regeneration, and failure to achieve the typical wound healing process [20] (**Figure 1.2**).

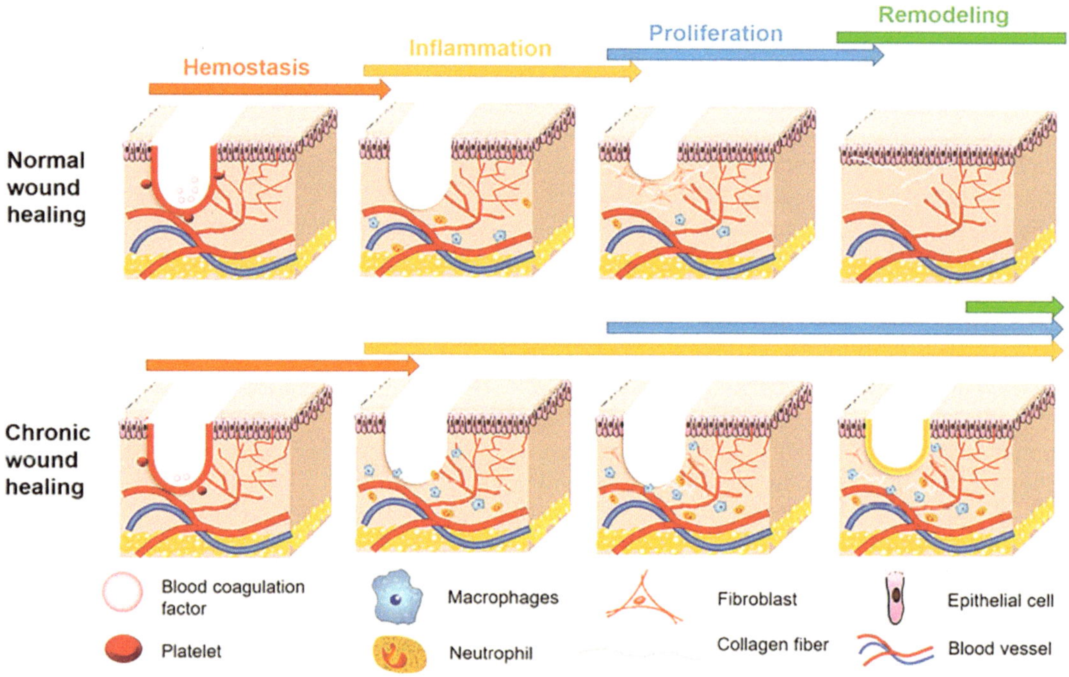

FIGURE 1.2 Comparison of normal wound and chronic wound healing process. Reprinted under Creative Commons Attribution (CC-BY) [23].

Prolonged inflammation is part of the progression of wounds where inflammation persists for weeks or months due to contributing factors such as persistent infections and impaired immune response. In the case of chronic wounds, there is often an imbalance between the enzymes and protein breakdown and their inhibitors [21]. Another cause of wound progression is biofilm formation. Biofilms contribute to chronicity by protecting bacteria from antibiotics and immune cells, allowing persistent infections [22]. Systemic factors also contribute to wound progression. High blood sugar levels impair immune function, reduce blood flow, and promote neuropathy, which leads to diabetic wounds. Vascular insufficiency, conditions such as venous insufficiency and peripheral arterial disease, reduce blood flow to the wound, depriving tissues of oxygen and nutrients [23].

1.4 CONVENTIONAL WOUND MANAGEMENT STRATEGIES

Conventional wound management strategies have proven effective in addressing both acute and chronic wound conditions. These encompass a range of methodologies, including the utilization of amniotic membrane application [24], hydrogel-based dressings, and the application of topical negative pressure [25]. Conventional wound care approaches concentrate on establishing the best possible environment to support the body's natural healing mechanisms while lowering the possibility of problems like infection. The main constituents of these techniques include wound cleaning, debridement, moisture balancing, infection management, and the application of suitable dressings and adjunct therapies. Within the context of burn wound management, where bacterial colonization poses a significant threat, conventional interventions such as hydrotherapy, debridement, and the application of topical antimicrobial agents play a pivotal role in mitigating infection risks. However, the practical implementation of cultured autografts and growth factors in burn wound treatment remains an ongoing area of exploration [26]. Infection control is critical to the management of wounds. Keeping a sterile atmosphere while changing dressings and applying topical antiseptics or antibiotics as needed are examples of preventive actions. When treating more serious or systemic infections, systemic antibiotics can be necessary. It's also critical to recognize and treat underlying illnesses like diabetes or immunological deficits that may make a person more susceptible to infection. Recent clinical trials have underscored the effectiveness of modern wound dressings and skin substitutes as a practical and economically viable approach to managing chronic wounds [27]. Adjunct therapies, including the use of growth factors, bioengineered skin substitutes, and negative pressure wound therapy, can enhance the healing process, especially in chronic or complex wounds. Growth factors, such as platelet-derived growth factor, can stimulate cellular activities critical for wound repair. Bioengineered skin substitutes provide a scaffold that supports cell proliferation and differentiation. Negative pressure wound therapy, also known as vacuum-assisted closure, involves the application of subatmospheric pressure to the wound through a sealed dressing, which helps remove exudate, reduce edema, and promote granulation tissue formation. Despite these advancements, it is pertinent to acknowledge that existing wound management strategies can be resource-intensive, both in terms of time and labor. The advancement of multidisciplinary methodologies within the field of wound care is poised to alleviate these challenges, thereby fostering improved clinical outcomes and enhancing the overall quality of life for patients [25, 28, 29].

1.5 ENGINEERED NANOMATERIALS FOR WOUND HEALING

Nanotechnology is a multidisciplinary field that involves the manipulation, study, and exploitation of structures and systems at the nanoscale, typically less than 100 nanometers [30]. It focuses on understanding and controlling matter at this scale, leading to the development of novel materials, devices, and products with unique properties. This technology has several applications, including in materials and manufacturing, medicine, energy, biotechnology, and national security [31]. Engineered nanomaterials for wound healing are specialized materials crafted at the nanoscale to facilitate the wound healing process. These materials are designed with the specific intention of aiding in the healing of wounds. They are engineered to possess distinct properties that enhance their effectiveness in treating acute to chronic wounds. Their formulation enables precise delivery of therapeutic agents to the wound site [32, 33].

1.6 CLASSIFICATION OF NANOTECHNOLOGY

Nanotechnology is classified into various categories according to their nature, use of materials, and applications (**Figures 1.3 and 1.4**).

The major type of nanoparticulate systems used for treatment of wounds is nanoparticles, their various types, nanocomposites, and scaffolds. Nanomaterials such as nanospheres nanocapsules, and nanoemulsions are mainly used for tissue regeneration [34]. Nanoparticulate systems used for wound therapy involve two types of systems: one that possesses intrinsic characteristics and the other with delivery vectors for therapeutic agents. The former is divided into metal and their oxide nanomaterials. Nanotechnology has brought about significant advancements in wound healing, offering a wide range of engineered nanomaterials that hold promise for treating both minor and severe wounds [35]. These specially designed nanomaterials can be used to deliver therapeutic agents that speed up the healing process of damaged skin tissues [36]. They come in various forms, as seen above, such as metallic and polymeric nanomaterials, each with unique features ideal for promoting WH and preventing infections [37].

1.7 NANOTECHNOLOGY-BASED METALLICS

Metallic nanoparticles are extremely small particles made from metals or metal alloys. They have unique physical, chemical, and electronic properties due to tiny size, larger surface area, and quantum effects [38]. This sets them apart from larger metallic particles. Metallic nanoparticles,

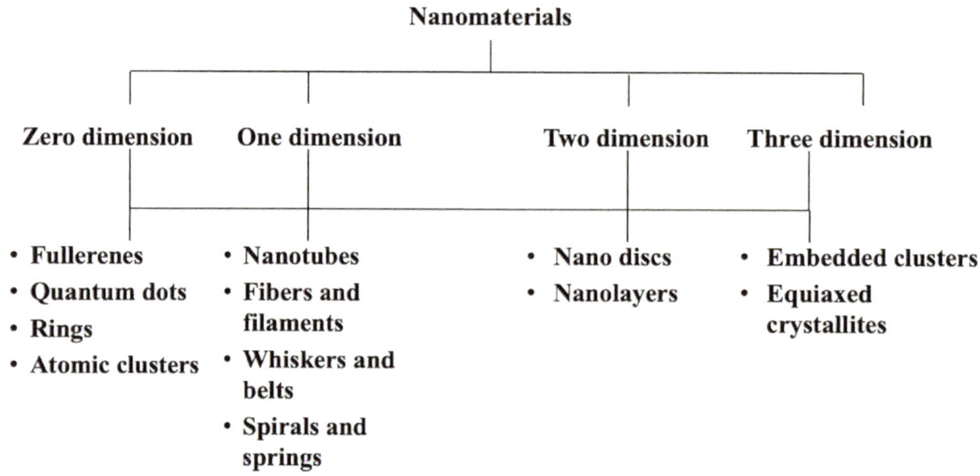

FIGURE 1.3 Classification of nanomaterials based on their nature.

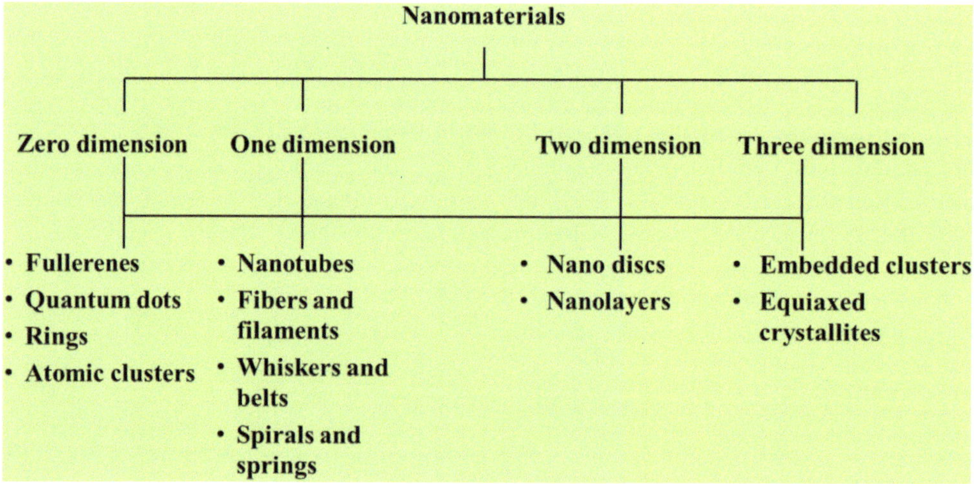

FIGURE 1.4 Classification based on dimensions of nanoparticles.

or MNPs, typically measure between 1 and 100 nanometers in diameter. Because of their small scale, these nanoparticles exhibit behaviors such as high reactivity, unusual optical properties, and enhanced catalytic activity [39]. Talking about organic nanoparticles, organic nanoparticles refer to materials formed when organic compounds undergo a transformation at the nanoscale. Examples of these materials include liposomes, dendrimers, micelles, and ferritin. Nanocapsules, micelles, and liposomes, categorized as nontoxic biodegradable nanoparticles, possess interior voids and exhibit sensitivity to thermal degradation, electromagnetic radiation, and light. Dendrimers, distinguished by their surfaces adorned with multiple chain ends conducive to specific chemical reactions, are employed in molecular recognition, nanosensing, light harvesting, and optoelectrochemical systems [40]. Moreover, their three-dimensional structure features internal voids capable of accommodating extra molecules, making them promising for drug delivery. Inorganic nanoparticles, lacking carbon atoms, are typically classified into metal-based and metal oxide-based categories [41]. Metallic nanoparticles are synthesized via constructive and destructive processes. Metal oxide nanomaterials consist of positive metallic ions and negative oxygen ions, with examples including silicon dioxide, titanium oxide, zinc oxide, and aluminum oxide [42–46]. Additionally, semiconductor nanomaterials, ceramic nanomaterials, and lipid-based nanomaterials are also classified as inorganic nanoparticles. Carbon nanomaterials encompass a diverse array of carbon allotropes, including zero-dimensional structures like fullerenes and quantum dots, one-dimensional carbon nanotubes, two-dimensional graphene, and three-dimensional nanodiamonds and nano-horns [47]. These materials find widespread application across various industries owing to their remarkable physical and chemical characteristics. Additionally, carbon-based nanoparticles offer exceptional electrical conductivity, thermal stability, and mechanical strength, making them ideal for use in electronics, energy storage, catalysis, and biomedical applications. Furthermore, carbon nanomaterials exhibit potential for environmental remediation, sensing, and drug delivery [48]. Gold nanoparticles (AuNPs) are well-regarded for their stability and unique optical properties. They have shown remarkable promise in wound healing applications due to their anti-inflammatory, antibacterial, and antioxidant properties, as well as their ability to speed up the healing process [49, 50]. Research indicates that AuNPs can contribute to faster wound closure, reduced inflammation, smaller wound size, increased tissue regeneration, and collagen production. They can also promote hair follicle formation and angiogenesis and display antioxidant and antibacterial effects [51]. Employing strategies like acidic microenvironments or photosensitizer molecules in conjunction with AuNPs may enhance their wound healing effectiveness [52]. Furthermore, metallic nanoparticles, including those made of gold, serve as an effective platform for developing low-cost, targeted treatments for chronic wounds, with the ability to deliver drugs, combat infections, and accelerate healing. Overall, AuNPs are a promising area for advancing wound care therapies. Silver nanoparticles (AgNPs) are valued for their antimicrobial properties and are widely used in healthcare and various consumer products [53]. Their antibacterial effects are primarily due to their ability to block respiratory enzyme pathways, alter microbial DNA, and affect cell walls. The use of silver in WH is advantageous because it is effective for a wide range of bacteria, including those that are resistant to multiple antibiotics and those that form biofilms. Silver compounds such as silver nitrate and silver sulfadiazine are commonly used to treat chronic wounds and burns. However, silver-based nanomaterials can cause side effects, such as blue-gray discoloration of the skin with prolonged use, and there is a risk of developing silver-resistant bacteria [54]. Selenium nanoparticles (SeNPs) have shown great potential in wound healing. Studies suggest that SeNPs can enhance antibacterial effects, promote tissue regeneration, and exhibit antioxidant properties [55]. SeNPs have been used in various structures and nanosystems to combat bacterial infections, including drug-resistant strains like MRSA and *E. coli*. Additionally, SeNPs have been incorporated into composite hydrogels to offer anti-inflammatory, pro-proliferative, and pro-migratory benefits, aiding in the healing of infected wounds. The combination of SeNPs with other materials like chitosan, polyethylenimine, and red blood cell membranes has shown promising results in improving wound healing and tackling antibiotic resistance [56]. Copper nanoparticles play a role in all stages of wound healing because they affect multiple cells and influence various cytokines and growth factor pathways. Copper is an essential metal required in small amounts for numerous metabolic processes [57]. Zinc oxide nanoparticles also show effective application ion WH due to their biocompatibility and permeability in dermal tissues. In rat models, these nanoparticles have shown regenerative properties by encouraging keratinocyte migration, tissue granulation, collagen fiber deposition, and re-epithelialization [58]. Several techniques are used to create metallic nanoparticles, such as chemical reduction, physical methods, biological synthesis, and electrochemical processes. In terms of properties, MNPs can be characterized based on various factors:

Optical Properties: They can show localized surface plasmon resonance, which results in unique color and light absorption characteristics

Catalytic Activity: Their high surface area means more active sites for catalytic reactions.

Magnetic Properties: Certain metallic nanoparticles, such as those made of iron, can display unique magnetic behaviors, useful in medical imaging and data storage.

Electrical Conductivity: Due to their metallic composition, these nanoparticles often exhibit high conductivity.

Despite their potential, metallic nanoparticles can pose risks and challenges. Their small size and high reactivity can lead to toxicity, especially within biological systems. Another challenge with metallic nanoparticles is their tendency to clump together, which can alter their expected behavior and reduce their effectiveness. This instability may also impact their applications [59]. Given these risks, it's crucial to carefully assess the safety and environmental impact of metallic nanoparticles to ensure their safe and effective use. Ongoing research into the behavior, safety, and potential risks associated with these nanoparticles will be key to unlocking their full potential while minimizing adverse effects. With responsible use, metallic nanoparticles could play a significant role across a range of industries, from healthcare to environmental science [54].

1.8 ENGINEERED NANOMATERIALS IN MANAGEMENT OF WOUND

Nanomaterials offer diverse strategies, including accelerating wound closure and aiding healing progression. The antibacterial, antifungal, and anti-inflammatory properties of nanoparticles make them crucial in wound healing. For instance, nanofibers have proven highly effective in creating wound dressings that act as nanoscaffolds, facilitating tissue repair [60]. The integration of nanomaterials into dressings has facilitated the release of growth factors and bioactive agents, enhancing the healing process for chronic wounds [61]. Nanotechnology-based engineered polymeric nanomaterials have shown significant advancement in treating both acute and chronic wounds. These advanced wound dressings utilize nanofibers, nanocomposites, and polymeric nanoparticles to enhance WH processes, prevent infections, and provide sustained drug release. Polymeric nanofibers have emerged as effective scaffolds for skin regeneration, while metal nanoparticles exhibit intrinsic antibacterial properties, making them valuable components in wound dressings [62]. Additionally, nanocarriers encapsulating drugs, biomolecules, and growth factors offer innovative treatment modalities for chronic wounds, further highlighting the potential of nanotechnology in wound care. The integration of nanomaterials in wound dressings not only promotes healing but also reduces healthcare costs and improves patient outcomes, revolutionizing wound management. Metal nanoparticles like silver, gold, zinc, and copper have shown promising results in promoting wound healing and controlling infections. Researchers have developed nanotechnology-based therapeutic agents to address specific challenges encountered in chronic wound healing, providing hope for more effective treatments in the future [63]. Gold nanoparticles and chitosan together dramatically improved wound healing in diabetic mouse research by encouraging angiogenesis and hastening ulcer closure (**Figure 1.5**) [7].

Abhijeet et al. illustrated the remarkable antioxidant and antibacterial properties of selenium nanoparticles derived from the bark of *Terminalia arjuna*, along with their potential in combating cancer. The utilization of TA-SeNPs (*Terminalia arjuna* selenium nanoparticles) in biomedical gel formulations was also explored. TA-SeNPs were found to exhibit significant antioxidant characteristics, rendering them suitable for integration into antioxidant therapies aimed at addressing ailments associated with oxidative stress. Moreover, the antimicrobial efficacy of TA-SeNPs implies their prospective use in the creation of antimicrobial treatments for infections. Also, the cytotoxic impact of TA-SeNPs on cancer cells, particularly the MCF-7 breast cancer cell line, shows promise in cancer treatment. The integration of TA-SeNPs into a gel matrix preserved their antimicrobial attributes, indicating their potential application in wound healing and topical drug administration systems. The biocompatibility and environmentally friendly synthesis method of TA-SeNPs position them as favorable candidates for diverse biomedical purposes, encompassing drug delivery, tissue regeneration, and diagnostic imaging [64]. Julalak et al. delve into the production of silver nanoparticles

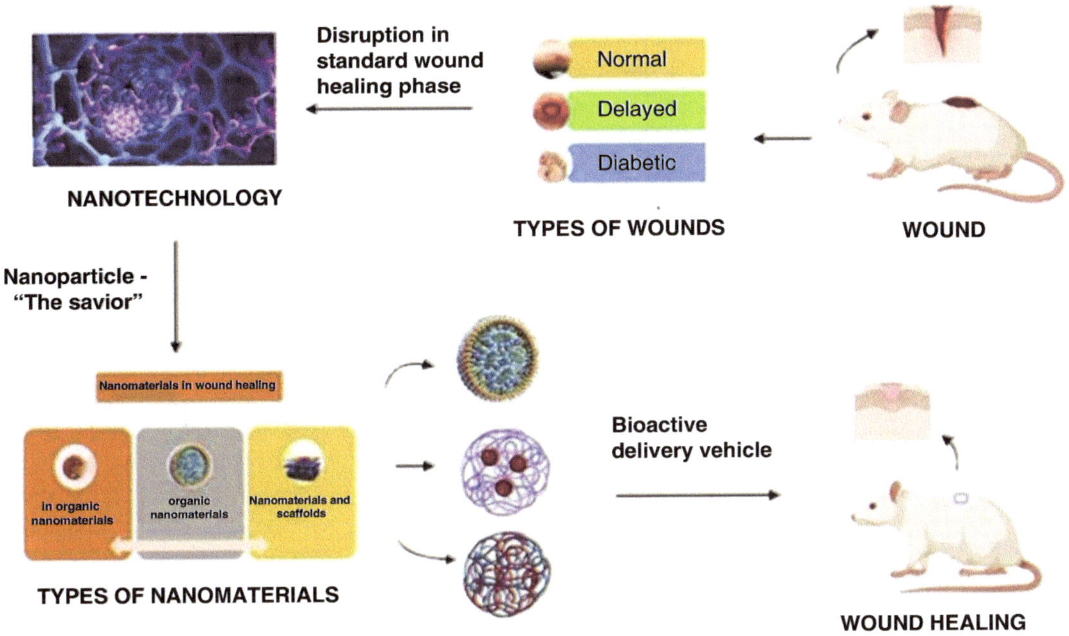

FIGURE 1.5 Role of nanotechnology for management of acute to chronic wounds. Reproduced from [7] under Creative Commons licenses.

through the combination of Rhodomyrtus tomentosa leaf extract and silk sericin, highlighting their antiseptic qualities when applied to topical gels. It underscores the environmentally friendly way stable silver nanoparticles can be rapidly produced: an approach fit for both pharmaceutical and cosmetic sectors due to the beneficial properties of silk sericin and R. tomentosa leaf extract that act as reducing agents. This study goes further to demonstrate how effective these silver nanoparticles are in wound care, showing high antimicrobial activity toward bacteria and fungi (common pathogens involved in wound infections), thus paving the way for alternative antiseptics with broader spectrum activity than available ones that only target surface cleanliness. There is potential for using these silver nanoparticles in hydrogels to accelerate wound closure, primarily because of their antimicrobial abilities. The growth of microorganisms that would otherwise be detrimental is effectively curtailed by these nanoparticles; this helps in two ways: preventing infections and ensuring that the area around the wound is clean. Antimicrobial activity can be sustained when stable and uniform silver nanoparticles are put into wound dressings or topical formulations. This also contributes to proper wound handling, since such components as infection risks are managed with ease. The synthesized silver nanoparticles have properties: antimicrobial, stability, and uniformity. These make them well-suited additives for wound care products, which play a significant role not only in expediting wound recovery but also in promoting healthy healing processes [65]. Sudarshan et al. developed biodegradable wound dressings that will use nanostructured materials for effective antibacterial as well as wound healing applications. The dressings exhibit high antibacterial properties along with a low inflammatory response and controlled release features, making them ideal candidates for treating infected wounds. With the application of green synthesized silver nanostructured materials in wound dressings, hydrophilic alternatives stand out as promising options: an alternative to traditional hydrophobic dressings. They present ease of hemostasis and good biocompatibility. The study focused on biodegradable wound dressings made of hypromellose, highlighting their excellent physicomechanical properties, controlled silver release, and broad-spectrum antimicrobial activity against wound pathogens and pathogenic fungi. When tested on healthy dogs, the hypromellose nanocomposites were found to significantly accelerate wound healing compared to the control group, indicating their effectiveness in promoting healing and controlling infection. The controlled-release properties of these dressings ensure a steady delivery of active ingredients, optimizing wound healing outcomes. These hybrid biodegradable dressings show promise for effectively treating infected wounds and reducing inflammatory responses. A key feature of the nanostructured wound dressing is its composition, which includes polyethylene with xanthan gum, hypromellose, or sodium carboxymethylcellulose, enhancing its effectiveness. The dressing exhibits a maximum absorption wavelength at420 nm and a zeta potential of approximately -35 mV, indicating stability and strong antimicrobial properties. It significantly inhibits the growth of major wound pathogens such as Acinetobacter baumannii, Escherichia coli, and Staphylococcus aureus, achieving over 99 % reduction in colony-forming units per milliliter. The hypromellose nanocomposite demonstrates excellent bactericidal activity, reducing bacterial growth by 99.9 % within an hour of treatment. Additionally, it shows inhibitory activity against free radical scavengers and reduces the inflammatory response of macrophages, underscoring its potential for wound healing applications [66]. Dwi et al. introduce a novel technology that enhances the antimicrobial properties of surgical nylon sutures by coating them with biogenic silver nanoparticles. This coating not only reduces surface roughness, crucial for preventing wound infections and promoting healing, but also provides potent bactericidal activity against various wound pathogens. The sutures maintain stable antimicrobial activity over time and are compatible with human red blood cells and fibroblasts, ensuring patient safety and reducing the risk of complications. The research demonstrates that the antimicrobial efficacy of the coated sutures lasts for at least 12 weeks, underscoring their long-term effectiveness in treating wound infections. Additionally, the coated sutures support wound healing and tissue regeneration without altering their mechanical properties, preserving their structural integrity. The antimicrobial coating significantly reduces bacterial load by more than 99.9 %, effective against both Gram-positive and Gram-negative pathogens. Compatibility tests with L929 fibroblasts revealed that the coated sutures promote cell viability and migration, enhancing the wound healing process. This study highlights the innovative use of silver nanoparticles on nylon monofilament sutures as a promising approach for treating wound infections with improved healing [67]. Dwi et.al demonstrated that coating surgical sutures with silver nanoparticles (AgNPs) synthesized from Eucalyptus camaldulensis extract significantly enhances their antibacterial properties and tensile strength while reducing surface roughness. This makes them highly effective against surgical site infections and managing wound infections. The AgNP-coated sutures exhibited strong bacteriostatic and bactericidal effects against common pathogens, including *Staphylococcus aureus*, *Acinetobacter baumannii*, *Escherichia coli*, *Klebsiella pneumoniae*, and *Pseudomonas aeruginosa*. Their antibacterial efficacy remained stable for up to 12 weeks, indicating long-lasting potential in preventing and managing surgical site infections. Moreover, despite the gradual release of silver, the coated sutures maintained biocompatibility with HaCaT keratinocyte cells, ensuring their safety for medical use. This stability in antibacterial properties over 12 weeks suggests that AgNP-coated sutures are a reliable strategy for managing bacterial infections associated with surgical wounds [68].

1.9 SCAFFOLD FORTIFIED WITH ENGINEERED NANOMATERIALS IN MANAGEMENT OF WOUND HEALING

Scaffolds serve an important role in tissue engineering and WH because they provide a structural framework that facilitates cell adhesion, proliferation, and differentiation. Scaffolds need to be porous and biocompatible for cells to adhere to and pass through them. These factors must be addressed while developing a scaffold for wound healing applications while maintaining high production and low costs. Nanomaterial-based scaffolds are built using numerous methods such as phase separation, self-assembly, and electrospinning. These techniques effectively produce polymeric nanofibers having pores, which stimulate fibroblasts in wounds and have structural and physical properties similar to the extracellular matrix. Nanofibrous scaffolds feature a high surface-to-volume ratio, which is predicted to enhance cell attachment. Several considerations must be taken into account while designing scaffolds for tissue engineering applications. Tissue engineering is a multidisciplinary field that combines engineering and life science principles to create biological substitutes that restore, maintain, and improve tissue function. Biomaterials play an important part in tissue engineering because they serve as three-dimensional matrices for proliferation, cellular growth, and new tissue synthesis [69–72]. To strengthen scaffolds, a variety of nanomaterials are used, each with its own set of advantages. Silver nanoparticles, which have significant capabilities, aid to prevent wound infections. Gold nanoparticles can promote cell proliferation and differentiation, whereas carbon nanotubes have high mechanical strength and electrical conductivity. Nanofibers, with their high surface area-to-volume ratio, can imitate the extracellular matrix, resulting in improved cell adhesion and tissue integration. These nanoparticles can also be functionalized with growth hormones, antibiotics, or other therapeutic substances to increase their effectiveness [73]. Fu et. al. showcased swift wound healing through the use of electrospun fibrous mats loaded with curcumin, made from poly(ε-caprolactone)-poly(ethylene glycol)-poly(ε-caprolactone). The electrospun scaffold shows significant antioxidant activity with lower cytotoxicity using *in vitro* studies, as well as enhanced WH using *in vivo* studies [74]. Shahveri et al. successfully employed silk fibroin and PLGA conjugate scaffolds to cultivate fibroblast cells for enhancing diabetic WH. The results from animal studies and histopathological evaluations supported the *in vitro* findings, demonstrating that the residual wound area treated with the scaffold is significantly smaller compared to the PLGA and control groups [75]. Another study using in vivo evaluation in Wistar rats revealed the potency of electrospun nanofiber membranes with AgNPs, showing decreased inflammation, enhanced wound healing, long-lasting antibacterial action, and low cytotoxicity. Despite the exciting promise of nanomaterial-enhanced scaffolds, several obstacles remain. Concerns about the long-term safety and potential toxicity of specific nanomaterials must be addressed. In conclusion, the fortification of scaffolds with nanomaterials represents a significant advancement in the management of wound healing. By enhancing the structural and functional properties of scaffolds, nanomaterials offer new possibilities for effective and accelerated wound care.

1.10 CLINICAL TRIALS REPORTED ON ENGINEERED NANOMATERIALS

Nanomaterials are emerging as a reliable approach in the management of wound healing due to their characteristic features such as antimicrobial activity, enhanced cellular interactions, and controlled drug delivery. Clinical studies of these nanomaterials have shown better therapeutic results in terms of efficacy and safety. Numerous studies have been reported to evaluate the use of nanomaterials in wound dressings. These revolutionary materials are being tested in clinical trials to determine their safety, effectiveness, and long-term benefits. Our objective is to create nanomaterial-based treatments that speed up healing and enhance patients' quality of life while also reducing complications. Ahia et al. conducted a randomized study on diabetic wound healing, comprising silver-based nanomaterials loaded into hydrogel wound dressing. The study involved sixty patients who had diabetes (type II) and diabetic foot wounds. The study's findings offered valuable insights on utilizing hydrogel/nano silver dressings for treating diabetic foot wounds [76]. Melamed et al. performed a study on copper oxide-integrated wound healing. The copper oxide incorporated microparticles were prepared and used as wound dressing. The goal of this study was to present early investigations on the efficacy of the dressing, especially in noninfected wounds. Wounds that did not respond well to traditional healing methods were treated with dressings infused with copper oxide. The results were analyzed by comparing them with previously reported studies on animals and in vitro models, which emphasized the role of copper in tissue regeneration and angiogenesis. The findings reveal that copper oxide-impregnated wound dressings have the potential for aiding WH, especially in challenging cases. Although it is clinically recommended to use an antimicrobial dressing, there is no unanimous opinion on the most effective way to apply these agents [77]. Miller et al. performed comparative studies to evaluate the antimicrobial potential of nanocrystalline silver and cadexomer iodine. For both antimicrobials, the number of cured wounds and the total healing rate were comparable. On the other hand, larger, older lesions with more exudate healed more quickly after the first two weeks of treatment when silver compounds were used. This study offers insightful data regarding the circumstances in which one product might be selected [78].

1.11 CHALLENGES OF ENGINEERED NANOMATERIALS

Incorporating engineered nanomaterials into wound healing treatments shows great potential, but also presents various obstacles. Biocompatibility and safety are among the main concerns. Extensive biocompatibility studies are necessary to ensure nanomaterials do not cause adverse immune responses, toxicity, or uncertain long-term effects [79]. Regulatory obstacles are also a significant challenge, as securing approval for these innovative materials requires thorough safety and effectiveness assessments, which can be lengthy and intricate. Scalability and manufacturing pose an additional challenge. It can be challenging to consistently produce nanomaterials on a large scale without compromising their quality and functionality. Inconsistent clinical outcomes can occur due to variability in production processes, which undermines the reliability of these therapies [80]. Furthermore, the high production and integration costs of nanomaterials into wound healing products could limit their accessibility and widespread adoption. Stability and controlled degradation of nanomaterials within the wound site are critical for their safe and effective use. Ensuring that nanomaterials remain stable and degrade in a controlled manner, avoiding prolonged exposure and potential side effects, is essential for successful clinical outcomes. Nanomaterials have great potential to advance wound healing therapies despite these obstacles. Due to their well-known antibacterial qualities, AgNPs have demonstrated promise in lowering infection rates and accelerating the healing of chronic wounds. AuNPs have exhibited anti-inflammatory and wound healing characteristics, augmenting the proliferation of fibroblasts and keratinocytes, which are crucial for tissue regeneration [6].

1.12 CONCLUSION AND FUTURE PROSPECTS

Engineered nanomaterials have shown significant potential in enhancing wound healing processes by promoting hemostasis, anti-inflammatory effects, and angiogenesis. The current landscape of nanomaterials in wound healing is marked by rapid advancements and promising applications through advancements in future-based products such as customized wound bandages, fortification in advanced drug delivery, improved biocompatibility, enhanced wound closure, improved antimicrobial activity, reduced scarring, and standardization and clinical trials. Therefore, by addressing the challenges associated with engineered nanomaterials and building upon current advancements, engineered nanomaterials have the potential to revolutionize wound healing treatments and improve patient outcomes.

In conclusion, the emergence of nanotechnology has revolutionized wound healing strategies by offering innovative solutions to overcome the limitations of conventional treatments. Engineered nanomaterials, including metallic nanoparticles like silver, gold, zinc, copper, and polymeric nanomaterials, have shown remarkable potential in managing acute and chronic wounds. By leveraging the unique properties of nanomaterials, such as enhanced antimicrobial activity, controlled drug delivery, and promotion of tissue regeneration, nanotechnology offers promising avenues for more efficient wound management. Moreover, developing a scaffold fortified with nanomaterials presents a promising approach for enhancing wound healing outcomes. Transdermal films incorporating metallic nanoparticles and engineered polymeric nanomaterials offer targeted delivery of therapeutic agents, ensuring sustained release and improved efficacy in wound healing. Incorporating nanotechnology into wound management strategies addresses the challenges associated with conventional treatments and opens up new possibilities for personalized and effective wound care. As research in this field continues to advance, it is evident that nanomaterial-based approaches hold great promise for revolutionizing the future of wound healing, offering hope for faster, more efficient, and scar-free recovery for patients worldwide.

REFERENCES

1. Guo S, Dipietro LA (2010) Factors Affecting Wound Healing. *Journal of Dental Research* 89 (3): 219–229. https://doi.org/10.1177/0022034509359125
2. Mallick S, Nag M, Lahiri D, Pandit S, Sarkar T, Pati S, Nirmal NP, Edinur HA, Kari ZA, Ahmad Mohd Zain MR, Ray RR (2022) Engineered Nanotechnology: An Effective Therapeutic Platform for the Chronic Cutaneous Wound. *Nanomaterials* 12 (5):778.
3. Kushwaha A, Goswami L, Kim BS (2022) Nanomaterial-Based Therapy for Wound Healing. *Nanomaterials* 12 (4):618.
4. Kumar A, Shah SR, Jayeoye TJ, Kumar A, Parihar A, Prajapati B, Singh S, Kapoor DU (2023) Biogenic Metallic Nanoparticles: Biomedical, Analytical, Food Preservation, and Applications In Other Consumable Products. *Frontiers in Nanotechnology* 5. https://doi.org/10.3389/fnano.2023.1175149
5. Patil S, Chandrasekaran R (2020) Biogenic Nanoparticles: A Comprehensive Perspective in Synthesis, Characterization, Application and Its Challenges. *Journal of Genetic Engineering and Biotechnology* 18 (1):67. https://doi.org/10.1186/s43141-020-00081-3
6. Sangnim T, Puri V, Dheer D, Venkatesh DN, Huanbutta K, Sharma A (2024) Nanomaterials in the Wound Healing Process: New Insights and Advancements. *Pharmaceutics* 16 (3):300.
7. Nandhini J, Karthikeyan E, Rajeshkumar S (2024) Nanomaterials for Wound Healing: Current Status and Futuristic Frontier. *Biomedical Technology* 6:26–45. https://doi.org/10.1016/j.bmt.2023.10.001
8. Wilkinson HN, Hardman MJ (2020) Wound Healing: Cellular Mechanisms and Pathological Outcomes. *Open Biology* 10 (9):200223. https://doi.org/10.1098/rsob.200223
9. Bowler PG, Duerden BI, Armstrong DG (2001) Wound Microbiology and Associated Approaches to Wound Management. *Clinical Microbiology Reviews* 14 (2):244–269. https://doi.org/10.1128/cmr.14.2.244-269.2001
10. Zhang P, Zou B, Liou Y-C, Huang C (2021) The Pathogenesis and Diagnosis of Sepsis Post Burn Injury. *Burns & Trauma* 9. https://doi.org/10.1093/burnst/tkaa047
11. Bowler PG (2002) Wound Pathophysiology, Infection and Therapeutic Options. *Annals of Medicine* 34 (6):419–427. https://doi.org/10.1080/078538902321012360
12. Chidrawar VR, Singh S, Jayeoye TJ, Dodiya R, Samee W, Chittasupho C (2023) Porous Swellable Hypromellose Composite Fortified with Eucalyptus camaldulensis Leaf Hydrophobic/Hydrophilic Phenolic-rich Extract to Mitigate Dermal Wound Infections. *Journal of Polymers and the Environment* 31 (9):3841–3856. https://doi.org/10.1007/s10924-023-02860-8
13. Singh S, Chidrawar VR, Hermawan D, Dodiya R, Samee W, Ontong JC, Ushir YV, Prajapati BG, Chittasupho C (2023) Hypromellose Highly Swellable Composite Fortified with Psidium Guajava Leaf Phenolic-rich Extract for Antioxidative, Antibacterial, Anti-inflammatory, Anti-melanogenesis, and Hemostasis Applications. *Journal of Polymers and the Environment* 31 (7):3197–3214. https://doi.org/10.1007/s10924-023-02819-9
14. Singh S, Chidrawar VR, Hermawan D, Nwabor OF, Olatunde OO, Jayeoye TJ, Samee W, Ontong JC, Chittasupho C (2023) Solvent-Assisted Dechlorophyllization of Psidium Guajava Leaf Extract: Effects on the Polyphenol Content, Cytocompatibility, Antibacterial, Anti-Inflammatory, and Anticancer Activities. *South African Journal of Botany* 158:166–179. https://doi.org/10.1016/j.sajb.2023.04.029
15. Syukri DM, Singh S, Nwabor OF, Ontong JC, Dejyong K, Sunghan J, Dejyong K, Lethongkam S, Voravuthikunchai SP (2024) Prevention of Post-operative Bacterial Colonization on Mice Buccal Mucosa Using Biogenic Silver Nanoparticles-Coated Nylon Sutures. *Regenerative Engineering and Translational Medicine*. https://doi.org/10.1007/s40883-024-00335-3
16. Huang J, Heng S, Zhang W, Liu Y, Xia T, Ji C, Zhang L-j (2022) Dermal Extracellular Matrix Molecules in Skin Development, Homeostasis, Wound Regeneration and Diseases. *Seminars in Cell & Developmental Biology* 128:137–144. https://doi.org/10.1016/j.semcdb.2022.02.027
17. Piipponen M, Li D, Landén NX (2020) The Immune Functions of Keratinocytes in Skin Wound Healing. *International Journal of Molecular Sciences* 21 (22):8790.
18. Lipsky BA, Silverman MH, Joseph WS (2016) A Proposed New Classification of Skin and Soft Tissue Infections Modeled on the Subset of Diabetic Foot Infection. Open Forum *Infectious Diseases* 4 (1). https://doi.org/10.1093/ofid/ofw255

19. Falanga V, Isseroff RR, Soulika AM, Romanelli M, Margolis D, Kapp S, Granick M, Harding K (2022) Chronic Wounds. *Nature Reviews Disease Primers* 8 (1):50. https://doi.org/10.1038/s41572-022-00377-3
20. Goldberg SR, Diegelmann RF (2020) What Makes Wounds Chronic. *Surgical Clinics of North America* 100 (4):681–693. https://doi.org/10.1016/j.suc.2020.05.001
21. Sachdeva C, Satyamoorthy K, Murali TS (2022) Microbial Interplay in Skin and Chronic Wounds. *Current Clinical Microbiology Reports* 9 (3):21–31. https://doi.org/10.1007/s40588-022-00180-4
22. Versey Z, da Cruz Nizer WS, Russell E, Zigic S, DeZeeuw KG, Marek JE, Overhage J, Cassol E (2021) Biofilm-Innate Immune Interface: Contribution to Chronic Wound Formation. *Frontiers in Immunology* 12. https://doi.org/10.3389/fimmu.2021.648554
23. Burgess JL, Wyant WA, Abdo Abujamra B, Kirsner RS, Jozic I (2021) Diabetic Wound-Healing Science. *Medicina* 57 (10):1072.
24. Ingraldi AL, Audet RG, Tabor AJ (2023) The Preparation and Clinical Efficacy of Amnion-Derived Membranes: A Review. *Journal of Functional Biomaterials* 14 (10). https://doi.org/10.3390/jfb14100531
25. Martin C, Low WL, Amin MCIM, Radecka I, Raj P, Kenward K (2013) Current Trends in the Development of Wound dressings, Biomaterials and Devices. *Pharmaceutical Patent Analyst* 2 (3): 341–359. https://doi.org/10.4155/ppa.13.18
26. Walter PH (1993) Burn Wound Management. *AACN Advanced Critical Care* 4 (2):378–387. https://doi.org/10.4037/15597768-1993-2015
27. Mirhaj M, Labbaf S, Tavakoli M, Seifalian AM (2022) Emerging Treatment Strategies in Wound Care. *International Wound Journal* 19 (7):1934–1954. https://doi.org/10.1111/iwj.13786
28. Kumar A, Mohite P, Munde S, Puri A, Edet HO, Ochoule DO, Chidrawar VR, Nagime PV, Uddin MJ, Singh S (2024) Revolutionizing the Biomedical and Environmental Clean-up Through Polymeric Nano-Sponges: A Review. *Nano-Structures & Nano-Objects* 39:101193. https://doi.org/10.1016/j.nanoso.2024.101193
29. Mohite P, Asane G, Rebello N, Munde S, Ade N, Boban T, Damiri F, Singh S (2024) Polymeric Hydrogel Sponges for Wound Healing Applications: A Comprehensive Review. *Regenerative Engineering and Translational Medicine*. https://doi.org/10.1007/s40883-024-00334-4
30. Ingle AP, Golińska P, Yadav A, Razzaghi-Abyaneh M, Patel M, Patel R, Plekhanova Y, Reshetilov A, Rai M (2020) Nanotechnology. In *Nanobiotechnology in Diagnosis, Drug Delivery, and Treatment* (pp. 1–24). https://doi.org/10.1002/9781119671732.ch1
31. Alberti T, S Coelho D, Voytena A, Pitz H, de Pra M, Mazzarino L, Kuhnen S, M Ribeiro-do-Valle R, Maraschin M, Veleirinho B (2017) Nanotechnology: A Promising Tool Towards Wound Healing. *Current Pharmaceutical Design* 23 (24):3515–3528.
32. Stern D, Cui H (2019) Crafting Polymeric and Peptidic Hydrogels for Improved Wound Healing. *Advanced Healthcare Materials* 8 (9):1900104.
33. Nagime PV, Singh S, Shaikh NM, Gomare KS, Chitme H, Abdel-Wahab BA, Alqahtany YS, Khateeb MM, Habeeb MS, Bakir MB (2023) Biogenic Fabrication of Silver Nanoparticles Using Calotropis Procera Flower Extract with Enhanced Biomimetics Attributes. *Materials* 16 (11):4058.
34. Niska K, Zielinska E, Radomski MW, Inkielewicz-Stepniak I (2018) Metal Nanoparticles in Dermatology and Cosmetology: Interactions with Human Skin Cells. *Chemico-Biological Interactions* 295:38–51. https://doi.org/10.1016/j.cbi.2017.06.018
35. Blanco-Fernandez B, Castano O, Mateos-Timoneda MÁ, Engel E, Pérez-Amodio S (2021) Nanotechnology Approaches in Chronic Wound Healing. *Advances in Wound Care* 10 (5):234–256.
36. Parani M, Lokhande G, Singh A, Gaharwar AK (2016) Engineered Nanomaterials for Infection Control and Healing Acute and Chronic Wounds. *ACS Applied Materials & Interfaces* 8 (16):10049–10069. https://doi.org/10.1021/acsami.6b00291
37. Mihai MM, Dima MB, Dima B, Holban AM (2019) Nanomaterials for Wound Healing and Infection Control. *Materials* 12 (13):2176.
38. Hayat M, Nawaz A, Chinnam S, Muzammal M, Latif MS, Yasin M, Ashique S, Zengin G, Farid A (2023) Formulation Development and Optimization of Herbo Synthetic Gel. In Vitro Biological Evaluation and in Vivo Wound Healing Studies. *Process Biochemistry* 130 (July 1):116–126.
39. Afzal O, Altamimi AS, Nadeem MS, Alzarea SI, Almalki WH, Tariq A, Mubeen B, Murtaza BN, Iftikhar S, Riaz N (2022) Nanoparticles in Drug Delivery: From History to Therapeutic Applications. *Nanomaterials* 12 (24):4494.
40. Pandey P, Dahiya M (2016) A Brief Review on Inorganic Nanoparticles. *Journal of Critical Reviews* 3 (3):18–26.

41. Kwiatkowska A, Drabik M, Lipko A, Grzeczkowicz A, Stachowiak R, Marszalik A, Granicka LH (2022) Composite Membrane Dressings System with Metallic Nanoparticles as an Antibacterial Factor in Wound Healing. *Membranes* 12 (2):215.
42. Nandhini SN, Sisubalan N, Vijayan A, Karthikeyan C, Gnanaraj M, Gideon DAM, Jebastin T, Varaprasad K, Sadiku R (2023) Recent Advances in Green Synthesized Nanoparticles for Bactericidal and Wound Healing Applications. *Heliyon* 9 (2).
43. Nwabor OF, Singh S, Paosen S, Vongkamjan K, Voravuthikunchai SP (2020) Enhancement of Food Shelf Life with Polyvinyl Alcohol-Chitosan Nanocomposite Films from Bioactive Eucalyptus Leaf Extracts. *Food Bioscience* 36:100609. https://doi.org/10.1016/j.fbio.2020.100609
44. Nwabor OF, Singh S, Wunnoo S, Lerwittayanon K, Voravuthikunchai SP (2021) Facile Deposition of Biogenic Silver Nanoparticles on Porous Alumina Discs, an Efficient Antimicrobial, Antibiofilm, and Antifouling Strategy for Functional Contact Surfaces. *Biofouling* 37 (5):538–554. https://doi.org/10.1080/08927014.2021.1934457
45. Nwabor OF, Singh S, Ontong JC, Vongkamjan K, Voravuthikunchai SP (2021) Valorization of Wastepaper Through Antimicrobial Functionalization with Biogenic Silver Nanoparticles, a Sustainable Packaging Composite. Waste and Biomass Valorization 12 (6):3287–3301. https://doi.org/10.1007/s12649-020-01237-5
46. Jayeoye TJ, Eze FN, Olatunde OO, Singh S, Zuo J, Olatunji OJ (2021) Multifarious Biological Applications and Toxic Hg2+ Sensing Potentiality of Biogenic Silver Nanoparticles Based on Securidaca Inappendiculata Hassk Stem Extract. *International Journal of Nanomedicine* 16:7557–7574. https://doi.org/10.2147/IJN.S325996
47. Hossain N, Mobarak MH, Mimona MA, Islam MA, Hossain A, Zohur FT, Chowdhury MA (2023) Advances and Significances of Nanoparticles in Semiconductor Applications–A Review. *Results in Engineering*:101347.
48. Mekuye B, Abera B (2023) Nanomaterials: An Overview of Synthesis, Classification, Characterization, and Applications. *Nano Select* 4 (8):486–501. https://doi.org/10.1002/nano.202300038
49. Jayeoye TJ, Nwude EF, Singh S, Prajapati BG, Kapoor DU, Muangsin N (2024) Sustainable Synthesis of Gold Nanoparticles for Drug Delivery and Cosmeceutical Applications: A Review. *BioNanoScience*. https://doi.org/10.1007/s12668-024-01436-7
50. Jayeoye TJ, Eze FN, Singh S, Olatunde OO, Benjakul S, Rujiralai T (2021) Synthesis of Gold Nanoparticles/Polyaniline Boronic Acid/Sodium Alginate Aqueous Nanocomposite Based on Chemical Oxidative Polymerization for Biological Applications. *International Journal of Biological Macromolecules* 179:196–205. https://doi.org/10.1016/j.ijbiomac.2021.02.199
51. Selmani A, Jeitler R, Auinger M, Tetyczka C, Banzer P, Kantor B, Leitinger G, Roblegg E (2023) Investigation of the Influence of Wound-Treatment-Relevant Buffer Systems on the Colloidal and Optical Properties of Gold Nanoparticles. *Nanomaterials* 13 (12):1878.
52. Chen W, Chu R, Li H, Hua T, Chen H, Li R, Zhou D, Cao S, Ye S, Li H (2023) A Novel Wound Dressing Based on a Gold Nanoparticle Self-Assembled Hydrogel to Promote Wound Healing. *Materials Advances* 4 (14):2918–2925. https://doi.org/10.1039/D3MA00130J
53. Nagarajan R (2008) *Nanoparticles: Building Blocks for Nanotechnology*. ACS Publications.
54. A Systematic Review of Metallic Nanoparticles: Synthesis, Biological Activities & Applications (2023). *Journal of Pharmaceutical Negative Results*:2525–2533. https://doi.org/10.47750/pnr.2023.14.S02.297
55. Ruan Q, Yuan L, Gao S, Ji X, Shao W, Ma J, Jiang D (2023). Development of ZnO/Selenium Nanoparticles Embedded Chitosan-Based Anti-Bacterial Wound Dressing for Potential Healing Ability and Nursing Care after Paediatric Fracture Surgery. *International Wound Journal* 20 (6):1819–1831. https://doi.org/10.1111/iwj.13947
56. Fang M, Zhang H, Wang Y, Zhang H, Zhang D, Xu P (2023). Biomimetic Selenium Nanosystems for Infectious Wound Healing. *Engineered Regeneration* 4 (2):152–160. https://doi.org/10.1016/j.engreg.2023.01.004
57. Mendes C, Thirupathi A, Corrêa MEAB, Gu Y, Silveira PCL (2022) The Use of Metallic Nanoparticles in Wound Healing: New Perspectives. *International Journal of Molecular Sciences* 23 (23):15376.
58. M G (2023) Biogenesis and Green Synthesis of Metal Nanoparticles and Their Pharmacological Applications. *Plant Science Today*. https://doi.org/10.14719/pst.2417
59. Cai YY, Choi YC, Kagan CR (2023) Chemical and Physical Properties of Photonic Noble-Metal Nanomaterials. *Advanced Materials* 35 (34):2108104
60. Sankar S, Kodiveri Muthukaliannan G (2024) Deciphering the Crosstalk Between Inflammation and Biofilm in Chronic Wound Healing: Phytocompounds Loaded Bionanomaterials as Therapeutics. *Saudi Journal of Biological Sciences* 31 (4):103963. https://doi.org/10.1016/j.sjbs.2024.103963

61. Prakashan D, Roberts A, Gandhi S. Recent Advancement of Nanotherapeutics in Accelerating Chronic Wound Healing Process for Surgical Wounds and Diabetic Ulcers. *Biotechnology and Genetic Engineering Reviews*:1–29. https://doi.org/10.1080/02648725.2023.2167432
62. Jiang Z, Zheng Z, Yu S, Gao Y, Ma J, Huang L, Yang L (2023) Nanofiber Scaffolds as Drug Delivery Systems Promoting Wound Healing. *Pharmaceutics* 15 (7):1829.
63. Vijayaram S, Razafindralambo H, Sun Y-Z, Vasantharaj S, Ghafarifarsani H, Hoseinifar SH, Raeeszadeh M (2024) Applications of Green Synthesized Metal Nanoparticles—A Review. *Biological Trace Element Research* 202 (1):360–386.
64. Puri A, Mohite P, Patil S, Chidrawar VR, Ushir YV, Dodiya R, Singh S (2023) Facile Green Synthesis and Characterization of Terminalia Arjuna Bark Phenolic–Selenium Nanogel: A Biocompatible and Green Nano-Biomaterial for Multifaceted Biological Applications. *Frontiers in Chemistry* 11. https://doi.org/10.3389/fchem.2023.1273360
65. Ontong JC, Singh S, Nwabor OF, Chusri S, Voravuthikunchai SP (2020) Potential of Antimicrobial Topical Gel with Synthesized Biogenic Silver Nanoparticle Using Rhodomyrtus Tomentosa Leaf Extract and Silk Sericin. *Biotechnology Letters* 42 (12):2653–2664. https://doi.org/10.1007/s10529-020-02971-5
66. Singh S, Nwabor OF, Sukri DM, Wunnoo S, Dumjun K, Lethongkam S, Kusolphat P, Hemtanon N, Klinprathum K, Sunghan J, Dejyong K, Lertwittayanon K, Pisuchpen S, Voravuthikunchai SP (2022) Poly (vinyl alcohol) Copolymerized with Xanthan Gum/Hypromellose/Sodium Carboxymethyl Cellulose Dermal Dressings Functionalized with Biogenic Nanostructured Materials for Antibacterial and Wound Healing Application. *International Journal of Biological Macromolecules* 216:235–250. https://doi.org/10.1016/j.ijbiomac.2022.06.172
67. Syukri DM, Nwabor OF, Singh S, Voravuthikunchai SP (2021) Antibacterial Functionalization of Nylon Monofilament Surgical Sutures Through in Situ Deposition of Biogenic Silver Nanoparticles. *Surface and Coatings Technology* 413:127090. https://doi.org/10.1016/j.surfcoat.2021.127090
68. Syukri DM, Nwabor OF, Singh S, Ontong JC, Wunnoo S, Paosen S, Munah S, Voravuthikunchai SP (2020) Antibacterial-Coated Silk Surgical Sutures by Ex Situ Deposition of Silver Nanoparticles Synthesized with Eucalyptus Camaldulensis Eradicates Infections. *Journal of Microbiological Methods* 174:105955. https://doi.org/10.1016/j.mimet.2020.105955
69. Adithya SP, Sidharthan DS, Abhinandan R, Balagangadharan K, Selvamurugan N (2020) Nanosheets-Incorporated Bio-Composites Containing Natural and Synthetic Polymers/Ceramics for Bone Tissue Engineering. *International Journal of Biological Macromolecules* 164:1960–1972. https://doi.org/10.1016/j.ijbiomac.2020.08.053
70. Smith LA, Ma PX (2004) Nano-fibrous Scaffolds for Tissue Engineering. *Colloids and Surfaces B: Biointerfaces* 39 (3):125–131. https://doi.org/10.1016/j.colsurfb.2003.12.004
71. T A, Prabhu A, Baliga V, Bhat S, Thenkondar ST, Nayak Y, Nayak UY (2023) Transforming Wound Management: Nanomaterials and Their Clinical Impact. *Pharmaceutics* 15 (5):1560.
72. Negut I, Dorcioman G, Grumezescu V (2020) Scaffolds for Wound Healing Applications. *Polymers* 12 (9):2010
73. Deepika B, Gopikrishna A, Girigoswami A, Banu MN, Girigoswami K (2022) Applications of Nanoscaffolds in Tissue Engineering. *Current Pharmacology Reports* 8 (3):171–187. https://doi.org/10.1007/s40495-022-00284-x
74. Fu S-Z, Meng X-H, Fan J, Yang L-L, Wen Q-L, Ye S-J, Lin S, Wang B-Q, Chen L-L, Wu J-B, Chen Y, Fan J-M, Li Z (2014) Acceleration of Dermal Wound Healing by Using Electrospun Curcumin-Loaded Poly(ε-caprolactone)-poly(ethylene glycol)-poly(ε-caprolactone) Fibrous Mats. *Journal of Biomedical Materials Research Part B: Applied Biomaterials* 102 (3):533–542. https://doi.org/10.1002/jbm.b.33032
75. Shahverdi S, Hajimiri M, Esfandiari MA, Larijani B, Atyabi F, Rajabiani A, Dehpour AR, Gharehaghaji AA, Dinarvand R (2014) Fabrication and Structure Analysis of Poly(lactide-co-glycolic acid)/Silk Fibroin Hybrid Scaffold for Wound Dressing Applications. *International Journal of Pharmaceutics* 473 (1):345–355. https://doi.org/10.1016/j.ijpharm.2014.07.021
76. Yahia EA, El-Sharkawey AE, Bayoumi MM (2021) Quantitative Evaluation of Diabetic Foot Wound Healing Using Hydrogel Nanosilver Based Dressing vs. Traditional Dressing: A Prospective Randomized Control Study. *Pakistan Journal of Medical & Health Sciences* 15:1571–1574.
77. Essa MS, Ahmad KS, Zayed ME, Ibrahim SG (2023) Comparative Study Between Silver Nanoparticles Dressing (SilvrSTAT Gel) and Conventional Dressing in Diabetic Foot Ulcer Healing: A Prospective Randomized Study. *The International Journal of Lower Extremity Wounds* 22 (1):48–55.
78. Miller CN, Newall N, Kapp SE, Lewin G, Karimi L, Carville K, Gliddon T, Santamaria NM (2010) A Randomized-Controlled Trial Comparing Cadexomer Iodine and Nanocrystalline Silver

on the Healing of Leg Ulcers. *Wound Repair and Regeneration* 18 (4):359–367. https://doi.org/10.1111/j.1524-475X.2010.00603.x

79. Khan S, Rai V, Srivastava S, Agarwal S, Bano N, Gupta Y (2024) Nano Medicine in Healing Chronic Wounds: Opportunities and Challenges. *Journal for Research in Applied Sciences and Biotechnology* 3 (1):12–16.
80. Gowda B, Mohanto S, Singh A, Bhunia A, Abdelgawad M, Ghosh S, Ansari M, Pramanik S (2023) Nanoparticle-Based Therapeutic Approaches for Wound Healing: A Review of the State-of-the-Art. *Materials Today Chemistry* 27:101319.

2 Nanocellulose
Advanced Application in Wound Healing

Sagnik Nag,[1] Dinesh R. Rao,[2] Kedlaya Srikrishna H. Damodar,[3] Sourav Mohanto,[4] Vetriselvan Subramaniyan,[1] and Mohammed Gulzar Ahmed[4]*

[1] Jeffrey Cheah School of Medicine and Health Sciences, Monash University, Malaysia, Jalan Lagoon Selatan, Bandar Sunway, 47500 Selangor Darul Ehsan, Malaysia.

[2] School of Bio-Sciences and Technology (SBST), Vellore Institute of Technology (VIT), Vellore, Tamil Nadu—632014, India.

[3] Department of Applied Bio-Sciences, Bhandarkars' Arts and Science College, Kundapura, Karnataka—576201, India.

[4] Department of Pharmaceutics, Yenepoya Pharmacy College & Research Centre, Yenepoya (Deemed to be University), Mangalore, Karnataka, 575018, India.

* Corresponding author: Vetriselvan Subramaniyan

2.1 INTRODUCTION

In the multiplex phraseology of human anatomy, skin stands as a sentinel, guarding our body against unpropitious agents—just like a soldier guarding a nation! Skin forms the largest organ in the human body. It is a testament to the marvelous bioengineering where multiple processes occur within a spur of a moment and have many physiological actions. Skin is an important organ that is involved in many functions including protective, metabolic, and immunoregulatory roles that are crucial facets for healthy living. Broadly, skin can be categorized into three main layers—the epidermis, dermis, and the hypodermis. Microscopic studies show many more layers within these, but this classification is based upon the location and physiological functions of skin (Abdo et al., 2020). The epidermis forms the outermost layer of the skin. It is characterized by dead connective tissues that are germ-proof and water-proof. It forms the first line of defense in the human body by extending its role as a physiological barrier for pathogens to enter our body. The middle layer is the dermis, which contains nerve endings, responsible for stimuli, and glands that secrete oils, called sebum, and sweat. The hypodermis, which is also called the subcutaneous tissue, is made of adipose tissue. It contains connective tissues and blood vessels for circulation. Together, these layers form an intricate interplay of the army that fosters various roles such as hydration, prevention of water loss, synthesis of important biomolecules like vitamin D, immune functions, and mechanical support. The skin interacts uniquely with several growth factors, cytokines, and bodily factors that provide a rough and tough support to the shape and integrity of the body, as well as extend immune functions (Tottoli et al., 2020). Beyond its tactile wonders, there lies a deeper narrative—one of resilience, adaptation, and the ceaseless battle against adversity. Every day, our skin confronts a barrage of challenges, from environmental pollutants to potent damaging factors. Yet, miraculously, it persists, continuously renewing itself in a silent oeuvre of cellular renewal. But the mightiest factor

DOI: 10.1201/9781003605966-2

that disturbs the integrity of skin and pesters normal cellular functions is wound—the demon in biological terms! A wound is any factor that disintegrates the layers of skin and causes injury. It disrupts the fragile equilibrium of skin and breaches the protective barrier to expose tissues vulnerable to infectious agents. In these moments of vulnerability, the body marshals its forces, initiating a cascade of events known as wound healing—a process as intricate as it is essential for our survival (Hayat et al., 2023; Gowda et al., 2023).

The technique of wound healing has been employed to assist the body in wound repair since time immemorial. Man has tried and applied new techniques to protect and heal wounds since the age of cavemen. The care of wounds has evolved from magic incantations, potions, and ointments to a general text written by Sushruta, Hippocrates, and Celsus on wound management. For all, including ancient Chinese, Korean, Egyptian, and African healers, the science of wound healing has been a matter of concern due to the frequent battles being fought during those days. There are two separate chapters on the recovery of such wounds in Sushruta Samhita, which describes the use of more than a hundred plants, in single or in combination, to effectively heal wounds. Sushruta has even discussed the approach to employ different plant-based drugs to clean wounds under *Vrana Shodhan* and the drugs to heal wounds under *Vrana Ropana* (Kumar et al., 2021). The ancient Indian surgical technique, popularly called as *Shalya Tantra*, describes eight ways to treat surgical wounds, which is collectively known as *Asthavidha Shastrakrama*, that put forth the ways in which wounds should be treated. It includes guidelines to various important procedures in first aid like disinfecting the wounds, cleaning, removing foreign materials, care, and methods to cover wounds, and so on, to effectively prevent infections and promote faster wound healing (Dudhamal, 2023).

Traditional Chinese Medicine (TCM) also has many perceptions on concepts dealing with wound care and healing. It employs the ancient concept of Yin and Yang to understand the dynamic balance within the body. Yin represents darkness, cold, and stillness, while Yang symbolizes light, warmth, and activity. TCM harmonizes these opposing forces to promote holistic healing, emphasizing balance and interconnectedness in health and wellness practices (Fu et al., 2021). While traditional methods in wound healing, like poultices and herbal remedies are widely popular and are deemed to be culturally significant, most of them often lack scientific validation and standardized application. Moreover, their limited efficacy and slow outcomes deter the modern acceptance. Additionally, they pose challenges such as contamination risk and inconsistent quality control, which hinder their widespread adoption in modern healthcare settings.

While contemporary methods like the use of topical ointments, stitches, band-aids, ice packs, and dressings are employed for wound care and healing, these methods come with many setbacks. The prevailing standard of care for wounds consists of swabbing for infection detection and examination, disinfecting, cleaning, dressing, and, in some cases, debridement of the wound. Common limitations in these techniques include recurring changes in dressings leading to re-injury, loopholes in techniques leading to infection, scarring, longer healing times, patient discomfort, and increased hospital care costs. To overcome all these, advances in nanotechnology are being promoted, which guarantee to resolve these issues by formulating patient-centric modern approaches in wound care. Even within this field, the use of nanobiomaterials is the neoteric way to deal with wound care. Biological nanomaterials include various naturally occurring and synthetically developed nanomaterials that are restructured and engineered in different forms, such as advanced nanomaterial-containing band-aids, hydrogels, interactive dressings, sponges, and bioactive dressings, which transform the wound by actually interacting with wound tissues and inducing stimuli for faster recovery, along with maintaining a conducive environment for tissue regeneration. These form the state-of-the-art innovation in wound care technology by bringing modern materials into play to heal wounds faster and more accurately (Kolimi et al., 2022; Pramanik et al., 2021).

This chapter deals with advances in nanocellulose, a nano-engineered biomaterial employed in wound healing. The chapter shall delve into the fine intricacies that lie among the techniques to synthesize nanocellulose, its different forms, and the mechanism of action that aids in the magical healing of even complicated wounds. Further, this chapter also provides scope to explore the future of nanocellulose in advanced wound healing, paired with technological and medical advancements.

2.2 BIOMATERIALS IN WOUND HEALING: RESEARCH GAPS AND PROSPECTS

The disruptions or breaks in the skin and tissues caused by injury, trauma, surgery, or medical conditions are called wounds. They are common and range from minor cuts to severe lacerations or surgical incisions (Tottoli et al., 2020). Wounds can be classified into acute and chronic wounds based on their pathogenicity, nuances, healing duration, and physiological characteristics. Acute wounds are sudden and heal normally by undergoing molecular steps without any complications. However, chronic wounds are long-lasting wounds with prolonged and recurring inflammation, necrosis, and underlying health issues, requiring specialized treatment and medical attention (Nag, Mitra, Bhattacharjee, et al., 2024; Raziyeva et al., 2021). Depending on the causes of injury, wounds are further classified into different types—traumatic, burn, surgical, ulcers, and infectious wounds. Infections occur when pathogenic microorganisms invade wound tissues, resulting in inflammation. Proper detection and diagnosis of these wounds are essential for effective wound management and treatment (Irfan-Maqsood, 2016). Wound healing is a dynamic process of tissue repair, and it involves several sequential and overlapping stages: hemostasis, inflammation, proliferation, and remodeling. It is aided by several factors, including cells, cytokines, and growth factors. Hemostasis stabilizes the wound site by clotting blood; inflammation has a role in immune activation and pathogen elimination. Proliferation induces vital processes like collagen synthesis and angiogenesis for healing. At last, remodeling concludes wound healing by restoring tissue architecture and strengthening the wound site. Throughout these stages, cellular and molecular interactions echo the music of healing, leading to the rehabilitation of tissue integrity and functions (Almadani et al., 2021; G. Kaur et al., 2022; Mohanto et al., 2024).

The conventional methods of wound healing rely upon a combination of methods like debridement, wound irrigation, the use of topical antiseptics or antimicrobials to prevent infections, and dressings. These methods are quite effective, but they come with challenges like recurring infections, scarring, delayed healing, and discomfort (Cancio, 2021). Infections lead to chronic wounds and complications. Scarring affects aesthetics and might adversely affect the functions of the tissue. Delays in healing discomfort patients and lead to increased healthcare costs. Addressing these issues is essential to develop patient-centric outcomes and to reduce healthcare burdens. This is possible with explorations leading to the use of modern techniques, especially those involving biomaterials, which offer promising approaches to enhance wound healing processes. Advanced biomaterials come with advantages like the removal of debris from the wound site, preventing infections, biocompatibility, facilitating oxygen transport, maintaining adequate moisture conditions, and precisely delivering therapeutic agents, which optimize the healing process (G. Kaur et al., 2022).

Over time, the definition of biomaterials has evolved to match advancements in medicine. Originally, they were described as nonliving materials intended to interact with biological systems. Later, this definition was refined to exclude naturally derived active substances. The current understanding recognizes them as substances that interact with our body and are biocompatible, meaning they can perform their functions to promote tissue responses without any harmful effects (Sultana et al., 2024). According to the American National Institute of Health, biomaterials are any synthetic or natural substances that enhance or replace bodily tissues, organs, or functions to improve the quality of life. There are varieties of biomaterials available, each with unique and distinguishing properties. With technological progress and bioengineering, these can be customized for various biomedical applications like wound healing and tissue regeneration. Biomaterials are chemically and structurally stable, thus maintaining integrity during wound healing. They are biocompatible—ensuring fewer adverse effects. They regulate moisture levels, promote cell adhesion and proliferation, and aid in the controlled and precise release of therapeutic agents and drugs, thus furnishing an optimal healing environment with minimized challenges and adverse effects, proving them to be better than conventional healing methods (Bianchera et al., 2020; Murray et al., 2019).

In the context of wound healing, biomaterials are commonly classified into three categories: synthetic polymers, natural polymers, and bioactive dressings. Synthetic polymers like silicones, polyesters, polyethylene glycol (PEG), and polylactic acid (PLA) are known for their versatile beneficial biocompatibility and lower degradation rates. Natural polymers like cellulose, collagen, chitosan, and alginate are derived from biological sources, and they possess outstanding biocompatibility and biodegradability. Bioactive dressings are biomaterials containing active agents like growth factors, antimicrobial agents, or nanoparticles to actively soothe the wound environment and promote healing. These biomaterials play crucial roles in providing mechanical support, promoting cell adhesion and proliferation, regulating inflammation, and facilitating tissue regeneration in the wound healing process (Bianchera et al., 2020; M. Li et al., 2023). The mechanism of action of biomaterials that aid in wound healing lies in their interaction with the cells and immune system. They act as scaffolds that mimic the extracellular matrix of the tissues, and over time they influence the topography, chemistry, and physiological properties of the tissues. Different forms of biomaterials aim to mimic normal tissue compositions. Microtubular biomaterials aid in the conduction of body fluids by mimicking blood vessels and capillaries. Polysaccharidic and fibrous biomaterials act as templates for the generation of the cell matrix, thus influencing cellular behavior and the tissue remodeling process. Biomaterials are also involved in inducing mechanical forces and signal stimuli in the wound tissues that aid in faster cell proliferation and quick regeneration of tissue (Gholap et al., 2024). The mechanical stimulus also plays a role in stiffening and giving tensile strength to the injured tissue, providing scope for faster recovery. Hydrogels aid in mimicking natural tissue and provide optimal moisture and texture at the site of injury. Recent advancements in biomaterial design focus on integrating growth factors, antimicrobial agents, and nanotechnology to enhance wound healing and regenerative medicine (Mohanto et al., 2023). Self-healing biomaterials employ microcapsules or dynamic chemical bonds within the matrix. Microcapsules rupture upon damage, releasing healing agents to form new bonds (Ansari et al., 2022). Some biomaterials possess intrinsic self-healing properties, like shape memory or regenerative capacity, facilitating repair. These mechanisms enable advanced biomaterials for tissue engineering, drug delivery, and medical devices. Smart and adaptable biomaterials can be engineered for hemostasis, antibacterial purposes, anti-inflammatory effects, promoting vascularization, regulating the microenvironment, self-healing capabilities, self-removal properties, monitoring functions, scar management, and many such versatile functions (M. Li et al., 2023; Y.-Y. Li et al., 2024; P. Wang et al., 2024). Many researches are focusing on incorporating growth factors like platelet-derived growth factor (PDGF) or transforming growth factor-beta (TGF-β) into biomaterials to stimulate active cell proliferation and tissue generation at the wound site. Additionally, integrating antimicrobial agents, antimicrobial biopeptides, and nanoparticles like silver nanoparticle impregnation can help prevent infections and promote a sterile wound environment conducive to healing (Hani et al., 2023; Nag, Mitra, Tripathi, et al., 2024). Advanced biomaterials like self-healing materials and adaptive biomaterials mechanically interact with tissues. They can be engineered to interact with tissues and develop neurons for quicker healing. Biodegradable biomaterials gradually degrade over time, providing scope for integration of new tissues and minimizing trapping and infections with foreign bodies. Biomaterial surface, topography, wettability, and stiffness are known to regulate the phenotype of macrophages, neutrophils, and T cell activation, thus having a greater effect on immune ecology. Furthermore, nanotechnology has enabled the development of biomaterials with precise control over structure and function, allowing for targeted delivery of therapeutic agents and enhanced interactions with biological systems. The use of biomaterials in wound healing objectifies patient-centric outcomes and bench-to-bedside applications, reducing healthcare burdens. With their versatility, biocompatibility, and ability to interact with biological systems, biomaterials offer tailored solutions for various medical needs, ultimately advancing wound healing therapies and regenerative medicine strategies (Agnihotri & Dhiman, 2017; Deng et al., 2022; Y.-Y. Li et al., 2024). While biomaterials seem to be advantageous in wound healing, they come with a variety of challenges involving immunogenicity, where some biomaterials may activate pro-inflammatory mechanisms leading to chronic inflammation (S. Das et al., 2024). Some biomaterials, like chitosan, face regulatory hurdles as they are sourced majorly from the exoskeletons of crustaceans and insects. Some biomaterials may be toxic or might become toxic during preparation, making them unusable for biomedical applications. There is a gap and demand in novel biomaterial research to discover ideal biomaterials for applications at ease (Deng et al., 2022; Smets & Rüdelsheim, 2018).

Recently, nanocellulose and its derivatives have emerged as promising biomaterials in aiding wound healing due to their unique properties, including biocompatibility, biodegradability, availability, ideal physicochemical properties, and also economic ease. They are obtained from "vegan" sources like plant materials and are processed into nanofibers or nanoparticles with distinguishing mechanical and biological properties. The biochemical properties of nanocellulose confer its suitability as wound dressing material to promote wound closure, tissue regeneration, and to provide a scaffold for cell growth. Nanocellulose-based dressings also offer antimicrobial properties, supporting aseptic wound environments. In recent research, nanocellulose has been proven to be a versatile and sustainable biomaterial with an excellent ability to interact with biological systems seamlessly. Nanocellulose swoops in as a "superhero" to aid wound healing, battle infections, and promote tissue regeneration, basically saving the day for patients in need (Bacakova et al., 2019; Yang et al., 2024).

2.3 CHARACTERISTICS OF NANOCELLULOSE

Nanocellulose, a versatile biomaterial sourced mainly from plant sources, has a multitude of potential applications because of its phenomenal physiochemical properties and advantages. Nanostructured cellulose, widely known as nanocellulose ($(C_6H_{10}O_5)_n$), is a long-chain, unbranched homopolysaccharide of identical β-D-glucose monomeric units connected by β (1→4) glycosidic linkage. Each monomer is twisted by an angle of 180° while forming glycosidic linkage. The repeating $C_{12}H_{22}O_{11}$ units in the structure are known as cellobiose. Cellulose can be structurally elucidated with a complex $^4C_1$ β-glucopyranose chair conformation. Being the most abundant renewable compound found in the biosphere, cellulose is widely available from plant sources. It is approximated that cellulose derived from plant wood sources contains about 15000 monomeric units (P. Kaur et al., 2021; Thakur et al., 2021). The repeating cellobiose units join together to form a linear chain. Adjacent chains are held together by means of hydrogen bonding and van der Waals forces of attraction to form an elementary fibril, which is a stacked chain structure. These fibrils are further bundled to form microfibrils, which are grouped together to get cellulose fiber. Cellulose fibers are composed of two regions within their structure—crystalline and amorphous regions. The crystalline region, made of the glucopyranose backbone, portrays ordered molecular arrangements which offer high chemical stability and less reactivity. These regions are characterized by tightly bound structures with high strength and rigidity. The amorphous region is brought about by the reactive chemical side chains bound to glucopyranose units. They are disorderly arranged and comparatively reactive with less chemical stability. The principle behind the preparation of nanocellulose synthetically often relies upon the need to selectively disrupt amorphous regions from cellulosic fibers and preserve their crystalline arrangements (P. Kaur et al., 2021; Yang et al., 2024).

Nanocellulose can be appropriately quoted as cellulose molecules with dimensions between 1 and 100 nanometers. It has gained popularity in nanotechnology due to its ability to create uniform materials with improved physicochemical and mechanical properties through size reduction. It is important to note that nanocellulose, obtained from downsized cellulose fibers, exhibits enhanced properties compared to its bulk form, making it promising for diverse applications. Traditional sources of nanocellulose include cellulose from various origins like hardwood, softwood, agricultural biomass, bacteria, algae, and tunicates. While the cellulose composition remains consistent, structural differences exist (P. Kaur et al., 2021). Plant-derived cellulose typically has diameters of 13 to 22 micrometers and crystallinity of 44% to 65%, whereas bacterial cellulose comprises nanoscale fibrils with higher crystallinity, around 90%. Nanocellulose size varies with raw material; for example, cotton, wood, and microcrystalline cellulose yield narrow particle size distributions, while bacterial and tunicate sources yield wider distributions. Based on their molecular structure and appearance, nanocellulose exists in various forms including cellulose nanocrystals (CNC), cellulose nanofibrils (CNF), cellulose nanosheets (CNS), spherical nanocellulose (SNC), microfibrillated cellulose (MFC), and nanocrystalline cellulose (NCC), each with specific synthesis methods and applications. CNCs are rod- or needle-shaped nanocrystals with high crystallinity and a rigid structure. They are 4 to 70 nm in diameter and usually less than 500 nm in length. They can be prepared by mechanical methods, acid hydrolysis, biological methods, using ionic liquids, and by eutectic solvent preparation methods. They have potent pharmaceutical and chemical applications. CNFs are reticulate or filamentous nanocellulose that have a flexible structure and tensile properties. They occur in fibrils of 2 to 60 nm in diameter and a few microns in length. They can be prepared by advanced methods like microjet methods, TEMPO oxidation, and ionic liquid methods. CNFs are widely used in the paper and food industries. CNSs are flaky, thin, and smooth sheets of nanocellulose existing in a monolayer of about 4 nm thick. They are prepared mechanically by ball milling and have wide biomedical applications. SNCs are biomedically important nanocellulose that are spherical in nature and possess high thermal stability. They are typically around 20 to 600 nm in size. They can be prepared by acid and enzymatic hydrolytic methods. Microfibrillated cellulose (MFC) is a nanocellulose material comprising fibrils with diameters typically ranging from 5 to 100 nanometers. The nanocellulose materials that are synthesized chemically are called nanocrystalline cellulose, which have high amounts of preserved crystalline state in their structure (P. Kaur et al., 2021; Yang et al., 2024). Bacterial cellulose (BC) is a nanocellulose material produced extracellularly during bacterial fermentation by numerous species within the *Rhizobium*, *Agrobacterium*, *Gluconacetobacter*, and *Sarcina* genera. Its primary advantage lies in its pure composition, devoid of other undesirable polymers often present in cellulose from plant sources, such as hemicelluloses and lignin (Gupta & Shukla, 2020). Unlike plant-derived cellulose, BC is synthesized extracellularly in the form of a dense network of nanofibrils through bacterial fermentation of sugars. It exhibits exceptional purity, high crystallinity (typically around 90%), and unique properties such as high mechanical strength, water retention capacity, and biocompatibility (P. Kaur et al., 2021; Yang et al., 2024).

Nanocellulose showcases impressive physical properties. Its robust mechanical strength ensures the creation of durable materials, while thermal stability guarantees resilience across diverse temperatures. With a large surface area (250 m g^{-1}) and aspect ratio, it facilitates extensive interactions in applications. The optical properties of nanocellulose render it adaptable for optoelectronics and transparent films. Furthermore, its rheological properties allow for tailored viscosity and flow behavior, enhancing its efficacy in formulations and coatings. The zeta potential determines nanocellulose surface charge, with negative values indicating abundant hydroxyl groups (-OH). Various researchers have found that the zeta potential of nanocellulose always exists in negative values, which confirms the presence of hydroxyl groups on its surface. The surface hydroxyl groups of nanocellulose enable strong interactions, making it an effective reinforcement in polymer composites with significantly increased mechanical strength and enhanced optical transparency (P. Kaur et al., 2021).

Microscopic techniques like field emission scanning electron microscopy (FESEM), transmission electron microscopy (TEM), and atomic force microscopy (AFM) enable dimensional analysis of nanocellulose, revealing right-handed chirality at fibril and bundle levels (P. Kaur et al., 2021). AFM additionally probes topography, morphology, and mechanical properties, favored by researchers for its resolution and adaptability, allowing investigations of sample changes in different environments. The transverse elastic modulus of CNCs and CNFs determines their structural and tensile features, which can be measured by AFM. Dynamic light scattering (DLS) determines the hydrodynamic size of nanocellulose, complementing other techniques for tracking consistency and size changes over time. Density and porosity can be measured using a helium pycnometer. The density of pure native celluloses like cotton and tunicin was reported to be around 1.60 g/cm^3 (Barbash et al., 2019; Daicho et al., 2020). Elemental analysis and functional group analysis of nanocellulose can be performed using X-ray photoelectron spectroscopy, while NMR analysis can be conducted for purity checks and to determine molecular structure. The signals from C4 and C6 can be analyzed to distinguish crystalline and noncrystalline phases within nanocellulose structures. Additionally, the ratio of C4 crystalline to noncrystalline phases can be used to calculate the crystallinity index (CI) value of nanocellulose (Daicho et al., 2020; P. Kaur et al., 2021). The study of rheological properties in nanocellulose aqueous suspensions is crucial for industrial applications, affecting preparation, processing, and material combinations. Nanocellulose's shape, size, and surface composition significantly influence its rheological behavior. Viscosity measurements of cellulose nanocrystal (CNC) dispersions at 1 wt% concentration and shear rates of 0.1 to 1 s revealed viscosities ranging from 10 to 100 mPas. Various methods such as Couette-type rheometers and rotational viscometers are commonly employed to assess nanocellulose suspension viscosity, considering wall depletion effects (Hubbe et al., 2017).

The unique properties of nanocellulose, stemming from its nanoscale dimensions and exceptional chemistry, enable a wide array of beneficial applications, effectively addressing sustainability concerns. Its stiffness, flexibility, and mechanical robustness make it advantageous for nanomaterial-based conductive applications. Additionally, its water retention capability, high aspect ratio, and gelation properties at lower temperatures render it suitable for cosmetic formulations. Nanocellulose's nanosized structure, flexibility, biocompatibility, and stability further enhance its utility in various optical material productions. Moreover, its adsorption properties, low toxicity, and ability to self-assemble make it ideal for biomedical applications, including the development of hydrogels, scaffolds, and wound dressings for tissue regeneration (P. Kaur et al., 2021; Patil et al., 2022). These attributes underscore nanocellulose's potential across diverse industries, from biomedicine to sustainable packaging, as it continues to advance through ongoing research and technological innovations (Al Abdallah et al., 2024; M. R. Khan et al., 2024).

2.4 SYNTHESIS TECHNIQUES OF NANOCELLULOSE

Cellulose is a natural polymer that can further degrade to form nanocellulose. Generally, the polymers can be synthesized by the addition of a long chain of monomers, where cellulose can be produced via etherification, esterification, through some ionic liquids such as 1-ethyl-3-methylimidazolium acetate or 1-butyl-3-methylimidazolium chloride, and some bacteria such as *Acetobacter xylinum*. Thus, based on the region of cellulose being disrupted and the agents used for the synthesis process, it can be classified into physical synthesis, chemical synthesis, and biological synthesis.

2.4.1 Physical Synthesis

In this mechanical method, the amorphous region of the cellulose will be derailed by physical actions. It includes grinding, homogenization, fibrillation, and extrusion (Yang et al., 2024).

2.4.1.1 Grinding

Grinding is a mechanical process that involves the deterioration of the complex structure of cellulose into nanoscale particles. This method is based on the application of mechanical shearing, where it employs mechanical shear forces and friction forces to break down the inter- and intramolecular hydrogen bonds existing in cellulose fibers (Yang et al., 2024). Through mechanical grinding (1500 rpm), the size of cellulose is lowered from the microscale to the nanoscale range. This reduction of size elevates the surface area-to-volume ratio and intensifies reactivity; the dimensions of cellulose particles are intricately linked to both the duration and velocity of the milling process (Phanthong et al., 2016).

This grinding process starts with the selection of a cellulose source, such as wood pulp, hemp, corn stalks, straw, or other biomass material, and then it is cleaned; all the impurities are dislodged to facilitate the grinding (Mohite et al., 2024). Various types of grinding machinery are available, such as ball mills, attrition mills, and planetary ball mills; depending on the source and application, suitable equipment is utilized to synthesize nanocellulose. Loading of cellulose fibers will be another step in this process, where fibers are loaded into the grinding chamber of the opted machinery with the grinding media, which comprises steel balls, ceramic beads, and other abrasive materials competent to generate the mechanical forces required to collide and squeeze out the fiber raw materials, leading to the reduction of the polymerization degree of cellulose and the formation of a nanocellulose suspension, which can be further characterized for relevant properties, recovered, and processed for further applications (Qi et al., 2023).

2.4.1.2 Homogenization

The process where cellulose is subjected to high pressure (50-2000 MPa) to reduce it into nanoscale fibers via the cavitation effect. The first step in this process is the preparation of a slurry by dispersing the cellulose fibers in a specific solvent. Prior to slurry formation, the cellulose fibers are pretreated to remove all the impurities to foster homogenization (Nasir et al., 2017). A homogenizer is the primary tool used in this process, which consists of a pump for the generation of pressure, a valve, and control for pressure and flow rate regulation. Using a homogenizer, cellulose slurry is passed through a close service or orifice at very high pressure, which triggers "bubble collapse" within the slurry due to the expeditious change in pressure and velocity that stimulates mechanical shearing forces, advocating the reduction in the dimension of cellulose fibers. Based on the required fiber size and properties, the cellulose suspension can be passed through the orifice multiple times. To safeguard the stability of the cellulose slurry from excessive heat produced from the frictional forces, a cooling system is used (Wu et al., 2021). Once the homogenization process is completed, the morphology, stability, or other properties are characterized and studied. Additional treatment, such as filtration and centrifugation, will be carried out as a posthomogenization treatment for the specific nanocellulose-based advanced applications (Y. Wang et al., 2015).

2.4.1.3 Fibrillation

The mechanical process where cellulose fibers are fragmented to nanofibrils is called fibrillation. Fibrillation is based on the disruption of hydrogen bonds present in the cellulose molecule (L. Wang et al., 2021). Removal of the impurities present in the cellulose fibers initiates the process of fibrillation, which is followed by the disintegration of cellulose fibers through grinding, milling, or refining (Pere et al., 2020). In the next step, cellulose slurry or suspension is formed by suspending the disintegrated fibers in a particular solvent. Fibrillation involves two kinds of treatments, namely fibrillation stage 1 (pretreatment) and fibrillation stage 2 (mechanical treatment). In the pretreatment, modification of surface properties takes place, which helps to untie the structure using chemicals and enzymes. Mechanical treatment deals with fibrillation and fragmentation of pretreated cellulose fibers via microfluidization and high-pressure homogenization (Jonoobi et al., 2015).

During the mechanical treatment, dependent variables such as pressure, shear rate, temperature, and number of passes should be controlled to upgrade the fibrillation with proper fiber size and low energy consumption. Thus, nanofibrils are then analyzed and characterized with the assistance of microscopy, spectroscopy, and rheology (Lengowski et al., 2020).

2.4.1.4 Extrusion

The process of extrusion is based on the principle of conversion of solid or semisolid materials into a specific shape or profile along with a die having a particular cross-sectional area by employing mechanical, thermal, and rheological attributes. In this method, cellulose suspension of proper consistency and concentration is prepared, and the suspension which is to be extruded is termed as feedstock (Debiagi et al., 2018).

The feedstock is then fed to the extruder, a device equipped for the extrusion procedure having a specialized component termed as a constricted orifice or die made up of a metal plate, which aims at providing specific shape and dimensions of fabricated materials. The constricted opening of the die forms a shear atmosphere. Along with the die, extruders also have a barrel and plunger (Debiagi et al., 2021). Feedstock is fed into the barrel, where it is presented to heating, and the plunger regulates the pressure. Under the influence of applied temperature and pressure, the feedstock will be plasticized, softened, and homogenized, which will further pass into the die to obtain the extruded nanocellulose fibers of specific dimensions. Thus, the extruded nanofibrils undergo cooling and solidification using water baths to stabilize structures and to prevent clumping or aggregation, which is followed by sizing of the extruded nanofiber by slicing down to the desired length and shape (J. Wang, Liu, et al., 2019).

2.4.2 Chemical Synthesis

The chemical method is employed to synthesize nanocellulose by the disruption of the crystalline region of cellulose using various chemical agents. This includes acid hydrolysis, oxidative hydrolysis, enzymatic hydrolysis, mechanochemical methods, and ionic liquid-assisted nitration (Mishra et al., 2018).

2.4.2.1 Acid Hydrolysis

Acid hydrolysis can be defined as the breaking down of polysaccharides into simple sugars using acid solutions. The selection of the cellulose source will be the primary step in the acid hydrolysis, which is followed by the elimination of impurities, drying, and grinding to allow the acid to penetrate and react. Cellulose fibers may be exposed to pretreatment for better accessibility and reactivity through modification of structural and surface morphology by mercerization and oxidation. Pretreated cellulose fibers are presented to acids for hydrolysis in the next step where they are immersed in sulfuric acid or hydrochloric acid of specific concentration. Heat the immersed fibers in the acid solution at 50°C to 120°C for 5 to 10 minutes with continuous stirring to prevent agglomeration (Wulandari et al., 2016). Fragmentation of cellulose fibers to nanofibrils takes place due to the breakage of glycosidic bonds between glucose units by the acid, leading to disruption of the crystalline structure of cellulose (Amezcua-Allieri et al., 2017). The hydrolysis reaction is then neutralized to stabilize the nanocellulose suspension with sodium hydroxide or calcium hydroxide. The nanosuspension is then passed through post treatments including filtration to expel the residual acids and drying to obtain dry powder or solid form of nanocellulose based on the applications, which will be further analytically analyzed using X-ray diffraction, spectroscopy, and thermal analysis (Niu et al., 2017).

2.4.2.2 Oxidative Methods

In this method, the cellulose is exposed to the oxidation process under controlled reaction parameters to obtain nano-dimension fibrils. For example, cellulose under some oxidative agents oxidizes hydroxyl groups to carboxyl groups, leading to depolymerization and formation of acetal bonds. Sodium hypochlorite (NaOCl), sodium periodate ($NaIO_4$), and TEMPO (2,2,6, 6-tetramethylpiperidin-1-oxyl) are prominent oxidizing agents used in the synthesis of nanocellulose (Barbash et al., 2022). The initial step in the oxidative method is to immerse pretreated (1–3 days) cellulose suspension in an oxidative solution to ease uniform oxidation. The oxidizing agent acts on hydroxyl groups, thereby disrupting glycosidic bonds and affixing other functional groups like COOH and CHO under the reaction temperature of 25°C–70°C, respectively. This reaction is catalyzed by NaBr to promote the rate of oxidation. Thus, the obtained nanocellulose is then filtered and washed with deionized water to remove traces of secondary impurities, followed by the characterization of the nanocellulose suspension (Liu et al., 2016).

2.4.2.3 Ionic Liquid-Assisted Nitration

The chemical process that employs ionic liquids for the introduction of the NO_2 group via substitution reaction is termed ionic liquid assisted nitration. The prominent nitration reagents used are a combination of nitric acid (HNO_3) and sulfuric acid (H_2SO_4). Sulfuric acid acts as a catalyst that eases the activation of nitric acid. The homogeneous solution is prepared by dissolving cellulose fibers into solvents such as N-methylmorpholine N-oxide (NMMO), ionic liquids, and dimethyl sulfoxide (DMSO), which assist in the dissolution of cellulose. Mix the prepared nitration reagent with the cellulose solution to initiate the process of nitration, which is followed by the formation of reaction bubbles between hydroxyl (-OH) groups of cellulose and nitric acids or ionic liquids, which directs the substitution of (-NO2) groups. Before isolating the nitrocellulose from the reaction mixture, add ample amounts of quenching agent to stop and stabilize the product (Stovbun et al., 2016). During the harvesting of nanocellulose, the nitrocellulose is hydrolyzed to yield nanocellulose with the elimination of nitro groups from the nitrocellulose backbone, which will further be washed, dried, and characterized to study the specific properties and biomedical applications (Budaeva et al., 2019).

2.4.2.4 Chemo-Mechanical Methods

The synergistic effect formed by mechanical forces and chemical reagents fabricates the nanocellulose fibrils with less consumption of energy, forming one of the cost-effective methods for the synthesis of nanocellulose. To consume less energy, the cellulose fibers are pretreated with chemical methods like acid hydrolysis, oxidation, and nitration for the initial disintegration of fibers. This chemically mediated treatment aids in the cleavage of bonds and disturbance in the surface area of cellulose fibers (Bharimalla et al., 2023). Thus, the pretreated cellulose fibers are then exposed to various physical methods for further processing of cellulose fibrils. Mechanical activation is mainly carried out by ball milling, microjet treatment, grinding, fibrillation, and high-pressure homogenization in this chemo-mechanical synthesis. This kind of synergistic effect eases the synthesis of nanocellulose fibrils, which are compatible, chemically stable, and biomedically potent for the field of nanomedicine and biomaterial-based therapeutics (Bharimalla et al., 2023; Yang et al., 2024).

2.4.2.5 Enzyme Hydrolysis

The catalytic process disrupts the amorphous region of the cellulose under the influence of the cellulase enzyme through surface deterioration. Cellulase is a significant hydrolytic enzyme found in microorganisms that helps in the cleavage of plant cell walls, aiding patho-invasion. Cellulase is composed of endoglucanases, exoglycanases, and β-glucosidases, each contributing to cellulose degradation. The external glucanase expels glucose units from the nonreducing end, while the internal glucanase interconverts crystalline to amorphous cellulose, finally producing oligosaccharides and glucose monomers (Nasir et al., 2017). Thus, in this method, specific cellulase is employed for the fragmentation of cellulose fibrils to nanoscale dimensions based on the requirements. This hydrolysis method is based on the principle of cleavage of glycosidic bonds via cellulase enzymes (Yang et al., 2024).

2.4.3 Biological Synthesis

Nanocellulose can be prepared through biological methods, which yield nanofibrils of exceptionally high crystallinity and thermal stability. These prominent applications have achieved remarkable milestones in the field of biomedical science and pharmacodynamics. Microbiological methods and cellulolytic enzyme-mediated nitration are the two prominent methods employed for the biological synthesis of nanocellulose.

2.4.3.1 Microbiological Method

The microbial biocatalytic method is based on the synthesis of nanocellulose by microbial metabolism. Bacterial cellulose (BC) is one kind of nanocellulose procured using bacteria such as Acetobacter, Komagataeibacter, and Gluconacetobacter. The synthesis of bacterial cellulose involves various steps such as selection of strains, culture media preparation, inoculation and incubation, cellulose synthesis, assembly of cellulose microfibrils, followed by maturation and harvesting (Yang et al., 2024).

Bacterial cellulose production starts with the selection of specific bacterial stains that can yield a good variety of cellulose. For example, a widely used bacterium in the preparation of bacterial cellulose is *Acetobacter xylose*. Suitable culture media must be prepared to aid the growth of bacterial populations by providing a carbon source, nitrogen source, mineral salts, and vitamins. Optimal conditions should be provided in the media for growth and cellulose production. Once the media is prepared, the bacterial strain is inoculated and incubated at the optimum temperature (28°C-30°C) with proper agitation (Ribeiro et al., 2019). During the augmentation phase, bacteria secrete cellulose as an extracellular polymer by the metabolism of the carbon source. The cellulose is formed in the cell membrane of bacteria, where the cellulose synthase complex catalyzes the glucose polymerization, followed by the assembly of gel-like cellulose microfibrils having exceptional mechanical resilience and thermal durability. The assembly of fibrils leads to the last step known as maturation, harvesting, and postprocessing based on the biomedical applications (Choi et al., 2022).

Another important method is microbial biocatalytic nitration, where microorganisms are employed as biological catalysts to assist in the nitration of cellulose fibers, which involves the addition of a nitro group with the elimination of a hydroxyl group using nitric acid and nitrogen dioxide. Many microorganisms use enzymatic machinery to procure nanocellulose under optimum conditions. Under the catalytic action of the enzyme nitro reductase, nitrate ions (NO^{3-}) are reduced to nitrite ions (NO^{2-}), leading to the formation of nitrocellulose, which is then hydrolyzed to remove the nitro group from the cellulose framework by the reaction with acids, resulting in the production of stable nanocellulose fibrils (J. Wang, Tavakoli, et al., 2019).

Several investigations have also been successful in the production of nanocellulose fibrils using a mixed culture of microbes that have the capacity to generate the highly complex polysaccharide, cellulose, via fermentation (Sharma & Bhardwaj, 2019). Another spectacular process that led to the development of nanocellulose-based wound dressing is the production of celluloses through genetically engineered microorganisms by incorporating cellulose synthase-encoding genes (Moradi et al., 2021).

2.4.3.2 Cellulolytic Enzyme-Mediated Nitration

Group of enzymes that cleaves cellulose into simpler glucose subunits and takes them as their carbon source are known as cellulolytic enzymes. These enzymes are produced by a variety of microbes that have a wide range of applications in the synthesis of nanocellulose on an industrial scale. It was also found that even termites and some marine invertebrates are capable of producing cellulolytic enzymes and (Fernandes et al., 2023) synthesized bacterial cellulose using the same extracted enzyme from these organisms. The cellulolytic enzymes are accompanied by some accessory enzymes like lytic polysaccharide monooxygenases (LPMOs), which aid in the oxidative-assisted glycosidic bond rupturing, and cellulose binding modules (CBMs), which are helpful in the formation of the enzyme-substrate complex for cellulose bifurcation (Ribeiro et al., 2019; Yang et al., 2024).

The cellulolytic-based nitration is based on the cellulose precursor molecule, which can bind with the substrate to form an enzyme-substrate bubble for the disintegration of the surface morphology of cellulose and the intramolecular hydrogen bonding, which is further carried out by the nitration agents, mainly nitric acid and sulfuric acid or acetic anhydride. Under the controlled enzyme degradation mediated with nitration, the cellulose will be converted to nitrocellulose, and on acid hydrolysis, it emerges as fine nanofibers after the drying and post-nitration processing, with extensive durability (Arantes et al., 2020). The various synthesis techniques and characterization processes of nanocellulose via advanced analytical instruments have been summarized in **Figure 2.1**.

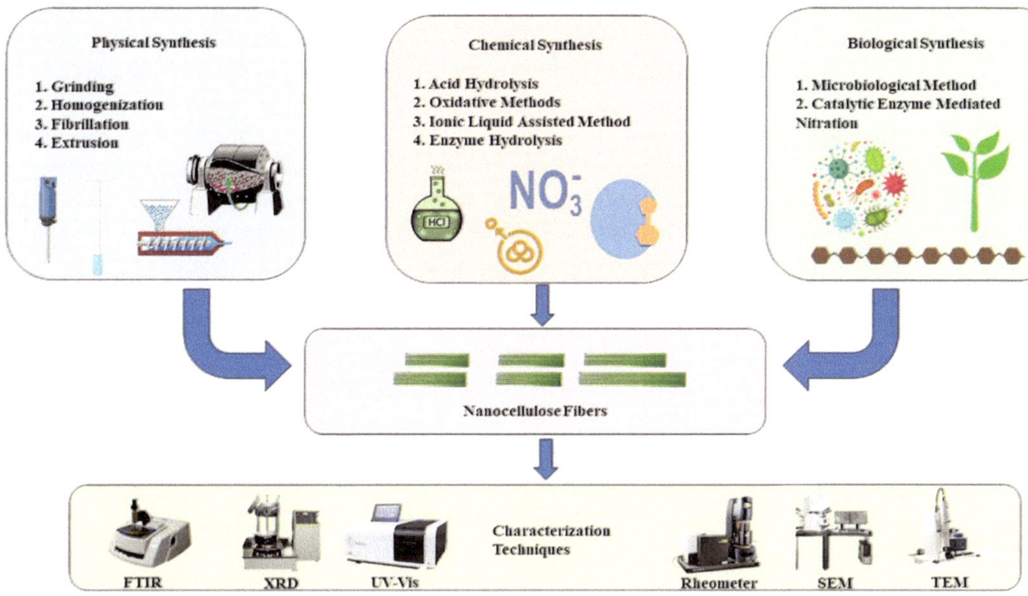

FIGURE 2.1 Glimpse of various synthesis techniques and characterization of nanocellulose fibers via advanced analytical techniques.

2.5 APPLICATIONS OF NANOCELLULOSE IN WOUND HEALING

Nanocellulose-based biomaterials are widely appreciated and renowned in pharmaceutical and medicinal applications—all thanks to their striking and magnificent properties like excellent biocompatibility, biodegradability, and biomimetic properties. Their unique physicochemical properties and the ability to interact with tissues make them a fabulous choice to adapt them in wound healing and tissue regeneration mechanisms. Various forms of nanocellulose-based biomaterials are adapted in wound healing, such as different forms of acetate celluloses, bacterial cellulose, cellulosic hydrogels, and their derivatives.

2.5.1 Acetate Cellulose in Wound Dressings

Cellulose acetate, or acetate cellulose (AC), are ester derivatives of cellulose that are extensively used in pharmaceuticals for their tissue regenerative and recuperative properties. AC comes in different forms, in films and sponges, based upon their structure and methods of synthesis (Tudoroiu et al., 2021; Yang et al., 2024). Cellulose acetate-based sponges can be made by incorporating mixtures of cellulosic nanofibers, cellulose acetoacetate, and other similar derivatives, which have a remarkable application in tissue recovery. The sponges are very light and less dense in nature, which makes them ideal to use as a scaffold for tissue regenerative and wound-mending purposes (Aderibigbe, 2024). In a recent study conducted by Xiao et al., cellulose acetate sponges with incorporated antibacterial agents like gentamicin via enamine bond have been determined to be effective against the growth of opportunistic pathogens like S. aureus and E. coli, with remarkable bacteriostatic properties, which proves these advanced bioactive sponges to be efficacious in inhibiting infections at the wound site by controlled and long-time release of drugs and bioactive agents like growth factors or corporate anti-inflammatory agents, which aids in smooth wound recovery with fewer adverse side effects like scarring and pain (Xiao et al., 2018).

Cellulose acetates are known for their extensive biocompatibility. Their sophisticated interaction with tissues makes them reconcilable with natural tissues with less risk of rejection. AC is hydrophilic in nature, which makes them an excellent agent in maintaining moist conditions near the site of trauma, which aptly aids in a soothing effect and healing. AC can dissolve in most organic solvents, which makes it a better choice to incorporate other bioactive materials like antibiotics, drugs, growth factors, and healing agents easily. These properties make it possible to use them as enteric-coated drug delivery devices and as band-aids having active pharmaceutical agents, which can ensure optimal drug release rates and a higher possibility for faster recovery. Improved drug utilization by cellulose acetate is highly advantageous in wound healing applications, as it ensures slow and steady drug release for a longer period of time. AC has a long water retention time, which improves the moist environment around the region of injury. This property is calculated to produce a conductive setting for faster tissue recuperation (Tudoroiu et al., 2021; Yang et al., 2024).

Usually, acetate celluloses are prepared from electrospinning methods. This makes it highly porous, which can ensure easy exchange of gases and drugs between the tissues. The large specific surface area of nanocellulose acetates is advantageous in a wide range of applications in wound healing. Along with a unique 3D network structure, cellulose acetates provide stability and rigidity, which makes them effective in providing mechanical strength to wound sites. These properties are also known to conduce mechanical and neural stimulus to cope with faster tissue regeneration and recovery. The porous structure, surface area, and highly networked structure mimic the extracellular matrix of the tissues, ensuring smooth compatibility as a curative and effective biotherapeutic drug delivery agent (Yang et al., 2024). A prominent study by Chen et al. describes the advancements in the production of acetate nanocellulose with different thymols by electrospinning. It was perceived that this method had enhanced water healing capacity and porous surface of AC, which provides scope for improved gaseous exchange between the tissues. It was established that initially these developed acetate cellulosic structures had a slow rate of drug release, but later, longer drug-releasing effects were observed. This had a significant impact on controlled drug-releasing events that dilatorily cause healing actions with enhanced properties like improved tissue interactions, pain-relieving activity, constant drug delivery, and enhanced biocompatibility. It was also seen that these acetate celluloses had a significant inhibitory bacteriostatic effect on *S. aureus* and *E. coli*, preventing infections. The porous structure and controlled drug release action also play a role in promoting the growth of L929 cells, specifically fibroblast cell lines, which are markers for squamous and cell differentiation, ensuring good cell compatibility of acetate celluloses and their decisive role in wound healing (Chen et al., 2020).

Furthermore, Gomaa et al. studied novel interactive wound dressing mats composed of polylactic acid/cellulose acetate (PLA/CA), which is described in **Figure 2.2 (i).** Thymoquinone (TQ) was incorporated in the scaffolds to exhibit its antibacterial property. These newly fabricated TQ-loaded PLA/CA wound dressings had several benefits, such as imitating the extracellular matrix (ECM) through a 3D nanofibrous structure and facilitating cell proliferation via the hydrophilicity and bioactivity of CA.

In addition, the wound dressings avoided infection by bacteria at an early stage due to the presence of TQ, as well as sustaining drug release for 9 days, which kept the bacterial load in the wound area to its minimal level. The wound healing process was significantly enhanced by *in vivo* evaluation of TQ-loaded PLA: CA (7:3) scaffolds, with increased re-epithelialization and controlled formation of granulation tissue. These results suggest that the developed TQ-loaded PLA/CA nanofibrous mats could be considered a good candidate for wound dressing applications (Gomaa et al., 2017).

Recently, Elsayed et al. developed acetate cellulose and its derivatives commingled with sulfonamide derivatives to create biodegradable acetate celluloses. Prepared by many methods, primarily by electrospinning, these developed acetate celluloses were also found to be highly potent against the growth of pathogenic microorganisms. These structures can be further consolidated with polylactic acid, which itself is a renowned bioactive agent with therapeutic benefits. When combined with acetate cellulose, it imparts enhanced interaction with cells, especially fibroblasts, and enables water and other molecules to easily diffuse through the matrix, which is crucial for

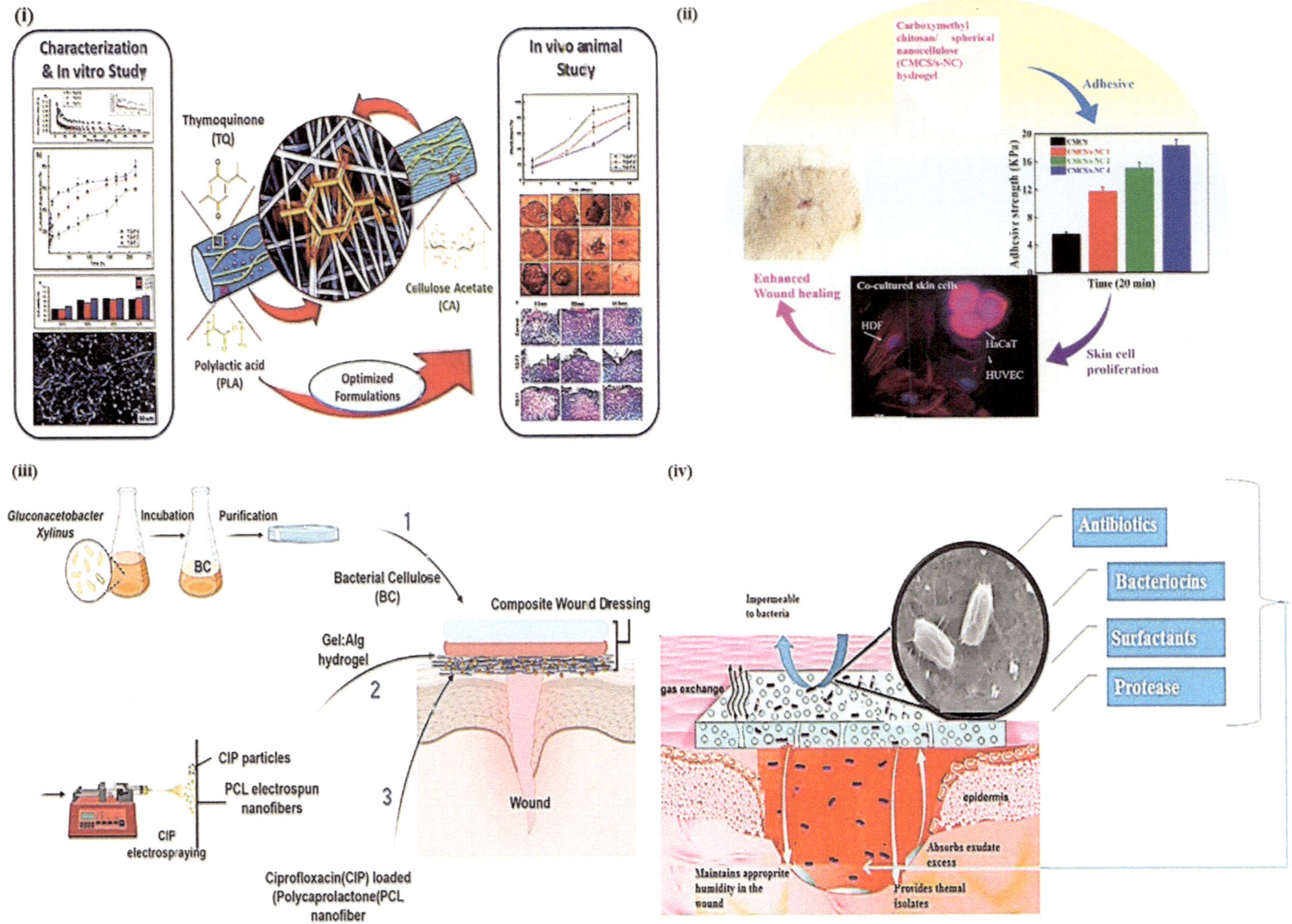

FIGURE 2.2 **(i)** Diagrammatic representation of thymoquinone-loaded polylactate/cellulose acetate-based interactive nanofibers depicting histopathological inferences showing enhanced *in vitro* and *in vivo* wound healing and tissue regeneration, reproduced with permission from Gomaa et al. (2017), Copyright 2017, Elsevier. **(ii)** Graphical representation of spherical nanocellulose-based functional hydrogels with exceptional adhesive properties and exhibiting enhanced skin cell proliferation aiding in effective wound healing, reproduced with permission from Patel et al. (2022), Copyright 2022, Elsevier. **(iii)** Schematic representation of the synergistic action of multilayered bacterial cellulose hydrogels with drug-loaded nanofiber composite as an advanced wound dressing material, reproduced with permission from Mohaghegh et al. (2024), Copyright 2024, Elsevier. **(iv)** Understanding of a versatile BC/BS-based biocomposite which accelerates wound repair and inhibits infection at the wound site (Savitskaya et al., 2019).

fluidic and gaseous exchange, aiding in the coping of injured cutaneous structures. The controlled release of anti-inflammatory agents by this acetate cellulose, sulfonamide, and polylactic acid-induced biomaterial avoids bacterial infections, sepsis, and scarring of the skin around the injured site. It is also known to induce premature granulation, tissue formation, and epithelialization, which are substantial stages in wound healing and tissue repair mechanisms. These biomaterials minimize inflammation, promote regeneration of blood vessels, support rapid healing, and provide pain relief, confronting patient-centric modulations and seamless healing therapy (Elsayed et al., 2020). The water retention property, cellular interactive nature, tissue regenerative capacities, porosity, exchange mediating property, and relieving effects make acetate cellulose forms ideal for healing burns, trauma, and chronic injuries (Yang et al., 2024).

2.5.2 Hydrogel Systems for Dressings

In the case of deep burn wounds, chronic injuries, and irregular wounds, the major complications during wound treatment are pain, burning sensation, prolonged healing times, re-injury during dressing changes, and injury due to separation of exudates from dehydration. To address these issues, hydrogels emerge as a "superstar" in wound healing therapies. Hydrogels are colloidal substances composed of about 90% water and insoluble polymers that remain within, forming a complex porous structure. Synthetic hydrogels are made of polyvinylpyrrolidone (PVP) and polymethyl acrylate (PMA), while some hydrogels are composed of alginate dressings. Hydrogels are popular for their high-water retention property, hydrophilicity, and 3D porous structure, all of which are ideal characteristics that position them among competitors for an ideal wound healing material (P. Wang et al., 2024; Yang et al., 2024).

The water retention property of hydrogel lies within its porous structure, whereas the hydrophilicity depends upon the cross-linking polar functional groups through which polymers are interconnected to each other. The shape and structure of hydrogels are dynamic—they can expand and contract freely as required. They are nonsticky and hence easy to handle. These properties of hydrogels make them fit for use in dressings to heal wounds (Aderibigbe, 2024). Hydrogel-based dressings have good biocompatibility. Their porous structure resembles the extracellular matrix of the tissues, rendering biocompatibility. Hydrogels are advantageous in wound healing because of their advanced functions like antimicrobial response, adhesive properties, hemostasis, anti-inflammatory and antioxidant behavior, substance delivery agent, stimulus response, and wound monitoring (Liang et al., 2021).

Nanocellulose-based hydrogels can be prepared by cross-linking the derivatives of cellulose like carboxymethyl chitosan (CMC) with dialdehyde modified cellulose nanocrystals (DACNC). The cross-linking is brought about by Schiff bases between the amine group of CMC and the aldehydic group of DACNC. Nanocellulose-based hydrogels are known for various benefits—(i) providing an ideal moist environment around the wound by their water retention property, (ii) absorbing and retaining wound exudates by their absorption capabilities, which prevents subsequent injuries, infections, and pain, (iii) providing good mechanical strength and tensile stimulus to the injured site by their dynamic omni-form shape, and (iv) acting as a good barrier against pathogens and preventing infections (Yang et al., 2024). Hydrogels aid in hemostasis not only by sealing the wound physically but also by enriching the coagulation factors through the absorption of wound extracts (Liang et al., 2021). Recent advances in hydrogel formulations have proceeded in the direction of loading bioactive agents into the hydrogel structure, enhancing its potency and activity. Erdagi I et al. have reported results of an investigation into diosgenin-modified interpenetrating networks by gelatin cross-linked nanocellulose hydrogels. It was observed that E. coli and S. aureus were significantly inhibited by neomycin and diosgenin, with inhibition ranging from 50% to 88%. The swelling behavior of the hydrogels was affected by the concentrations of nanocellulose and diosgenin, with pore diameters reported between 10 to 30 micrometers. Beyond these, many exceptional properties like increased fibroblast activity to reduce inflammation and noncytotoxicity to L929 and fibroblast cell lines in mammals, have also been tested in vivo for nanocellulose-based drug-loaded hydrogels (Ilkar Erdagi et al., 2020).

Nanocellulose-based hydrogels come in various forms based on their chemical and biological properties. They can be cationic polymers, like quaternized hydroxyethyl nanocellulose, which aids in hemostasis, or they can be self-healing hydrogels, like oxidized cellulose nanocrystals. Some nanocellulose hydrogels are involved in stimulus response induction. Researchers like Goh et al. have prepared these by loading cytokinin, which promotes angiogenesis, or by loading factors like epithelial growth factors (EGFs), CD31, and many others that promote tissue regeneration and stimulus response. In recent times, there has been a trend in synthesizing hydrogels with dynamic bonding through cross-linking in situ polymers and interpenetrating networks, which can provide good advantages in terms of rigidity, stability, and tensile strength (Goh et al., 2016).

Antibacterial nanocellulose-based hydrogels synthesized as a result of loading inorganic metal nanoparticles are also in trend. Maliha et al. has worked on incorporating metal NPs like silver, gold, zinc, and copper, which have already been tested for their excellent antibacterial, bactericidal, and bacteriostatic potencies, into hydrogels. Hydrogels containing inorganic complexes like phenyl bisphosphate bismuth (III) were reported to inhibit the growth of potential pathogens like *Pseudomonas aeruginosa* and *Acinetobacter baumannii* (Maliha et al., 2021). Many such hydrogels have also been tested against methicillin-resistant *Staphylococcus aureus* and vancomycin-resistant *Enterobacteriaceae* and have proven to be successful inhibitors of such pathogens (Yang et al., 2024). Recently, in a study conducted by Basu et al., metal ion cross-linking nanocellulose-based hydrogels with copper and calcium ions have also been proven to be antibacterial in nature. Calcium ions inhibit the growth of biofilm-forming *Pseudomonas*, and copper ions inhibit the growth of *Staphylococcus epidermidis*, thus preventing chronic wounds and infections (Basu et al., 2018).

According to a study conducted by Patel et al., tissue regeneration complications in wound repair can be addressed by hydrogels that are reinforced with spherical nanocellulose (s-NC) and carboxymethyl chitosan, using better adhesion, injectability, and conductivity (**Figure 2.2 (ii)**). They outperform cellulose nanocrystals in terms of biological activity. With the help of such improved antibacterial properties, these hydrogels tested on rat wound models provide insights into a prospect for quicker regeneration of skin. Since they can be given any shape according to requirements, these materials are malleable. As far as skin is concerned, this advancement represents a major breakthrough in biomaterials that could be useful for wound repair in tissue engineering (Patel et al., 2022). Hydrogel-based sponges, characterized by hydrophilic gels with a 3D network structure, are the recent highlights in cellulose-based biomaterials in wound repair. They are found to expand rapidly in water. This property holds significant promise for wound healing applications, facilitating increased water retention capacities and providing a supportive environment for tissue regeneration, thereby potentially accelerating the healing process (Mohite et al., 2024).

Since most of the fraction of hydrogel contains water, it provides a cooling effect to the site of injury. This thermoregulation aids in the anti-inflammatory response and pain-relieving action. The exchange of water brought about by hydrogel dressings provides for a numbing effect against burns and pain. Being easy to handle and self-healing, hydrogels can also be directly injected into large and irregular wounds for utmost coverage and maximum benefits. All these properties of hydrogels make them an archetypal material for constructing dressings for wound healing benefits (Aderibigbe, 2024; Liang et al., 2021).

2.5.3 Bacterial Cellulose in Wound Dressing

Bacterial cellulose (BC) is an excellent and promising bioactive material that is effective against complex wounds (Yang et al., 2024). Originating naturally from Southeast Asia, where cellulose is produced by anaerobic bacteria of the genus *Komagataeibacter* spp. by absorbing nutrients from coconut water, it was consumed as a popular dessert—*Nata de coco*. Little did the people there know that this dessert could be a boon in medical and pharmaceutical applications. Bacterial cellulose is produced by various bacteria of the genera *Agrobacterium*, *Rhizobium*, *Salmonella*, and *Pseudomonas*. The most efficient producer is *Komagataeibacter xylinus*, which produces a fibrillar nanobiocellulosic structure anaerobically. BC membranes are produced synthetically by static culturing, and spherical or fibrous BC can be synthesized by agitated culture techniques. It is one of the nontoxic biomaterials of huge medicinal importance (Horue et al., 2023).

BC is a felicitous biomaterial for tissue regeneration and tissue engineering uses. It has moisture retention capacity; has a mesh-like networked structure, giving it strong supporting force and a porous matrix that allows penetration and exchange of liquids and gases through and within its fibrillar structure. With these advantageous properties, BC is naturally nontoxic and has good biocompatibility—making it an apt choice for wound dressings. BC is available commercially in wound dressings under different brands like Bionext®, Membracell®, and Xcell®, which mimic the extracellular matrix of the cells to improve and foster epithelialization. BC is proven to have high *in vivo* biocompatibility and the ability to replenish cell fixation and wound healing by providing a 3D substrate, quintessential for tissue regeneration and proliferation. BC is known for its rapid epithelialization properties and enhanced tissue generation rates, which extend its use in the dressings of complicated wounds like diabetic foot wounds, chronic wounds, and severe burns (Horue et al., 2023). BC-based biomaterials are preferred over synthetic nanocellulose-based dressings because of a higher range of biocompatibility than those that are synthetically produced (Yang et al., 2024).

While BC is definitely a beneficial biomaterial, it also comes with many drawbacks. It is observed that bacterial cellulose lacks antimicrobial properties. To overcome this drawback, antimicrobial agents are impregnated into fibers within the BC structures by physical or chemical loading methods (Horue et al., 2023; Yang et al., 2024). Recently, Asanarong et al. have prepared BC incorporated with papain solution by dipping it, which enhances cutaneous cell regenerative properties. Papain, obtained from papaya latex, is naturally known to possess bioactive activities like antibacterial efficacy and antioxidant properties, which contribute to wound healing processes when combined with bacterial cellulose (Asanarong et al., 2021). Another prominent study by Wan et al., states that BC can also be combined with inorganic nanoparticles like silver nanowires and nanoparticles to intensify its antibacterial properties (Wan et al., 2020). Another drawback of BC is that it possesses a relatively shorter drying time—that is, it dries out soon. Thus, BC-based dressings need to be applied in two layers, where the outer second layer should be replaced once every few weeks. This further increases the chances of re-injury during the changing of the dressings or increases the risk of chronic inflammation leading to delayed healing. It also causes hospital costs for patients. Study and innovations on increasing the liquid retention properties of BC have now become a prominent area of research in pharmaceutics to develop BC-based composites in wound healing with enhanced properties, making it suitable for use. There are various ways in which the liquid retention properties of BC can be amplified. One method is that during the biosynthesis, some other composites of biopolymers may be added to enhance the liquid retention property. These biopolymers may be chitosan or alginate or their derivatives. Some of these impart antibacterial activities to BC. The production of BC is limited. As it is a metabolite produced by the bacteria, it is difficult to produce. The product needs to be properly purified to remove all biological and chemical impurities, which leads to fewer finished products. BC produced in this manner cannot meet the demands of the industry. Modified BC is available where alginate is evenly distributed within the fibrillar structure of BC, which renders its availability, as well as providing a longer drying time than neat BC. It also enhances drug release rates, conferring value-added benefits in the production and properties of BC (Yang et al., 2024). **Table 2.1** further summarizes various sources, synthesis techniques, and biomedical applications of nanocellulose fibers.

Recent advances in BC-based dressings include the development of state-of-the-art BC-based dressings with aggrandized biocompatibility and therapeutic properties.

In recent research conducted by Mohaghegh et al., several-layered wound dressings that contain bacterial cellulose hydrogel, gelatin/alginate hydrogel, and ciprofloxacin-laden polycaprolactone nanofibers were fabricated (Figure 2.2 (iii)). Such dressings showed good performance in terms of

TABLE 2.1
Summarization of Various Types of Nanocellulose Derivatives From Different Sources, Synthesis Techniques, and Their Biomedical Applications

Sources	Types	Methods of Isolation	Dimensions of Isolated Nanocellulose	Applications	References
Industrial cotton waste	Cellulose nanofibers (CNF)	Acid hydrolysis	Length (100–200 nm), width (10–20 nm)	Water remediation and antibacterial nanocellulose	Vismara et al. (2021)
Miscanthus sacchariflorus	Cellulose nanocrystals (CNCs)	Sulfuric acid hydrolysis	Length (680 ± 61 nm), diameter (18 ± 4 nm)	Reinforcing agents	El Achaby et al. (2018)
Corncob wastage	Esterified cellulose nanofibers (E-CNFs)	Mechano-chemical esterification	Length (10–100 nm), width (1.5–2.8 nm)	Electronics and optical devices	Kang et al. (2017)
Nypa fruticans trunk, coconut husk fiber, rice husk	Cellulose nanocrystals (CNCs)	Acid hydrolysis	Length (200–500 nm), width (10–15 nm)	Optoelectronics, packaging, mechanically reinforced polymer composites, tissue scaffolds, environmental remediation	Nang An et al. (2020)
Dunchi fiber	Cellulose Nanocrystals (CNCs)	Alkali treatment, Chlorine free bleaching, Sulphuric acid hydrolysis	Length (202.87 ± 79.87 nm), Width (20.67± 7.39nm)	Reinforcement materials in composite and packaging	(M. N. Khan et al., 2020)
Sugar beet molasses	Bacterial cellulose nanocrystals (BCNCs)	Sulfuric acid hydrolysis	Length (306 ± 112 nm), diameter (25 ± 5 nm)	Bio-nanocomposite films, food packaging	Salari et al. (2019)
Cannabis sativa (hemp)	Cellulose nanofibers (CNFs)	Chemo-mechanical method	Length (659–1056 nm), diameter (64–76 nm)	Reinforcing agents	Rana & Gupta (2020)
Wheat bran	Cellulose nanocrystals (CNCs)	Sulfuric acid hydrolysis	Length (644.77 ± 225.20 nm), diameter (33.80 ± 9.83 nm)	Reinforcing agents	Xiao et al. (2019)
Ramie fibers (*Boehmeria nivea*)	Cellulose nanocrystals (CNCs)	Sulfuric acid hydrolysis, sonication, and homogenization	Length (145.61 nm), diameter (6.67 nm)	Reinforcing agents	Kusmono et al. (2020)
Walnut shells	Cellulose nanocrystals (CNCs)	Acid hydrolysis	Diameter (130 nm)	Reinforcement materials in composite and packaging	Hemmati et al. (2018)

water uptake, controlled drug release, and faster healing of wounds, which indicated possibilities for their application in medicine (Mohaghegh et al., 2024). Another development is the assimilation of synthetic polymers like polydopamine (PDA) and poly-ε-lysine (PL) with BC to produce an advanced BC composite, as prepared by Wahid et al. in their study. This is known as BCP@ε-PL, and it is renowned for its antibacterial activity, enhanced hemocompatibility, and cytocompatibility (Wahid et al., 2021). Further innovation in the pursuit of BC is the development of flexible BC by incorporating polycaprolactone, which is done by Kotcharat et al. This is further integrated with antibiotics like gentamicin and streptomycin to confer excellent antibacterial activity and outstanding biocompatibility. This BC-based composite is also found to be nontoxic to BHK-21 cells *in vivo*, the cell lines of fibroblasts from the kidneys of baby hamsters, which are markers for providing a structural framework for tissues and cells, thus proving BC composites to be advantageous in wound healing (Kotcharat et al., 2021).

Furthermore, Figure 2.2 (iv) deals with the research conducted by Savitskaya et al., where a biocomposite comprising bacterial cellulose (BC) gel-film and *Bacillus subtilis* (BS) cells was fabricated and assessed for potential biomedical applications. *Staphylococcus aureus*, *Staphylococcus epidermidis*, *Escherichia coli*, and *Pseudomonas aeruginosa* are common pathogens that cause wound infections. The immobilized *Bacillus subtilis* (BS) cells demonstrated significant antagonistic activity against these pathogens. Applying the BC/BS biocomposite to excision wounds of animal models sped up the healing process and promoted reparative processes (Savitskaya et al., 2019). Ultrastructural studies of BC-based composites were studied by Das et al., reveals that they provide a uniformly distributed scaffold for drug release in the wound site, which helps in healing (M. Das et al., 2022). BC composites can also be doped with zinc oxides and titanium oxides, which have improved their mechanical properties, enhanced their cell adhesive properties, and played a vital role in accelerating burn healing in mouse models. These studies reveal that upon combining BC with suitable agents like biopolymers, chemicals, or drugs, their drawbacks could be combated and their properties can be enhanced (Horue et al., 2023). There is a lot of scope for discovering BC-based composites in dressings to enhance their properties and to develop a patient-centric and medically ideal dressing material for faster wound healing and tissue repair (Horue et al., 2023; Yang et al., 2024).

2.6 CONCLUSION AND FUTURE PROSPECTS

It is found that the nanocellulose-based wound dressing definitely introduces a remarkable platform in tissue engineering and biomaterial science due to its ideal biological, physicochemical properties, and molecular structure. The rejuvenation of damaged tissue can be done with nanocellulose-based wound dressings through different overlapping phases of tissue repair, in which remodeling will play a significant role in the process of healing. In recent years, the biocompatibility, biodegradability, porosity, reduced scarring, flexibility, and conformability of nanocellulose-based wound dressings contribute to rapid and better tissue regeneration. Nowadays, biomaterials promise efficient wound care with reduced sepsis. Hydrogels are one among those biomaterials where the high water content (90%) and insoluble polymers aid in better wound care and management. Many research studies are ongoing to improve the mechanical strength and bioactivity of hydrogels by merging them with nanoparticles, nanomaterials, antimicrobial agents, etc. Along with hydrogels, electrospun nanofibrous scaffolds and 3D bioprinting technologies help the tissue reintegrate with its normal structure and function.

Due to their high precision and resolution, 3D and 4D bioprinting help in the accumulation of biomaterials, biological cells, and bioink with exceptional accuracy. Multimaterial printing is one of the significant applications of 3D and 4D bioprinting, which enables the embedment of multiple cells and bioactive molecules within the same casting print. This aids in the development of heterogeneous tissues and organs with explicit features such as vascular networks, gradients of growth factors, and mechanical properties. The integration of stimuli-responsive materials into 4D bioprints helps in the fabrication of adaptive tissues that ease self-organization and morphogenesis, thereby promising an ideal bioconstruct that will remodel, repair, and respond to harsh physiological cues. The incorporation of artificial intelligence and machine learning into nanocellulose-based wound healing fosters material characterization and prediction of various physicochemical features of nanocellulose using AI or ML algorithms. AI can analyze some reaction parameters such as temperature, pH, agitation speed, etc., for improved quality and effectiveness. Machine learning contributes to tailoring the material design based on wound size, depth, anatomical site, and exudate level, which can suit different types of wounds. The development of predictive models using AI/ML for the efficacy of wound healing and biomaterials promotes rapid healing with fewer complications. By incorporating AI sensors into nanocellulose wound dressings, individuals can get real-time wound status and can detect the moisture level and microbial load around the wound to prevent sepsis and further complications.

The current study of nanocellulose-based wound healing lays a foundation in the field of tissue engineering due to its exceptional healing properties. Nanocellulose can be synthesized using physical, chemical, and biological methods, depending on the part of cellulose disrupted and the agents used for the disruption. Nanocellulose can be combined with various wound dressings, either alone or with other biomaterials, for the advanced healing of the wound. Nanocellulose-based wound dressings aid in creating a nurturing environment for cell migration, proliferation, and cell differentiation, which facilitates rapid wound closure and a shorter healing time period. Due to more systematic, coordinated, and well-regulated tissue regeneration, nanocellulose-based wound dressings decrease scar formation with proficient cosmetic outcomes of wound healing. It has been claimed that nanocellulose scaffolds can be combined with various kinds of antibacterial and anti-inflammatory agents to prevent sepsis at the origin site of the wound, as well as some growth factors that support the healing process. Another ideal property of nanocellulose is its ability to confirm the optimal interface between the wound dressing and damaged tissue, which is flexibility and conformability, making it more effective as the basic principle of tissue regeneration. Nanocellulose can be procured from renewable sources, leading to eco friendly scaffolds that can replace synthetic wound healing materials.

Most pathophysiological conditions associated with wounds involve tissue decay or mortification resulting from injury or lack of blood supply. To overcome this biological threat, we can opt for nanocellulose-based wound dressings due to the highly porous structure of nanocellulose, which permits the smooth exchange of oxygen, nutrients, and developmental factors in and around the wound, essential for tissue recovery achieved via the facilitation of cellular infiltration. Recently, it was found that silver nanoparticles (AgNPs) combined with nanocellulose tend to have inbuilt antimicrobial properties that can inhibit pathogens from invading wound sites. The AgNPs, in association with the bacterial cell membrane, disturb DNA replication and help program the death of pathogenic bacteria. The AgNPs form a perfect and stable suspension due to the high surface area of nanocellulose, resulting in a synergistic effect. Along with these, there are different forms of nanocellulose that have significant effects on wound healing. Acetate cellulose and bacterial cellulose are examples of nanocellulose that have been extensively used to regenerate damaged tissue. Acetate cellulose, a derivative, has a higher ability to absorb water and can maintain a moist wound area necessary for the healing process. Cellulose acetate is tolerable to the human body, which guarantees that there won't be significant adverse effects, and it is a biodegradable wound dressing with a higher point of recovery and patient safety. Another kind of bioactive material for complex wounds is bacterial cellulose, obtained naturally from the bacterial genera *Komagataeibacter*, *Acetobacter*, and *Gluconacetobacter*, respectively, which assist in the reintegration of normal cellular components for the regeneration of damaged tissue, which is essential for normal function and tissue activities.

Tissue engineering in the wound healing domain will introduce remarkable advanced biomedical applications with nanocellulose-based biomaterials. More research is unfolding on nanocellulose-based wound dressings for better wound healing and management. In the future, immunomodulatory biomaterials can be applied to modulate immune response and aid regenerative healing by avoiding the formation of fibrotic tissue. Nanocellulose-incorporated drugs can be effective against particular pathogenic conditions, representing a novel method for drug delivery for infections associated with complex wounds. Nanocellulose-based biomaterials can be customized according to various wound characteristics such as wound size, anatomical position, and severity, which are purely based on personalized medicine, providing patient-centric care and enhancing the patient quality of life.

2.7 LIST OF KEYWORDS

Wound healing; Nanocellulose; Biomaterials; Wound dressing; Anti-oxidant; Anti-inflammatory.

REFERENCES

Abdo, J. M., Sopko, N. A., & Milner, S. M. (2020). The applied anatomy of human skin: A model for regeneration. *Wound Medicine*, *28*, 100179. https://doi.org/10.1016/j.wndm.2020.100179

Aderibigbe, B. A. (2024). Application of nanocellulose for wound dressings. In *Nanocellulose* (pp. 193–219). Wiley. https://doi.org/10.1002/9781394172825.ch8

Agnihotri, S., & Dhiman, N. K. (2017). Development of nano-antimicrobial biomaterials for biomedical applications. In *Advances in Biomaterials for Biomedical Applications* (pp. 479–545). Springer Nature Singapore. https://doi.org/10.1007/978-981-10-3328-5_12

Al Abdallah, H., Tannous, J. H., & Abu-Jdayil, B. (2024). Cellulose and nanocellulose aerogels, their preparation methods, and potential applications: A review. *Cellulose*. https://doi.org/10.1007/s10570-024-05743-w

Amezcua-Allieri, M. A., Sánchez Durán, T., & Aburto, J. (2017). Study of chemical and enzymatic hydrolysis of cellulosic material to obtain fermentable sugars. *Journal of Chemistry*, *2017*, 1–9. https://doi.org/10.1155/2017/5680105

Ansari, M. J., Rajendran, R. R., Mohanto, S., Agarwal, U., Panda, K., Dhotre, K., Manne, R., Deepak, A., Zafar, A., Yasir, M., & Pramanik, S. (2022). Poly(N-isopropylacrylamide)-based hydrogels for biomedical applications: A review of the state-of-the-art. In *Gels* (Vol. 8, Issue 7). https://doi.org/10.3390/gels8070454

Arantes, V., Dias, I. K. R., Berto, G. L., Pereira, B., Marotti, B. S., & Nogueira, C. F. O. (2020). The current status of the enzyme-mediated isolation and functionalization of nanocelluloses: Production, properties, techno-economics, and opportunities. *Cellulose*, *27*(18), 10571–10630. https://doi.org/10.1007/s10570-020-03332-1

Asanarong, O., Minh Quan, V., Boonrungsiman, S., & Sukyai, P. (2021). Bioactive wound dressing using bacterial cellulose loaded with papain composite: Morphology, loading/release and antibacterial properties. *European Polymer Journal*, *143*, 110224. https://doi.org/10.1016/j.eurpolymj.2020.110224

Bacakova, L., Pajorova, J., Bacakova, M., Skogberg, A., Kallio, P., Kolarova, K., & Svorcik, V. (2019). Versatile application of nanocellulose: From industry to skin tissue engineering and wound healing. *Nanomaterials*, *9*(2), 164. https://doi.org/10.3390/nano9020164

Barbash, V. A., Yashchenko, O. V, Gondovska, A. S., & Deykun, I. M. (2022). Preparation and characterization of nanocellulose obtained by TEMPO-mediated oxidation of organosolv pulp from reed stalks. *Applied Nanoscience*, *12*(4), 835–848. https://doi.org/10.1007/s13204-021-01749-z

Barbash, V. A., Yashchenko, O. V, & Vasylieva, O. A. (2019). Preparation and properties of nanocellulose from *Miscanthus x giganteus*. *Journal of Nanomaterials*, *2019*, 1–8. https://doi.org/10.1155/2019/3241968

Basu, A., Heitz, K., Strømme, M., Welch, K., & Ferraz, N. (2018). Ion-crosslinked wood-derived nanocellulose hydrogels with tunable antibacterial properties: Candidate materials for advanced wound care applications. *Carbohydrate Polymers*, *181*, 345–350. https://doi.org/10.1016/j.carbpol.2017.10.085

Bharimalla, A. K., Deshmukh, S. P., Patil, S., Nadanathangam, V., & Saxena, S. (2023). Development of energy efficient nanocellulose production process by enzymatic pretreatment and controlled temperature refining of cotton linters. *Cellulose*, *30*(2), 833–847. https://doi.org/10.1007/s10570-022-04959-y

Bianchera, A., Catanzano, O., Boateng, J., & Elviri, L. (2020). The place of biomaterials in wound healing. In *Therapeutic Dressings and Wound Healing Applications* (pp. 337–366). Wiley. https://doi.org/10.1002/9781119433316.ch15

Budaeva, V. V, Gismatulina, Y. A., Mironova, G. F., Skiba, E. A., Gladysheva, E. K., Kashcheyeva, E. I., Baibakova, O. V, Korchagina, A. A., Shavyrkina, N. A., Golubev, D. S., Bychin, N. V, Pavlov, I. N., & Sakovich, G. V. (2019). Bacterial nanocellulose nitrates. *Nanomaterials*, *9*(12), 1694. https://doi.org/10.3390/nano9121694

Cancio, L. C. (2021). Topical antimicrobial agents for burn wound care: History and current status. *Surgical Infections*, *22*(1), 3–11. https://doi.org/10.1089/sur.2020.368

Chen, Y., Qiu, Y., Chen, W., & Wei, Q. (2020). Electrospun thymol-loaded porous cellulose acetate fibers with potential biomedical applications. *Materials Science and Engineering: C*, *109*, 110536. https://doi.org/10.1016/j.msec.2019.110536

Choi, S. M., Rao, K. M., Zo, S. M., Shin, E. J., & Han, S. S. (2022). Bacterial cellulose and its applications. *Polymers*, *14*(6), 1080. https://doi.org/10.3390/polym14061080

Daicho, K., Fujisawa, S., Kobayashi, K., Saito, T., & Ashida, J. (2020). Cross-polarization dynamics and conformational study of variously sized cellulose crystallites using solid-state 13C NMR. *Journal of Wood Science*, *66*(1), 62. https://doi.org/10.1186/s10086-020-01909-9

Das, M., Zandraa, O., Mudenur, C., Saha, N., Sáha, P., Mandal, B., & Katiyar, V. (2022). Composite scaffolds based on bacterial cellulose for wound dressing application. *ACS Applied Bio Materials*, *5*(8), 3722–3733. https://doi.org/10.1021/acsabm.2c00226

Das, S., Mukherjee, T., Mohanty, S., Nayak, N., Mal, P., Ashique, S., Pal, R., Mohanto, S., & Sharma, H. (2024). Impact of NF-κB signaling and sirtuin-1 protein for targeted inflammatory intervention. *Current Pharmaceutical Biotechnology*. https://doi.org/10.2174/0113892010301469240409082212

Debiagi, F., Faria-Tischer, P. C. S., & Mali, S. (2018). *Cellulose and Nanocellulose Produced from Lignocellulosic Residues by Reactive Extrusion* (pp. 227–242). https://doi.org/10.1021/bk-2018-1304.ch012

Debiagi, F., Faria-Tischer, P. C. S., & Mali, S. (2021). A green approach based on reactive extrusion to produce nanofibrillated cellulose from oat hull. *Waste and Biomass Valorization*, *12*(2), 1051–1060. https://doi.org/10.1007/s12649-020-01025-1

Deng, X., Gould, M., & Ali, M. A. (2022). A review of current advancements for wound healing: Biomaterial applications and medical devices. *Journal of Biomedical Materials Research Part B: Applied Biomaterials*, *110*(11), 2542–2573. https://doi.org/10.1002/jbm.b.35086

Dudhamal, T. S. (2023). Review of grey literature on Ayurveda wound healing formulations and procedures—A systematic review. *Journal of Ayurveda and Integrative Medicine*, *14*(4), 100779. https://doi.org/10.1016/j.jaim.2023.100779

El Achaby, M., El Miri, N., Hannache, H., Gmouh, S., Trabadelo, V., Aboulkas, A., & Ben Youcef, H. (2018). Cellulose nanocrystals from Miscanthus fibers: Insights into rheological, physico-chemical properties and polymer reinforcing ability. *Cellulose*, *25*(11), 6603–6619. https://doi.org/10.1007/s10570-018-2047-1

Elsayed, R. E., Madkour, T. M., & Azzam, R. A. (2020). Tailored-design of electrospun nanofiber cellulose acetate/poly(lactic acid) dressing mats loaded with a newly synthesized sulfonamide analog exhibiting superior wound healing. *International Journal of Biological Macromolecules*, *164*, 1984–1999. https://doi.org/10.1016/j.ijbiomac.2020.07.316

Fernandes, M., Alves, C., Melro, L., Fernandes, R. D. V, Padrão, J., Salgado, A. J., & Zille, A. (2023). Modification of nanocellulose. In *Handbook of Biomass* (pp. 1–39). Springer Nature Singapore. https://doi.org/10.1007/978-981-19-6772-6_35-1

Fu, R., Li, J., Yu, H., Zhang, Y., Xu, Z., & Martin, C. (2021). The yin and yang of traditional Chinese and Western medicine. *Medicinal Research Reviews*, *41*(6), 3182–3200. https://doi.org/10.1002/med.21793

Gholap, A. D., Kapare, H. S., Pagar, S., Kamandar, P., Bhowmik, D., Vishwakarma, N., Raikwar, S., Garkal, A., Mehta, T. A., Rojekar, S., Hatvate, N., & Mohanto, S. (2024). Exploring modified chitosan-based gene delivery technologies for therapeutic advancements. *International Journal of Biological Macromolecules*, *260*, 129581. https://doi.org/10.1016/j.ijbiomac.2024.129581

Goh, M., Hwang, Y., & Tae, G. (2016). Epidermal growth factor loaded heparin-based hydrogel sheet for skin wound healing. *Carbohydrate Polymers*, *147*, 251–260. https://doi.org/10.1016/j.carbpol.2016.03.072

Gomaa, S. F., Madkour, T. M., Moghannem, S., & El-Sherbiny, I. M. (2017). New polylactic acid/cellulose acetate-based antimicrobial interactive single dose nanofibrous wound dressing mats. *International Journal of Biological Macromolecules*, *105*, 1148–1160. https://doi.org/10.1016/j.ijbiomac.2017.07.145

Gowda, B. H. J., Mohanto, S., Singh, A., Bhunia, A., Abdelgawad, M. A., Ghosh, S., Ansari, M. J., & Pramanik, S. (2023). Nanoparticle-based therapeutic approaches for wound healing: A review of the state-of-the-art. *Materials Today Chemistry*, *27*, 101319. https://doi.org/10.1016/j.mtchem.2022.101319

Gupta, G. K., & Shukla, P. (2020). Lignocellulosic biomass for the synthesis of nanocellulose and its eco-friendly advanced applications. *Frontiers in Chemistry*, *8*. https://doi.org/10.3389/fchem.2020.601256

Hani, U., Gowda, B. H. J., Haider, N., Ramesh, K., Paul, K., Ashique, S., Ahmed, M. G., Narayana, S., Mohanto, S., & Kesharwani, P. (2023). Nanoparticle-based approaches for treatment of hematological malignancies: A comprehensive review. *AAPS PharmSciTech*, *24*(8), 233. https://doi.org/10.1208/s12249-023-02670-0

Hayat, M., Nawaz, A., Chinnam, S., Muzammal, M., Latif, M. S., Yasin, M., Ashique, S., Zengin, G., & Farid, A. (2023). Formulation development and optimization of herbo synthetic gel: In vitro biological evaluation and in vivo wound healing studies. *Process Biochemistry*, *130*, 116–126.

Hemmati, F., Jafari, S. M., Kashaninejad, M., & Barani Motlagh, M. (2018). Synthesis and characterization of cellulose nanocrystals derived from walnut shell agricultural residues. *International Journal of Biological Macromolecules*, *120*, 1216–1224. https://doi.org/10.1016/j.ijbiomac.2018.09.012

Horue, M., Silva, J. M., Berti, I. R., Brandão, L. R., Barud, H. da S., & Castro, G. R. (2023). Bacterial cellulose-based materials as dressings for wound healing. *Pharmaceutics*, *15*(2), 424. https://doi.org/10.3390/pharmaceutics15020424

Hubbe, M. A., Tayeb, P., Joyce, M., Tyagi, P., Kehoe, M., Dimic-Misic, K., & Pal, L. (2017). Rheology of nanocellulose-rich aqueous suspensions: A review. *BioResources*, *12*(4), 9556–9661. https://doi.org/10.15376/biores.12.4.Hubbe

Ilkar Erdagi, S., Asabuwa Ngwabebhoh, F., & Yildiz, U. (2020). Genipin crosslinked gelatin-diosgenin-nanocellulose hydrogels for potential wound dressing and healing applications. *International Journal of Biological Macromolecules*, *149*, 651–663. https://doi.org/10.1016/j.ijbiomac.2020.01.279

Irfan-Maqsood, M. (2016). Classification of wounds: Know before research and clinical practice. *Cell Therapy and Regenerative Medicine Journal*, *1*(2), 79. https://doi.org/10.15562/ctrm.21

Jonoobi, M., Oladi, R., Davoudpour, Y., Oksman, K., Dufresne, A., Hamzeh, Y., & Davoodi, R. (2015). Different preparation methods and properties of nanostructured cellulose from various natural resources and residues: A review. *Cellulose*, *22*(2), 935–969. https://doi.org/10.1007/s10570-015-0551-0

Kang, X., Sun, P., Kuga, S., Wang, C., Zhao, Y., Wu, M., & Huang, Y. (2017). Thin cellulose nanofiber from corncob cellulose and its performance in transparent nanopaper. *ACS Sustainable Chemistry & Engineering*, *5*(3), 2529–2534. https://doi.org/10.1021/acssuschemeng.6b02867

Kaur, G., Narayanan, G., Garg, D., Sachdev, A., & Matai, I. (2022). Biomaterials-based regenerative strategies for skin tissue wound healing. *ACS Applied Bio Materials*, *5*(5), 2069–2106. https://doi.org/10.1021/acsabm.2c00035

Kaur, P., Sharma, N., Munagala, M., Rajkhowa, R., Aallardyce, B., Shastri, Y., & Agrawal, R. (2021). Nanocellulose: Resources, physio-chemical properties, current uses and future applications. *Frontiers in Nanotechnology*, *3*. https://doi.org/10.3389/fnano.2021.747329

Khan, M. N., Rehman, N., Sharif, A., Ahmed, E., Farooqi, Z. H., & Din, M. I. (2020). Environmentally benign extraction of cellulose from dunchi fiber for nanocellulose fabrication. *International Journal of Biological Macromolecules*, *153*, 72–78. https://doi.org/10.1016/j.ijbiomac.2020.02.333

Khan, M. R., Wasim, M., Farooq, A., Naeem, M. A., Mushtaq, M., Liu, J., Chen, C., & Wei, Q. (2024). A review study on derivation of nanocellulose to its functional properties and applications in drug delivery system, food packaging, and biosensing devices. *Polymer Bulletin*. https://doi.org/10.1007/s00289-024-05190-4

Kolimi, P., Narala, S., Nyavanandi, D., Youssef, A. A. A., & Dudhipala, N. (2022). Innovative treatment strategies to accelerate wound healing: Trajectory and recent advancements. *Cells*, *11*(15), 2439. https://doi.org/10.3390/cells11152439

Kotcharat, P., Chuysinuan, P., Thanyacharoen, T., Techasakul, S., & Ummartyotin, S. (2021). Development of bacterial cellulose and polycaprolactone (PCL) based composite for medical material. *Sustainable Chemistry and Pharmacy*, *20*, 100404. https://doi.org/10.1016/j.scp.2021.100404

Kumar, V., Nesari, T., Ghildiyal, S., & Sherkhane, R. (2021). Pharmacodynamic appraisal of wound-healing herbs of *Sushruta Samhita. AYU (An International Quarterly Journal of Research in Ayurveda)*, *42*(1), 1. https://doi.org/10.4103/ayu.AYU_34_20

Kusmono, Listyanda, R. F., Wildan, M. W., & Ilman, M. N. (2020). Preparation and characterization of cellulose nanocrystal extracted from ramie fibers by sulfuric acid hydrolysis. *Heliyon*, *6*(11), e05486–e05486. https://doi.org/10.1016/j.heliyon.2020.e05486

Lengowski, E. C., Bonfatti Júnior, E. A., Simon, L., de Muñiz, G. I. B., de Andrade, A. S., Nisgoski, S., & Klock, U. (2020). Different degree of fibrillation: Strategy to reduce permeability in nanocellulose-starch films. *Cellulose*, *27*(18), 10855–10872. https://doi.org/10.1007/s10570-020-03232-4

Li, M., Xia, W., Khoong, Y. M., Huang, L., Huang, X., Liang, H., Zhao, Y., Mao, J., Yu, H., & Zan, T. (2023). Smart and versatile biomaterials for cutaneous wound healing. *Biomaterials Research*, *27*(1), 87. https://doi.org/10.1186/s40824-023-00426-2

Li, Y.-Y., Ji, S.-F., Fu, X.-B., Jiang, Y.-F., & Sun, X.-Y. (2024). Biomaterial-based mechanical regulation facilitates scarless wound healing with functional skin appendage regeneration. *Military Medical Research*, *11*(1), 13. https://doi.org/10.1186/s40779-024-00519-6

Liang, Y., He, J., & Guo, B. (2021). Functional hydrogels as wound dressing to enhance wound healing. *ACS Nano*, *15*(8), 12687–12722. https://doi.org/10.1021/acsnano.1c04206

Liu, C., Li, B., Du, H., Lv, D., Zhang, Y., Yu, G., Mu, X., & Peng, H. (2016). Properties of nanocellulose isolated from corncob residue using sulfuric acid, formic acid, oxidative and mechanical methods. *Carbohydrate Polymers*, *151*, 716–724. https://doi.org/10.1016/j.carbpol.2016.06.025

Maliha, M., Brammananth, R., Dyson, J., Coppel, R. L., Werrett, M., Andrews, P. C., & Batchelor, W. (2021). Biocompatibility and selective antibacterial activity of a bismuth phosphinato-nanocellulose hydrogel. *Cellulose*, *28*(8), 4701–4718. https://doi.org/10.1007/s10570-021-03835-5

Mishra, R. K., Sabu, A., & Tiwari, S. K. (2018). Materials chemistry and the futurist eco-friendly applications of nanocellulose: Status and prospect. *Journal of Saudi Chemical Society*, *22*(8), 949–978. https://doi.org/10.1016/j.jscs.2018.02.005

Mohaghegh, H., Assadi, Z., Derakhshan, A., & Masaeli, E. (2024). Accelerating full-thickness wound healing with bacterial cellulose-based multilayer composites. *Journal of Pharmaceutical Sciences*, *113*(3), 754–763. https://doi.org/10.1016/j.xphs.2023.09.018

Mohanto, S., Biswas, A., Gholap, A. D., Wahab, S., Bhunia, A., Nag, S., & Ahmed, M. G. (2024). Potential biomedical applications of terbium-based nanoparticles (TbNPs): A review on recent advancement. *ACS Biomaterials Science & Engineering*. https://doi.org/10.1021/acsbiomaterials.3c01969

Mohanto, S., Narayana, S., Merai, K. P., Kumar, J. A., Bhunia, A., Hani, U., Al Fatease, A., Gowda, B. H. J., Nag, S., Ahmed, M. G., Paul, K., & Vora, L. K. (2023). Advancements in gelatin-based hydrogel systems for biomedical applications: A state-of-the-art review. *International Journal of Biological Macromolecules*, *253*, 127143. https://doi.org/10.1016/j.ijbiomac.2023.127143

Mohite, P., Asane, G., Rebello, N., Munde, S., Ade, N., Boban, T., Damiri, F., & Singh, S. (2024). Polymeric hydrogel sponges for wound healing applications: A comprehensive review. *Regenerative Engineering and Translational Medicine*. https://doi.org/10.1007/s40883-024-00334-4

Moradi, M., Jacek, P., Farhangfar, A., Guimarães, J. T., & Forough, M. (2021). The role of genetic manipulation and in situ modifications on production of bacterial nanocellulose: A review. *International Journal of Biological Macromolecules*, *183*, 635–650. https://doi.org/10.1016/j.ijbiomac.2021.04.173

Murray, R. Z., West, Z. E., Cowin, A. J., & Farrugia, B. L. (2019). Development and use of biomaterials as wound healing therapies. *Burns & Trauma*, *7*. https://doi.org/10.1186/s41038-018-0139-7

Nag, S., Mitra, O. P. S., Bhattacharjee, A., Mohanto, S., Gowda, B. H. J., Kar, S., Ramaiah, S., Anbarasu, A., & Ahmed, M. G. (2024). Exploring the emerging trends in the synthesis and theranostic paradigms of cerium oxide nanoparticles (CeONPs): A comprehensive review. *Materials Today Chemistry*, *35*, 101894. https://doi.org/10.1016/j.mtchem.2023.101894

Nag, S., Mitra, O. P. S., Tripathi, G., Adur, I., Mohanto, S., Nama, M., Samanta, S., Gowda, B. H. J., Subramaniyan, V., Sundararajan, V., & Kumarasamy, V. (2024). Nanomaterials-assisted photothermal therapy for breast cancer: State-of-the-art advances and future perspectives. *Photodiagnosis and Photodynamic Therapy*, *45*, 103959. https://doi.org/10.1016/J.PDPDT.2023.103959

Nang An, V., Chi Nhan, H. T., Tap, T. D., Van, T. T. T., Van Viet, P., & Van Hieu, L. (2020). Extraction of high crystalline nanocellulose from biorenewable sources of Vietnamese agricultural wastes. *Journal of Polymers and the Environment*, *28*(5), 1465–1474. https://doi.org/10.1007/s10924-020-01695-x

Nasir, M., Hashim, R., Sulaiman, O., & Asim, M. (2017). Nanocellulose. In *Cellulose-Reinforced Nanofibre Composites* (pp. 261–276). Elsevier. https://doi.org/10.1016/B978-0-08-100957-4.00011-5

Niu, F., Li, M., Huang, Q., Zhang, X., Pan, W., Yang, J., & Li, J. (2017). The characteristic and dispersion stability of nanocellulose produced by mixed acid hydrolysis and ultrasonic assistance. *Carbohydrate Polymers*, *165*, 197–204. https://doi.org/10.1016/j.carbpol.2017.02.048

Patel, D. K., Ganguly, K., Hexiu, J., Dutta, S. D., Patil, T. V, & Lim, K.-T. (2022). Functionalized chitosan/spherical nanocellulose-based hydrogel with superior antibacterial efficiency for wound healing. *Carbohydrate Polymers*, *284*, 119202. https://doi.org/10.1016/j.carbpol.2022.119202

Patil, T. V, Patel, D. K., Dutta, S. D., Ganguly, K., Santra, T. S., & Lim, K.-T. (2022). Nanocellulose, a versatile platform: From the delivery of active molecules to tissue engineering applications. *Bioactive Materials*, *9*, 566–589. https://doi.org/10.1016/j.bioactmat.2021.07.006

Pere, J., Tammelin, T., Niemi, P., Lille, M., Virtanen, T., Penttilä, P. A., Ahvenainen, P., & Grönqvist, S. (2020). Production of high solid nanocellulose by enzyme-aided fibrillation coupled with mild mechanical treatment. *ACS Sustainable Chemistry & Engineering*, *8*(51), 18853–18863. https://doi.org/10.1021/acssuschemeng.0c05202

Phanthong, P., Guan, G., Ma, Y., Hao, X., & Abudula, A. (2016). Effect of ball milling on the production of nanocellulose using mild acid hydrolysis method. *Journal of the Taiwan Institute of Chemical Engineers*, *60*, 617–622. https://doi.org/10.1016/j.jtice.2015.11.001

Pramanik, S., Mohanto, S., Manne, R., Rajendran, R. R., Deepak, A., Edapully, S. J., Patil, T., & Katari, O. (2021). Nanoparticle-based drug delivery system: The magic bullet for the treatment of chronic pulmonary diseases. *Molecular Pharmaceutics*, *18*(10), 3671–3718. https://doi.org/10.1021/acs.molpharmaceut.1c00491

Qi, Y., Guo, Y., Liza, A. A., Yang, G., Sipponen, M. H., Guo, J., & Li, H. (2023). Nanocellulose: A review on preparation routes and applications in functional materials. *Cellulose*, *30*(7), 4115–4147. https://doi.org/10.1007/s10570-023-05169-w

Rana, S. S., & Gupta, M. K. (2020). Isolation of nanocellulose from hemp (Cannabis sativa) fibers by chemo-mechanical method and its characterization. *Polymer Composites*, *41*(12), 5257–5268. https://doi.org/10.1002/pc.25791

Raziyeva, K., Kim, Y., Zharkinbekov, Z., Kassymbek, K., Jimi, S., & Saparov, A. (2021). Immunology of acute and chronic wound healing. *Biomolecules*, *11*(5), 700. https://doi.org/10.3390/biom11050700

Ribeiro, R. S. A., Pohlmann, B. C., Calado, V., Bojorge, N., & Pereira, N. (2019). Production of nanocellulose by enzymatic hydrolysis: Trends and challenges. *Engineering in Life Sciences*, *19*(4), 279–291. https://doi.org/10.1002/elsc.201800158

Salari, M., Sowti Khiabani, M., Rezaei Mokarram, R., Ghanbarzadeh, B., & Samadi Kafil, H. (2019). Preparation and characterization of cellulose nanocrystals from bacterial cellulose produced in sugar beet molasses and cheese whey media. *International Journal of Biological Macromolecules*, *122*, 280–288. https://doi.org/10.1016/j.ijbiomac.2018.10.136

Savitskaya, I. S., Shokatayeva, D. H., Kistaubayeva, A. S., Ignatova, L. V, & Digel, I. E. (2019). Antimicrobial and wound healing properties of a bacterial cellulose based material containing B. subtilis cells. *Heliyon*, *5*(10), e02592–e02592. https://doi.org/10.1016/j.heliyon.2019.e02592

Sharma, C., & Bhardwaj, N. K. (2019). Biotransformation of fermented black tea into bacterial nanocellulose via symbiotic interplay of microorganisms. *International Journal of Biological Macromolecules*, *132*, 166–177. https://doi.org/10.1016/j.ijbiomac.2019.03.202

Smets, G., & Rüdelsheim, P. (2018). Biotechnologically produced chitosan for nanoscale products. A legal analysis. *New Biotechnology*, *42*, 42–47. https://doi.org/10.1016/j.nbt.2018.02.005

Stovbun, S. V, Nikol'skii, S. N., Mel'nikov, V. P., Mikhaleva, M. G., Litvin, Y. A., Shchegolikhin, A. N., Zlenko, D. V, Tverdislov, V. A., Gerasimov, D. S., & Rogozin, A. D. (2016). Chemical physics of cellulose nitration. *Russian Journal of Physical Chemistry B*, *10*(2), 245–259. https://doi.org/10.1134/S199079311602024X

Sultana, R., Mohanto, S., Bhunia, A., Biswas, A., Akhtar, M. S., Mishra, V., Modi, D., Aljabali, A. A., Tambuwala, M., & Faiyazuddin, M. (2024). Current progress and emerging role of essential oils in drug delivery therapeutics. *Current Drug Delivery*. https://doi.org/10.2174/0115672018287719240214075810

Thakur, V., Guleria, A., Kumar, S., Sharma, S., & Singh, K. (2021). Recent advances in nanocellulose processing, functionalization and applications: A review. *Materials Advances*, *2*(6), 1872–1895. https://doi.org/10.1039/D1MA00049G

Tottoli, E. M., Dorati, R., Genta, I., Chiesa, E., Pisani, S., & Conti, B. (2020). Skin wound healing process and new emerging technologies for skin wound care and regeneration. *Pharmaceutics*, *12*(8), 735. https://doi.org/10.3390/pharmaceutics12080735

Tudoroiu, E.-E., Dinu-Pîrvu, C.-E., Albu Kaya, M. G., Popa, L., Anuța, V., Prisada, R. M., & Ghica, M. V. (2021). An overview of cellulose derivatives-based dressings for wound-healing management. *Pharmaceuticals*, *14*(12), 1215. https://doi.org/10.3390/ph14121215

Vismara, E., Bertolini, G., Bongio, C., Massironi, N., Zarattini, M., Nanni, D., Cosentino, C., & Torri, G. (2021). Nanocellulose from cotton waste and its glycidyl methacrylate grafting and allylation: Synthesis, characterization and adsorption properties. *Nanomaterials*, *11*(2), 476. https://doi.org/10.3390/nano11020476

Wahid, F., Zhao, X.-J., Zhao, X.-Q., Ma, X.-F., Xue, N., Liu, X.-Z., Wang, F.-P., Jia, S.-R., & Zhong, C. (2021). Fabrication of bacterial cellulose-based dressings for promoting infected wound healing. *ACS Applied Materials & Interfaces*, *13*(28), 32716–32728. https://doi.org/10.1021/acsami.1c06986

Wan, Y., Yang, S., Wang, J., Gan, D., Gama, M., Yang, Z., Zhu, Y., Yao, F., & Luo, H. (2020). Scalable synthesis of robust and stretchable composite wound dressings by dispersing silver nanowires in continuous bacterial cellulose. *Composites Part B: Engineering*, *199*, 108259. https://doi.org/10.1016/j.compositesb.2020.108259

Wang, J., Liu, X., Jin, T., He, H., & Liu, L. (2019). Preparation of nanocellulose and its potential in reinforced composites: A review. *Journal of Biomaterials Science, Polymer Edition*, *30*(11), 919–946. https://doi.org/10.1080/09205063.2019.1612726

Wang, J., Tavakoli, J., & Tang, Y. (2019). Bacterial cellulose production, properties and applications with different culture methods—A review. *Carbohydrate Polymers*, *219*, 63–76. https://doi.org/10.1016/j.carbpol.2019.05.008

Wang, L., Li, K., Copenhaver, K., Mackay, S., Lamm, M. E., Zhao, X., Dixon, B., Wang, J., Han, Y., Neivandt, D., Johnson, D. A., Walker, C. C., Ozcan, S., & Gardner, D. J. (2021). Review on nonconventional fibrillation methods of producing cellulose nanofibrils and their applications. *Biomacromolecules*, *22*(10), 4037–4059. https://doi.org/10.1021/acs.biomac.1c00640

Wang, P., Cai, F., Li, Y., Yang, X., Feng, R., Lu, H., Bai, X., & Han, J. (2024). Emerging trends in the application of hydrogel-based biomaterials for enhanced wound healing: A literature review. *International Journal of Biological Macromolecules*, *261*, 129300. https://doi.org/10.1016/j.ijbiomac.2024.129300

Wang, Y., Wei, X., Li, J., Wang, F., Wang, Q., Chen, J., & Kong, L. (2015). Study on nanocellulose by high pressure homogenization in homogeneous isolation. *Fibers and Polymers*, *16*(3), 572–578. https://doi.org/10.1007/s12221-015-0572-1

Wu, C., McClements, D. J., He, M., Zheng, L., Tian, T., Teng, F., & Li, Y. (2021). Preparation and characterization of okara nanocellulose fabricated using sonication or high-pressure homogenization treatments. *Carbohydrate Polymers*, *255*, 117364. https://doi.org/10.1016/j.carbpol.2020.117364

Wulandari, W. T., Rochliadi, A., & Arcana, I. M. (2016). Nanocellulose prepared by acid hydrolysis of isolated cellulose from sugarcane bagasse. *IOP Conference Series: Materials Science and Engineering*, *107*, 12045. https://doi.org/10.1088/1757-899X/107/1/012045

Xiao, Y., Liu, Y., Wang, X., Li, M., Lei, H., & Xu, H. (2019). Cellulose nanocrystals prepared from wheat bran: Characterization and cytotoxicity assessment. *International Journal of Biological Macromolecules*, *140*, 225–233. https://doi.org/10.1016/j.ijbiomac.2019.08.160

Xiao, Y., Rong, L., Wang, B., Mao, Z., Xu, H., Zhong, Y., Zhang, L., & Sui, X. (2018). A light-weight and high-efficacy antibacterial nanocellulose-based sponge via covalent immobilization of gentamicin. *Carbohydrate Polymers*, *200*, 595–601. https://doi.org/10.1016/j.carbpol.2018.07.091

Yang, C., Zhu, Y., Tian, Z., Zhang, C., Han, X., Jiang, S., Liu, K., & Duan, G. (2024). Preparation of nanocellulose and its applications in wound dressing: A review. *International Journal of Biological Macromolecules*, *254*, 127997. https://doi.org/10.1016/j.ijbiomac.2023.127997

3 Advancements in Nanomaterial-Based Drug Delivery Systems for Wound Healing

Arun Kumar,[1] Vikanksha,[1] Jatinder Singh,[1] Ab Waheed Wani,[1] and Glomeraz Abdi[2]*

[1] School of Agriculture, Lovely Professional University, Phagwara, Punjab, 144411, India

[2] Department of Biotechnology, Persian Gulf Research Institute, Persian Gulf University. Bushehr, 75169, Iran

* Corresponding author: Glomeraz Abdi

3.1 INTRODUCTION

Wound healing is a multistep process that incorporates numerous phases to repair the integrity of injured tissue. This process is both complicated and dynamic. Hemostasias, which is the procedure of stopping bleeding by activating blood clotting processes, is the first step in the method. After this, there occurs a period of inflammation, which is defined by the migration of antibodies to the wound site to avoid infection and clean debris accumulated there. The subsequent period, known as the phase of proliferation, is responsible for the production of new blood vessels and contributes to the formation of new tissues. The remodeling phase is responsible for ensuring that the newly produced tissue receives the necessary strength and functioning (Nour *et al.*, 2019). Each step is essential, and it is imperative that they take place in a coordinated way to guarantee optimal recovery. When it comes to wound healing, the conventional medication delivery methods depend mostly on topical treatments such as ointments, creams, and dressings. As a result of their straightforwardness and ease of implementation, these techniques have seen widespread adoption. On the other hand, they have a number of important restrictions. In many cases, topical treatments have a limited ability to penetrate deeper layers of the skin, which results in therapeutic effects that are less than ideal (Zhang *et al.*, 2013). Furthermore, these systems have the potential to demonstrate inconsistency in drug release, which may lead to either an insufficient therapeutic impact or the possibility of toxicity as a consequence of excessive exposure. Traditional drug delivery methods have a number of drawbacks, which highlights the need for developing more modern drug delivery mechanisms that are able to overcome these deficiencies. In comparison to more traditional approaches, advanced drug delivery systems, especially those that make use of nanotechnology, provide a number of distinctive benefits (Patra *et al.*, 2018). It is possible to develop nanomaterials to exhibit distinctive features, such as increased permeability and retention effects, which enables them to penetrate tissues more effectively. Furthermore, they may be constructed to produce a regulated and prolonged release of therapeutic substances, which guarantees a dose of the medicine that is both consistent and efficacious at the site of the wound. Within the realm of wound healing, the development of drug delivery systems that are based on nanomaterials represents a major achievement. There are

DOI: 10.1201/9781003605966-3

many different kinds of nanomaterials that are used in such systems. These nanomaterials include polymeric nanoparticles, metallic nanoparticles, and carbon-based nanomaterials (Fritea *et al.*, 2021). Every one of these nanomaterials has unique features that may be adapted to suit specific purposes. For example, polymeric nanoparticles are biocompatible and may be made to disintegrate at regulated rates, which allows the medicine to be released over a prolonged period of time. Not only do metallic nanoparticles, such as those manufactured from silver or gold, have the ability to carry medications, but they also possess inherent antibacterial qualities, which are very helpful in the prevention of wound infections. It is possible to utilize carbon-based nanoparticles, such as graphene and carbon nanotubes, to fortify the wound site while providing restorative medicines. These nanostructures give extraordinary mechanical strength (Maleki *et al.*, 2020). The incorporation of nanotechnology into medication delivery systems for the purpose of wound healing has resulted in the development of creative tactics that go above the straightforward delivery of drugs. In reaction to certain stimuli, such as pH, temperature, or the existence of certain enzymes, smart nanomaterials are able to adapt to the environment of the wound and release medications. By ensuring that the medicine is only released when it is required, this reactivity helps to maximize the therapeutic effectiveness while simultaneously limiting the adverse effects. It is also possible to develop customized means of administration to distribute medications selectively to cells that are engaged in the method of wound healing (Saghazadeh *et al.*, 2018). This will enhance the treatment's accuracy as well as its efficacy.

In spite of these developments, the development of medication delivery systems for wound healing that are based on nanomaterials is not without its obstacles. To guarantee the security of these systems, it is essential to address the fundamental problems of biocompatibility and toxicity. When attempting to determine the impact that nanomaterials have on the body over the long term, it is necessary to do extensive preclinical and clinical research (Bourquin *et al.*, 2018). The procedure for obtaining authorization for novel therapies based on nanomaterials entails intensive testing and validation to verify that the treatments fulfill safety and effectiveness criteria. Regulatory concerns additionally have a significant part in this method. The potential influence that medication administration methods built around nanomaterials might have on the healing process of wounds is significant. This innovative technology has the potential to dramatically enhance the process of healing, which will ultimately lead to improved results for patients. It does this by transcending the constraints of standard medication delivery techniques (Zhou *et al.*, 2022). The capability of administering medications in a more efficient and accurate manner leads to a reduction in the likelihood of problems and speeds up the healing process. In addition, the inclusion of antimicrobial characteristics into certain types of nanomaterials has the potential to safeguard against diseases, which are a typical consequence that occur during the healing process of wounds.

3.2 FUNDAMENTALS OF NANOMATERIALS

Because of their high surface area-to-volume ratio and quantum effects, nanomaterials, which are characterized by their nanoscale dimensions (ranging from 1 to 100 nanometers), display property characteristics that are different from those of their bulk counterparts. These substances are available in a variety of forms, such as nanoparticles, nanofibers, and nanocomposites, each of which is tailor-made for a particular use. Enhanced permeability, regulated drug release, and the capacity to distribute drugs in a targeted manner are all qualities that nanomaterials provide in the context of drug delivery for wound healing (Hamad *et al.*, 2020). Nanoparticles made of polymers supply biological compatibility and controlled degradation, while nanoparticles made of metallic materials, such as silver and gold, have antibacterial capabilities. As a result of their outstanding mechanical strength and drug-loading capability, carbon-based nanomaterials, such as graphene, are very flexible instruments in the field of enhanced wound care.

3.2.1 Synthesis and Characterization of Nanomaterials

Creating materials with dimensions that generally range from one to one hundred nanometers is what is involved in the process of nanomaterial synthesis. The synthesis of these materials may be accomplished by a variety of processes, which can be classified as either bottom-up or top-down procedures.

- Bottom-up Synthesis: In this approach, nanomaterials are built from atomic or molecular precursors. Common methods are listed as follows:
  - **Sol-Gel Synthesis:** For the purpose of obtaining nanomaterials, this technique includes first transforming a solution or colloidal suspension into a solid gel phase, then drying and calcining the gel. The production of metal oxides, silica nanoparticles, and hybrid organic-inorganic compounds are all commonly utilized applications of the sol-gel synthesis technique (Guo *et al.*, 2016).
  - **Chemical Vapor Deposition (CVD):** Through a chemical response of gas precursors, chemical vapor deposition (CVD) is a process that includes the application of very thin films or coatings onto a substrate. The production of carbon nanotubes, graphene, and thin-film materials are all popular applications for this material (Shah & Tali, 2016).
- **Top-Down Synthesis:** In this approach, bulk materials are broken down into smaller nanosized structures. Common methods include:
  - **Lithography:** It is possible to shape structures at the nanoscale via the use of lithography methods such as photolithography and electron beam lithography. Production of integrated circuits and nano electronic devices requires the use of this approach (Liddle & Gallatin, 2011).
  - **Milling:** Utilizing mechanical milling processes, such as ball milling and attrition milling, it is possible to decrease the size of large quantities to the nanoscale. Nanoparticles and nanocomposites are both obtained via the use of this technique (Yadav *et al.*, 2012). To adjust the dimension, form, structure, and surface features of nanomaterials, it is vital to have precise oversight over the production variables. These factors include temperature, pressure, reaction time, and precursor concentration.

3.2.2 Characterization Techniques

The process of characterization of nanomaterials comprises the use of a variety of analytical methods to explore the structural, chemical, and physical characteristics of the nanomaterials. The chemical makeup, shape, coating chemistry, and effectiveness of nanomaterials may all be better understood with the use of these methodologies.

- **Microscopy Techniques:** Microscopy techniques, such as transmission electron microscopy (TEM), scanning electron microscopy (SEM), and atomic force microscopy (AFM), are used to visualize the morphology, size, and shape of nanomaterials with high spatial resolution (Das, 2017).
- **Spectroscopic Techniques:** Spectroscopic techniques, including Fourier-transform infrared spectroscopy (FTIR), UV-visible spectroscopy, and X-ray photoelectron spectroscopy (XPS), are used to analyze the chemical composition, functional groups, and electronic structure of nanomaterials.
- **Particle Sizing Techniques:** Dynamic light scattering (DLS) and laser diffraction are commonly used for measuring the size distribution of nanoparticles in suspension.
- **Crystallographic Techniques**: X-ray diffraction (XRD) is employed to determine the crystal structure, phase composition, and crystallite size of nanomaterials (Krishna & Philip, 2022).

3.3 NANOMATERIAL-BASED DRUG DELIVERY SYSTEMS

By encapsulating therapeutic compounds into nanosized carriers, drug delivery systems that are based on nanomaterials provide precise and targeted treatment (**Table 3.1**). The use of these systems allows for controlled release, increased penetration, and targeted distribution to particular areas, hence reducing the likelihood of systemic collateral effects. The development of stimuli-responsive nanocarriers, which made it possible to release drugs on request, and bioactive dressings that integrated antibacterial capabilities for wound care are examples of recent advancements. In general, the administration of drugs using nanomaterials has a tremendous potential to revolutionize therapies, provide enhanced effectiveness, and reduce the number of unwanted effects (Chenthamara *et al.*, 2019).

TABLE 3.1
Fundamental Aspects of Different Nanomaterials in Wound Healing, Including Their Type, Advantages, and Size

Nanomaterial	Advantage	Size	References
Graphene	Promotes tissue regeneration by enhancing angiogenesis and cell proliferation	Nanoscale	Yosida *et al.* (2023)
Carbon nanotubes	Provides structural support for cell growth and tissue ingrowth	Nanoscale	Zhou *et al.* (2023)
Silver nanoparticles	Exhibits potent antimicrobial properties, preventing infections in wounds	Typically <100 nm	Cheng *et al.* (2022)
Liposomes	Enables targeted and controlled drug delivery, optimizing therapeutic efficacy	Variable size	Manzari *et al.* (2021)
Polymeric nanoparticles	Offers tunable properties for drug encapsulation and controlled release	Typically <200 nm	Liu *et al.* (2020)
Chitosan nanoparticles	Biocompatible and biodegradable, enhances wound healing and tissue regeneration	Typically <200 nm	Shalaby *et al.* (2022)
Zinc oxide nanoparticles	Possesses antimicrobial and anti-inflammatory properties, aiding in wound healing	Typically <100 nm	Huang *et al.* (2020)
Gold nanoparticles	Facilitates controlled release of therapeutic agents, promoting wound healing	Typically <100 nm	Hu *et al.* (2021)
Cerium oxide nanoparticles	Exhibits antioxidant properties, reducing oxidative stress in wounds	Typically <100 nm	Dong *et al.* (2021)
Quantum dots	Enables precise imaging and monitoring of wound healing processes	Typically <10 nm	Johnson *et al.* (2021)
Iron oxide nanoparticles	Facilitates magnetic targeting of therapeutic agents to wound sites	Typically <100 nm	Yang *et al.* (2022)

(*Continued*)

TABLE 3.1 *(Continued)*
Fundamental Aspects of Different Nanomaterials in Wound Healing, Including Their Type, Advantages, and Size

Nanomaterial	Advantage	Size	References
Hydrogel nanoparticles	Offers high water content and biocompatibility, promoting wound hydration	Typically <200 nm	Xiang *et al.* (2020)
Dendrimers	Enables multifunctional drug delivery and targeting in wound healing	Variable scale	Yin *et al.* (2021)
Mesoporous silica nanoparticles	Facilitates controlled release of therapeutic agents, enhancing wound healing	Typically <100 nm	Augustine *et al.* (2020)
Polymer-lipid hybrid nanoparticles	Provides a versatile platform for drug delivery and wound healing	Typically <200 nm	Catanzano *et al.* (2021)

3.3.1 Mechanisms of Drug Delivery

For the purpose of precisely managing the pharmacokinetics, biodistribution, and release of therapeutic drugs inside the body, drug delivery systems comprise a wide variety of different ways. These mechanisms can be broadly categorized into several strategies:

1. **Passive Diffusion:** In this process, medicines gradually permeate through biological boundaries such as the membranes of cells or the blood-brain barrier. This is the most straightforward method currently in existence. There are a number of parameters that influence the rate of diffusion, including the concentration of the medication gradient, the size of the molecule, and the ability to dissolve off different lipids (Seelig, 2007).
2. **Active Transport:** Active transport is a process in which specialized transport proteins enhance the flow of medicines through the membranes of cells regardless of the concentration gradients that are present. The ATP-binding cassette (ABC) transporters and the solute carrier (SLC) transporters are two examples of transport carriers. This particular mechanism enables selective absorption or efflux of medications, and it has the potential to greatly impact both the distribution and effectiveness of therapeutic agents (Drozdzik *et al.*, 2021).
3. **Endocytosis and Exocytosis:** Various vesicular transport systems, including phagocytosis, pinocytosis, and receptor-mediated endocytosis, are involved in the process of endocytosis, which is the process of bringing external materials into cells. The process of exocytosis, on the other hand, is characterized by the release of chemicals from inside cells into the extracellular environment. The use of nanocarriers or liposomes allows for the tailored delivery of drugs via these mechanisms (Foroozandeh *et al.*, 2021).
4. **Carrier-Mediated Transport:** For the purpose of facilitating the movement of pharmaceuticals across biological membranes, carrier-mediated transport makes use of certain carrier proteins. When it comes to transferring polar or charged compounds that are unable to efficiently permeate across lipid bilayers, this method is especially significant. Examples of this comprise the continuous movement of transmitters by vesicular transporters and the facilitation of glucose diffusion via the transportation of glucose by glucose transporters (Smith *et al.*, 2014).

5. **Nanotechnology-Based Delivery Systems:** In the realm of medication delivery, nanotechnology presents a number of novel techniques, such as nanoparticles, liposomes, dendrimers, and micelles. For the purpose of encapsulating medications, protecting them from degradation, and facilitating targeted administration to certain tissues or cells, these nanocarriers are a useful tool. Furthermore, stimuli-responsive nanomaterials make it possible to regulate the release of drugs in response to either external stimuli or physiological circumstances (Mohapatra *et al.*, 2018).
6. **Implantable Devices and Controlled Release Systems:** Technologies that are inserted, such as drug-eluting stents, patches, and microchips, are able to deliver medications in a continuous distribution over an extended amount of time. To modulate the kinetics of drug release, systems with controlled release make use of a variety of processes, including diffusion, erosion, and osmosis. This allows for the optimization of therapeutic effectiveness while simultaneously limiting adverse effects (Kaur *et al.*, 2022).

3.3.2 Nanocarriers and Their Types

Nanocarriers are specialized structures that make it possible to encapsulate and transport therapeutic materials, such as medications or genes, to certain target areas inside the body. They provide a multitude of benefits, including improved stability of medications, controlled administration, and targeted distribution, all while lowering the incidence of undesirable side effects. Various types of nanocarriers have been developed, each with unique properties and applications:

3.3.2.1 Liposomes

Polymeric nanoparticles are nanoparticles that are made up of biodegradable or biocompatible polymers, such as chitosan, polylactic acid (PLGA), or polyethylene glycol (PEG). Personalized drug delivery systems are made possible by the fact that these nanoparticles have attributes that can be altered, such as their size, surface charge, and the rate at which they release drugs (Kumari *et al.*, 2012). It is possible for polymeric nanoparticles to encapsulate pharmaceuticals by either chemical coupling or physical trapping. This provides safeguards against deterioration and makes it possible to determine regulated release patterns. There are several uses for polymeric nanoparticles, namely cancer treatment, gene transfer, and vaccine administration. This is because polymeric nanoparticles are versatile and have the potential to target particular tissues or cells. Any alterations made to their surfaces have the potential to significantly improve their biological acceptability and aiming abilities (Sánchez *et al.*, 2020).

3.3.2.2 Polymeric Nanoparticles

Polymeric nanoparticles are nanoparticles that are made up of biodegradable or biocompatible polymers, such as chitosan, polylactic acid (PLGA), or polyethylene glycol (PEG). Personalized drug delivery systems are made possible by the fact that these nanoparticles have attributes that can be altered, such as their size, surface charge, and the rate at which they release drugs (Alvarado *et al.*, 2019). It is possible for polymeric nanoparticles to encapsulate pharmaceuticals by either chemical coupling or physical trapping. This provides protection against degradation and makes it possible to determine regulated release patterns. There are several uses for polymeric nanoparticles, namely cancer treatment, gene transfer, and vaccine administration. This is because polymeric nanoparticles are versatile and have the potential to target particular tissues or cells. Any alterations made to their surfaces have the potential to significantly improve their biological compatibility and aiming abilities.

3.3.2.3 Dendrimers

Dendrimers are macromolecules that are extensively branching and tree-like in structure. They have a central core structure and many peripheral functional groups. They have structures that are accurately specified, in addition to having a regulated size and molecular weight, and they provide a high drug-loading capacity as well as flexibility in surface modification. Through electrostatic interactions or covalent conjugation, dendrimers have the ability to encapsulate pharmaceuticals inside their interiors or bind medicinal molecules to their surface. Given their flexibility, dendrimers are considered to be interesting prospects for applications in the fields of imaging, diagnostics, and drug administration (Svenson & Tomalia, 2012). The fact that they are able to target certain tissues or cells, in addition to the fact that they have the capacity to modify surfaces in a multifunctional manner, contributes to their rising significance in nanomedicine.

3.3.2.4 Micelles

Self-assembling of amphiphilic substances in water-based solutions results in the formation of micelles, which are colloidal nanostructures. They have a hydrophobic core and a hydrophilic shell, which allows for the solubilization of hydrophobic pharmaceuticals inside the center and the stability of the medication in aqueous settings (Lombardo *et al.*, 2020). It is possible to functionalize micelles with targeting ligands or stimuli-responsive moieties to improve medication delivery to certain tissues or to activate drug release in reaction to signals from the outside world. Micelles, due to their ease of manufacture, high drug-loading capacity, and plasticity in modifying the surface, are highly desirable nanocarriers for drug delivery applications. These applications include cancer treatment and intracellular administration of therapeutic substances (Zorkina *et al.*, 2020)

3.4 APPLICATIONS OF NANOMATERIALS IN WOUND HEALING

Because of their one-of-a-kind features, nanomaterials offer a great deal of potential for use in applications related to wound healing. They have the ability to reduce inflammation, enable the administration of drugs to specific locations, and stimulate tissue regeneration (Julier *et al.*, 2017). The use of nanofibers for scaffolds that promote cell development, nanoparticles for controlled release of growth factors and antibacterial agents, and nanocomposites for bioactive wound dressings are some examples of nanomaterials found in medical applications. Furthermore, nanoparticles make it possible to create smart wound dressings that are able to sense, monitor, and react to the microenvironment of the wound because of their capabilities. In general, techniques that are based on nanomaterials provide novel solutions that may speed up the method of wound healing and improve the results of acute as well as long-term wounds (**Table 3.2**).

3.4.1 Antimicrobial Properties

The use of nanomaterials that possess antimicrobial qualities is essential for the healing of wounds, the prevention of infections, the reduction of inflammation, and the promotion of tissue regeneration. Nanoparticles of metals such as silver, copper, and zinc oxide release ions that damage the membranes of microorganisms (Pormohammad *et al.*, 2021; Hayat *et al.*, 2023). This disrupts enzyme function and induces oxidative stress, which ultimately results in the death of microorganisms due to oxidative stress. Infused with antimicrobial chemicals such as chitosan or polyvinyl alcohol, electrospun nanofibers imitate the extracellular matrix. This allows them to promote cell adhesion and migration while simultaneously releasing antimicrobial compounds into the wound bed. Nanocomposites, such as hybrids of graphene oxide and silver, have the ability to provide synergistic antibacterial properties and controlled release, which speeds up the rehabilitation process. With nanoscale topographies or antimicrobial nanoparticles, coatings applied to medical devices may inhibit the development of biofilm, hence lowering the number of infections that are linked with implants. As a result, nanomaterials provide treatments that are both targeted and efficient for wound care, hence improving healing results and reducing the likelihood of developing antibiotic resistance (Mihai *et al.*, 2019).

TABLE 3.2
Potential Applications of Nanomaterials in Wound Healing

Nanomaterial	Application	Mechanism	References
Silver nanoparticles	Antibacterial agent	Disruption of bacterial cell membranes	Lin *et al.* (2021)
Zinc oxide nanoparticles	Anti-inflammatory agent	Modulation of inflammatory pathways	Moustaqil *et al.* (2021)
Gold nanoparticles	Drug delivery carrier	Controlled release of therapeutic agents	Lopes *et al.* (2022)
Chitosan nanoparticles	Hemostatic agent	Promotion of blood clotting	Fan *et al.* (2021)
Hydrogel nanoparticles	Wound dressing	Moist wound environment maintenance	Aisa *et al.* (2022)
Nanofibers	Tissue regeneration scaffold	Mimicking extracellular matrix for cell growth	Amorim *et al.* (2021)
Liposomes	Growth factor delivery	Encapsulation and targeted release	Yammine *et al.* (2024)
Quantum dots	Wound imaging	Fluorescence for visualization of wound healing processes	Zheng *et al.* (2021)
Cerium oxide nanoparticles	Antioxidant	Scavenging of reactive oxygen species	Apak *et al.* (2022)
Carbon nanotubes	Electrical stimulation	Enhancement of cellular proliferation and migration	Sukhanov *et al.* (2021)

3.4.2 Anti-inflammatory Properties

Through the reduction of inflammation, the alleviation of pain, and the promotion of tissue repair, nanomaterials that contain anti-inflammatory characteristics play a crucial role in the process of wound healing. Through the targeted administration of anti-inflammatory drugs, they are able to effectively regulate localized swelling while simultaneously limiting the adverse effects on the system. For example, nanoparticles loaded with medications such as corticosteroids or nonsteroidal anti-inflammatory drugs (NSAIDs) allow for regulated release, which ensures stable therapeutic levels at the site of the lesion. In a similar manner, electrospun nanofibers that contain anti-inflammatory substances provide comfort that is both widespread and targeted, thus facilitating healing while simultaneously alleviating pain (Wei *et al.*, 2023). Furthermore, nanocomposite materials mix anti-inflammatory nanoparticles with other bioactive compounds to influence inflammation in a synergistic manner and to boost tissue regeneration. In general, techniques that are based on nanomaterials provide potential avenues for fighting inflammation and enhancing the results of wound healing (Nqakala *et al.*, 2021).

3.4.3 Promotion of Tissue Regeneration

Establishing an atmosphere that is favorable for the proliferation and differentiation of cells is one of the most important roles that nanomaterials play in the process of fostering tissue regeneration. Nanomaterials enhance interactions with cells by imitating the structure and biochemistry of the extracellular matrix. This allows nanomaterials to promote adhesion, migration, and proliferation of cells. They preserve bioactive compounds and development factors by means of tailored drug delivery, which guarantees prolonged release and increased bioavailability to facilitate tissue healing. In addition to this, scaffolds made of nanomaterials provide a framework for the growth of tissue, which helps to direct the process of rehabilitation. The use of nanomaterial carriers, which safeguard and encourage the incorporation of stem cells into host tissues, is an additional advantage of stem cell treatment (Zhang *et al.*, 2022).

3.5 POLYMERIC NANOMATERIALS IN WOUND HEALING

Polymeric nanoparticles are a potential avenue in the field of wound healing because they provide a wide range of solutions that may be individualized to meet the requirements of the process of wound healing. The use of these materials, which include electrospun nanofibers and nanoparticles, makes a substantial contribution to the processes of tissue regeneration, infection control, and wound healing alike (Memic *et al.*, 2019). Through the development of scaffolds, polymeric nanofibers provide structural support for the adhesion and proliferation of cells, which in turn promotes the healing of wounds and the regeneration of tissue. In addition, their function in medication delivery makes it possible to administer therapeutic drugs, including antimicrobials and growth factors, to the location of the wound in a targeted and regulated manner, which in turn promotes healing while simultaneously treating infections. In addition, bioactive dressings may include the incorporation of polymeric nanoparticles, which can improve wound treatment by producing an environment that is conducive to the regeneration of tissue. Polymeric nanoparticles have the potential to significantly advance wound care methods due to their biocompatibility and the adjustable features they possess (Tiwari *et al.*, 2023).

3.5.1 Types of Polymeric Nanomaterials

Polymeric nanomaterials encompass a diverse array of materials with nanosized dimensions, offering tailored solutions for various biomedical applications, including wound healing (**Figure 3.1**). Here are several types of polymeric nanomaterials commonly employed in wound healing:

- **Polymeric Nanofibers:** When it comes to uses related to wound healing, electrospun nanofibers are among the polymeric nanomaterials that have received the most research attention. The fabrication of these fibers is accomplished by the use of electrospinning, a technique that results in the production of continuous ultrafine fibers with diameters in the nanometer range. Nanofibers made of polymeric materials have a substantial surface area-to-volume ratio (Islam *et al.*, 2019), which allows them to imitate the structure of the extracellular matrix. This allows them to promote cell adhesion, proliferation, and tissue regeneration.
- **Polymeric Nanoparticles:** The sizes of polymeric nanoparticles generally range from one to one thousand nanometers. These particles are colloidal in shape. Polymers that are biocompatible and biodegradable, such as poly(lactic-co-glycolic acid) (PLGA), chitosan, or polyethylene glycol (PEG), provide the basis for their synthesis. In the process of drug delivery, polymeric nanoparticles operate as carriers, encapsulating therapeutic chemicals that include antimicrobial agents, growth stimulants, or anti-inflammatory medications (D'souza & Shegokar, 2016). In addition to better therapeutic effectiveness, they provide regulated discharge kinetics, extended drug absorption at the wounded site, and extended drug retention.
- **Polymeric Hydrogels:** These interconnected networks of hydrophilic polymers are known as polymeric hydrogels, and they have the ability to consume significant volumes of water. To preserve a humid atmosphere that is beneficial to wound healing, these materials infuse the wound bed with moisture. Polymeric hydrogels have the ability to encapsulate bioactive compounds and facilitate their prolonged release, providing a regulated distribution of therapeutic molecules to the site of the wound (Jacob *et al.*, 2021). In addition, they have mechanical qualities that are comparable to those of real tissues, which helps to reduce patient pain and encourages patient compliance.

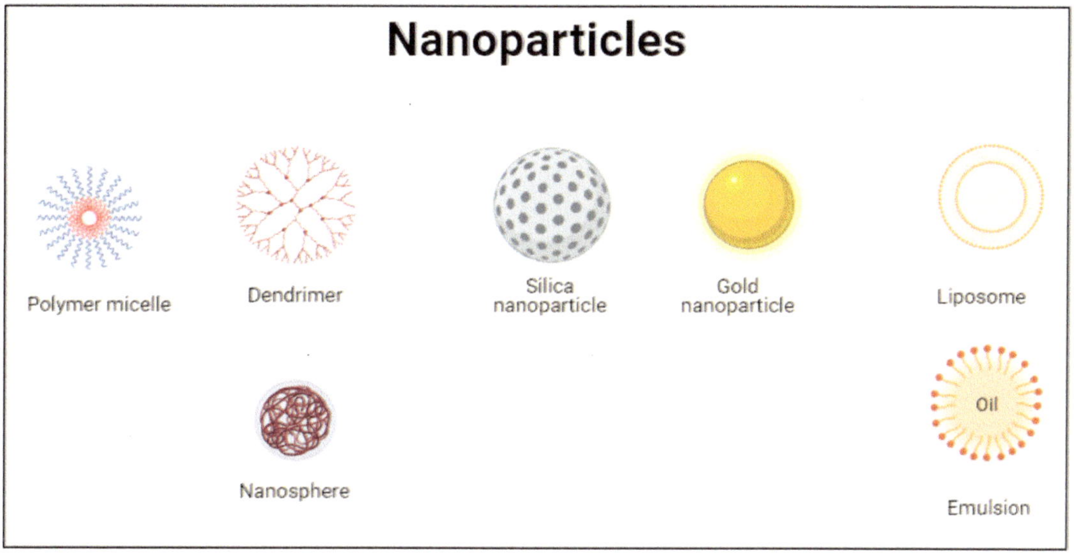

FIGURE 3.1 Types of nanoparticles employed.

- **Polymeric Nanocomposites:** To improve the mechanical, thermal, and biological characteristics of polymers, polymeric nanocomposites use nanoparticles or nanofillers in conjunction with polymers. When compared with pure polymers, these materials provide stronger and more stable properties, as well as enhanced utility (Kumar *et al.*, 2019). Polymeric nanocomposites have the potential to integrate antibacterial nanoparticles, bioactive compounds, or growth factors into wound dressings throughout the healing process. This may result in the creation of multipurpose wound dressings that have increased therapeutic effectiveness.
- **Polymeric Micelles:** Within an aqueous solution, amphiphilic block copolymers are responsible for the formation of polymeric micelles, which are colloidal entities that self-assemble. These nanocarriers have the ability to enclose hydrophobic pharmaceuticals inside their core, therefore preventing the medications from degrading. As a result of their enhanced drug solubility, stability, and bioavailability, polymeric micelles are becoming more interesting as possibilities for use in wound healing (Parra *et al.*, 2021).

3.5.2 Drug Loading and Release Mechanisms

A crucial component of polymeric nanoparticles that are used in drug delivery systems is the processes that are responsible for drug loading and discharge. The manner in which pharmaceuticals are encapsulated inside the nanomaterials and then discharged at the intended spot is determined by these processes (Chandrakala *et al.*, 2022). The release of the medicine takes place when the polymer undergoes swelling and expansion, enabling water to enter and dissolve the drug. The pace at which the release occurs is determined by the contraction kinetics of the polymer and the chemical reactions between the medication and the polymer. Here's an overview of drug loading and release mechanisms in polymeric nanomaterials:

1. **Physical Encapsulation:** The process of mechanically encapsulating pharmaceuticals involves physically entrapping them inside the fundamental component of nanoparticles or the polymeric matrix (Sánchez *et al.*, 2020). This may take place during the process of nanoparticle generation, which involves the dissolution or dispersion of medicines in a

polymer solution or suspension medium. The medicine is able to get entrapped inside the polymer matrix when the solvent evaporates or when the nanoparticles solidify.

2. **Chemical Conjugation:** It is possible for some medications to develop chemical conjugations or covalent bonds with the chains of polymers. By tethering the drug particles to the polymer vertebral column, this method makes it possible to regulate the introduction and discharge of the medication (Utterström *et al.*, 2021). It is possible to improve medication durability using chemical conjugation, which also allows for controlled release kinetics and enables tailored distribution.
3. **Adsorption:** Electrostatic interactions, hydrophobic interactions, and hydrogen bonding are all potential mechanisms for the adsorption of drugs upon the outermost layer of polymeric nanoparticles. The drug-polymer combination could possess varying degrees of equilibrium, despite the fact that this approach is reasonably straightforward and has the potential to reach substantial drug loading capabilities (Lübtow *et al.*, 2019).

Drug Release Mechanisms:

1. **Diffusion-Controlled Release:** Diffusion-controlled release involves the release of pharmaceuticals from polymeric nanoparticles by means of dispersion throughout the polymer matrix. The pace at which the release occurs is affected by parameters such as the porous nature of the polymer, the variation of drug quantity, and the chemical reactions between the polymer and the drug (Ganetri *et al.*, 2021). Dispersal-controlled discharge generally adheres to Fick's law of diffusion.
2. **Swelling-Controlled Release:** Certain polymeric nanomaterials, namely hydrogels, experience an increase in size when they come into contact with water-based environments.
3. **Degradation-Controlled Release:** Biodegradable polymers degrade under physiological settings, causing the slow erosion of the polymer matrix and eventual dissolution of drugs. The pace at which drugs are released is regulated by the speed at which the polymer breaks down. This breakdown rate may be adjusted by changing the chemical structure or molecular mass of the polymers (Kumar *et al.*, 2023).
4. **Stimuli-Responsive Release:** Certain polymeric nanoparticles have stimulus-responsive characteristics, whereby the release of drugs is activated by certain environmental stimuli, such as pH, temperature, light, or enzymes that are found at the intended location. Stimuli-responsive polymers experience structural changes in reaction to various stimuli, resulting in the liberation of encapsulated medicines (Handa *et al.*, 2022).

3.6 METALLIC NANOMATERIALS IN WOUND HEALING

Metallic nanomaterials, including silver nanoparticles, zinc oxide nanoparticles, and gold nanoparticles, possess exceptional antibacterial characteristics, making them very helpful in the context of wound healing. They counteract infections by breaking the membranes of microorganisms and preventing their development, hence reducing the likelihood of complications in wounds. Moreover, metallic nanoparticles have the ability to enhance tissue regeneration by promoting healing processes such as stimulating the growth of new blood vessels and regulating the body's inflammatory reactions. Due to their adaptability and effectiveness, they are highly attractive prospects for the development of new wound dressings and treatments (Kolimi *et al.*, 2022). These materials have the potential to expedite healing and enhance results in wound care.

3.6.1 Gold and Silver Nanoparticles

Gold and silver nanoparticles are two prominent metallic nanomaterials with unique properties that find diverse applications in wound healing.

3.6.1.1 Gold Nanoparticles (AuNPs)

Gold nanoparticles (**Figure 3.1**) have remarkable biocompatibility, chemical stability, and surface plasmon resonance (SPR) qualities, which make them very beneficial for wound healing applications. Their adjustable optical characteristics provide straightforward identification and visualization inside biological systems. Gold nanoparticles are being investigated for their ability to kill many types of microorganisms, such as bacteria, fungi, and viruses, in the process of wound healing. They cause disruption of microbial membranes, inhibition of enzyme function, and induction of oxidative stress, resulting in the death of microbial cells. Moreover, gold nanoparticles aid tissue regeneration by augmenting angiogenesis, collagen formation, and cell proliferation. Due to their capacity to regulate inflammatory reactions and facilitate wound healing, they show great potential as candidates for the development of sophisticated wound care products and pharmacological substances (Öhnstedt *et al.*, 2019).

3.6.1.2 Silver Nanoparticles (AgNPs)

Silver nanoparticles are well recognized for their strong antibacterial capabilities, which are essential in the process of wound healing. They demonstrate wide-ranging effectiveness against bacteria, including those that are resistant to antibiotics, as well as fungi and viruses. Silver nanoparticles cause disruption of microbial cell membranes, interference with cellular functions, and induction of oxidative stress, ultimately leading to the death of microorganisms (Adeyemi *et al.*, 2020). Silver nanoparticles not only possess antibacterial properties, but they also facilitate wound healing by diminishing inflammation, promoting the growth of new blood vessels, and expediting the regeneration of tissue. Additionally, they have anti-inflammatory characteristics that help reduce excessive inflammation at the location of the lesion. Wound dressings, coatings, and hydrogels often include silver nanoparticles that prevent diseases and facilitate wound healing (Diniz *et al.*, 2020). Nevertheless, the possibility of their ability to cause cell death and the possible negative impact on the surroundings need careful evaluation before their use. Gold and silver nanoparticles have distinct benefits in the process of wound healing, such as their ability to combat microbial activity, promote tissue regeneration, and exhibit compatibility with living organisms. Due to their versatility and adjustable features, they are highly attractive prospects for the development of enhanced wound care treatments that effectively tackle the difficulties associated with infection oversight, inflammation management, and tissue restoration (Khondakar *et al.*, 2024). Nevertheless, further investigation is required to comprehensively comprehend the processes by which they operate, refine their compositions, and guarantee their safety and effectiveness in clinical environments.

3.6.2 Mechanisms of Action

The actions of gold and silver nanoparticles in wound healing include a complex interaction of antibacterial, anti-inflammatory, and tissue-regenerative activities. These nanoparticles cause damage to microbial membranes, trigger oxidative stress, and disrupt biological processes. This ultimately results in the death of microbial cells and prevents invasion at the injury site (Mammari *et al.*, 2022). Furthermore, they regulate inflammatory reactions by decreasing the production of pro-inflammatory cytokines and fostering a well-balanced immune response that supports the healing of wounds. Moreover, gold and silver nanoparticles promote tissue regeneration by augmenting angiogenesis, collagen production, and cell proliferation, hence expediting wound healing and epithelialization.

3.6.3 Clinical Applications and Efficacy

Gold and silver nanoparticles have shown potential in several therapeutic uses for wound healing owing to their antimicrobial, anti-inflammatory, and tissue-regenerative characteristics. These nanoparticles are used in medical procedures to improve the effectiveness of wound dressings, coatings, and topical preparations. Clinical applications and efficacy include (Rajendran *et al.*, 2018):

1. **Infection Control:** Gold and silver nanoparticles have potent antimicrobial activity against a diverse array of pathogens, including bacteria, fungi, and viruses. Clinical trials have shown their effectiveness in lowering the amount of harmful microorganisms and avoiding diseases in both short-term and long-term wounds, as well as in diabetic ulcers and burns (Ding *et al.*, 2022).
2. **Accelerated Healing:** Gold and silver nanoparticles facilitate the healing of wounds and the growth of new tissue by promoting the formation of blood vessels, the production of collagen, and the multiplication of cells. Studies conducted in clinical trials have shown that wounds treated with nanoparticle-based dressings exhibit faster rates of wound healing and enhanced formation of epithelial tissue, in comparison to traditional treatment methods (Berthet *et al.*, 2017).
3. **Inflammation Modulation:** Gold and silver nanoparticles demonstrate anti-inflammatory properties by decreasing the levels of pro-inflammatory cytokines and alleviating excessive inflammation at the site of injury. Empirical data indicates that treatments using nanoparticles stimulate a harmonized immune reaction, resulting in enhanced healing results and decreased problems (Rybka *et al.*, 2022).
4. **Pain Management:** Nanoparticle-infused wound dressings deliver pain management by forming a shielding layer, decreasing contact, and limiting wound discharge. Nanoparticle-containing treatments have been shown to result in decreased pain ratings and enhanced satisfaction for patients in clinical investigations, as opposed to normal dressings (Omar *et al.*, 2021).
5. **Minimization of Scarring:** The possibility of gold and silver nanoparticles to reduce scarring and enhance aesthetically pleasing wound healing results has been studied. Empirical evidence indicates that treatments using nanoparticles have the potential to diminish the growth of scars and enhance the visual appearance of treated wounds (Salvo *et al.*, 2022).

3.7 CARBON-BASED NANOMATERIALS IN WOUND HEALING

Carbon-based nanomaterials, such as carbon nanotubes, graphene, and carbon dots, possess distinctive characteristics that make them suitable for use in applications in wound healing. Due to their exceptional properties such as large surface area, strong durability, and compatibility with living organisms, they are well-suited for applications in tissue engineering, drug delivery, and bioimaging (Nguyen *et al.*, 2020). Carbon nanoparticles enhance the attachment, growth, and specialization of cells, expediting the healing of wounds and the restoration of tissues. Moreover, they have antibacterial characteristics, effectively fighting against infections in wounds.

3.7.1 Graphene and Carbon Nanotubes

Carbon nanotubes are tubular structures made of coiled graphene sheets. These materials exhibit exceptional mechanical strength, flexibility, and conductivity, which makes them very adaptable for the purpose of wound healing. Carbon nanotubes (CNTs) enhance the attachment, growth, and movement of cells, hence aiding in the process of tissue regeneration. They possess antibacterial characteristics, which hinder the proliferation of germs and fungi in wounds. In addition, scaffolds made from carbon nanotubes (CNTs) provide structural support for the proliferation of cells and development of tissue, hence improving wound closure and epithelialization (Pormohammad *et al.*, 2021). Moreover, carbon nanotubes (CNTs) have the capability to undergo functionalization with bioactive compounds or growth factors, which may enhance the specific delivery of drugs and expedite the healing of wounds.

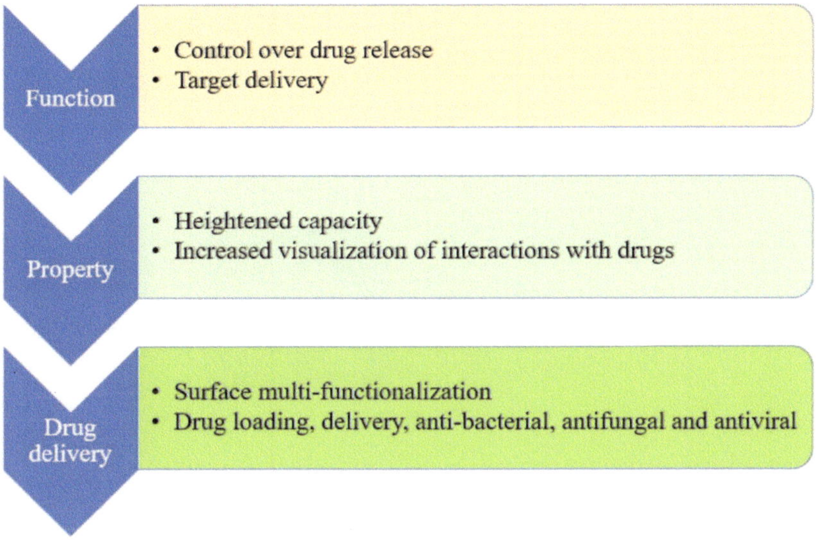

FIGURE 3.2 Functions, properties, and drug delivery components of nanotechnology.

3.7.2 Drug Delivery and Healing Mechanisms

Graphene and carbon nanotubes play crucial roles in drug delivery and wound healing through various mechanisms (**Figure 3.2**):

Drug Delivery Mechanisms:

1. **High Drug Loading Capacity:** Graphene and carbon nanotubes have extensive surface areas, enabling them to accommodate enormous amounts of drugs. Drugs have the ability to adhere to the surfaces of materials or be enclosed inside their molecular makeup, allowing for effective transportation to the location of the wound (Finbloom *et al.*, 2020).
2. **Controlled Release:** Nanomaterials provide a mechanism for precise and regulated release of medicinal substances, limiting sudden release and providing a consistent and prolonged distribution of drugs. Surface alterations and integration allow for the control of discharge kinetics, hence improving the effectiveness of therapeutic treatment (Patra *et al.*, 2018).
3. **Targeted Delivery:** Targeted medication distribution to certain cells or tissues may be achieved by functionalizing drugs with targeting ligands or stimuli-responsive moieties. This reduces unintended negative consequences and increases the concentration of the medicine at the location of the wound, resulting in better treatment results (Veith *et al.*, 2019).

*Healing Mechanisms (**Figure 3.3**):*

1. **Cell Adhesion and Proliferation:** Graphene and carbon nanotubes enhance cell adhesion, proliferation, and migration, hence promoting the healing of tissues. Their distinct surface characteristics and topographies closely resemble those of the matrix that surrounds cells, making them an optimal foundation for cellular adhesion and proliferation (Tudureanu *et al.*, 2022).
2. **Angiogenesis Stimulation:** These nanoparticles promote the growth of newly formed blood vessels, known as angiogenesis, by increasing the reproduction and movement of endothelial cells. Enhanced blood circulation to the injured area expedites the process of tissue healing and regrowth (Kaur *et al.*, 2022).

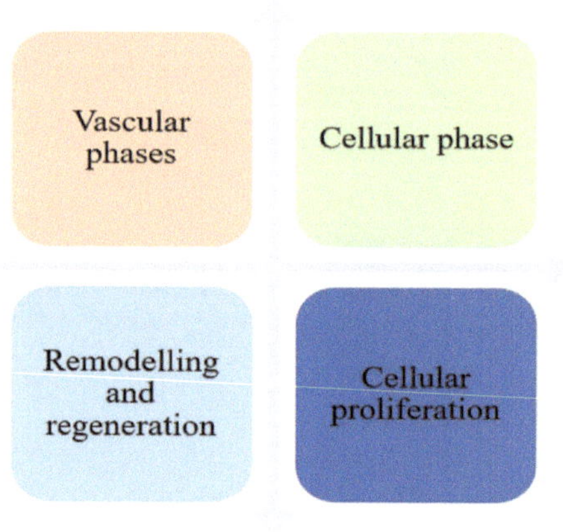

FIGURE 3.3 Stages of wound healing.

3. **Anti-inflammatory Effects:** Graphene and carbon nanotubes possess anti-inflammatory characteristics, which result in a decrease in the generation of pro-inflammatory cytokines and the alleviation of chronic swelling at the exact location of the lesion. This fosters a beneficial milieu for the process of healing and mitigates the occurrence of problems (Muire *et al.*, 2022).
4. **Antimicrobial Activity:** Both materials have intrinsic antibacterial qualities, which prevent the development of germs and fungus in wounds. This measure serves to avoid infections and foster a sterile atmosphere that is beneficial to the process of healing (Dubey *et al.*, 2022).
5. **Enhanced Wound Closure:** Graphene and carbon nanotubes enhance cell proliferation and tissue proliferation, promoting wound healing and epithelial growth. Their robustness and pliability enhance the effectiveness of wound healing (Amiryaghoubi *et al.*, 2021).

3.8 BIOCOMPATIBILITY AND SAFETY OF NANOMATERIALS

When developing nanomaterial-based systems to deliver drugs for wound healing, it is crucial to prioritize biocompatibility and stability. Nanomaterials such as graphene and carbon nanotubes have distinct possibilities for precise and regulated administration of drugs to specific wound locations. Nevertheless, it is important to conduct a comprehensive assessment of their impact on biological structures to guarantee their effectiveness and safety in clinical settings (Albahri *et al.*, 2023).

Graphene and carbon nanotubes provide favorable biocompatibility characteristics, as shown by research showing compliance with many kinds of cells that participate in the process of wound healing, including fibroblasts, keratinocytes, and endothelial cells (Sadat *et al.*, 2022). Nanomaterials have a crucial role in supporting vital cellular processes such as adhesion, proliferation, and differentiation, which in turn promote tissue regeneration. Additionally, the biocompatibility of these materials may be improved by surface modifications and functionalization techniques, which reduce unspecific interactions and boost material-cell connections (Chandy *et al.*, 2020).

Although graphene and carbon nanotubes show promise for medication delivery in wound healing, there are ongoing worries about their possible toxicity. Excessive amounts or incorrect methods of administration might result in negative health consequences, such as inflammation, fibrosis, or genotoxicity. Therefore, conducting thorough toxicity evaluations is essential for assessing the safety characteristics of drug delivery systems that utilize nanomaterials. Comprehending the diffusion and elimination mechanisms of nanomaterials is crucial for evaluating their safety inside a living organism. Research has shown that graphene and carbon nanotubes may be eliminated from the human body by renal excretion after being broken down by hepatic enzymes. This process helps to reduce their buildup and any harmful effects. Nevertheless, comprehensive long-term research is necessary to completely clarify their impact on overall bodily functions and guarantee their secure use in clinical environments. Strict adherence to regulatory requirements is essential for successfully transitioning nanomaterial-based drug delivery devices into clinical trials. Regulatory bodies, such as the FDA and EMA, rigorously oversee the safety and effectiveness of these treatments to safeguard the well-being of patients. To adhere to regulatory requirements and guarantee the health of patients, it is essential to conduct thorough preclinical and clinical investigations, which include biological compatibility tests and long-term evaluations (Bernard *et al.*, 2018).

3.9 CONCLUSION

Nanomaterial-based drug delivery technologies are bringing about a breakthrough age in the field of wound healing. These advanced technologies use the distinctive characteristics of nanoparticles to tackle decades-old problems in wound care. Nanomaterial-based systems provide a very precise method for delivering therapeutic chemicals, allowing for focused and regulated treatment of infections, regulation of inflammation, and stimulation of tissue regeneration. Although continuous endeavors are being made to guarantee biocompatibility and safety, the prospective advantages of these advances are significant. As scientific investigation progresses in the field of nanomedicine, the possibility of faster healing, better results for patients, and higher standards of life for those with wounds becomes more imminent.

REFERENCES

Adeyemi, O. S., Shittu, E. O., Akpor, O. B., Rotimi, D., & Batiha, G. E. S. (2020). Silver nanoparticles restrict microbial growth by promoting oxidative stress and DNA damage. *EXCLI Journal*, *19*, 492.

Aisa, J., & Parlier, M. (2022). Local wound management: A review of modern techniques and products. *Veterinary Dermatology*, *33*(5), 463–478.

Albahri, A. S., Duhaim, A. M., Fadhel, M. A., Alnoor, A., Baqer, N. S., Alzubaidi, L., . . . Deveci, M. (2023). A systematic review of trustworthy and explainable artificial intelligence in healthcare: Assessment of quality, bias risk, and data fusion. *Information Fusion*, *96*, 156–191.

Alvarado, Y., Muro, C., Illescas, J., & Riera, F. (2019). Polymer nanoparticles for the release of complex molecules. *Materials for Biomedical Engineering*, 135–163.

Amiryaghoubi, N., Fathi, M., Barzegari, A., Barar, J., Omidian, H., & Omidi, Y. (2021). Recent advances in polymeric scaffolds containing carbon nanotube and graphene oxide for cartilage and bone regeneration. *Materials Today Communications*, *26*, 102097.

Amorim, S., Reis, C. A., Reis, R. L., & Pires, R. A. (2021). Extracellular matrix mimics using hyaluronan-based biomaterials. *Trends in Biotechnology*, *39*(1), 90–104.

Apak, R., Calokerinos, A., Gorinstein, S., Segundo, M. A., Hibbert, D. B., Gülçin, İ., & Arancibia-Avila, P. (2022). Methods to evaluate the scavenging activity of antioxidants toward reactive oxygen and nitrogen species (IUPAC Technical Report). *Pure and Applied Chemistry*, *94*(1), 87–144.

Augustine, R., Rehman, S. R. U., Ahmed, R., Zahid, A. A., Sharifi, M., Falahati, M., & Hasan, A. (2020). Electrospun chitosan membranes containing bioactive and therapeutic agents for enhanced wound healing. *International Journal of Biological Macromolecules*, *156*, 153–170.

Berthet, M., Gauthier, Y., Lacroix, C., Verrier, B., & Monge, C. (2017). Nanoparticle-based dressing: The future of wound treatment?. *Trends in Biotechnology*, *35*(8), 770–784.

Bernard, M., Jubeli, E., Pungente, M. D., & Yagoubi, N. (2018). Biocompatibility of polymer-based biomaterials and medical devices–regulations, in vitro screening and risk-management. *Biomaterials Science*, *6*(8), 2025–2053.

Bourquin, J., Milosevic, A., Hauser, D., Lehner, R., Blank, F., Petri-Fink, A., & Rothen-Rutishauser, B. (2018). Biodistribution, clearance, and long-term fate of clinically relevant nanomaterials. *Advanced Materials*, *30*(19), 1704307.

Catanzano, O., Quaglia, F., & Boateng, J. S. (2021). Wound dressings as growth factor delivery platforms for chronic wound healing. *Expert Opinion on Drug Delivery*, *18*(6), 737–759.

Chandrakala, V., Aruna, V., & Angajala, G. (2022). Review on metal nanoparticles as nanocarriers: Current challenges and perspectives in drug delivery systems. *Emergent Materials*, *5*(6), 1593–1615.

Chandy, T. (2020). Biocompatibility of materials and its relevance to drug delivery and tissue engineering. In *Biointegration of medical implant materials* (pp. 297–331). Woodhead Publishing.

Cheng, S., Wang, H., Pan, X., Zhang, C., Zhang, K., Chen, Z., . . . Qi, X. (2022). Dendritic hydrogels with robust inherent antibacterial properties for promoting bacteria-infected wound healing. *ACS Applied Materials & Interfaces*, *14*(9), 11144–11155.

Chenthamara, D., Subramaniam, S., Ramakrishnan, S. G., Krishnaswamy, S., Essa, M. M., Lin, F. H., & Qoronfleh, M. W. (2019). Therapeutic efficacy of nanoparticles and routes of administration. *Biomaterials Research*, *23*(1), 20.

D'souza, A. A., & Shegokar, R. (2016). Polyethylene glycol (PEG): A versatile polymer for pharmaceutical applications. *Expert Opinion on Drug Delivery*, *13*(9), 1257–1275.

Das, S. K. (2017). Nanoparticles advanced characterization techniques: A view point. *Journal of Atoms and Molecules*, *7*(4), 1091.

Ding, X., Tang, Q., Xu, Z., Xu, Y., Zhang, H., Zheng, D., & Chen, J. (2022). Challenges and innovations in treating chronic and acute wound infections: From basic science to clinical practice. *Burns & Trauma*, *10*, tkac014.

Diniz, F. R., Maia, R. C. A., de Andrade, L. R. M., Andrade, L. N., Vinicius Chaud, M., da Silva, C. F., . . . Severino, P. (2020). Silver nanoparticles-composing alginate/gelatine hydrogel improves wound healing in vivo. *Nanomaterials*, *10*(2), 390.

Dong, R., & Guo, B. (2021). Smart wound dressings for wound healing. *Nano Today*, *41*, 101290.

Drozdzik, M., Drozdzik, M., & Oswald, S. (2021). Membrane carriers and transporters in kidney physiology and disease. *Biomedicines*, *9*(4), 426.

Dubey, S. K., Parab, S., Alexander, A., Agrawal, M., Achalla, V. P. K., Pal, U. N., . . . Kesharwani, P. (2022). Cold atmospheric plasma therapy in wound healing. *Process Biochemistry*, *112*, 112–123.

Fan, Q., Bai, J., Shan, H., Fei, Z., Chen, H., Xu, J., & Wang, C. (2021). Implantable blood clot loaded with BMP-2 for regulation of osteoimmunology and enhancement of bone repair. *Bioactive Materials*, *6*(11), 4014–4026.

Finbloom, J. A., Sousa, F., Stevens, M. M., & Desai, T. A. (2020). Engineering the drug carrier biointerface to overcome biological barriers to drug delivery. *Advanced Drug Delivery Reviews*, *167*, 89–108.

Foroozandeh, P., Jusoh, S. A. M., & Shamsuddin, S. (2021). Passive drug delivery, mechanisms of uptake, and intracellular trafficking. In *Organelle and molecular targeting* (pp. 129–152). CRC Press.

Fritea, L., Banica, F., Costea, T. O., Moldovan, L., Dobjanschi, L., Muresan, M., & Cavalu, S. (2021). Metal nanoparticles and carbon-based nanomaterials for improved performances of electrochemical (Bio) sensors with biomedical applications. *Materials*, *14*(21), 6319.

Ganetri, I., Essamlali, Y., Amadine, O., Danoun, K., Aboulhrouz, S., & Zahouily, M. (2021). Controlling factors of slow or controlled-release fertilizers. In *Controlled release fertilizers for sustainable agriculture* (pp. 111–129). Academic Press.

Guo, X., Zhang, Q., Ding, X., Shen, Q., Wu, C., Zhang, L., & Yang, H. (2016). Synthesis and application of several sol–gel-derived materials via sol–gel process combining with other technologies: A review. *Journal of Sol-Gel Science and Technology*, *79*, 328–358.

Hamad, A., Khashan, K. S., & Hadi, A. (2020). Silver nanoparticles and silver ions as potential antibacterial agents. *Journal of Inorganic and Organometallic Polymers and Materials*, *30*(12), 4811–4828.

Handa, M., Singh, A., Flora, S. J. S., & Shukla, R. (2022). Stimuli-responsive polymeric nanosystems for therapeutic applications. *Current Pharmaceutical Design*, *28*(11), 910–921. https://doi.org/10.2174/1381612827666211208150210. PMID: 34879797.

Hayat, M., Nawaz, A., Chinnam, S., Muzammal, M., Latif, M. S., Yasin, M., Ashique, S., Zengin, G., & Farid, A. (2023). Formulation development and optimization of herbo synthetic gel: In vitro biological evaluation and in vivo wound healing studies. *Process Biochemistry*, *130*, 116–126.

Hu, B., Gao, M., Boakye-Yiadom, K. O., Ho, W., Yu, W., Xu, X., & Zhang, X. Q. (2021). An intrinsically bioactive hydrogel with on-demand drug release behaviors for diabetic wound healing. *Bioactive Materials*, *6*(12), 4592–4606.

Huang, W. C., Ying, R., Wang, W., Guo, Y., He, Y., Mo, X., . . . Mao, X. (2020). A macroporous hydrogel dressing with enhanced antibacterial and anti-inflammatory capabilities for accelerated wound healing. *Advanced Functional Materials*, *30*(21), 2000644.

Islam, M. S., Ang, B. C., Andriyana, A., & Afifi, A. M. (2019). A review on fabrication of nanofibers via electrospinning and their applications. *SN Applied Sciences*, *1*, 1–16.

Jacob, S., Nair, A. B., Shah, J., Sreeharsha, N., Gupta, S., & Shinu, P. (2021). Emerging role of hydrogels in drug delivery systems, tissue engineering and wound management. *Pharmaceutics*, *13*(3), 357.

Johnson, K. B., Wei, W. Q., Weeraratne, D., Frisse, M. E., Misulis, K., Rhee, K., . . . Snowdon, J. L. (2021). Precision medicine, AI, and the future of personalized health care. *Clinical and Translational Science*, *14*(1), 86–93.

Julier, Z., Park, A. J., Briquez, P. S., & Martino, M. M. (2017). Promoting tissue regeneration by modulating the immune system. *Acta Biomaterialia*, *53*, 13–28.

Kaur, G., Narayanan, G., Garg, D., Sachdev, A., & Matai, I. (2022). Biomaterials-based regenerative strategies for skin tissue wound healing. *ACS Applied Bio Materials*, *5*(5), 2069–2106.

Khondakar, K. R., Tripathi, D., Mazumdar, H., Ahuja, K., & Kaushik, A. (2024). Tailored MXene and graphene as efficient telemedicine platforms for personalized health wellness. *Materials Advances*.

Kolimi, P., Narala, S., Nyavanandi, D., Youssef, A. A. A., & Dudhipala, N. (2022). Innovative treatment strategies to accelerate wound healing: Trajectory and recent advancements. *Cells*, *11*(15), 2439.

Krishna, D. N. G., & Philip, J. (2022). Review on surface-characterization applications of X-ray photoelectron spectroscopy (XPS): Recent developments and challenges. *Applied Surface Science Advances*, *12*, 100332.

Kumar, A. R. S., Padmakumar, A., Kalita, U., Samanta, S., Baral, A., Singha, N. K., & Qiao, G. G. (2023). Ultrasonics in polymer science: Applications and challenges. *Progress in Materials Science*, *136*, 101113.

Kumar, A. R. S., Sharma, K., & Dixit, A. R. (2019). A review of the mechanical and thermal properties of graphene and its hybrid polymer nanocomposites for structural applications. *Journal of Materials Science*, *54*(8), 5992–6026.

Kumari, A., Kumar, V., & Yadav S. (2012). Nanotechnology: A tool to enhance therapeutic values of natural plant products. *Trends in Medical Research*, *7*, 34–42.

Liddle, J. A., & Gallatin, G. M. (2011). Lithography, metrology and nanomanufacturing. *Nanoscale*, *3*(7), 2679–2688.

Lin, L., Chi, J., Yan, Y., Luo, R., Feng, X., Zheng, Y., & Pan, X. (2021). Membrane-disruptive peptides/peptidomimetics-based therapeutics: Promising systems to combat bacteria and cancer in the drug-resistant era. *Acta Pharmaceutica Sinica B*, *11*(9), 2609–2644.

Liu, Y., Yang, G., Jin, S., Xu, L., & Zhao, C. X. (2020). Development of high-drug-loading nanoparticles. *ChemPlusChem*, *85*(9), 2143–2157.

Lombardo, D., Calandra, P., Pasqua, L., & Magazù, S. (2020). Self-assembly of organic nanomaterials and biomaterials: The bottom-up approach for functional nanostructures formation and advanced applications. *Materials*, *13*(5), 1048.

Lopes, J. H., Tabary, N., & Hernandez-Montelongo, J. (2022). Biomacromolecules Systems Applied to Medical Implants for the Release of Therapeutic Agents. *Frontiers in Bioengineering and Biotechnology*, *10*, 910203.

Lübtow, M. M., Haider, M. S., Kirsch, M., Klisch, S., & Luxenhofer, R. (2019). Like dissolves like? A comprehensive evaluation of partial solubility parameters to predict polymer–drug compatibility in ultrahigh drug-loaded polymer micelles. *Biomacromolecules*, *20*(8), 3041–3056.

Maleki, M., Zarezadeh, R., Nouri, M., Sadigh, A. R., Pouremamali, F., Asemi, Z., . . . Yousefi, B. (2020). Graphene oxide: A promising material for regenerative medicine and tissue engineering. *Biomolecular Concepts*, *11*(1), 182–200.

Mammari, N., Lamouroux, E., Boudier, A., & Duval, R. E. (2022). Current knowledge on the oxidative-stress-mediated antimicrobial properties of metal-based nanoparticles. *Microorganisms*, *10*(2), 437.

Manzari, M. T., Shamay, Y., Kiguchi, H., Rosen, N., Scaltriti, M., & Heller, D. A. (2021). Targeted drug delivery strategies for precision medicines. *Nature Reviews Materials*, *6*(4), 351–370.

Memic, A., Abdullah, T., Mohammed, H. S., Joshi Navare, K., Colombani, T., & Bencherif, S. A. (2019). Latest progress in electrospun nanofibers for wound healing applications. *ACS Applied Bio Materials*, *2*(3), 952–969.

Mihai, M. M., Dima, M. B., Dima, B., & Holban, A. M. (2019). Nanomaterials for wound healing and infection control. *Materials*, *12*(13), 2176.

Mohapatra, S., Ranjan, S., Dasgupta, N., Thomas, S., & Mishra, R. K. (Eds.). (2018). *Applications of targeted nano drugs and delivery systems: Nanoscience and nanotechnology in drug delivery*. Elsevier.

Moustaqil, M., Ollivier, E., Chiu, H. P., Van Tol, S., Rudolffi-Soto, P., Stevens, C., & Gambin, Y. (2021). SARS-CoV-2 proteases PLpro and 3CLpro cleave IRF3 and critical modulators of inflammatory pathways (NLRP12 and TAB1): Implications for disease presentation across species. *Emerging Microbes & Infections*, *10*(1), 178–195.

Muire, P. J., Thompson, M. A., Christy, R. J., & Natesan, S. (2022). Advances in immunomodulation and immune engineering approaches to improve healing of extremity wounds. *International Journal of Molecular Sciences*, *23*(8), 4074.

Nguyen, E. P., Silva, C. D.C. C., & Merkoçi, A. (2020). Recent advancement in biomedical applications on the surface of two-dimensional materials: From biosensing to tissue engineering. *Nanoscale*, *12*(37), 19043–19067.

Nour, S., Baheiraei, N., Imani, R., Khodaei, M., Alizadeh, A., Rabiee, N., & Moazzeni, S. M. (2019). A review of accelerated wound healing approaches: Biomaterial-assisted tissue remodeling. *Journal of Materials Science: Materials in Medicine*, *30*, 1–15.

Nqakala, Z. B., Sibuyi, N. R., Fadaka, A. O., Meyer, M., Onani, M. O., & Madiehe, A. M. (2021). Advances in nanotechnology towards development of silver nanoparticle-based wound-healing agents. *International Journal of Molecular Sciences*, *22*(20), 11272.

Öhnstedt, E., Lofton Tomenius, H., Vågesjö, E., & Phillipson, M. (2019). The discovery and development of topical medicines for wound healing. *Expert Opinion on Drug Discovery*, *14*(5), 485–497.

Omar, M. M., Laprise-Pelletier, M., Chevallier, P., Tuduri, L., & Fortin, M. A. (2021). High-sensitivity permeation analysis of ultrasmall nanoparticles across the skin by positron emission tomography. *Bioconjugate Chemistry*, *32*(4), 729–745.

Parra, A., Jarak, I., Santos, A., Veiga, F., & Figueiras, A. (2021). Polymeric micelles: A promising pathway for dermal drug delivery. *Materials*, *14*(23), 7278.

Patra, J. K., Das, G., Fraceto, L. F., Campos, E. V. R., Rodriguez-Torres, M. D. P., Acosta-Torres, L. S., & Shin, H. S. (2018). Nano based drug delivery systems: Recent developments and future prospects. *Journal of Nanobiotechnology*, *16*, 1–33.

Pormohammad, A., Monych, N. K., Ghosh, S., Turner, D. L., & Turner, R. J. (2021). Nanomaterials in wound healing and infection control. *Antibiotics*, *10*(5), 473.

Rajendran, N. K., Kumar, S. S. D., Houreld, N. N., & Abrahamse, H. (2018). A review on nanoparticle based treatment for wound healing. *Journal of Drug Delivery Science and Technology*, *44*, 421–430.

Rybka, M., Mazurek, Ł., & Konop, M. (2022). Beneficial effect of wound dressings containing silver and silver nanoparticles in wound healing—from experimental studies to clinical practice. *Life*, *13*(1), 69.

Sadat, Z., Farrokhi-Hajiabad, F., Lalebeigi, F., Naderi, N., Gorab, M. G., Cohan, R. A., . . . Maleki, A. (2022). A comprehensive review on the applications of carbon-based nanostructures in wound healing: From antibacterial aspects to cell growth stimulation. *Biomaterials Science*, *10*(24), 6911–6938.

Saghazadeh, S., Rinoldi, C., Schot, M., Kashaf, S. S., Sharifi, F., Jalilian, E., . . . Khademhosseini, A. (2018). Drug delivery systems and materials for wound healing applications. *Advanced Drug Delivery Reviews*, *127*, 138–166.

Salvo, J., & Sandoval, C. (2022). Role of copper nanoparticles in wound healing for chronic wounds: Literature review. *Burns & Trauma*, *10*, tkab047.

Sánchez, A., Mejía, S. P., & Orozco, J. (2020). Recent advances in polymeric nanoparticle-encapsulated drugs against intracellular infections. *Molecules*, *25*(16), 3760.

Seelig, A. (2007). The role of size and charge for blood–brain barrier permeation of drugs and fatty acids. *Journal of Molecular Neuroscience*, *33*, 32–41.

Shah, K. A., & Tali, B. A. (2016). Synthesis of carbon nanotubes by catalytic chemical vapour deposition: A review on carbon sources, catalysts and substrates. *Materials Science in Semiconductor Processing*, *41*, 67–82.

Shalaby, M. A., Anwar, M. M., & Saeed, H. (2022). Nanomaterials for application in wound Healing: Current state-of-the-art and future perspectives. *Journal of Polymer Research*, *29*(3), 91.

Smith, D., Artursson, P., Avdeef, A., Di, L., Ecker, G. F., Faller, B., . . . Testa, B. (2014). Passive lipoidal diffusion and carrier-mediated cell uptake are both important mechanisms of membrane permeation in drug disposition. *Molecular Pharmaceutics*, *11*(6), 1727–1738.

Sukhanov, S., Higashi, Y., Yoshida, T., Mummidi, S., Aroor, A. R., Russell, J. J., & Chandrasekar, B. (2021). The SGLT2 inhibitor Empagliflozin attenuates interleukin-17A-induced human aortic smooth muscle cell proliferation and migration by targeting TRAF3IP2/ROS/NLRP3/Caspase-1-dependent IL-1β and IL-18 secretion. *Cellular Signalling*, *77*, 109825.

Svenson, S., & Tomalia, D. A. (2012). Dendrimers in biomedical applications—reflections on the field. *Advanced Drug Delivery Reviews*, *64*, 102–115.

Tiwari, N., Kumar, D., Priyadarshani, A., Jain, G. K., Mittal, G., Kesharwani, P., & Aggarwal, G. (2023). Recent progress in polymeric biomaterials and their potential applications in skin regeneration and wound care management. *Journal of Drug Delivery Science and Technology*, *82*, 104319.

Tudureanu, R., Handrea-Dragan, I. M., Boca, S., & Botiz, I. (2022). Insight and recent advances into the role of topography on the cell differentiation and proliferation on biopolymeric surfaces. *International Journal of Molecular Sciences*, *23*(14), 7731.

Utterström, J., Naeimipour, S., Selegård, R., & Aili, D. (2021). Coiled coil-based therapeutics and drug delivery systems. *Advanced Drug Delivery Reviews*, *170*, 26–43.

Veith, A. P., Henderson, K., Spencer, A., Sligar, A. D., & Baker, A. B. (2019). Therapeutic strategies for enhancing angiogenesis in wound healing. *Advanced Drug Delivery Reviews*, *146*, 97–125.

Wei, J., Mu, J., Tang, Y., Qin, D., Duan, J., & Wu, A. (2023). Next-generation nanomaterials: Advancing ocular anti-inflammatory drug therapy. *Journal of Nanobiotechnology*, *21*(1), 282.

Xiang, J., Shen, L., & Hong, Y. (2020). Status and future scope of hydrogels in wound healing: Synthesis, materials and evaluation. *European Polymer Journal*, *130*, 109609.

Yadav, T. P., Yadav, R. M., & Singh, D. P. (2012). Mechanical milling: A top down approach for the synthesis of nanomaterials and nanocomposites. *Nanoscience and Nanotechnology*, *2*(3), 22–48.

Yammine, J., Chihib, N. E., Gharsallaoui, A., Ismail, A., & Karam, L. (2024). Advances in essential oils encapsulation: Development, characterization and release mechanisms. *Polymer Bulletin*, *81*(5), 3837–3882.

Yang, L., Patel, K. D., Rathnam, C., Thangam, R., Hou, Y., Kang, H., & Lee, K. B. (2022). Harnessing the therapeutic potential of extracellular vesicles for biomedical applications using multifunctional magnetic nanomaterials. *Small*, *18*(13), 2104783.

Yin, M., Wu, J., Deng, M., Wang, P., Ji, G., Wang, M., . . . Huang, P. (2021). Multifunctional magnesium organic framework-based microneedle patch for accelerating diabetic wound healing. *ACS Nano*, *15*(11), 17842–17853.

Yoshida, Y. G., Yan, S., Xu, H., & Yang, J. (2023). Novel metal nanomaterials to promote angiogenesis in tissue regeneration. *Engineered Regeneration*.

Zhang, T., Lin, R., Wu, H., Jiang, X., & Gao, J. (2022). Mesenchymal stem cells: A living carrier for active tumor-targeted delivery. *Advanced Drug Delivery Reviews*, *185*, 114300.

Zhang, Z., Tsai, P. C., Ramezanli, T., & Michniak-Kohn, B. B. (2013). Polymeric nanoparticles-based topical delivery systems for the treatment of dermatological diseases. *Wiley Interdisciplinary Reviews: Nanomedicine and Nanobiotechnology*, *5*(3), 205–218.

Zheng, K., Tong, Y., Zhang, S., He, R., Xiao, L., Iqbal, Z., . . . Li, Y. (2021). Flexible bicolorimetric polyacrylamide/chitosan hydrogels for smart real-time monitoring and promotion of wound healing. *Advanced Functional Materials*, *31*(34), 2102599.

Zhou, L., Min, T., Bian, X., Dong, Y., Zhang, P., & Wen, Y. (2022). Rational design of intelligent and multifunctional dressing to promote acute/chronic wound healing. *ACS Applied Bio Materials*, *5*(9), 4055–4085.

Zhou, X., Qian, Y., Chen, L., Li, T., Sun, X., Ma, X., & He, C. (2023). Flowerbed-inspired biomimetic scaffold with rapid internal tissue infiltration and vascularization capacity for bone repair. *ACS Nano*, *17*(5), 5140–5156.

Zorkina, Y., Abramova, O., Ushakova, V., Morozova, A., Zubkov, E., Valikhov, M., . . . Chekhonin, V. (2020). Nano carrier drug delivery systems for the treatment of neuropsychiatric disorders: Advantages and limitations. *Molecules*, *25*(22), 5294.

4 Artificial Intelligence for Wound Healing (I.E., Real-Time Monitoring, Image-Based, Bioinformatics, and Precision Regulation)

Karishma Rathi,[1] *Shraddha Gupta,*[2] *Gayatri Korade,*[2] *Gulshan Rathi,*[3] *and S. M. Firdous**

[1] Assistant Professor of the Department of Pharmacy Practice at Dr. D.Y. Patil Institute of Pharmaceutical Sciences and Research Pimpri, Pune

[2] Pharm D Intern, Department of Pharmacy Practice at Dr. D.Y. Patil Institute of Pharmaceutical Sciences and Research Pimpri, Pune

[2] Student Pharm D, Department of Pharmacy Practice at Dr. D.Y. Patil Institute of Pharmaceutical Sciences and Research Pimpri, Pune

[3] Assistant Professor, Department of Quality Assurance at VSS Institute of Pharmacy, Badnapur

* Head of Department of Pharmacology at Calcutta Institute of Pharmaceutical Technology & AHS, Uluberia, Howrah

Corresponding author: Dr. S M Firdous
Professor and Head
Calcutta Institute of Pharmaceutical Technology and ASH
Uluberia, Howrah, West Bengal, India

4.1 INTRODUCTION TO WOUND HEALING AND AI

A skin wound occurs when there is a breakdown in the integrity of the epidermal layer (1). Any injury to tissue that involves disruption to anatomical integrity and results in functional loss is considered a wound. Wound healing primarily refers to the natural physiological response to skin damage. Wound healing is a complicated process that involves the intricate interaction of different cell types, cytokines, mediators, and the vascular system, despite its seeming simplicity. The sequence begins with initial vasoconstriction and platelet aggregation to halt bleeding, followed by the infiltration of inflammatory cells, starting with neutrophils. These cells promote angiogenesis, thrombosis, and epithelialization by releasing mediators and cytokines. By adding extracellular components, fibroblasts aid in the healing process by creating a scaffold (1). The inflammatory phase, which is marked by chemotaxis, hemostasis, and increased vascular permeability, functions to seal the wound, get rid of germs and detritus from the cell, and promote cellular migration. Typically lasting several days, this phase marks the beginning of wound healing after an injury to the epidermal layer. The entire process is dynamic, encompassing well-organized cellular, humoral, and molecular mechanisms, and may extend over the course of several years (2).

 DOI: 10.1201/9781003605966-4

The three interlinked phases of wound healing are inflammation, proliferation, and remodeling. Any disturbance in this process can result in abnormal wound healing (3, 4).

4.1.1 Phases of Wound Healing

The sequence of phases of wound healing is shown in **Figure 4.1** and is described in detail below.

1. Phase I-Hemostasis
   At the onset of wound development, tissue loss triggers bleeding, exposing blood to extracellular matrix (ECM) components. To prevent further bleeding, vasoconstriction and platelet plugging occur, constricting blood vessels and aggregating platelets. This process, referred to as hemostasis, leads to the formation of fibrin. Fibrin facilitates cell motility by functioning as a transient matrix and is subsequently dissolved through fibrinolysis, paving the way for the next stage of wound healing (5).
2. Phase II-Inflammation
   Fibrin breakdown initiates inflammation, making vessels permeable. This allows plasma release, enabling neutrophils (within 24–48 hrs.) to phagocytize debris, defending against infection. Macrophages (24–36 hrs. later) contribute to bacterial removal and release growth factors to promote granulation, re-epithelialization, and the development of new blood vessels (6).
3. Phase III-Proliferation:
   Granulation tissue forms and keratinocytes migrate, promoting epithelial tissue regeneration for epidermal layer continuity. Collagen, elastin, glycosaminoglycans, fibrous proteins, and fibrin and fibronectin are all present in the new tissue. Extracellular proteins produced

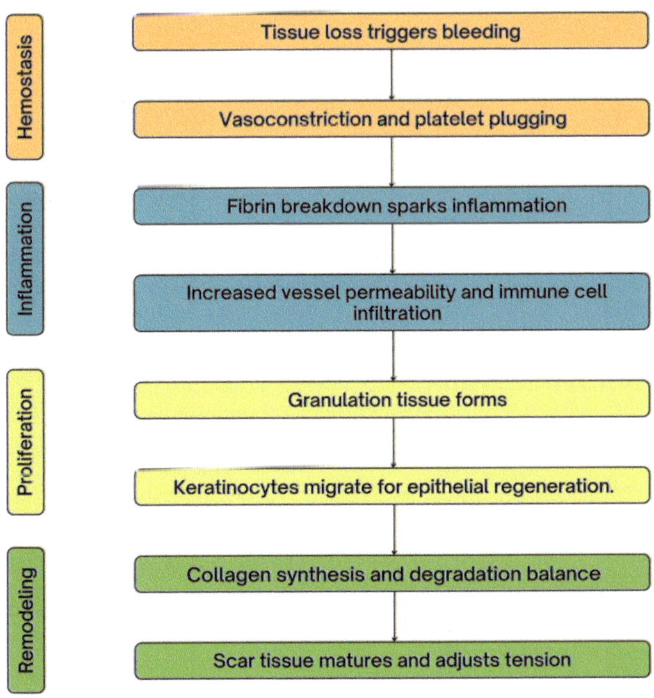

FIGURE 4.1 Phases of wound healing.

by fibroblasts control angiogenesis and cell division, employing growth and angiogenic factors (6).
4. Phase IV Remodeling:

When collagen production and breakdown are balanced and the wound closes, the remodeling phase starts. In this phase, the body produces new, stronger tissue to replace the temporary tissue created during the inflammation and proliferation stages. Additionally, it improves the scar's appearance and modifies the scar tissue's tension so that it matches the surrounding tissue (7).

4.1.2 Wound Healing Classification

1. Primary Healing:
   Uncomplicated healing of a noninfected, well-approximated wound, is termed primary healing, e.g., surgical wounds (8).
2. Secondary Healing:

When the healing process is disrupted by factors like infection, dehiscence, hypoxia, or immune dysfunction, the wound enters the secondary healing stage. This involves granulation tissue formation and epithelialization over the new tissue (8). Wounds in this category, like venous leg ulcers or extensive tissue loss, as in fourth-degree burns, are more susceptible to infections and exhibit poorer healing outcomes.

Approximately 8 million individuals globally suffer from wounds, whether infected or not (1). In the United States, chronic wounds impact 2% of the total population, and comparable figures are observed in European nations (2, 3). Chronic wound prevalence rises with age, with diabetic and obese individuals at heightened risk due to factors like hyperglycemia, impaired vascular function, and neuropathy (9).

4.1.3 Artificial Intelligence

John McCarthy, recognized as the father of artificial intelligence for coining the term in 1956, defined AI as the fusion of science and engineering to create intelligent devices for human welfare. He envisioned AI as an intellect surpassing the best human brains across various fields, including computer science and linguistic logic. AI encompasses philosophy, computer science, mathematics, linguistics, biology, neuroscience, sociology, and more (10).

Artificial General Intelligence (AGI) aims to emulate human-like intellectual behavior, involving learning, perception, problem-solving, and adaptive solutions. It incorporates linguistic logic and reasoning. AI is categorized as follows:

1. Weak AI, where machines mimic intelligence through programmed actions,
2. Strong AI, where machines independently perform calculations and make predictions, exemplified by IBM's supercomputer "WATSON." (10)

Artificial intelligence (AI) has the ability to completely transform the way healthcare is delivered by providing new methods for data collection and storage, administrative ease, patient evaluation, and treatment efficiency. AI, particularly machine learning (ML), employs various algorithms for programming or "learning," allowing it to adapt to experiences and inputs, mimicking human-like tasks. While AI has already found widespread adoption in finance and information technology, its utilization in healthcare is also expanding. In the realm of wound healing, there is a growing interest in leveraging AI models. A study conducted by Cho et al. highlights an artificial intelligence algorithm that forecasts the healing of chronic wounds in 12 weeks. The model demonstrated acceptable accuracy levels, emphasizing that wound characteristics, rather than patient demographics, were more precise predictors of recovery (11). Different studies reported promising results in using artificial intelligence for predicting pressure ulcers.

Beyond predicting wound healing, AI shows potential as a valuable tool for wound assessment. D.M. Anisuzzaman et al. study reveals that image-based AI software holds promise for providing accurate data on wound assessment (12). In comparison to manual assessments, which are prone to observer variability, computer-based image analysis significantly enhances wound assessment by eliminating observer bias. This improvement has the potential to enhance wound care outcomes and increase patient satisfaction.

4.2 CHALLENGES IN WOUND HEALING

Obstacles in the wound healing process arise from various factors. Specific biological markers delineate the nonhealing nature of chronic wounds, influenced by both local and systemic elements. Despite good clinical practices, eliminating or reducing the impact of these factors is deemed impossible. Alongside the local and systemic factors that hinder healing, potential biomarkers for chronic wounds include diminished tissue growth factors, heightened proteolytic enzymes like matrix metalloproteinases causing extracellular matrix degradation, increased inflammatory mediators (notably, excessive neutrophil infiltration in pressure ulcers), and the existence of senescent cells (referenced as potential biomarkers for chronic wounds in source00013).

Table 4.1 provides an extensive overview of the factors influencing wound healing, categorizing them into local and systemic components for clarity and thorough examination (14):

Beyond these health-related challenges, the fundamental issue lies in the perception of wounds. Chronic wounds are not often recognized as a disease. Nurses and doctors use a variety of treatment modalities depending on their experiences and perspectives, which vary greatly between settings and geographical areas (14).

Divergence in approaches across different settings and regions underscores the challenge of not perceiving wounds as a distinct disease. This viewpoint exacerbates serious problems with the wound care system as it exists now, encompassing (15):

1. Limited robustness and reproducibility in research and clinical results might result in limitations on meta-analysis and incorrect interpretation of wound healing data.

TABLE 4.1
Classification of Factors Affecting Wound Healing

Local Factors	Systemic Factors
Foreign body	Conditions include diabetes, fibrosis, keloids, jaundice, uremia, and inherited healing abnormalities.
• Venous sufficiency	Medication includes chemotherapy, nonsteroidal anti-inflammatory medications, and glucocorticoids.
Tissue maceration	• Age and gender
Biofilm	• Immunocompromised conditions: cancer, AIDS
Hypoxia	• Alcoholism and smoking
Ischemia	Malnutrition
Oxygenation	• Sex hormones
Infection	• Stress

2. Insufficient knowledge of the biomolecular signals that coincide with the wound environment, especially in exudates.
3. Reliance on antiquated instruments and techniques for classifying and quantifying different kinds of wounds.
4. Inadequate coordination between stakeholders.
5. An overcrowded, ineffective market full of identical goods.
6. The lack of a Centre of Excellence for the Treatment of Chronic Wounds. The fragmented community lacks a central hub to coordinate multidisciplinary collaboration among surgeons, internal medicine practitioners, infectious disease specialists, specialty nurses, engineers, basic scientists, social workers, and health economists.
7. Education: Traditionally, nursing has taken the lead in wound care, but formal training is lacking in basic nursing education (16). The patients' lack of understanding is a major problem in the treatment of chronic wounds, with a shortage of educational materials designed for them.
8. The requirement for modifications to FDA guidelines to simplify and clarify PMA and 510(k) approval processes and limit the entry of new "me too" goods within the crowded chronic wound market.
9. Challenges in insurance and reimbursement, where approval delays and "standard of care" requirements impede advanced therapies, lead to disease progression and increased costs for preventive treatments, causing patient suffering.
10. The expanding aging population without a concurrent increase in clinician expertise, with a scarcity of qualified professionals with experience in wound care, such as nurses and geriatricians.
11. Commercial management firms drive wound care centers, funding and overseeing operations. Their crucial role in billing and regulatory compliance underscores the complex decision-making in patient treatment strategies.

4.3 TRADITIONAL APPROACHES FOR WOUND HEALING

Wound management starts with identifying the cause of the wound and addressing systemic and lifestyle factors. Wounds are irrigated with saline or an antibacterial solution, and a customized dressing is applied (17). Debridement, the removal of nonviable tissue, is the first step in wound care, shown to halve the healing time of chronic wounds. Moist dressings, such as moist compresses, are commonly used, but a variety of dressings are also used:

- Hydrogel dressings
- Hydrocolloid dressings
- Gauze
- Dressings containing silver or alginates
- Films
- Foam dressings

The purpose of the dressings is to keep the wound from becoming infected and to drain excess fluid (18). Adjuncts to wound healing aim to tackle and overcome the barriers hindering the process of wound healing. These adjuncts include:

- Bioengineered Skin: Living bioengineered skin, like Apligraft, provides growth factors, collagens, and a collagen matrix. Its exact mechanism isn't fully understood, but it stimulates wound healing processes like re-epithelialization, angiogenesis, and immune cell attraction, making it effective for both acute and chronic wounds (19, 20).

- Electrostimulation: Electrostimulation mimics the skin's natural electrical currents post-injury, potentially accelerating wound healing by enhancing the migration of key healing cells like neutrophils, macrophages, and fibroblasts. Studies demonstrate increased collagen production, tensile strength, and improved blood flow with electrostimulation, bolstering the wound healing process (20).
- Negative pressure wound therapy: Negative pressure wound therapy utilizes an airtight dressing connected to a pump to continuously remove fluid from the wound, promoting healing by enhancing blood flow and maintaining moisture. While commonly used in hospitals for chronic or large open wounds, it can also be used at home. However, its effectiveness lacks strong evidence, and the efficacy of other therapies like shockwave, ozone, or light therapy remains uncertain (18).
- Ultrasound therapy: Ultrasound therapy applies sound waves to tissues, inducing a warming effect. Its therapeutic effects stem from both thermal and nonthermal properties. Ultrasound converts electrical energy at frequencies exceeding 20,000 Hz, with penetration depth determined by frequency. Traditionally used for muscular spasm relief and musculoskeletal pain, the thermal component has also been explored to enhance scar outcomes (18, 21).
- Hyperbaric oxygen therapy: Hyperbaric oxygen therapy exposes individuals to high-pressure oxygen environments, increasing wound oxygen levels and potentially enhancing healing by boosting blood oxygen concentration and improving blood supply (18, 21).
- Lasers: Low-energy lasers utilized in wound management can slightly raise tissue temperature. These lasers increase cellular activity in wounded skin, although the precise mechanism remains incompletely understood. Proposed mechanisms include stimulation of ascorbic acid uptake, activation of photoreceptors in mitochondria, alterations in cellular ATP or cyclic AMP levels, and stabilization of cell membranes (21, 22).

To manage pain generated from wounds like mechanical injuries or surgical wounds, NSAIDs are commonly prescribed. However, due to concerns about systemic exposure, they cannot be used throughout the entire healing process. For chronic wounds causing persistent pain, acetaminophen or ibuprofen can provide relief. Stronger painkillers are available by prescription if needed. In cases of primary hyperalgesia, a combination of NSAIDs and mild opioids may be used in hospitals to reduce inflammation and modulate pain at the spinal cord level (23).

Wounds infected with bacteria may hinder effective healing, necessitating antibiotics in severe cases, applied topically using ointments or compresses. Topical delivery requires smaller drug volumes, ensuring high and sustained antibiotic concentrations while minimizing systemic side effects. Combined with proper nutrition, dressing, and removal of dead tissue, this approach helps reduce microbial infection (23). Antisepsis is crucial in chronic wound care, with agents like iodine, chlorhexidine, silver, and polyhexamethylene biguanide available in various forms such as solutions, gels, and dressings (24).

4.4 ROLE OF ARTIFICIAL INTELLIGENCE IN WOUND HEALING

4.4.1 Overview of Artificial Intelligence

Artificial intelligence (AI), an area of computer science, traces its roots back to the 1950s (25). The term "artificial intelligence" (AI) was coined in a proposal for a conference at Dartmouth College in 1955 (26). Emphasizing the ambition to mimic and automate human thinking processes through the application of computer algorithms, this historical origin underscores the ongoing development of AI, dedicated to advancing computational methods for replicating and enhancing human cognitive abilities.

Machine learning in AI is predominantly driven by artificial neural networks (ANNs) and fuzzy logic. The pervasive influence of AI on modern technology and daily life is evident, with "deep learning," a subtype of ANNs, playing a pivotal role as a catalyst for this impact (27–29). Inspired by the human brain, ANN architecture mirrors interconnected neurons processing complex information. Comprising interconnected layers of artificial neurons or nodes that have been educated on data and produced using mathematical techniques, ANNs excel in making inferences or predictions when faced with new information. An example of deep learning is represented by deep neural networks, which include millions of artificial neurons arranged in several linked layers. The algorithm is enabled to extract intricate and concealed information from training data under this layer depth. Consequently, deep learning models prove invaluable for tasks like object detection and predicting outcomes, surpassing the competitive capabilities of human experts (29).

In the early 1970s, AI began to make its mark in healthcare, signifying an early integration of this technology into practical fields (25). Ethical and safety considerations likely contributed to the cautious adoption of new technologies by healthcare professionals, especially when human lives are at stake (30). Despite this cautious approach, industry experts foresee widespread AI utilization enhancing care quality and patient outcomes by reducing human errors (31). Moreover, clinicians may gain more time for complex responsibilities as AI takes over routine tasks. However, factors such as trust, perceived usefulness, and privacy remain crucial in the acceptance and adoption of AI in healthcare (8, 9). The AI in healthcare market is projected to grow significantly, from generating \$6.9 billion in 2021 to an estimated \$67.4 billion by 2027, with a compound annual growth rate of 46.2% during 2021–2027. The increasing focus on cost optimization, data management, public-private alliances, and regional budget allocations are key drivers fueling demand for AI in the healthcare sector (32).

AI-driven image processing techniques have emerged as promising computerized tools for precise wound assessment, offering objectivity and noninvasiveness. In recent times, a wide range of predictive techniques have been developed that outperform human wound specialists by utilizing diverse data processing systems, prediction algorithms, machine learning, and statistical models. In wound care, deep learning in particular has gained prominence and produced ground-breaking results.

4.4.2 Application of AI in Wound Healing

Artificial Intelligence (AI) plays a significant role in various aspects of healthcare, including wound healing. Here are some applications of AI in wound healing:

1. Smart Dressing:
   Smart wound dressings represent a novel approach to wound management, leveraging built-in sensors and intelligent materials to interact with wounds, monitor their condition, and adapt to environmental changes. Each stage of the wound healing process has unique characteristics and physiological requirements, and smart dressings are designed to address these needs effectively. Many types of smart wound dressings have been created, such as self-healing dressings for mobile wounds, biomechanical dressings, stimuli-responsive dressings, self-removable dressings, and monitoring dressings (33). These dressings play a crucial role in interacting with the wound environment, tracking changes throughout the healing process, managing wound infections, preventing chronic wound development, and promoting autonomous wound healing.
2. Monitoring Progress:
   Machine learning models were created by Berezo et al. to identify patients who are at risk of their wounds not healing. Using electronic health record (EHR) data, machine learning models were created to identify patients at risk of wounds that would not heal within 4, 8, and 12 weeks of the commencement of therapy (34). The inputs used to train these models

included pertinent variables related to chronic wounds, such as size, depth, location, and infection presence, in addition to patient data including age, gender, and medical conditions. The machine learning models discovered patterns and associations through the analysis of these input variables, which helped them predict how long chronic wounds wound take to heal. These models' output provides an approximate timeline for wound healing (29).

3. Predictive Analysis:
Digital photos of wounds are used as the main source of data for the AI models. Features that are essential for wound evaluation and diagnosis are extracted from these pictures through analysis, including the size, depth, color, texture, presence of necrotic or granulation tissue, and indications of infection. The AI models provide automated wound evaluation, wound type classification, stage or severity estimation, complication identification, and time-series tracking of healing. The purpose of these outputs is to supply doctors with standardized data to support treatment planning and oversight (35).
4. Mobile-based Wound Analysis Systems:
Numerous smartphone applications have been created expressly for measuring and detecting wounds. Manuel Dujovny examined one such app, MOWA, which measures the wound edge and can evaluate ulcer size, wound color (black, yellow, and red), and three other tissue colors. Users need to draw the wound edge manually for size calculation (36). Additionally, the app offers therapeutic treatment advice based on international guidelines Similarly, Friesen et al. developed an Android app for managing chronic wounds that lets users keep track of clinical wound history, add new patients, update wound evaluations, save patient data, and compare week-to-week outcomes (37).
5. Data Integration and Electronic Health Records (EHR) and Clinical Decision Support:
With the advent of AI, accurate disease classification and reclassification based on individual characteristics are now achievable goals. This technology enables the integration of data from electronic health records, incorporating information about a patient's medical history, medications, and previous treatments, swift analysis of medical images and data stored in Electronic Medical Records (EMRs), facilitating prompt and precise service delivery. The integration of AI into medical platforms has given rise to innovative services aimed at enhancing medical quality and enabling real-time health management (38). AI-powered Electronic Health Records (EHRs) function as intelligent aides for healthcare providers by offering evidence-based recommendations and alerts. By scrutinizing patient data and medical literature, AI can propose potential diagnoses and treatment options and alert doctors to possible drug interactions or allergies, thereby aiding in informed decision-making.
6. Wound Assessment and Diagnosis:
Traditionally, for wound assessment, the ruler method is employed, where wounds are measured manually for length, width, and depth using cotton swabs. However, recent studies have explored the use of artificial intelligence (AI) to overcome the limitations of manual measurement. AI has the capacity to create completely automated wound assessment system tools capable of accurately assessing various wound features, aiding in treatment planning (9). However, the development of AI-assisted tools necessitates high-quality datasets with a large volume of data. Despite this, approaches such as superpixel segmentation, region growing, and expert systems offer alternatives that don't require extensive training data (32, 39, 40). Convolutional neural networks (CNNs) have emerged as a promising method for automatic wound segmentation, offering higher accuracy, quicker processing times, and standardized protocols compared to traditional methods. Various convolutional neural network (CNN) architectures have been introduced for image analysis tasks, among which are MobileNet, U-Net, ResNet, and EfficientNet. Liu et al. developed WoundSeg specifically for wound segmentation, leveraging architectures such as MobileNet and VGG-16 (40).

7. Remote Monitoring and Telemedicine:
   Telemedicine provides efficient and cost-effective healthcare delivery for patients with various medical conditions. It includes real-time video consultations and asynchronous methods like forwarding wound images for review. It's widely used for chronic (e.g., diabetic foot ulcers, pressure injuries) and acute wounds (e.g., postsurgical). Telemedicine enables remote wound assessment, patient consultation, and collaboration among medical professionals, especially between primary and tertiary care providers (41). For example, community nurses can change dressings and send images for remote assessment by specialists, while primary care nurses can seek advice on wound management using digital tools (42).

4.5 AI TECHNIQUES IN WOUND HEALING

4.5.1 Machine Learning Algorithms

Machine learning, a prominent application of artificial intelligence, coined by Arthur Samuel, entails endowing computers with the capability to learn autonomously without explicit programming. This learning process enables machines to continuously enhance their performance by assimilating real-time data and feedback (43). When contemplating concepts like big data, data science, and analysis, artificial intelligence and machine learning are often intertwined. These sophisticated machines, sometimes referred to as "Humanoids," exhibit remarkable precision and efficacy in their operations (10).

Machine learning distinguishes itself by using learning algorithms derived from data rather than handcrafted rules. Two fundamental categories of machine learning algorithms prevail: supervised and unsupervised (12, 43). Classification of machine learning is depicted in **Figure 4.2**. Labeled data, encompassing both input and related output, is used by supervised algorithms for their operations. They autonomously discern patterns to map input data to their respective outputs. This category boasts a plethora of popular algorithms, including k-nearest neighbors, neural networks, logistic regression, decision trees, random forests, support vector machines, Bayesian networks, naïve Bayes, and discriminant analysis (44, 45). Conversely, unsupervised algorithms grapple with unlabeled data, discerning patterns and organizing data into clusters sans predefined outputs. Notable examples of unsupervised algorithms include decision trees, k-means, tree matching, Bayesian information criteria, hierarchical clustering, spectral clustering, independent component analysis, and principal component analysis (12, 43). In medical imaging, machine learning finds extensive application, with algorithms such as linear discriminant analysis, support vector machine, decision tree, random factor, and neural network being prominent choices for enhancing diagnostic and prognostic capabilities (12).

4.5.2 Deep Learning Models

Deep learning is a subset of machine learning. In 2006, "Deep Learning" (DL) was introduced by Hinton et al., based on the concept of artificial neural networks (ANNs) (46). Deep learning operates by extracting features from raw data through multiple layers, identifying various aspects relevant to the input data. Techniques in deep learning include convolutional networks, recurrent neural networks, and deep neural networks, with a particular emphasis on convolutional networks within artificial neural networks (47). Deep learning primarily focuses on two critical factors: nonlinear processing in multiple layers or stages and supervised or unsupervised learning. Nonlinear processing in multiple layers entails an algorithm where each layer utilizes the output of the preceding layer as its input (48). This hierarchical approach allows for the extraction of increasingly abstract features from the input data, leading to more nuanced and sophisticated representations.

FIGURE 4.2 Classification of machine learning.

Voulodimos et al. carried out a survey on deep learning methods in the computer vision field, highlighting several widely used techniques including stacked (denoising) autoencoders, deep belief networks, convolutional neural networks (CNNs), and deep Boltzmann machines (49). In their review, Bakator et al. explored the use of deep learning techniques in medical diagnosis, such as class structure-based deep convolutional neural networks (CSDCNNs), stacked denoising autoencoders (SDAEs), convolutional recurrent neural networks (CRNNs), self-advised support vector machines (SA-SVMs), convolutional neural networks (CNNs), and undirected graph recursive neural networks (UGRNNs) (50).

4.5.3 Natural Language Processing

Natural Language Processing is a branch of computer science, artificial intelligence, and linguistics concerned with the interactions between computers and human (natural) language. The goal of Natural Language Processing (NLP) is to enable algorithms or systems to specialize in one or more areas. The metric for assessing NLP algorithms or systems involves evaluating their capability for integrating language understanding and language generation (51).

Natural language processing (NLP) involves techniques to extract grammatical structure and meaning from input, enabling useful tasks. Natural language generation builds output based on target language rules and tasks. NLP finds applications in tutoring systems, duplicate detection, computer-supported instruction, and database interfaces, enhancing interactivity and productivity (52). The main applications of natural language processing (NLP) are listed below:

1) Speech Synthesis: NLP facilitates text-to-speech conversion (TTS), where text data is inputted. High-level modules for speech synthesis are utilized, along with sentence segmentation to handle punctuation marks via decision trees.
2) Speech Recognition: NLP aids automatic speech recognition systems by employing grammar. Context-free grammars represent syntax, assisting in dealing with spontaneous speech. Automatic summarization, including indexing, extracts the gist of speech transcriptions for information retrieval and dialogue system tasks. Along with this, natural language processing (NLP) finds applications in (51–53):

1. Text Summarization: Condensing lengthy text.
2. Sentiment Analysis: Understanding emotions in text.
3. Chatbots and Virtual Assistants: Providing automated assistance.
4. Machine Translation: Translating between languages.
5. Spam Detection: Identifying and filtering spam.
6. Frequently Asked Questions (FAQs): Automating responses to common queries.

4.5.4 Computer Vision

Computer vision, an integral facet of artificial intelligence, empowers computers to comprehend and scrutinize visual data, mirroring the perceptual abilities of humans in interpreting their surroundings. Utilizing machine learning methodologies, it enables the identification and classification of objects within digital images and videos, facilitating responsive actions based on visual input (54). Subdomains within computer vision encompass a broad array of tasks, including scene reconstruction, object detection, event detection, activity recognition, video tracking, 3D pose estimation, learning, indexing, motion estimation, visual servoing, 3D scene modeling, and image restoration (55).

This field encompasses diverse aspects, from capturing raw data to extracting intricate patterns and interpreting visual information, amalgamating principles from digital image processing, pattern recognition, artificial intelligence, and computer graphics. At its core, computer vision aims to extract meaningful insights and descriptions from input scenes, culminating in image understanding (54).

Prominent computer vision algorithms include SIFT, SURF, and Viola-Jones for object detection; adaptive thresholding for segmentation; eigenfaces using PCA for facial recognition; Lucas-Kanade for motion tracking; Kalman filter for state estimation; and mean shift for clustering. Junfeng Gao et al. reviewed several prevalent computer vision algorithms applied in medical imaging. These include segmentation techniques like ensemble of polyharmonic extreme learning machines (EPELM) and methods tailored for specific medical imaging tasks such as high-intensity focused ultrasound (HIFU), random walkers (RW), RWBayes, apparent diffusion coefficient map (ADC MAP), prostate diffusion-weighted imaging (DWI), and digital path approach despeckle filter (56).

4.6 AI APPLICATIONS IN WOUND ASSESSMENT

4.6.1 Image Analysis for Wound Classification

In the context of image analysis in AI for wound classification, wounds are typically classified based on their appearance and characteristics as captured in images. Here are some common classifications used in AI-based wound image analysis:

4.6.1.1 Wound Type

- Pressure Ulcers: Due to sustained pressure on the skin, often occurring in bedridden patients.
- Diabetic Foot Ulcers: Patients with diabetes commonly have foot ulcers, often on the feet due to poor circulation and neuropathy.
- Venous Ulcers: Caused by poor venous circulation, typically on the lower legs.
- Arterial Ulcers: Resulting from poor arterial circulation, often on the feet or lower legs.

4.6.1.2 Wound Stage

- Stage 1: Superficial wounds with intact skin but may show discoloration.
- Stage 2: Thickness of skin is partially lost, including epidermis and possibly the dermis.
- Stage 3: Complete thickness of skin is lost, involving damage or necrosis of subcutaneous tissue.
- Stage 4: Complete thickness of skin is lost with extensive tissue damage, including muscle, bone, or supporting structures.

4.6.1.3 Wound Size and Depth

Classification is based on the size and depth of the wound, often measured in square centimeters and millimeters, respectively.

4.6.1.4 Wound Bed Characteristics

- Necrotic: Presence of dead tissue in the wound bed, which can be categorized based on color (e.g., black, brown, and yellow).
- Granulating: Indication of healing, with the presence of granulation tissue (red or pink).
- Epithelializing: Wound bed covered with epithelial cells, indicating advanced healing.

4.6.1.5 Infection Status

- Infected: Signs of infection, such as increased redness, warmth, swelling, or pus.
- Noninfected: Wounds without signs of infection, showing normal healing progression.

The above types of wounds demonstrate the application of AI, specifically deep learning algorithms, in classifying various types of wounds based on image analysis, showcasing the potential for AI to enhance wound care management (57–59).

Image analysis for wound classification typically involves several steps, including image preprocessing, feature extraction, and classification using machine learning or deep learning algorithms.

Here's an overview of the process:

1. **Image Acquisition:** Wound images are captured using various imaging modalities, such as digital cameras, smartphones, or specialized wound imaging devices. These images are typically in the form of red, green, blue (RGB) color images or grayscale images.
2. **Image Preprocessing:** Preprocessing steps are applied to enhance the quality of the images and remove noise. They may include resizing, normalization, denoising, and color correction to ensure consistency across images.
3. **Feature Extraction:** Features relevant to wound classification are extracted from the preprocessed images. These features may include texture, shape, and color features. Texture features describe the spatial arrangement of pixel intensities, shape features characterize the geometric properties of the wound, and color features capture the color distribution within the wound area.
4. **Machine Learning/Deep Learning:** The extracted features are used as input to machine learning or deep learning algorithms for classification. Machine learning algorithms, for example, Support Vector Machines (SVM) or Random Forests, are commonly used for this purpose. Deep learning algorithms, specifically Convolutional Neural Networks (CNNs), have also shown promising output in wound classification tasks due to their capability to automatically learn relevant features from raw pixel data.
5. **Classification**: The machine learning or deep learning model is trained on a labeled dataset of wound images, where each image is associated with a specific wound type or category (e.g., diabetic foot ulcer, pressure ulcer). Once trained, the model can classify new unseen images of wounds into different categories, such as stages of healing or types of wounds. Machine learning algorithms, for example, support vector machines (SVMs), decision trees, or deep learning models like convolutional neural networks (CNNs), are commonly used for classification. These algorithms learn patterns from the extracted features and can accurately classify new wound images.
6. **Validation and Testing:** The effectiveness of the classification model is examined using validation and testing datasets to assess its accuracy, sensitivity, specificity, and other performance metrics. Cross-validation techniques are often used to ensure the robustness of the model.
7. **Deployment:** Once validated, the trained model can be deployed in clinical settings to help healthcare providers in the automated classification of wounds, thereby increasing the efficiency and accuracy of wound assessment and management.

Overall, image analysis for wound classification is a complicated process that needs expertise in both image processing and wound care. However, advancements in AI and machine learning have significantly improved the accuracy and efficiency of this process, making it a valuable tool in clinical practice (57, 58)

4.6.2 Automated Wound Measurement

Automated wound measurement is a critical component of modern wound care, offering advantages in accuracy, efficiency, and consistency over manual methods. This section provides an overview of automated wound measurement techniques, including image analysis and machine learning approaches, along with their applications in clinical practice.

4.6.2.1 Image Analysis for Wound Measurement

Image analysis techniques play a crucial role in automated wound measurement. These techniques involve extracting relevant features from wound images, such as size, depth, and tissue composition, to quantify wound characteristics. Pasero et al. developed an algorithm for the automatic segmentation of leg ulcers into "background" and "not background" classes using mobile phone images, based on pixel classification using RGB and HSV color map values (60). Sobel edge detection has been used to determine ulceration areas, providing information on the location, size, and shape of wounds. Thermal imaging has been utilized for wound and location detection, utilizing color-based thermal patterns (12).

4.6.2.2 Machine Learning for Wound Measurement

Machine learning algorithms, particularly deep learning models, in automated wound measurement tasks have shown promising results. These algorithms can learn complex patterns from wound images and provide accurate measurements. Wang et al. developed a cascaded two-stage classifier based on SVM for diabetic foot ulcer area determination, achieving accurate wound region segmentation (57). The robustness of the SVM classifier for wound segmentation and added textural features to enhance segmentation accuracy was investigated by Kolesnik et al. (61). A system for segmenting burnt wounds from images using Cr-Transformation, Luv-Transformation, and Fuzzy C-mean Clustering (FCM) algorithms was proposed by Wantanajittikul et al. (62)

4.6.2.3 Clinical Applications

Automated wound measurement techniques have several clinical applications, including wound assessment, monitoring wound healing progress, and optimizing treatment strategies. These techniques enable healthcare providers to accurately measure wound characteristics, such as size and depth, for better wound management decisions. They can also predict wound healing outcomes based on measured characteristics, helping tailor treatment plans to individual patients' needs.

In conclusion, automated wound measurement techniques, including image analysis and machine learning approaches, offer significant advancements in wound care, providing accurate and efficient methods for wound assessment and management in clinical practice.

4.6.3 Predictive Modeling for Wound Healing

Predictive modeling for wound healing involves the development of algorithms and models to forecast the healing progress of wounds. These models can be real-time and image-based, utilize bioinformatics data, and adhere to precision regulations to ensure accuracy and reliability.

4.6.4 Real Time

Predictive modeling for wound healing is a multifaceted approach that incorporates various technologies and methodologies to forecast the progression of wound healing. One key aspect is the real-time nature of these models, which involves continuously updating predictions based on new data. For example, as a wound heals, new images and patient data can be fed into the model to refine its predictions, enabling clinicians to make informed decisions about treatment adjustments or interventions.

Real-time predictive models leverage data collected over time to predict future outcomes. For example, Li et al. (2020) developed a machine learning model that used wound images and patient data to predict wound healing likelihood. This real-time prediction can help clinicians make timely decisions regarding treatment strategies (34).

4.6.5 Image-Based

Image-based predictive models play a crucial role in this process by extracting relevant features from wound images to predict healing outcomes. These models can analyze images captured at different stages of the healing process to identify patterns and trends that indicate how a wound is likely to progress. For instance, features such as wound size, depth, and tissue composition can be extracted from images and used to train machine learning algorithms to predict healing outcomes. Image-based predictive models utilize features extracted from wound images to predict healing outcomes. Wang et al. (2020) used a deep learning algorithm to predict diabetic foot ulcer healing based on image features. These models can provide valuable insights into wound characteristics that affect healing (57).

4.6.6 Bioinformatics

Bioinformatics-based predictive models integrate molecular and genetic data into the prediction process. By analyzing gene expression patterns or biomarker levels associated with wound healing, these models can provide an understanding of the underlying biological processes that influence healing. With this data, individualized treatment programs that take into account each patient's unique genetic profile and wound characteristics can be created. The study by Sood et al. (2014) investigated the role of microRNAs (miRNAs) in diabetic foot ulcers (DFUs). They found that miRNAs play a critical role in regulating the expression of genes involved in wound healing and identified specific miRNAs that were dysregulated in DFUs compared to healthy skin (63).

4.6.7 Precision Regulations

Precision regulations are essential to ensure that predictive models meet specific standards for accuracy, reliability, and safety. These regulations may include validation requirements, data quality standards, and guidelines for model transparency and interpretability. Adhering to these regulations ensures that predictive models are developed and deployed responsibly, minimizing the risk of errors or biases that could impact patient care. Precision regulations ensure that predictive models meet specific standards for accuracy and reliability. For example, the FDA's regulatory framework for AI in healthcare requires developers to validate models rigorously to ensure they meet safety and efficacy standards (64).

In conclusion, predictive modeling for wound healing combines real-time data analysis, image-based feature extraction, bioinformatics, and adherence to precision regulations to provide clinicians with valuable insights into wound healing processes. By leveraging these technologies and methodologies, clinicians can make more informed decisions about wound care, leading to improved results for patients.

4.7 AI-DRIVEN THERAPIES AND TREATMENTS IN WOUND

Artificial intelligence (AI) is revolutionizing wound healing by leveraging advanced algorithms and data analytics to enhance diagnosis, treatment, and monitoring. These AI-driven therapies offer personalized approaches, predictive insights, and remote monitoring capabilities, promising improved outcomes and efficiency in wound care management (**Table 4.2**).

4.7.1 Personalized Medicine

Personalized medicine relies on specific medical data to create tailored strategies and treatments for identifying and managing illnesses, with the goal of maintaining individuals' health (65). Physicians in this approach integrate a range of patient data sources, including symptoms, conventional test results, medical and family history, and genomic data. By combining this extensive information, healthcare professionals can accurately diagnose conditions and develop personalized treatment plans that address the unique needs and characteristics of each patient (66) AI plays a pivotal role in personalized medical services by swiftly analyzing this diverse medical data, encompassing a patient's disease history, treatment records, genetic profile, lifestyle factors, and dietary habits. Leveraging this comprehensive information, these services deliver customized diagnoses and treatment plans tailored to the specific requirements of each individual. This approach also enables the

TABLE 4.2
Different AI Tools With Descriptions Used in Drug Discovery

AI Tool	Description	Website
DeepChem	An MLP model employing Python-based AI for finding suitable drug candidates. (72)	Website: [DeepChem] (https://github.com/deepchem/deepchem
DeepTox	Software predicting the toxicity of 12,000 drugs.	Website: [DeepTox] (www.bioinf.jku.at/research/DeepTox)
] (	A Python-based system aiding in detecting molecular activity.	Website: [DeepNeuralNet-QSAR] (https://github.com/Merck/DeepNeuralNet-QSAR)
ORGANIC	A molecular generation tool for creating molecules with desired properties.	Website: [ORGANIC] (https://github.com/aspuru-guzik-group/ORGANIC)
Potential Net	Utilizes neural networks to predict ligand binding affinity	Website: [PotentialNet] (https://pubs.acs.org/doi/full/10.1021/acscentsci.8b00507)
Hit Dexter	Applies ML techniques to predict molecules that might respond to biochemical assays	Website: [HitDexter] (http://hitdexter2.zbh.uni-hamburg.de)
DeltaVina	A scoring function for rescoring drug-ligand binding affinity.	Website: [DeltaVina] (https://github.com/chengwang88/deltavina)
Neural Graph Fingerprint	Assists in predicting the properties of novel molecules.	Website: [Neural Graph Fingerprint] (https://github.com/HIPS/neural-fingerprint)
AlphaFold	Predicts 3D structures of proteins.	Website: [AlphaFold] (https://deepmind.com/blog/alphafold)

prediction of future diseases, exemplified by AI programs like IBM Watson, which utilize advanced algorithms to forecast potential health issues (38).

Virtual health assistants represent an innovative technology revolutionizing healthcare by aiding healthcare professionals. These systems are created to mimic human conversation, providing personalized patient care tailored to individual input (67). Digital assistants, powered by AI applications, chatbots, audio cues, and interfaces, are indispensable in personalized wound healing treatment plans. These virtual allies offer crucial support by analyzing symptoms to identify underlying issues, providing personalized medical advice, sending medication reminders, scheduling appointments, and monitoring vital signs. Additionally, they streamline communication by collecting daily health data and transmitting reports to healthcare providers, reducing their workload and ensuring timely interventions. By automating tasks and offering personalized care, virtual assistants enhance efficiency and improve patient outcomes in wound healing journeys. Furthermore, AI plays a pivotal role in developing personalized treatment strategies by examining individual patient characteristics and responses to previous therapies. Through the analysis of extensive datasets, AI can identify relevant patterns and recommend the most effective and personalized interventions, thereby optimizing wound healing outcomes.

4.7.2 Drug Discovery and Development

AI in medicinal chemistry accelerates drug discovery by predicting drug efficacy and toxicity, bypassing time-consuming experimentation. Traditional methods are costly and uncertain, but AI analyzes vast datasets to swiftly and accurately forecast the pharmacological properties and safety profiles of potential compounds. This efficiency enables researchers to explore a wider chemical space and identify promising candidates faster, ultimately leading to the discovery of safer and more effective drugs (68). Drug discovery involves two main steps: 1. identifying targetable processes or molecules responsible for specific clinical conditions and 2. designing agents to modulate them. AI methods are applied to both steps, swiftly analyzing data to pinpoint targets and facilitating the design and optimization of drug candidates, leading to more precise and effective treatments (69).

Predictive and generative tasks comprise the majority of AI's applications in drug development, including virtual screening, de novo drug creation, retrosynthesis and reaction prediction, de novo protein design, and more (70). Artificial intelligence is used in drug discovery at several phases, from target validation and identification to estimating therapeutic response. Molecular property prediction and molecule creation are two essential processes involved in lead identification, an important step. A crucial component of virtual screening is molecular property prediction, which forecasts a molecule's value based on its structure or learned representation. Drug-target interaction (DTI) prediction, toxicity prediction, drug-induced liver damage (DILI) prediction, and other reasons are all served by this prediction. (71) Here are examples of AI tools used in drug discovery:

The process of creating new molecules, or molecules with desired qualities, is known as "molecule generation." It consists of two levels of tasks: 1) realistic molecule generation, which creates molecules within the limitations set by chemical rules, and 2) goal-directed molecule generation, which creates chemically valid molecules with desired properties (69, 71).

In drug development, AI is increasingly utilized to aggregate, organize, and analyze big data to enhance trial performance and regulatory approval. AI methods screen potential clinical trial sites, analyze electronic health records to identify suitable candidates, and integrate real-world data to support trial performance. Additionally, AI assists in dynamic participant monitoring and early detection of adverse events, improving trial management (69).

4.7.3 Smart Wound Dressing

Smart dressings with integrated multiparametric sensing capabilities will revolutionize wound care by providing real-time wireless communication of wound data. Two major advantages of this breakthrough are that it speeds up the healing of wounds and makes remote patient monitoring possible (73) As a result, there will be less need for home visits, quicker recovery, fewer complications, shorter hospital stays, and fewer visits to general practitioners' offices and public health clinics. These enhancements will optimize healthcare budgets by reducing total costs associated with the provision of wound care (74). The next section will address the precise wound characteristics being felt, their functions in wound healing, and the available measurement techniques.

4.7.3.1 pH-Responsive Wound Dressings

Wound exudate contains information on the pH of the wound, with varying pH values indicating different stages of the healing process (75). An acidic pH (5–6) is typical in healing wounds, while chronic wounds (nonhealing) have a higher pH (7.15–8.90), indicating infection due to bacterial proliferation. Monitoring wound pH is crucial for early infection detection and wound characterization (73). Various strategies, such as flex circuit transducers, optical sensor-based dyes, pH-sensitive color dyes, pH-sensitive electrodes, and carbon dots, are employed to create wound dressings that can detect changes in pH levels (75, 76).

4.7.3.2 Moisture Responsive Wound Dressings

Currently, available techniques for identifying wound moisture include eye inspection, the application of indicators such as humidity sensors or electrical impedance moisture mapping, or the use of wound sense dipsticks. "Wound Sense," the first moisture-sensitive wound dressing to be sold commercially, was released in 2016. To create breathable dressings with integrated sensors, a variety of composite materials are used, including carbon nanotubes, graphene oxide, graphene, polyvinyl alcohol, and palladium and cerium oxide nanosheets (75, 77).

4.7.3.3 Temperature

Normal healing wounds maintain a temperature around 37.8°C, with deviations of 2.2°C signaling potential issues. Decreases suggest ischemia, while increases indicate infection. Temperature sensors utilize various detection systems, including infrared sensors, thermistors, and resistance temperature sensors. Elements such as nanofibers or nanomeshes of materials like polyurethane, graphene, platinum, and gold are commonly employed in sensor preparation (75, 78).

4.7.3.4 Pressure-Responsive Wound Dressings

Pressure-responsive wound dressings are crucial for managing pressure ulcers and diabetic foot ulcers, as they alleviate pressure on injured areas. Pressure sensor dressings facilitate easy monitoring of wounds. These sensors come in four types based on operation principles: piezoresistive, capacitive, triboelectric, and piezoelectric sensors. Ongoing developments focus on creating multiple flexible pressure sensors to enhance wound monitoring capabilities (73, 75).

4.7.3.5 Oxygen-Sensitive Dressing

To promote collagen deposition, epithelialization, fibroplasia, angiogenesis, and infection prevention, oxygen is essential to the wound healing process. Pulse oximetry, which depends on the oxygen-dependent characteristics of hemoglobin, and optical approaches that use sensor films and two-dimensional luminescence imaging to measure wound oxygenation are examples of noninvasive monitoring techniques. (73) Li et al. introduced an innovative approach involving phosphorescence emission captured by paint-on sensing bandages, enabling the generation of oxygen maps of underlying tissue, though not continuously (79). This method shows promise for potential use in wearable smart wound dressings.

4.7.3.6 Sustained Drug Releasing Wound Dressings

Stimuli-responsive wound dressings allow for controlled drug release at injury sites. Drugs are embedded within dressing layers, with hydrogels being commonly used for sustained drug release. These drug delivery dressings utilize polymers like polylactide-co-glycoside, polyvinyl pyrrolidone, polyvinyl alcohol, polyhydroxyalkyl-methacrylate, polyurethane foam, hydrocolloid, alginate, hyaluronic acid, collagen, and chitosan (75).

4.8 CHALLENGES AND LIMITATIONS OF AI IN WOUND HEALING

Artificial intelligence (AI) has shown great potential in advancing wound healing by enabling more accurate and efficient wound assessment, prediction, and treatment. However, several challenges and limitations need to be addressed to fully realize the benefits of AI in wound healing.

4.8.1 Data Quality and Quantity

One of the major challenges in applying AI to wound healing is the availability of high-quality and large-scale data. Wound images and patient data used for training AI models need to be accurately labeled and representative of diverse wound types and conditions. However, obtaining such datasets can be challenging due to privacy concerns, variability in wound characteristics, and the need for expert annotations. Furthermore, the quality of data can significantly impact the performance of AI models. Low-quality images, incomplete patient records, and inconsistent data collection practices can lead to biased or inaccurate AI predictions. Addressing these challenges requires efforts to improve data collection standards, enhance data sharing practices, and develop methods for data augmentation and synthesis to overcome data scarcity (80, 81).

4.8.2 Interpretability and Trustworthiness

The interpretability of AI models and the reliability of their predictions present another difficulty in using AI for wound healing. AI algorithms, particularly deep learning models, are sometimes referred to as "black boxes" because of their intricate designs and opaque decision-making processes. When it comes to clinical contexts, where physicians need explanations for AI suggestions to make well-informed judgments, this lack of interpretability may hinder the adoption of AI. Ensuring the trustworthiness of AI predictions is also crucial, as inaccurate or biased predictions can have serious consequences for patient care. Techniques for interpreting AI models, such as feature visualization, attention mechanisms, and explainable AI methods, are being developed to improve the transparency and trustworthiness of AI in wound healing. (82, 83)

4.8.3 Regulatory and Ethical Considerations

The regulatory landscape surrounding the use of AI in healthcare, including wound healing, is complex and evolving. Regulatory bodies, such as the Food and Drug Administration (FDA) in the United States, require rigorous validation and evaluation of AI algorithms before they can be used in clinical practice. Ensuring compliance with regulatory requirements adds to the complexity and cost of developing AI solutions for wound healing.

Ethical considerations also play a significant role in the deployment of AI in wound healing. Issues such as patient consent, information privacy, and algorithmic bias need to be carefully addressed to ensure that AI technologies are used responsibly and ethically. Collaborations between researchers, clinicians, regulators, and ethicists are essential to navigate these challenges and develop AI solutions that benefit patients while upholding ethical standards (31, 84).

Despite these challenges and limitations, AI holds immense promise for transforming wound healing by improving the accuracy, efficiency, and accessibility of wound care. Addressing the challenges related to data quality, interpretability, and regulatory considerations will require collaborative efforts from stakeholders across the healthcare ecosystem. AI has the potential to revolutionize wound healing and improve outcomes for patients with chronic and acute wounds by overcoming these challenges.

4.9 FUTURE DIRECTIONS AND EMERGING TRENDS IN AI FOR WOUND CARE

4.9.1 Integrating AI with Other Technologies

The future of wound care will likely see the integration of AI with other technologies to enhance diagnostic capabilities, improve treatment outcomes, and streamline healthcare delivery. One promising integration is the combination of AI with wearable sensors and smart wound dressings. These technologies can continuously monitor wound healing progress, detect early signs of infection, and provide real-time feedback to healthcare providers and patients. Furthermore, AI can be integrated with telemedicine platforms to enable remote wound assessment and consultation. This approach can improve access to specialized wound care in underserved areas and reduce healthcare costs associated with frequent clinic visits. (85)

Additionally, AI can be integrated with robotic systems for automated wound debridement and surgical procedures. Robotic systems equipped with AI can provide precise and consistent wound care, reducing the risk of human error and improving patient outcomes (86).

4.9.2 Advancements in Wound Care Devices

Advancements in AI are driving the development of innovative wound care devices. For example, AI-powered imaging devices can provide high-resolution images of wounds, allowing for more accurate and detailed assessment. These devices can also analyze wound images to track healing progress and predict outcomes. Similarly, AI algorithms can be integrated into smart wound dressings to detect changes in wound condition, such as infection or delayed healing, and deliver targeted therapies, such as antimicrobial agents or growth factors, to promote healing (87).

Another emerging trend is the use of 3D bioprinting technology combined with AI to create personalized tissue scaffolds for wound repair. These scaffolds can mimic the structure and function of native tissue, promoting faster and more effective wound healing (88).

4.9.3 Potential Impact on Healthcare Delivery

Artificial intelligence (AI) in wound care has the potential to revolutionize the way healthcare is delivered by increasing productivity, cutting expenses, and improving patient outcomes (89, 90) Healthcare professionals can concentrate more on patient care by using AI to automate repetitive processes like wound measurement and recordkeeping. AI can also assist healthcare professionals in making better judgments by assessing vast collections of clinical and wound imaging data to spot trends and forecast results. This can result in patients with complex or chronic wounds receiving more individualized and efficient treatment regimens. (91) Additionally, AI can improve communication and collaboration among healthcare providers by facilitating the sharing of wound images, treatment plans, and outcomes in real time. This can lead to more coordinated and efficient care, particularly for patients receiving care from multiple providers or in different healthcare settings (92).

In conclusion, the integration of AI with other technologies, advancements in wound care devices, and the potential impact on healthcare delivery are exciting developments that hold great promise for the future of wound care. By leveraging AI, healthcare providers can enhance diagnostic capabilities, improve treatment outcomes, and provide more personalized and efficient care to patients with wounds.

4.10 CONCLUSION AND OUTLOOK

4.10.1 Summary of Key Findings

The application of artificial intelligence (AI) in wound assessment and management has advanced significantly, providing automated and accurate solutions for wound classification, measurement, and prediction. Various methods, including machine learning algorithms and image processing techniques, have been developed for feature extraction and wound segmentation from 2D images. These methods have shown promising results in accurately identifying wound characteristics such as size, depth, and tissue composition. Additionally, AI algorithms have been used for predicting wound healing outcomes, guiding treatment decisions, and monitoring wound progression over time.

4.10.2 Implications for Clinical Practice

Enhancing the efficacy and precision of wound assessment through the use of AI in clinical practice has the potential to transform wound care completely. When making decisions about wound treatment and monitoring, healthcare providers can benefit from the use of automated wound assessment and classification. Moreover, AI-based solutions can help with remote wound monitoring, allowing for prompt interventions and lessening the demand for medical resources. Large datasets of clinical and wound image data can also be analyzed by AI algorithms to find trends and variables affecting wound healing, which can help patients receive individualized treatment plans.

4.10.3 Opportunities for Further Research

Despite the significant progress in AI applications for wound assessment, several opportunities for further research exist. Future studies should concentrate on enhancing the accuracy and robustness of AI algorithms for wound segmentation and classification. Additionally, the development of

AI-based systems for real time wound monitoring and feedback could improve patient outcomes. Furthermore, integrating AI with other technologies, such as wearable devices and telemedicine, could enhance the accessibility and effectiveness of wound care. Collaborative efforts between researchers, clinicians, and industry partners are essential to drive innovation and translate AI research into clinical practice.

REFERENCES

1. Ozgok Kangal MK, Regan JP. Wound healing. *StatPearls*. 2024 [cited 2024 Apr 21]. Available from: https://pubmed.ncbi.nlm.nih.gov/30571027/
2. Coger V, Million N, Rehbock C, Sures B, Nachev M, Barcikowski S, et al. Tissue concentrations of zinc, iron, copper, and magnesium during the phases of full thickness wound healing in a rodent model. *Biol Trace Elem Res*. 2019 Sep 15 [cited 2024 Apr 5];191(1):167. Available from: /pmc/articles/PMC6656798/
3. Bowden LG, Byrne HM, Maini PK, Moulton DE. A morphoelastic model for dermal wound closure. *Biomech Model Mechanobiol*. 2016 Jun 1 [cited 2024 Apr 5];15(3):663–81. Available from: https://pubmed.ncbi.nlm.nih.gov/26264498/
4. Cullen B, Smith R, Mcculloch E, Silcock D, Morrison L. Mechanism of action of PROMOGRAN, a protease modulating matrix, for the treatment of diabetic foot ulcers. *Wound Repair Regen*. 2002 [cited 2024 Apr 5];10(1):16–25. Available from: https://pubmed.ncbi.nlm.nih.gov/11983003/
5. Gale AJ. Current understanding of hemostasis. *ToxicolPathol*. 2011 [cited 2024 Apr 5];39(1):273. Available from: /pmc/articles/PMC3126677/
6. Baum CL, Arpey CJ. Normal cutaneous wound healing: Clinical correlation with cellular and molecular events. *Dermatol Surg*. 2005 Mar 21 [cited 2024 Apr 21];31(6):674–86. Available from: https://pubmed.ncbi.nlm.nih.gov/15996419/
7. Dalisson B, Barralet J. Bioinorganics and wound healing. *Adv Healthc Mater*. 2019 Sep 1 [cited 2024 Apr 5];8(18). Available from: https://pubmed.ncbi.nlm.nih.gov/31402608/
8. Hayat M, Nawaz A, Chinnam S, Muzammal M, Latif MS, Yasin M, Ashique S, Zengin G, Farid A. Formulation development and optimization of herbo synthetic gel: In vitro biological evaluation and in vivo wound healing studies. *Process Biochem*. 2023 Jul 1;130:116–26.
9. Chairat S, Chaichulee S, Dissaneewate T, Wangkulangkul P, Kongpanichakul L. AI-assisted assessment of wound tissue with automatic color and measurement calibration on images taken with a smartphone. *Healthcare*. 2023 Jan 1 [cited 2024 Apr 6];11(2). Available from: /pmc/articles/PMC9858639/
10. Rupali M, Amit P. A review paper on general concepts of "artificial intelligence and machine learning". *Int Adv Res J Sci*. 2017;4.
11. Cho SK, Mattke S, Gordon H, Sheridan M, Ennis W. Development of a model to predict healing of chronic wounds within 12 weeks. *Adv Wound Care (New Rochelle)*. 2020 Sep 1 [cited 2024 Apr 6];9(9):516–24. Available from: https://pubmed.ncbi.nlm.nih.gov/32941121/
12. Anisuzzaman DM, Wang C, Rostami B, Gopalakrishnan S, Niezgoda J, Yu Z. Image-based artificial intelligence in wound assessment: A systematic review. 2022 Sep 15 [cited 2024 Apr 6];11(12):687–709. Available from: www.liebertpub.com/doi/10.1089/wound.2021.0091
13. Diegelmann RF, Evans MC. Wound healing: An overview of acute, fibrotic and delayed healing. *Front Biosci*. 2004 [cited 2024 Apr 6];9:283–9. Available from: https://pubmed.ncbi.nlm.nih.gov/14766366/
14. Monika P, Chandraprabha MN, Rangarajan A, Waiker PV, Chidambara Murthy KN. Challenges in healing wound: Role of complementary and alternative medicine. *Front Nutr*. 2022 Jan 20 [cited 2024 Apr 6];8. Available from: https://pubmed.ncbi.nlm.nih.gov/35127787/
15. Mahmoudi M, Gould LJ. Opportunities and challenges of the management of chronic wounds: A multidisciplinary viewpoint. *Chronic Wound Care Management and Research*. 2020 Jul [cited 2024 Apr 6]; 7:27–36. Available from: www.tandfonline.com/action/journalInformation?journalCode=dcwc20
16. Zulkowski K, Ayello EA, Wexler S. Certification and education: Do they affect pressure ulcer knowledge in nursing? *Adv Skin Wound Care*. 2007 Jan [cited 2024 Apr 21];20(1):34–8. Available from: https://pubmed.ncbi.nlm.nih.gov/17195785/
17. Wilkinson HN, Hardman MJ. Wound healing: Cellular mechanisms and pathological outcomes. *Open Biol*. 2020 Jun 23 [cited 2024 Apr 6];10(9):341–70. Available from: https://pubmed.ncbi.nlm.nih.gov/32993416/
18. What are the treatment options for chronic wounds? 2018 Jun 14 [cited 2024 Apr 6]. Available from: www.ncbi.nlm.nih.gov/books/NBK326436/

19. Jimenez PA, Jimenez SE. Tissue and cellular approaches to wound repair. *Am J Surg*. 2004 [cited 2024 Apr 6];187(5 Suppl. 1):S56–64. Available from: https://pubmed.ncbi.nlm.nih.gov/15147993/
20. Broughton G, Janis JE, Attinger CE. Wound healing: An overview. *PlastReconstrSurg*. 2006 Jun [cited 2024 Apr 6];117(7 Suppl). Available from: https://pubmed.ncbi.nlm.nih.gov/16801750/
21. Broughton G, Janis JE, Attinger CE. Wound healing: An overview. *PlastReconstrSurg*. 2006 Jun [cited 2024 Apr 21];117(7 Suppl). Available from: https://pubmed.ncbi.nlm.nih.gov/16801750/
22. Low-level laser therapy—PubMed. [cited 2024 Apr 6]. Available from: https://pubmed.ncbi.nlm.nih.gov/15957498/
23. Deng X, Gould M, Ali MA. A review of current advancements for wound healing: Biomaterial applications and medical devices. *J Biomed Mater Res B Appl Biomater*. 2022 Nov 1;110(11):2542–73.
24. Falcone M, De Angelis B, Pea F, Scalise A, Stefani S, Tasinato R, et al. Challenges in the management of chronic wound infections. *J Glob Antimicrob Resist*. 2021 Sep 1 [cited 2024 Apr 6];26:140–7. Available from: https://pubmed.ncbi.nlm.nih.gov/34144200/
25. Kaplan A, Haenlein M. Siri, Siri, in my hand: Who's the fairest in the land? On the interpretations, illustrations, and implications of artificial intelligence. *Bus Horiz*. 2019 Jan 1;62(1):15–25.
26. Guo J, Li B. The application of medical artificial intelligence technology in rural areas of developing countries. *Health Equity*. 2018 [cited 2024 Apr 6];2(1):174. Available from: /pmc/articles/PMC6110188/
27. Jain AK, Mao J, Mohiuddin KM. Artificial neural networks: A tutorial. *Computer (Long Beach Calif)*. 1996 Mar;29(3):31–44.
28. What is a Neural Network? Definition, Types and How It Works. [cited 2024 Apr 6]. Available from: www.techtarget.com/searchenterpriseai/definition/neural-network
29. Le DTP, Pham TD. Unveiling the role of artificial intelligence for wound assessment and wound healing prediction. *Open Exploration*. 2023 Aug 31 [cited 2024 Apr 6];4(4):589–611. Available from: www.explorationpub.com/Journals/em/Article/1001163
30. Arora A. Conceptualising artificial intelligence as a digital healthcare innovation: An introductory review. *Med Devices (Auckl)*. 2020 [cited 2024 Apr 6];13:223–30. Available from: https://pubmed.ncbi.nlm.nih.gov/32904333/
31. Topol EJ. High-performance medicine: The convergence of human and artificial intelligence. *Nat Med*. 2019 Jan 7 [cited 2024 Apr 6];25(1):44–56. Available from: www.nature.com/articles/s41591-018-0300-7
32. Artificial Intelligence (AI) in Healthcare Market Size, Share, Industry Report, Statistics—2032. [cited 2024 Apr 21]. Available from: www.marketsandmarkets.com/Market-Reports/artificial-intelligence-healthcare-market-54679303.html
33. Dong R, Guo B. Smart wound dressings for wound healing. *Nano Today*. 2021 Dec 1 [cited 2024 Apr 6];41. Available from: www.researchgate.net/publication/355095271_Smart_wound_dressings_for_wound_healing
34. Berezo M, Budman J, Deutscher D, Hess CT, Smith K, Hayes D. Predicting chronic wound healing time using machine learning. *Adv Wound Care (New Rochelle)*. 2022 Jun 1 [cited 2024 Apr 6];11(6):281. Available from: /pmc/articles/PMC8982125/
35. Howell RS, Liu HH, Khan AA, Woods JS, Lin LJ, Saxena M, et al. Development of a method for clinical evaluation of artificial intelligence-based digital wound assessment tools. *JAMA Netw Open*. 2021 May 19 [cited 2024 Apr 6];4(5). Available from: https://pubmed.ncbi.nlm.nih.gov/34009348/
36. Pressure Injury Evolution: Mobile Wound Analyzer | Request PDF. [cited 2024 Apr 6]. Available from: www.researchgate.net/publication/325376834_Pressure_Injury_Evolution_Mobile_Wound_Analyzer
37. Friesen MR, Hamel C, McLeod RD. A mHealth application for chronic wound care: Findings of a user trial. *Int J Environ Res Public Health*. 2013 Nov 19 [cited 2024 Apr 6];10(11):6199. Available from: /pmc/articles/PMC3863895/
38. Lee S, Kim HS. Prospect of artificial intelligence based on electronic medical record. *J Lipid Atheroscler*. 2021 Sep 1 [cited 2024 Apr 6];10(3):282. Available from: /pmc/articles/PMC8473961/
39. Chang CW, Christian M, Chang DH, Lai F, Liu TJ, Chen YS, et al. Deep learning approach based on superpixel segmentation assisted labeling for automatic pressure ulcer diagnosis. *PLoS One*. 2022 Feb 1 [cited 2024 Apr 6];17(2). Available from: https://pubmed.ncbi.nlm.nih.gov/35176101/
40. Chairat S, Chaichulee S, Dissaneewate T, Wangkulangkul P, Kongpanichakul L. AI-assisted assessment of wound tissue with automatic color and measurement calibration on images taken with a smartphone. *Healthcare*. 2023 Jan 1 [cited 2024 Apr 21];11(2). Available from: /pmc/articles/PMC9858639/
41. Zhu X, Ren BFSJ, Lim VH, Wan L, Chen Y, Wang X, et al. Description and utilization of telewound monitoring services in primary care patients with acute wounds in Singapore: A retrospective study. *Adv Skin Wound Care*. 2022 Oct 1 [cited 2024 Apr 6];35(10):544–9. Available from: https://pubmed.ncbi.nlm.nih.gov/36125454/

42. Chang YY, Ang SY, Ong CE, Peng SS, Zulkifli H, Hashim N, et al. Right siting of complex acute wound management—preliminary study of teleconsultation wound services between acute and primary care in Singapore. *J Tissue Viability*. 2022 May 1 [cited 2024 Apr 21];31(2):353–7. Available from: https://pubmed.ncbi.nlm.nih.gov/34711417/
43. (PDF) Machine Learning Algorithms – A Review. [cited 2024 Apr 5]. Available from: www.researchgate.net/publication/344717762_Machine_Learning_Algorithms_-A_Review
44. Kotsiantis SB. Supervised machine learning: A review of classification techniques. *Informatica*. 2007;31:249–68.
45. Singh A, Thakur N, Sharma A. A review of supervised machine learning algorithms. International Conference on Computing for Sustainable Global Development. 2016.
46. Sarker IH. Deep learning: A comprehensive overview on techniques, taxonomy, applications and research directions. *SN Comput Sci*. 2021 Nov 1 [cited 2024 Apr 21];2(6):1–20. Available from: https://link.springer.com/article/10.1007/s42979-021-00815-1
47. Mishra RK, Reddy GYS, Pathak H. The understanding of deep learning: A comprehensive review. *Math Probl Eng*. 2021;2021.
48. (PDF) Deep Learning: A Review. [cited 2024 Apr 5]. Available from: www.researchgate.net/publication/328285467_Deep_Learning_A_Review
49. Voulodimos A, Doulamis N, Doulamis A, Protopapadakis E. Deep learning for computer vision: A brief review. *ComputIntellNeurosci*. 2018;2018.
50. Bakator M, Radosav D. Deep learning and medical diagnosis: A review of literature. *Multimodal Technologies and Interaction*. 2018 Sep 1;2(3).
51. Khurana D, Koli A, Khatter K, Singh S. Natural language processing: State of the art, current trends and challenges. *Multimed Tools Appl*. 2023 Jan 1 [cited 2024 Apr 5];82(3):3713–44. Available from: www.researchgate.net/publication/319164243_Natural_Language_Processing_State_of_The_Art_Current_Trends_and_Challenges
52. Reshamwala A, Mishra D, Pawar P. Review on natural language processing. *An International Journal (ESTIJ)*. 2013 [cited 2024 Apr 5];3(1):2250–3498. Available from: www.researchgate.net/publication/235788362
53. Mah PM, Skalna I, Muzam J. Natural language processing and artificial intelligence for enterprise management in the era of industry 4.0. *Applied Sciences*. 2022 Sep 14 [cited 2024 Apr 5];12(18):9207. Available from: www.mdpi.com/2076-3417/12/18/9207/htm
54. G KY, Ravichandra M. A review paper on computer vision and image processing. *Int J Adv Res Sci, Communication and Technology (IJARSCT)*. 2022 [cited 2024 Apr 21];2(2). Available from: www.ijarsct.co.in
55. Wiley V, Lucas T. Computer vision and image processing: A paper review. *Int J Artif Intell Res*. 2018 Jun 1 [cited 2024 Apr 5];2(1):22. Available from: www.researchgate.net/publication/323106814_Computer_Vision_and_Image_Processing_A_Paper_Review
56. Gao J, Yang Y, Lin P, Park DS. Editorial: Computer vision in healthcare applications. *J Healthc Eng*. 2018 [cited 2024 Apr 5];2018. Available from: www.researchgate.net/publication/323556158_Computer_Vision_in_Healthcare_Applications
57. Nagaraju S, Kumar KV, Rani BP, Lydia EL, Ishak MK, Filali I, et al. Automated diabetic foot ulcer detection and classification using deep learning. *IEEE Access*. 2023;11:127578–88.
58. Wang C, Anisuzzaman DM, Williamson V, Dhar MK, Rostami B, Niezgoda J, et al. Fully automatic wound segmentation with deep convolutional neural networks. *Sci Rep*. 2020 Dec 1 [cited 2024 Apr 3];10(1). Available from: https://pubmed.ncbi.nlm.nih.gov/33318503/
59. Anisuzzaman DM, Patel Y, Niezgoda JA, Gopalakrishnan S, Yu Z. A Mobile app for wound localization using deep learning. *IEEE Access*. 2020;10:61398–409.
60. Pasero E, Castagneri C. Application of an automatic ulcer segmentation algorithm. *International Forum on Research and Technologies for Society and Industry Leveraging a Better Tomorrow*. 2017 Oct 11;
61. Kolesnik M, Fexa A. Segmentation of wounds in the combined color-texture feature space. 2004 May 12 [cited 2024 Apr 3];5370:549–56. Available from: www.spiedigitallibrary.org/conference-proceedings-of-spie/5370/0000/Segmentation-of-wounds-in-the-combined-color-texture-feature-space/10.1117/12.535041.full
62. Wantanajittikul K, Auephanwiriyakul S, Theera-Umpon N, Koanantakool T. Automatic segmentation and degree identification in burn color images. *BMEiCON-2011–4th Biomedical Engineering International Conference*. 2011 [cited 2024 Apr 3];169–73. Available from: www.researchgate.net/publication/254022343_Automatic_segmentation_and_degree_identification_in_burn_color_images

63. Ma X, Lu C, Lv C, Wu C, Wang Q. The expression of miR-192 and its significance in diabetic nephropathy patients with different urine albumin creatinine ratio. *J Diabetes Res.* 2016 [cited 2024 Apr 3];2016. Available from: /pmc/articles/PMC4736386/
64. Mennella C, Maniscalco U, De Pietro G, Esposito M. Ethical and regulatory challenges of AI technologies in healthcare: A narrative review. *Heliyon.* 2024 Feb 29;10(4):e26297.
65. Kawamoto K, Lobach DF, Willard HF, Ginsburg GS. A national clinical decision support infrastructure to enable the widespread and consistent practice of genomic and personalized medicine. *BMC Med Inform Decis Mak.* 2009 [cited 2024 Apr 23];9(1). Available from: https://pubmed.ncbi.nlm.nih.gov/19309514/
66. Parekh ADE, Shaikh OA, Simran, Manan S, Hasibuzzaman Md Al. Artificial intelligence (AI) in personalized medicine: AI-generated personalized therapy regimens based on genetic and medical history: Short communication. *Annals of Medicine and Surgery.* 2023 Nov [cited 2024 Apr 23];85(11):5831. Available from: /pmc/articles/PMC10617817/
67. Curtis RG, Bartel B, Ferguson T, Blake HT, Northcott C, Virgara R, et al. Improving user experience of virtual health assistants: Scoping review. *J Med Internet Res.* 2021 Dec 1 [cited 2024 Apr 23];23(12). Available from: https://pubmed.ncbi.nlm.nih.gov/34931997/
68. Blanco-González A, Cabezón A, Seco-González A, Conde-Torres D, Antelo-Riveiro P, Piñeiro Á, et al. The role of AI in drug discovery: Challenges, opportunities, and strategies. *Pharmaceuticals.* 2023 Jun 1 [cited 2024 Apr 23];16(6):891. Available from: www.mdpi.com/1424-8247/16/6/891/htm
69. Liebman M. The role of artificial intelligence in drug discovery and development. *Chem Int.* 2022 Jan 1 [cited 2024 Apr 23];44(1):16–9. Available from: www.degruyter.com/document/doi/10.1515/ci-2022-0105/html
70. Mullard A. New drugs cost US$2.6 billion to develop. *Nat Rev Drug Discov.* 2014 Dec 1 [cited 2024 Apr 23];13(12):877–877. Available from: www.nature.com/articles/nrd4507
71. Deng J, Yang Z, Ojima I, Samaras D, Wang F. Artificial intelligence in drug discovery: Applications and techniques. *Brief Bioinform.* 2022 Jan 1 [cited 2024 Apr 23];23(1). Available from: https://pubmed.ncbi.nlm.nih.gov/34734228/
72. Paul D, Sanap G, Shenoy S, Kalyane D, Kalia K, Tekade RK. Artificial intelligence in drug discovery and development. *Drug Discov Today.* 2021 Jan 1 [cited 2024 Apr 23];26(1):80. Available from: /pmc/articles/PMC7577280/
73. O'Callaghan S, Galvin P, O'Mahony C, Moore Z, Derwin R. 'Smart' wound dressings for advanced wound care: A review. *J Wound Care.* 2020 Jul 2;29(7):394–406.
74. Guest JF, Ayoub N, McIlwraith T, Uchegbu I, Gerrish A, Weidlich D, et al. Health economic burden that wounds impose on the National Health Service in the UK. *BMJ Open.* 2015 [cited 2024 Apr 23];5(12). Available from: https://pubmed.ncbi.nlm.nih.gov/26644123/
75. Rani Raju N, Silina E, Stupin V, Manturova N, Chidambaram SB, Achar RR. Multifunctional and smart wound dressings—a review on recent research advancements in skin regenerative medicine. *Pharmaceutics.* 2022 Aug 1 [cited 2024 Apr 6];14(8). Available from: /pmc/articles/PMC9414988/
76. Omidi M, Yadegari A, Tayebi L. Wound dressing application of pH-sensitive carbon dots/chitosan hydrogel. *RSC Adv.* 2017 Feb 7 [cited 2024 Apr 6];7(18):10638–49. Available from: https://pubs.rsc.org/en/content/articlehtml/2017/ra/c6ra25340g
77. Lu SH, Samandari M, Li C, Li H, Song D, Zhang Y, et al. Multimodal sensing and therapeutic systems for wound healing and management: A review. *Sens Actuators Rep.* 2022 Nov 1;4:100075.
78. Tang N, Zheng Y, Haick H, Jiang X, Zhou C, Jin H, et al. Wearable sensors and systems for wound healing-related ph and temperature detection. *Micromachines (Basel).* 2021 Apr 1 [cited 2024 Apr 6];12(4). Available from: /pmc/articles/PMC8070747/
79. Brown MS, Ashley B, Koh A. Wearable technology for chronic wound monitoring: Current dressings, advancements, and future prospects. *Front BioengBiotechnol.* 2018 Apr 26;6:350710.
80. Kelly CJ, Karthikesalingam A, Suleyman M, Corrado G, King D. Key challenges for delivering clinical impact with artificial intelligence. *BMC Med.* 2019 Oct 29 [cited 2024 Apr 3];17(1):1–9. Available from: https://bmcmedicine.biomedcentral.com/articles/10.1186/s12916-019-1426-2
81. Jiang L, Wu Z, Xu X, Zhan Y, Jin X, Wang L, et al. Opportunities and challenges of artificial intelligence in the medical field: Current application, emerging problems, and problem-solving strategies. *J Int Med Res.* 2021 [cited 2024 Apr 3];49(3). Available from: https://pubmed.ncbi.nlm.nih.gov/33771068/
82. Razzak MI, Naz S, Zaib A. Deep learning for medical image processing: Overview, challenges and future.
83. Cabitza F, Rasoini R, Gensini GF. Unintended consequences of machine learning in medicine. *JAMA.* 2017 Aug 8 [cited 2024 Apr 3];318(6):517–8. Available from: https://pubmed.ncbi.nlm.nih.gov/28727867/

84. Char DS, Shah NH, Magnus D. Implementing machine learning in health care—addressing ethical challenges. *N Engl J Med.* 2018 Mar 3 [cited 2024 Apr 3];378(11):981. Available from: /pmc/articles/PMC5962261/
85. Kruse CS, Krowski N, Rodriguez B, Tran L, Vela J, Brooks M. Telehealth and patient satisfaction: A systematic review and narrative analysis. *BMJ Open.* 2017 Aug 1 [cited 2024 Apr 23];7(8). Available from: https://pubmed.ncbi.nlm.nih.gov/28775188/
86. Kim J, Neilipovitz D, Cardinal P, Chiu M, Clinch J. A pilot study using high-fidelity simulation to formally evaluate performance in the resuscitation of critically ill patients: The University of Ottawa critical care medicine, high-fidelity simulation, and crisis resource management I study. *Crit Care Med.* 2006 Aug [cited 2024 Apr 23];34(8):2167–74. Available from: https://journals.lww.com/ccmjournal/fulltext/2006/08000/a_pilot_study_using_high_fidelity_simulation_to.21.aspx
87. Oliveira FPM, Tavares JMRS. Medical image registration: A review. *Comput Methods Biomech Biomed Engin.* 2014 [cited 2024 Apr 23];17(2):73–93. Available from: https://pubmed.ncbi.nlm.nih.gov/22435355/
88. Gungor-Ozkerim PS, Inci I, Zhang YS, Khademhosseini A, Dokmeci MR. Bioinks for 3D bioprinting: An overview. *Biomater Sci.* 2018 May 5 [cited 2024 Apr 23];6(5):915. Available from: /pmc/articles/PMC6439477/
89. Sittig DF, Singh H. A new socio-technical model for studying health information technology in complex adaptive healthcare systems. *Qual Saf Health Care.* 2010 [cited 2024 Apr 23];19(Suppl 3):i68. Available from: /pmc/articles/PMC3120130/
90. Exudate Management Made Easy Update—Wounds International [Internet]. [cited 2024 Apr 23]. Available from: https://woundsinternational.com/made-easy/exudate-management-made-easy-update-wint/
91. Rippon MG, Fleming L, Chen T, Rogers AA, Ousey K. Artificial intelligence in wound care: Diagnosis, assessment and treatment of hard-to-heal wounds: A narrative review. *J Wound Care.* 2024 Apr 2 [cited 2024 Apr 23];33(4):229–42. Available from: www.magonlinelibrary.com/doi/10.12968/jowc.2024.33.4.229
92. Greenhalgh T, Shaw S, Wherton J, Vijayaraghavan S, Morris J, Bhattacharya S, et al. Real-world implementation of video outpatient consultations at macro, meso, and micro levels: Mixed-method study. *J Med Internet Res.* 2018 Apr 1 [cited 2024 Apr 23];20(4). Available from: https://pubmed.ncbi.nlm.nih.gov/29625956/

5 Role of the Immune System in Wound Healing and Strategies to Modulate Immune Responses to Promote Wound Healing

Phool Chandra,[1] Shahjad,[2] Anurag Verma,[1] and Neetu Sachan[3]*

[1] Teerthanker Mahaveer College of Pharmacy, Teerthanker Mahaveer University, Moradabad-244001, Uttar Pradesh, India

[2] Department of Pharmacology, Global Institute of Pharmaceutical Education and Research, Kashipur-244713, Uttarakhand, India

[3] Maharana Pratap College of Pharmacy, Mandhana, Kanpur-209217, Uttar Pradesh, India

* Corresponding author: Prof. Phool Chandra
Teerthanker Mahaveer College of Pharmacy,
Teerthanker Mahaveer University, Moradabad,
Uttar Pradesh, India-244001

5.1 INTRODUCTION

The immune system, a key component in wound healing, sets off a complex series of processes that restore tissue integrity and functionality. An injury triggers a cascade of immune responses that help repair damaged tissue, eliminate waste products, and protect the body from infections. This intricate process involves various immune cell types. The phases of wound healing that involve inflammation, proliferation, and remodeling are coordinated by the cooperative action of messenger molecules and cytokines. At the site of injury, immune cells such as neutrophils, macrophages, and lymphocytes are recruited to mount a defense against invading microbes and initiate tissue repair mechanisms. One of the initial response mechanisms is neutrophils acting to contain infection through phagocytosis and the release of antimicrobial agents. Subsequently, macrophages assume a central role in orchestrating the different stages of wound healing by phagocytosing debris, secreting growth factors, and modulating the inflammatory response. Furthermore, lymphocytes facilitate tissue regeneration and immune response regulation, which both aid in the healing process. In addition, the immune system facilitates angiogenesis, collagen deposition, and wound closure by intimately interacting with other biological constituents like fibroblasts and endothelial cells (1–7). Dysregulation of immune responses can lead to impaired wound healing, as seen in chronic wounds or conditions associated with immunodeficiency. Developing innovative treatment approaches to improve wound healing and reduce complications requires an understanding of the complex

DOI: 10.1201/9781003605966-5

interactions between immune cells and their signaling pathways. The diverse function of the immune system in wound healing is thoroughly examined in this review, which also highlights the important cellular and molecular mechanisms involved in each stage of the process (8–12). Apart from the widely recognized functions of immune cell types, including lymphocytes, neutrophils, and macrophages, contemporary studies have illuminated the contributions that other immune cell types and their subgroups make to the process of wound healing. For example, it has been demonstrated that regulatory T cells (Tregs), a specific fraction of T lymphocytes recognized for their immunosuppressive properties, are essential for controlling inflammation and encouraging tissue healing. Tregs work by inhibiting overreactions from the immune system, which helps avoid tissue damage and provide an environment that is favorable to recovery. Furthermore, new data points to the involvement of innate lymphoid cells (ILCs), a broad class of lymphocytes without antigen-specific receptors, in wound healing through cytokine production and interactions with nonimmune cells and antibodies (13–16). Recent advances in single-cell sequencing technologies have enabled the identification and characterization of previously unrecognized immune cell populations within the wound microenvironment, providing insights into their specific roles and interactions during different phases of wound healing. Additionally, the role of the microbiome in modulating immune responses and influencing wound healing outcomes has garnered increasing attention. Commensal microbes residing on the skin and mucosal surfaces can shape the local immune milieu through interactions with immune cells and the production of metabolites, eventually influencing the processes involved in wound healing. Comprehending the complex interactions among these heterogeneous immune cell populations, the gut microbes, and the wound microenvironment could potentially lead to the creation of novel immunomodulatory treatments that improve wound healing and reduce complications in a range of pathological circumstances (17–23).

5.2 IMMUNE SYSTEM AND WOUND HEALING

For wounds to heal, the immune system is essential. Phases of hemostasis, inflammation, proliferation, and remodeling are coordinated by a complex interaction of different cell types and signaling chemicals. For wounds to heal successfully, the shift from inflammation to proliferation is especially important because it establishes the equilibrium between tissue regeneration and scar formation. Impaired tissue regeneration and ongoing inflammation are common characteristics of chronic wounds (24–29).

5.2.1 Inflammatory Phase

To rid the wound site of pathogens and debris and prepare the tissue for repair, the inflammatory phase of wound healing is crucial. Through the recruitment and activation of several cell types and signaling molecules, the immune system plays a pivotal role throughout this phase. Among the principal immunological participants in the inflammatory phase are as follows (**Figure 5.1**).

Neutrophils are the initial cells to reach the site of the lesion, where they produce antimicrobial compounds and phagocytose bacteria. Additionally, they release chemokines and cytokines that draw additional immune cells to the wound. (30–33).

Macrophages are the most prevalent at the wound site and serve a variety of purposes. They have the ability to generate growth factors and cytokines that aid in tissue repair, phagocytose dead cells and debris, and change between pro- and anti-inflammatory (M1) and M2 morphologies to control the inflammatory response (34–37).

Lymphocytes: This group comprises natural killer (NK) cells, T cells, and B cells. NK cells provide immunity that adapts and controls the immune response. They can generate growth factors and cytokines that affect how quickly wounds heal (38–41). Recent studies have focused on the immune system during the inflammatory phase. The population of microorganisms that live on the skin, known as the epidermal microbiome, has the ability to affect several aspects of wound healing and immune response, including the recruitment and activation of immune cells, the synthesis

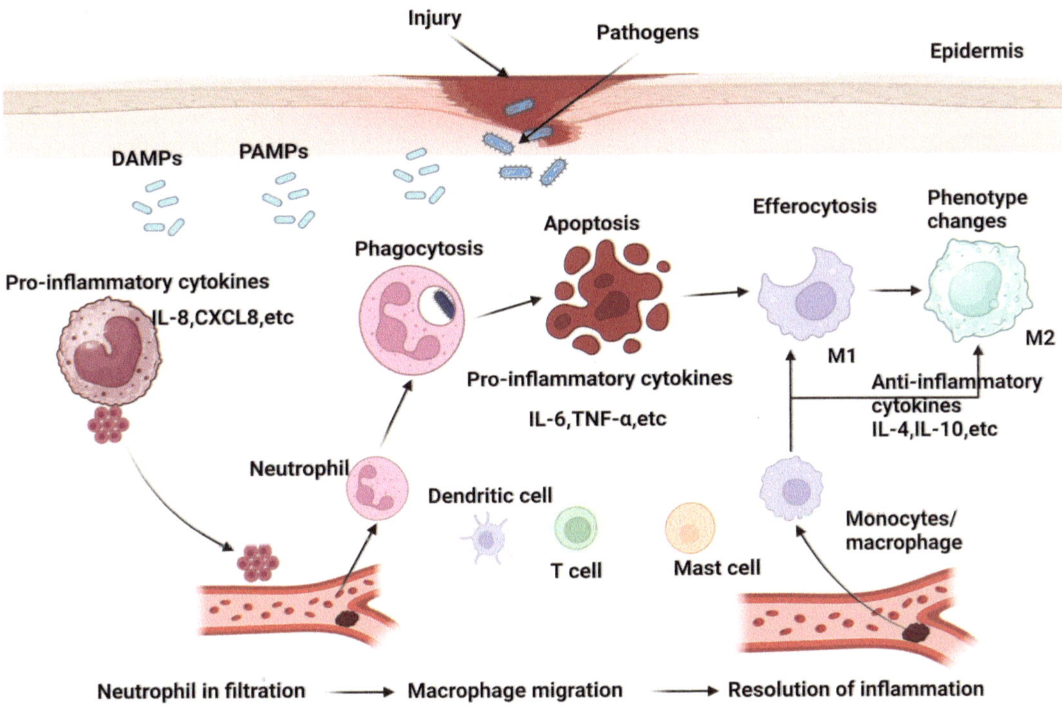

FIGURE 5.1 Immune system in the inflammatory phase.

of peptides that inhibit microbes, and the equilibrium between inflammation and tissue repair (42). Nanozymes, which are nanomaterials with enzyme-like activities, can modulate inflammation and promote skin regeneration by scavenging reactive oxygen species, inhibiting bacterial biofilms, and enhancing angiogenesis (43, 44). Immunomodulatory treatments, including stem cell therapy, growth factors, and cytokines, can improve wound healing and avoid chronicity by influencing the immune system, promoting tissue regeneration, and lowering the production of scars (45)

5.2.2 Proliferative Phase

The body's immune system is still vital in coordinating a number of processes that support tissue regeneration and repair throughout the proliferative stage of wound healing. This is a thorough examination of the role the immune system plays in the proliferative phase of wound healing. (46)

5.2.2.1 Angiogenesis and Immune Cells

The development of fresh blood vessels, or angiogenesis, is crucial for supplying oxygen and nutrients to the location of the wound and promoting tissue healing. Angiogenic factors, such as fibroblast growth factor (FGF) and vascular endothelial growth factor (VEGF), are secreted by immune cells, including T lymphocytes and macrophages, which encourage the development of fresh blood vessels. Furthermore, angiogenesis and neovascularization in the wound bed are facilitated by immune cells known as endothelial progenitor cells, as shown in **Figure 5.2** (47, 48).

5.2.2.2 Fibroblast Activation and Extracellular Matrix Deposition

Upon injury, quiescent fibroblasts become activated and undergo phenotypic changes to become myofibroblasts. A high level of proteins that contract, including α-smooth muscle actin (α-SMA), and higher synthetic activity for ECM components are characteristics of myofibroblasts. Numerous growth factors and cytokines, including transforming growth factor-beta (TGF-β), platelet-derived

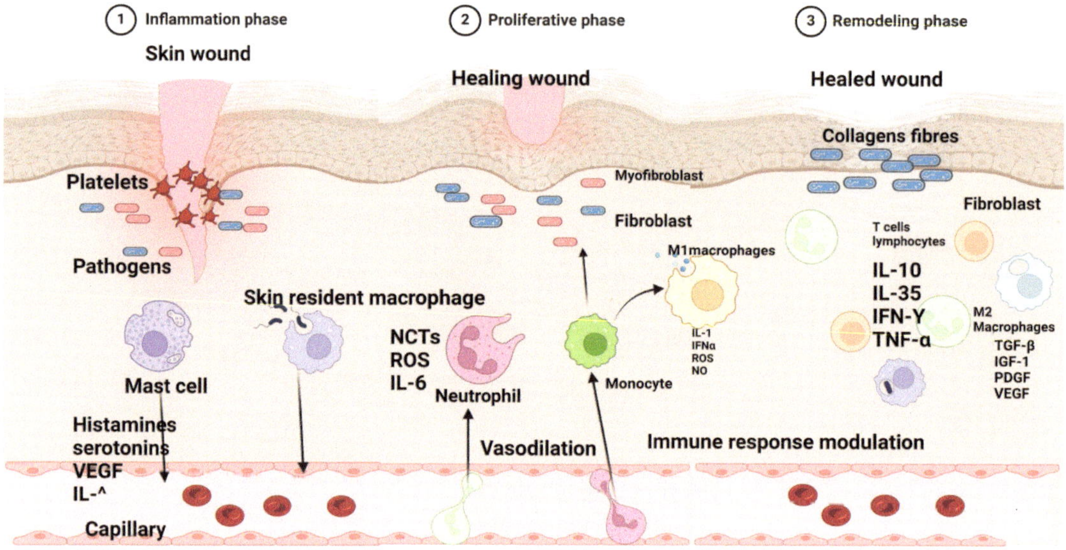

FIGURE 5.2 Immune system in proliferative and remodeling phase.

growth factor (PDGF), and fibroblast growth factor (FGF), are generated during the inflammatory phase of wound healing and can activate fibroblasts. Increased ECM production results from these substances stimulating fibroblast migration, proliferation, and conversion into myofibroblasts (49–51). The synthesis and secretion of different ECM proteins by fibroblasts are essential for the deposition of ECM. Collagen is the most abundant protein in the ECM and provides tensile strength to the wound tissue. Fibronectin and elastin contribute to the elasticity and flexibility of the ECM, while proteoglycans help maintain hydration and regulate cell signaling. The deposition of ECM components occurs in a highly organized manner, with collagen fibers aligning along the lines of tension to promote wound contraction and closure. Additionally, fibroblasts interact with other cell types, such as endothelial cells and immune cells, to coordinate ECM remodeling and tissue repair processes (52–54).

5.2.3 Remodeling Phase

As a way to coordinate the process of tissue repair and provide prospective approaches for regenerative therapy, the immune system is essential to the remodeling phase of wound healing. Key immune components, such as monocytes and lymphocytes, are particularly integral in facilitating wound closure and promoting scar-less healing. However, challenges such as inadequate drug delivery systems and safety concerns can hinder their clinical efficacy (55, 56). During the remodeling phase, the immune system is involved in the modulation of the extracellular matrix (ECM) and the activation of immune responses that are essential to tissue regeneration. The interplay between immune cells and the extracellular matrix (ECM) facilitates the process of regeneration and prevents scarring, as demonstrated in tissue-regenerating mice (57). Macrophages are among the most prominent immune cells that play a vital role in the regulation of immune system reactions during this phase. Their functions suggest a potential link between immunosuppressive processes in tumor growth and wound healing mechanisms. Additionally, studies show that innate cell populations from both the skin and the immune system are involved in the remodeling phase, responding to infections within the wound area and causing inflammation and tissue damage, which is a crucial step toward healing (58). Research has identified the need for more immunocompetent 3D dermis models to fully replicate the effects of the inflammatory response on ECM remodeling in wound healing, which has not been completely understood yet (59). Moreover, factors influenced by the immune system, such as growth factors, inflammation, and tensile forces, are known to impact the regenerative responses and determine the transition between scar formation and tissue regeneration (60). Another important component of the remodeling phase is lymphangiogenesis, which promotes tissue regeneration and helps to drain inflammatory chemicals, making it a possible target for treatment in chronic wounds (61).

5.3 IMMUNE DYSFUNCTION AND IMPAIRED WOUND HEALING

In the past few years, a great deal of research has been done on the role of immunological dysregulation in defective wound healing, highlighting its crucial impact on the healing process. Degradation of wound healing is a result of multiple causes, including immunological disarray, arterial neuropathy, and cellular senescence in certain disorders like diabetes. Contributing variables include inflammation, microbiota disruption, growth factor imbalances, and changes in fibroblast migration and proliferation. (62) Immune events involving various immunoregulatory cells are essential for orchestrating the normal wound healing process, which, when dysregulated, can result in sustained inflammation and subpar healing outcomes. (63) The interplay between immune responses and healing is not limited to humans. For example, in horses affected by pituitary pars intermedia dysfunction (PPID), alterations in the gingival tissue might suggest a link between PPID, age, and immune-related soft tissue changes. (64) The connective tissue sealing surrounding dental implants deteriorates in patients with Type 2 diabetes mellitus (T2DM), and this condition is exacerbated by immune inflammation, bacterial infestation, excessive matrix metalloproteinase activity, and oxidative stress—all of which emphasize the intimate connection between immune function and wound healing. (65) Among the immunological mechanisms affecting wound healing, the formation of neutrophil extracellular traps (NETs) has been implicated in promoting pathological inflammatory responses leading to tissue injury and thrombosis, thereby exacerbating diabetic complications, including wound healing impairment. (66) In infants with congenital heart disease, malnutrition can result in impaired wound healing and a weakened immune system, adversely affecting surgical outcomes both in short and long terms. (67) One study suggests that targeting heme oxygenase-1 (HO-1) could be a promising approach for treating diabetic foot ulcers, owing to its antioxidant, anti-inflammatory, pro-proliferative, angiogenic, and cytoprotective properties. (68) Patients with cirrhosis also present with increased perioperative risks due to immune dysfunction that predisposes them to infection and hampers wound healing due to compromised protein synthesis. (69) Furthermore, the COVID-19 pandemic highlighted the vulnerability of individuals with impaired wound healing, who reported not only worse mood and quality of life but also diminished immune fitness in the shadow of lockdown measures, disruptions in medical care, and negative effects on their condition. (70) This was also seen in additional research, whereby participants who claimed to have poor wound healing had significantly lower levels of perceived immunological fitness, symptoms of insomnia, exhaustion during the day, and quality of sleep. (71) There appears to be a significant correlation between perceived immune fitness and impaired wound healing, with individuals in the "COMBI" group exhibiting more immune-related complaints. (72) One tsRNA known as tRF-Gly-CCC-039 has been implicated in the negative regulation of endothelial cell activity during wound healing in the context of diabetes. This tsRNA inhibits critical processes like cell proliferation, migration, and tube formation. Moreover, diabetic mice showed delayed corneal wound healing and a subdued immune response following an injury, potentially linked to altered gut microbiome diversity and specific bacterial abundance changes. (73) Beyond natural processes, certain therapeutic interventions exhibit promise. For instance, Donepezil has been hypothesized to prevent high glucose-induced endothelial barrier dysfunction, promoting wound healing by preventing the destabilization of junctional proteins. (74) The potential of engineering serpin-loaded extracellular vesicles could prove beneficial in wound healing and regulation of immune cell signaling. The bountiful presence of immune cells and other components in the skin also substantiate its role as an

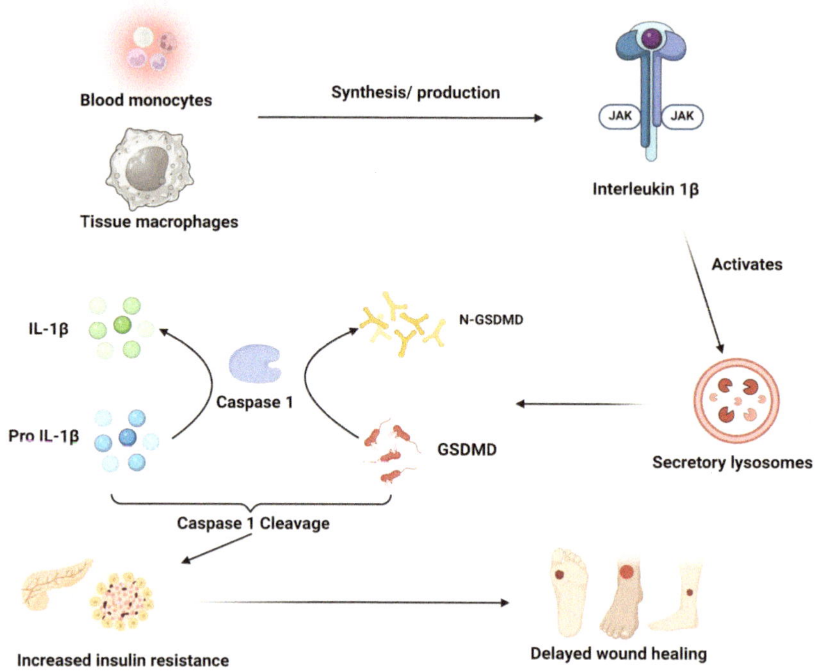

FIGURE 5.3 Immune dysfunction and delayed wound healing.

immunological barrier, which is fundamental for wound healing. (75) Monocytes, as well as lymphocytes, play a critical role in the change in wound macrophage phenotype from pro-inflammatory to anti-inflammatory, which is a critical phase in both normal and impaired healing processes, particularly in diabetes patients (76)m Finally, it has been demonstrated that both flavored and nicotine-containing e-cigarettes can hinder wound healing by raising oxidative stress and decreasing the bioavailability of nitric oxide, two factors necessary for the functioning of endothelial cells and wound healing, as shown in in **Figure 5.3** (77).

5.3.1 Chronic Wounds and Immune Dysregulation

Chronic injuries are unable to heal by going through regular stages in a timely and systematic manner. These wounds, which are frequently the consequence of underlying illnesses like diabetes, venous insufficiency, or pressure ulcers, may be trapped in the inflammatory phase. Because they are persistent and require sophisticated treatment, chronic wounds pose a substantial burden to patients and healthcare systems. A crucial component of the pathophysiology of chronic wounds is immune dysregulation. In a functional immune response, injury triggers an acute inflammatory response that clears pathogens and debris, leading to the proliferation and remodeling phases that culminate in wound closure. However, in chronic wounds, the immune system is disrupted, and several factors come into play, as presented in **Figure 5.4** (78–80).

Persistent inflammation is a crucial aspect of chronic wounds and plays a major role in their pathogenesis. Inflammation is often an essential component of the body's innate immune response, which aids in the defense against infections and starts the healing process after damage. A series of actions are involved in the inflammatory phase, such as the production of growth regulators and cytokines, the infiltration of immune cells such as neutrophils and macrophages into the site of injury, and the elimination of dead tissue and infections. However, the inflammatory response does

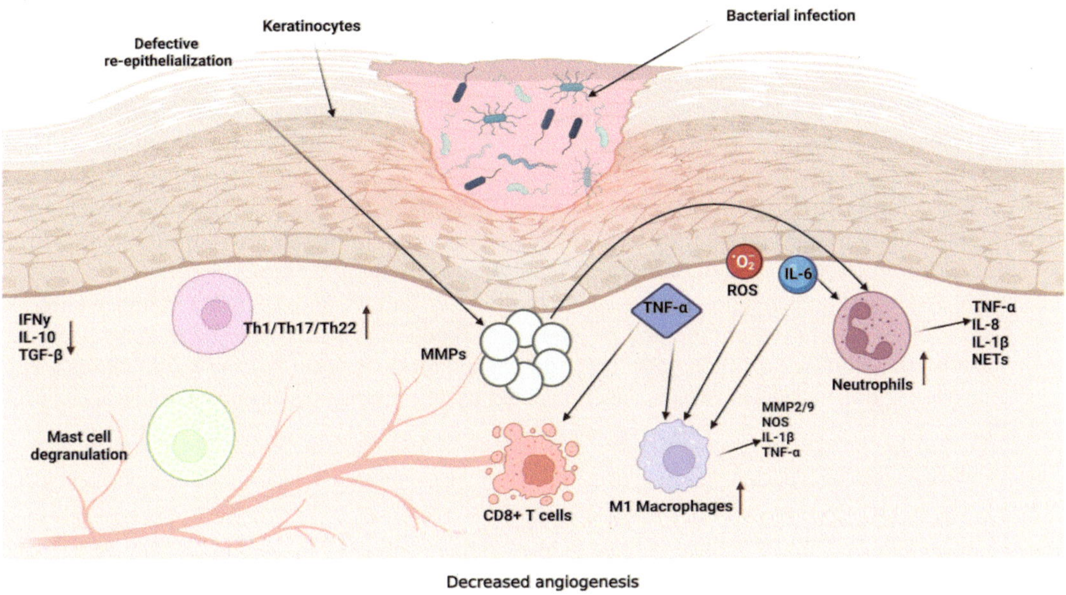

FIGURE 5.4 Chronic wounds and immune dysregulation.

not work as it should in the case of chronic wounds. The wound stays trapped in a prolonged state of inflammation rather than moving on to the proliferative and remodeling stages of healing. Several distinguishing characteristics of this dysregulated inflammation are: An imbalance of cytokines: Chronic injuries frequently display a disparity in cytokines, with high concentrations of pro-inflammatory cytokines that both initiate and maintain inflammation, such as tumor necrosis factor-alpha (TNF-α), interleukin-1 (IL-1), and interleukin-6 (IL-6). The presence of matrix metalloproteinases (MMPs) in excess: Extracellular matrix proteins are broken down by MMPs, an enzyme, throughout the course of normal wound healing. However, in chronic wounds, there is an overproduction of MMPs, which can lead to the destruction of growth factors and proteins necessary for tissue repair. Prolonged Presence of Immune Cells: Acute inflammation is marked by a transient influx of neutrophils, followed by macrophages that clear debris and facilitate the transition to repair. In chronic wounds, there is a persistent influx or activation of these cells without a successful transition to tissue repair. Altered Immune Cell Function: Immune cells in chronic wounds may exhibit impaired function, such as reduced ability to phagocytose bacteria, decreased responsiveness to signals that promote healing, and an inability to resolve inflammation. Biofilm Formation: Bacterial biofilms can form within chronic wounds, protecting bacteria from both the host immune response and antimicrobial treatments, leading to persistent inflammation and infection. The persistent inflammation in chronic wounds is detrimental because it disrupts the normal balance and timing of wound healing processes, leading to tissue destruction instead of repair. As a result, chronic wounds are typically challenging to treat, and addressing the underlying immune dysregulation is pivotal for successful wound management. To combat persistent inflammation, treatments may include the use of anti-inflammatory agents, protease inhibitors, dressings that manage wound exudate, and techniques to disrupt biofilms. Additionally, research is ongoing to identify new therapeutic targets and strategies to modulate the immune response and promote the resolution of inflammation in chronic wounds. (81–87)

Accumulation of senescent cells in the wound area can secrete factors that perpetuate inflammation and impede proper cell function necessary for healing. (88, 89)

Biofilm Formation: Chronic wounds often harbor bacterial biofilms, protective communities of microorganisms that are difficult to eradicate and can disrupt immune function and delay healing. (90, 91)

Immune dysregulation may prevent new blood vessels from forming, which is necessary to provide nutrition and oxygen to the injured area (92). Modified Cell Dynamics: Immune cell types such as neutrophils and macrophages, which are crucial to the healing process, as well as keratinocytes and fibroblasts, may exhibit dysfunction. Because of these complexities, research into chronic wounds and immune dysregulation involves understanding the molecular and cellular mechanisms at play to develop targeted therapies. Potential strategies include modulating the immune response, enhancing wound bed preparation, combating biofilm formation, and applying advanced wound care technologies like tissue-engineered products and regenerative medicine approaches. The latest research often focuses on bioengineering, growth factors, stem cell therapy, and personalized medicine to tackle the multifactorial issues presented by immune dysregulation in chronic wounds (93–95).

5.3.2 Factors Contributing to Impaired Wound Healing

Impaired wound healing is a complex process that is impacted by a number of variables that might make it more difficult for the body to heal wounds. This is a synopsis of the most recent studies that address the variables influencing poor wound healing. Diabetes affects the oral mucosa, which might result in delayed vascularization, decreased blood flow, decreased innate immunity, lower growth factor synthesis, and higher psychological stress. These variables likely contribute to altered wound healing mechanisms in diabetic individuals (96). Particularly, diabetic foot ulcers show reduced recruitment of endothelial progenitor cells, diminished angiogenesis, and impaired cellular responses to hypoxia. This is partially due to the instability and suppression of hypoxia-inducible factor 1-alpha (HIF-1α) brought on by hyperglycemia (97). Additionally, by improving cell connections and vascularization and by delivering therapeutic chemicals like growth factors, nanotherapeutics—engineered at the nanoscale—can expedite diabetic wound healing (98). Diabetic wounds are subject to the interaction of neutrophil extracellular traps and ferroptosis, two types of programmed cell death, which impede the healing process (99). Wound healing is greatly aided by enzymes such as plasmin and matrix metalloproteinases, which encourage the hydrolysis of substances necessary for cell migration (100). Other factors impeding the healing mechanism include enhanced inflammatory responses, restricted cell proliferation, restrained angiogenesis, limited collagen deposition, and infections (101). The insufficient induction of adipose tissue senescence following injury might be indicated by altered regulation of senescent mesenchymal cells and the cellular senescence-associated secretory phenotype in diabetic adipose tissue transplants. These modifications can also hinder recovery (102). Increased matrix metalloproteinase (MMP) production, extended reactive oxygen species (ROS) production, and excessive inflammation are all linked to delayed recovery in ischemic type 2 diabetes rats (103). Treatments such as platelet release can potentially improve healing by modulating the secretion of endogenous growth factors and restoring essential healing steps (104). Deficiencies in certain receptors, such as α3β1 integrin in keratinocytes, may lead to impaired angiogenesis (105). Platelets can also facilitate healing by reducing oxidative stress in endothelial cells, which decreases apoptosis and promotes neovascularization (106). Furthermore, the monitoring of wound fluid nitrate levels could predict healing success in diabetic foot ulcers (107). MicroRNAs within adipose-derived stem cells (ASCs) play significant roles in the pathogenesis of impaired wound healing mechanisms (108) while genetic deficiencies can also lead to poor angiogenesis and wound healing, such as in the case of WRN mutation affecting HGF insufficiency and PI3K/AKT dysregulation (109). A critical protein required for normal healing is the lysosomal trafficking regulator, which affects epithelialization and collagen deposition (110). Paeonol was shown to promote healing by affecting macrophage polarization and inflammatory responses in diabetic rats (111). In ischemia-impaired dermal ulcers, applying growth factors like KGF-2 topically can significantly promote re-epithelialization and granulation tissue formation, indicating a possible mesenchymal–epithelial interaction beneficial to wound healing (112). Disruptions in hemodynamic forces, such as disturbed shear stress, can impair angiogenesis and wound healing by downregulating VEGFB expression (113). Impairments can also arise from infections, unregulated inflammation, poor cellular functions, and enhanced protease activity (114). Lastly, the pathology of Crohn's disease includes impairments in enteric glial cells (EGCs) that hinder intestinal epithelial healing due to deficiencies in specific lipid metabolite production (115) and dysregulations in the molecular microenvironment of wounds can delay neutrophil recruitment and inflammation resolution (116). Wound healing in liver fibrosis, another complex condition, is influenced by the interplay of inflammation, oxidative stress, and autophagy (117).

5.3.3 Examples of Diseases Affecting Wound Healing

5.3.3.1 Diabetes and Autoimmune Diseases

Diabetes is frequently associated with impaired wound healing due to factors such as inflammation, impaired glucose metabolism, angiogenesis disorders, and delayed healing related to inflammatory cell infiltration and collagen deposition (118). Autoimmune disorders can similarly affect the healing process, often as a result of systemic inflammation and autoantibodies that target healthy tissues, disrupting normal regenerative processes (119).

5.3.3.2 Obesity and Cardiovascular Disease

Wounds in obese individuals can heal more slowly due to factors like reduced perfusion, chronic low-grade inflammation, and hormonal imbalances that impair tissue repair. Cardiovascular diseases, on the other hand, can hinder wound healing by compromising blood flow and, thus, nutrient and oxygen delivery necessary for tissue repair (120).

5.3.3.3 Malnutrition

Malnutrition can dramatically obstruct wound healing given that nutrients are fundamental building blocks of the repair process. Deficiencies in proteins, vitamins, and minerals can severely delay healing. (120)

5.3.3.4 Chronic Renal Disease and Cancers

Renal disease can alter wound healing through imbalances in electrolytes and hormones, such as calcium and phosphorus, which are critical to the repair of tissues, while cancers can disrupt healing through systemic effects such as cachexia and local effects such as impeding angiogenesis (120),

5.3.3.5 Chronic Liver Diseases

Liver fibrosis, which can result from chronic liver disease, impairs the liver's regenerative ability due to an imbalance in extracellular matrix (ECM) generation and degradation, where factors like inflammation and oxidative stress play significant roles in the pathogenesis (121).

5.3.3.6 COPD and Fibrotic Diseases

Chronic obstructive pulmonary disease (COPD) patients exhibit abnormal fibrogenesis in the lungs due to oxidative stress-induced cellular senescence, affecting healing mechanisms. In fibrotic diseases affecting various organs, excessive ECM deposition impairs normal function and healing processes (122).

5.3.3.7 Infections

Wounds can be complicated by infections, especially those resistant to conventional antibiotics. This has led to research into antimicrobial substances without the drawbacks of antibiotics, which include aloe vera, curcumin, and cinnamaldehyde, with applications in wound healing (123).

5.3.3.8 Keloid Disease

Abnormal healing processes can result in keloid disease, a benign condition marked by overgrown scar tissue extending beyond the wound's original margins, mediated by factors such as transforming growth factor beta (TGF-beta) (124).

5.3.3.9 Oculodentodigital Dysplasia (ODDD)

ODDD affects wound healing, likely through subclinical defects that have been explored through both human and animal models. The healing mechanism is orchestrated by a complex interplay of systemic and local factors, and its disruption due to diseases can lead to chronic nonhealing wounds, posing significant healthcare challenges. Treatments are being developed that target these underlying mechanisms, such as matrix metalloproteinase inhibition in diabetic ulcers or the potential of resveratrol to promote healing through anti-inflammatory, antioxidant, and antibacterial effects (125).

5.4 STRATEGIES TO MODULATE IMMUNE RESPONSES FOR WOUND HEALING

One of the strategies includes the use of immunomodulatory hydrogels that control macrophage polarization to improve chronic wound healing (126). The design of biomaterials with surface modifications impacts cell attachment, migration, and immune-inflammatory response, thereby enhancing bone healing and regeneration through optimized biomaterial properties (127). Bioactive materials can modulate cell behaviors across different stages of wound healing, including hemostasis, anti-inflammation, tissue regeneration and deposition, and matrix remodeling (128). Strategies to improve the effectiveness of biomaterials in tissue repair involve intentional activation and timing of monocytes and macrophages, as well as the modulation of their interactions with other cell types involved in the wound healing process (129). The potential of medicinal plants in modulating immune responses for wound healing is also recognized, with plants like aloe vera, *Calendula officinalis*, and *Curcuma longa* traditionally used for treatment and infection prevention (130) Nanomedicines that manipulate or re-educate macrophages during inflammation have been proposed as an effective strategy for wound management (131). Additionally, enhanced immune responses have been observed with the use of silver nanoparticle-containing hydrogels, indicating triggered immune responses in diabetic wound healing (132). Co-transplantation of multipotent stromal cells (MSCs) and fibroblasts has been suggested to educate the wound microenvironment and normalize aberrant healing processes in aged individuals (133). Novel strategies include creating an immune microenvironment conducive to wound healing by utilizing whole-course-repair systems based on neurogenesis-angiogenesis crosstalk and macrophage reprogramming (134). Other approaches involve shape-programmable hierarchical fibrous membranes designed to modulate the inflammatory microenvironment, promoting wound healing in diabetic patients (135). Activation of an adaptive immune response has been demonstrated to induce regenerative healing through the use of D-peptide cross-linked hydrogels (136) Furthermore, lymphangiogenesis is being targeted as a new therapeutic area for wound healing, breaking a cycle of inflammation and promoting tissue repair (137) Modulation of the immune response has wide-ranging potential applications, including preventing postsurgical adhesions, enhancing angiogenesis, and improving periapical healing in endodontically treated teeth (138, 139).

5.4.1 Immunomodulatory Agents

Chitosan-based dressings have anti-inflammatory and immunomodulatory properties that significantly contribute to the skin healing process postinjury. A systematic review emphasized the significance of such wound therapy (140). Silver nano formulations have been recognized for their wound healing capabilities due to their potential effectiveness as new agents in the treatment of chronic wounds, underscoring their role as promising immunomodulatory agents (141) Surface-active molecules like surfactants and biosurfactants are utilized in wound treatments due to their anti-inflammatory, antioxidant, and antibacterial properties (142). Antimicrobial peptides have been suggested to hold therapeutic potential for wound healing due to their engagement in various beneficial pathways (143). Nanotechnology-based medications, particularly for diabetic wounds, show promise by enhancing healing through slow, localized drug delivery, which stimulates various healing aspects—indicating expansion in this field (144). Postbiotics display beneficial properties like immunomodulatory, anti-inflammatory, and angiogenesis characteristics that expedite wound healing processes (145). Curcumin and silver nanoparticles decorated by silk sericin encapsulated in bioactive hydrogels can reduce bacterial infections and pro-inflammatory factors while promoting angiogenesis, thus enhancing wound healing (144) MicroRNAs (miRNAs) control proteins and pathways at the posttranscriptional level, showing significant potential for managing wounds by inducing favorable pathways (146) Temulawak extract, applied in ointment form, has been effective in promoting wound healing in Wistar rats by improving macroscopic appearances and histopathological factors like epithelialization (147). Placental-derived biomaterials exhibit immunomodulatory properties, creating a reparative environment conducive to wound healing (148). Kefir grains show multiple activities, including immunomodulatory effects, which may translate to wound healing benefits (149) Helichrysum italicum essential oils exhibit antioxidant, antimicrobial, and anti-inflammatory effects that significantly contribute to the various phases of wound healing (150).

5.4.1.1 Anti-inflammatory Drugs

Qingre Huoxue Decoction (QHD) is a Traditional Chinese Medicine preparation enhancing wound healing by modulating pathways related to blood vessel formation (angiogenesis) and inflammation (151).

Ozonated Sunflower Oil (OSO) is an oil treated with ozone that possesses antioxidant properties beneficial for treating inflammatory conditions of the skin like atopic dermatitis (152).

Chitosan is a natural polysaccharide known to suppress inflammatory enzymes and modulate pro-inflammatory mediators, beneficial in wound healing (153).

Cell-free supernatant (CFS) from *Lactobacillus johnsonii* LJO02, a probiotic derivative combined with vitamin D, can reduce inflammatory cytokines in human skin cell models (154).

Silk fibroin films are biomaterials derived from silk that, when enhanced with antimicrobial agents, can aid in anti-inflammatory responses and immune modulation during healing (155).

Backpacks are microparticles designed to carry anti-inflammatory agents directly to macrophages (immune cells), aiding in immune modulation and wound resolution (156).

Bioactive scaffolds and nanofibrous composites are engineered materials that influence immune cells like macrophages to adopt a healing-promoting (M2) phenotype (157).

Nanocomposite hydrogels are advanced gels infused with nanoparticles that can steer immune cells towards anti-inflammatory profiles, thereby aiding in wound healing (158).

Nanomedicines are tiny medicinal compounds that can target immune cells and alter their responses to promote healing (159)

Curcumin-encapsulated hydrogel is a gel-like delivery system containing the anti-inflammatory compound curcumin to reduce immune cell infiltration and inflammation at wound sites (160).

Carvacrol prodrugs/hyaluronic acid formulations are compounds made from thyme that are designed to release their active ingredient slowly, paired with hyaluronic acid to promote wound healing (161).

MSC-derived sEV-loaded hyaluronic acid hydrogels are hyaluronic acid hydrogels loaded with small extracellular vesicles (sEVs) from mesenchymal stem cells (MSCs) that can change macrophage behavior towards a regeneration-friendly state (162)

Immunomodulatory hydrogels are hydrogels designed to interact with immune cells, particularly macrophages, to improve chronic wound healing by reducing prolonged inflammation (163).

Plant-derived compounds targeting PI3K/AKT are compounds extracted from plants that can affect specific immune cell signaling pathways, which are crucial for immune responses and wound healing (164).

5.4.1.2 Growth Factors and Cytokines

The role of growth factors and cytokines is essential in modulating immune responses for wound healing. These signaling molecules orchestrate the inflammatory responses to eliminate pathogens and help recruit additional immune cells to the site of injury, thereby influencing the dynamics of wound healing. Various studies between 2020 and 2024 have shed light on different aspects of this complex process. Małgorzata Zielińska and colleagues highlight the significance of wound microbiota and its impact on wound healing through the coordination of immune responses (165). Similarly, Morio and others have shown that UV irradiation can induce a range of molecules, including growth factors, cytokines, and antimicrobial peptides, that contribute to the immune response modulation vital for wound healing (166). The crosstalk between keratinocytes and fibroblasts, mediated by these signaling molecules, is particularly important in the various stages of wound healing, from inflammation to extracellular matrix modulation (167). Transformation of growth factors such as TGF-β and matrix metalloproteinases (MMPs) has been linked to the immune response in wound healing and extracellular matrix remodeling, which can influence the severity of diseases like COVID-19 (168). Polymorphonuclear neutrophils (PMNs) also contribute to immune modulation via cytokines, chemokines, and growth factors, playing a role in wound healing, mucosal inflammation, and other processes (169). Moreover, aging can impact the immune responses associated with wound healing, affecting cells like gingival fibroblasts and consequently influencing diseases such as periodontal disease (170). In addition, biomaterials have been used to regulate macrophage cell fate, guiding immune responses toward a pro-regenerative environment conducive to healing (171). Specialized T cells such as gamma delta (γδ) T cells help regulate immune responses in the skin, aiding in wound healing through their secretion of growth factors and inflammatory factors (172). The use of human keratin matrices in wound healing applications suggests that cytokines and growth factors can be modulated to enhance wound closure and reduce wound size compared to other treatments (173). Electrical stimulation has been found to promote diabetic human skin fibroblast adhesion and growth, reducing pro-inflammatory cytokine secretion and supporting the secretion of growth factors integral to the healing process (174). Additionally, Chinese medicinal materials show potential by regulating cytokines and growth factors to accelerate wound healing and inhibit scar formation (175). Combinations of growth factors and antioxidants can stimulate cellular responses and manage inflammation, facilitating wound healing (176). For instance, quercetin has demonstrated the ability to accelerate wound healing in rats by modulating the expression of growth factors such as VEGF and TGF-β1, and cytokines like IL-10 (177). The sequential delivery of growth factors and cytokines has been studied to show improved wound healing and tissue regeneration, promoting granulation tissue formation and functional neovascularization (178). Incorporating these signaling molecules into biomaterials enhances wound healing by providing a sustained release and minimizing side effects (179). Biomaterials containing growth factor/cytokine combinations and stem cells can also improve cardiac tissue regeneration after injury, reducing inflammation and fibrosis. Furthermore, physiological fibrin hydrogels have been shown to modulate immune cells favorably, accelerating skin wound healing compared to more concentrated fibrin sealants. Understanding the intricate roles of growth factors and cytokines in wound healing also extends to the treatment of burns and the interactions with the skin microbiome, which can significantly influence the immune response and wound microenvironment. Lastly, advancements in understanding how mesenchymal stromal cells can be primed with proinflammatory cytokines to enhance wound healing through IL-6-dependent macrophage polarization provide a promising avenue for improved wound repair speed and quality. All data present in the **Table 5.1** (180–184)

TABLE 5.1
Growth Factors and Cytokines for Wound Healing

Factors	Source	Function	References
Platelet-derived growth factor	Platelet Macrophage Keratinocyte	Fibroblast proliferation Chemoattraction	(185)
Vascular endothelial growth factor	Epidermal cell Macrophage	Angiogenesis Increases vascular permeability	(186)
Epidermal cell growth factor	Platelet Macrophage Fibroblast	Cell migration Cell proliferation	(187)
Fibroblast growth factor	Macrophage Mast cell Endothelial cells Keratinocyte Fibroblast	Angiogenesis Fibroblast proliferation Keratinocyte migration	(188)
Transforming growth factor β1	Platelet Macrophage Lymphocyte Keratinocyte Fibroblast	Cell migration Chemoattraction Granulation tissue formation Re-epithelialization Extracellular matrix synthesis	(189)
Tumor necrosis factor-α	Neutrophil Macrophage	Growth factor expression Re-epithelialization	(190)
Interleukin-1	Neutrophil Macrophage Fibroblast	Initiation of inflammation Re-epithelialization	(191)
Interleukin-6	Neutrophil Macrophage Fibroblast	Re-epithelialization Formation of granulation tissue Angiogenesis	(192)
Interleukin-8	Neutrophil Macrophage	Re-epithelialization	(193)

5.4.2 Biomaterial-Based Approaches

These materials come in various forms such as hydrogels, films, nanofibers, foams, and patches. Their high biocompatibility allows them to interact well with the body. They can encapsulate different types of cells and biomolecules. Controlled release in various wound models makes them promising candidates for wound healing applications. Tissue Remodeling and Healing Mechanisms: Understanding tissue remodeling phenomena and wound healing mechanisms is essential. Factors like cell therapy, bioactive therapeutic delivery, and biomaterial-based approaches contribute to healing acceleration. These methods aim to create a supportive microenvironment, prevent microbial contamination, and deliver therapeutic agents (194, 195). One mechanism involves the use of immunomodulatory substances. Melatonin-enriched autologous bio gels, for instance, can restore immune tissue homeostasis and reduce inflammation (196). Immunomodulatory cytokines, when delivered through biomaterials, can control wound healing by influencing the fate of immune cells, as seen with biomaterials designed for the ischemic myocardium (197). Spermidine-functionalized biomaterials modulate the immune response by interacting with macrophages and other immune cells, enhancing wound healing processes and managing inflammation around implants (198). The immunomodulatory properties of placental-derived biomaterials could also address chronic wounds by altering immune responses in the wound microenvironment (199). Biomaterials influence chemokine signaling by modifying the body's immune reactions. This can involve the enhancement or suppression of pro-inflammatory and anti-inflammatory properties, which can dictate the outcome of healing and may depend on the timing and duration of biomaterial presence (200) Furthermore, biomaterials can regulate macrophage polarization, as seen in the advanced biomaterials that modulate macrophage transformation to promote healing, thereby addressing immune dysfunction and

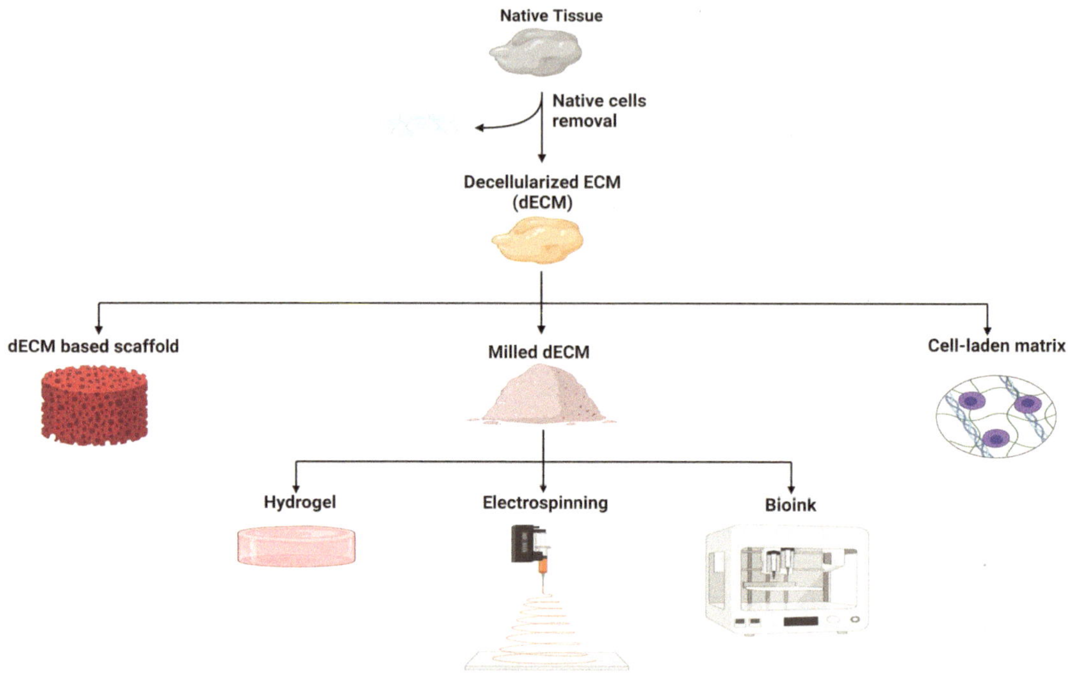

FIGURE 5.5 Biomaterial-based approaches.

creating conducive microenvironments for tissue repair (201). Surface modifications of biomaterials provide another mechanism by influencing the immune-inflammatory responses to promote healing and regulate inflammation. This is essential for optimal tissue regeneration due to the biomaterial surface properties influencing immune cell activity (202). An innovative approach involves the use of glycopeptide hydrogels to activate a pro-regenerative Type 2 immune response in wound healing. This is achieved by enhancing macrophage-T-cell crosstalk, thus leveraging the benefits of immune cell interaction to evoke strong T-cell immunity that supports wound repair (203). The physical and topographical properties of biomaterials are also significant, as they. can affect cell activities such as macrophage transformation and facilitate cell migration. This can lead to a better reparative immune response and promote epithelialization that is beneficial for wound healing (204). Lastly, optimized glycaemic control in diabetic patients is crucial to prevent immune dysfunction that can contribute to impaired wound healing. Here, biomaterials can play a role in providing structured and favorable environments that promote tissue repair despite sustained hyperglycemic conditions, as presented in **Figure 5.5** (205).

5.4.2.1 Scaffolds for Immune Cell Recruitment

Bioengineered Scaffolds for Cutaneous Wound Healing. These scaffolds play a pivotal role in wound repair and skin regeneration (**Figure 5.6**). Researchers focus on designing biomaterial scaffolds, including precursor development. Immunomodulation is a key aspect, where scaffolds interact with immune cells. By conjugating with bioactive molecules and stem cells, these scaffolds enhance wound healing. Collagen-Glycosaminoglycan (CG) Scaffold: One of the earliest tissue engineering scaffolds, the CG scaffold disrupts scar formation. It provides a template that fills the space to be regenerated, facilitating cell attachment and new tissue generation. Pro-regenerative scaffolds lead to complete skin structures, while traditional biological scaffolds often result in scarred skin. Hydrogel Scaffolds: Hydrogels offer a physiologic microenvironment for transplanted cells. They provide mechanical support and protection from native immune cells. Tailoring cell–hydrogel interactions based on tissue properties enhances cell therapy delivery (206–208). Spatial design and

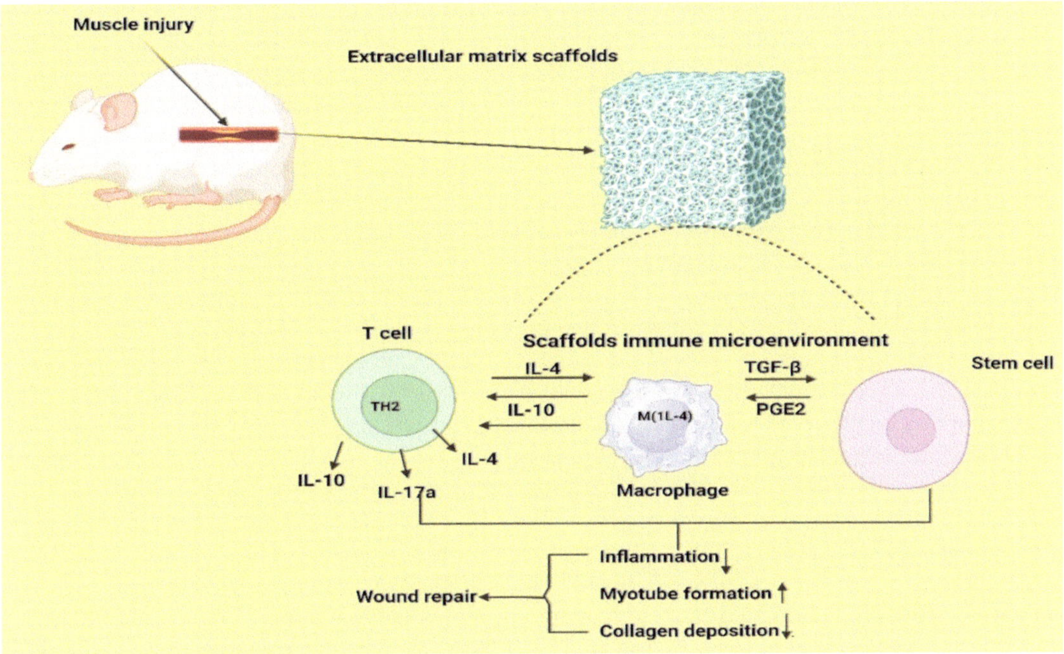

FIGURE 5.6 Bioengineered scaffolds for cutaneous wound healing.

chemotaxis induced by growth factors in hierarchically assembled nanofiber scaffolds with dual growth factor gradients can rapidly recruit immune cells, thus promoting skin wound healing, as demonstrated (209). Microporous annealed particle (MAP) scaffolds have been shown to improve wound healing and reduce inflammation by promoting a balanced pro-regenerative macrophage response, as mentioned (210). The deregulation of neutrophil and macrophage recruitment orchestrated by FOXM1 is said to impair wound healing in diabetic foot ulcers, indicating the significance of correct immune cell mobilization for healing (211). A study by (212) has illustrated that Angptl4 deficiency led to prolonged neutrophil presence and fewer monocyte-derived macrophages, which was controlled by an Angptl4-ifi202b axis during wound healing—highlighting the significance of managing immune cell dynamics. Hydrogel scaffolds create a physiologic microenvironment that encourages the recruitment of immune cells to the wound site, protecting transplanted cells from native immune attacks (213). Monocytes and lymphocytes are essential in diabetic wound healing, where their impaired transitions may hinder the healing process, as detailed (214). Elastin-based hydrogels enhance wound healing by promoting innate immune cell recruitment and angiogenesis according to (215) Bioengineered scaffolds composed of thermo-responsive polymers guide immune cell recruitment for wound healing through controlled drug release strategies, illustrating a sophisticated application of scaffolds (216). Mast cells have been identified as mediators of inflammatory effector cell recruitment, showing their importance in immune dysfunction within the context of wound healing (217). The role of Ag-doped bioactive glass-ceramic nanoparticles in skin substitutes includes promoting antibacterial activity, immune response modulation, and support for scarless cutaneous wound regeneration. Present in the (218)

5.4.2.2 Controlled Release Systems for Bioactive Molecules

Silk-based Particulate Systems: Silk fibroin and sericin, derived from silkworm production, serve as a unique biocompatible and biodegradable protein platform. These sustainable silk-based systems can control the release of pharmaceuticals and bioactive agents. By ensuring optimal bioavailability, they address challenges associated with conventional delivery methods. Electronically controlled drug release system: Biocompatible and electroactive polymers (EAP) form the basis of this system. Negatively and positively charged drugs are loaded into EAP systems through electro-polymerization. This approach promotes wound healing by precisely releasing therapeutic compounds. Novel Microenvironments: Researchers explore various strategies, including controlled-release materials, hydrogels, and scaffolds. Antimicrobial and growth factor delivery, electrospinning, 3D printing, and nanomaterial approaches contribute to enhanced wound repair (219–221). Sustained release of inhibitors from bionic scaffolds enhances skin tissue regeneration, leading to quicker and better wound healing compared to other systems with burst release limitations. This controlled release fosters not just wound healing but also hair follicle neogenesis (222). Controlled release systems modulate the immune system by optimizing the actions of immune cells and promoting the release of anti-inflammatory cytokines, thus supporting the healing of chronic skin wounds (223). Apigenin-loaded ointments have been designed to display anti-inflammatory and wound healing activities, demonstrating a successful drug-delivery system for wound healing. Systems providing spatiotemporal control over the delivery of substances like nitric oxide (NO), essential for immune responses and wound healing, showcase the capability of controlled release systems to manage immune dysfunction (224). Degradable hydrogel delivery systems target endogenous cells with bioactive molecules such as the intercellular adhesion molecule-1 (ICAM-1) to enhance protein retention and promote healing in chronic wounds (225). Bio nanocomposites for controlled release of antioxidants and other bioactive compounds, like prodigiosin, provide a strategy to improve the immune system and are suitable for treating wounds (226). Magnetic responsive release from microspheres derived from metal-organic frameworks (MOFs) enables controlled delivery of bioactive molecules like NO for the treatment of bacteria-infected cutaneous wounds, further affirming the role of controlled release systems in immune modulation (227). Liquid-crystal-infused porous surfaces that enable both pulsatile and continuous release modes of bioactive molecules have potential applications in immune dysfunction and wound healing (228). Investigations have also expanded to include the development of immunomodulatory biomaterials for regenerating infarcted myocardium, illustrating the broader implications of controlled release systems in tissues affected by immune dysfunction (229).

5.4.3 Cell-Based Therapies

Cell-based therapies for wound healing are emerging as promising alternatives to conventional treatments that often fail to achieve satisfactory outcomes. Cell-based therapies involve the use of various types of cells, such as stem cells, progenitor cells, or differentiated cells, to modulate the wound healing process and enhance tissue regeneration (230, 231). Cell-based therapies can be applied for both acute and chronic wounds, such as burns, diabetic ulcers, or surgical wounds. Cell-based therapies can improve wound healing by providing various beneficial effects, such as accelerating wound closure, promoting angiogenesis, reducing inflammation, stimulating re-epithelialization, and remodeling the extracellular matrix. Cell-based therapies can also reduce scar formation, donor site morbidity, and infection risk. Cell-based therapies can be delivered in different forms, such as cell suspensions, cell sheets, cell scaffolds, or cell-derived extracellular vesicles (232, 233). Cell-based therapies can be derived from various sources, such as adipose tissue, bone marrow, umbilical cord blood, or skin. Among them, mesenchymal stem cells (MSCs) are the most widely studied and applied cell type for wound healing due to their high proliferation and differentiation potential, easy isolation and expansion, low immunogenicity and tumorigenicity, and paracrine secretion of various growth factors and cytokines. MSCs can be obtained from different tissues, such as adipose tissue, bone marrow, umbilical cord, or placenta. MSCs can also secrete extracellular vesicles (EVs), which are membrane-bound structures that contain proteins, lipids, and nucleic acids. EVs can mimic the therapeutic effects of MSCs without the risk of cell engraftment or differentiation. EVs can also be engineered to enhance their targeting and delivery efficiency. Cell-based therapies for wound healing are still in the early stages of development and face several challenges, such as standardization of cell sources and products, optimization of cell delivery methods and doses, evaluation of long-term safety and efficacy, and regulation of clinical trials and applications. However, cell-based therapies have shown great potential and promise for improving wound healing outcomes and quality of life for patients with various types of wounds (234, 235).

5.4.3.1 Mesenchymal Stem Cells

Exosomes derived from MSCs (MSC-Exo) have therapeutic effects on wound repair and scar attenuation, thanks to their immunomodulation mechanisms, acting on different immune cells and molecules involved in wound healing and scar formation (236, 237). Umbilical cord-derived MSCs reportedly decrease blood-brain barrier permeability, modulate cytokine ratios, enhance tight junction proteins, and promote wound healing, suggesting their protective role against barrier dysfunction induced by surgical wounds in mice (238). MSCs orchestrate wound healing and regeneration through interactions with various ordered inflammatory processes, particularly with neutrophils, macrophages, and T cells, each of which plays critical roles in different phases of healing (236). Adipose-derived MSCs mediate immune modulation by releasing immune mediators and exosomes, which help resolve prolonged inflammatory responses crucial for the wound healing process (239). A specific molecular mechanism of MSC-exosome action has been proposed, where circ-Astn1 enhances SIRT1 expression through miR-138–5p adsorption, thereby improving endothelial precursor cell function and angiogenesis, which are vital for wound repair (240). Advances have been made in preconditioning methods for human MSCs in scaffolds, exhibiting enhanced wound healing in models such as porcine wound healing, and MSC treatments promote M2 macrophage polarization and growth factor production, proving critical for reducing inflammation and enhancing wound healing (241). Additionally, injectable peptide hydrogel encapsulation of MSCs has shown improved viability, stemness, and anti-inflammatory effects, leading to enhanced early-stage wound healing in vivo compared to traditional hyaluronic acid-based materials (242). These studies highlight the critical role of MSCs in wound healing by utilizing their immunomodulatory, anti-inflammatory, and pro-regenerative capabilities. The evidence points to MSCs and MSC-derived products as promising candidates for treating wounds complicated by immune dysfunction (**Figure 5.7**).

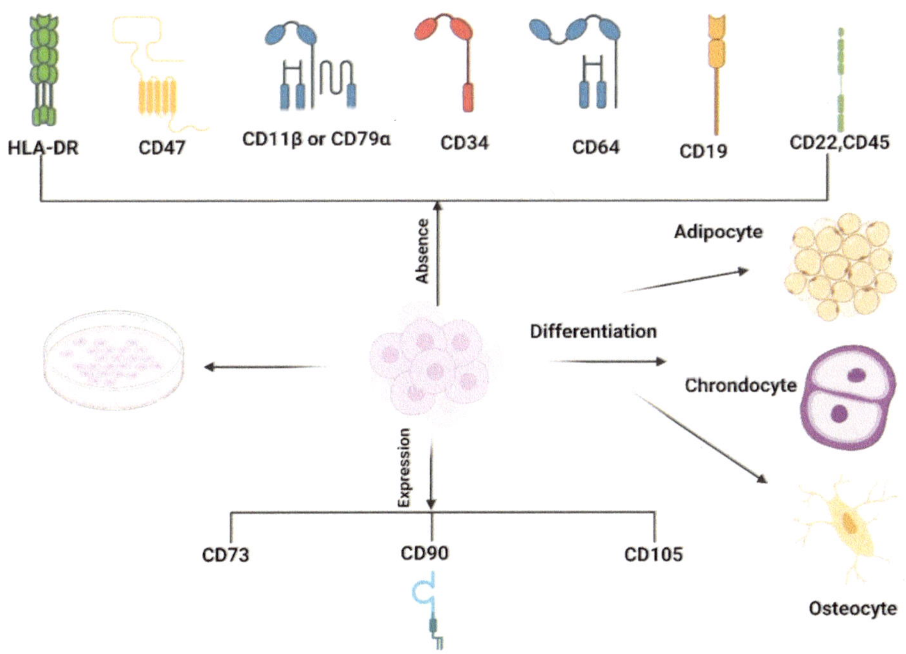

FIGURE 5.7 Mesenchymal stem cell therapy.

5.4.3.2 Immune Cell Therapies

Cell-based approaches for wound healing have utilized hydrogel scaffolds for the improved delivery of stem cells, providing a physiologic microenvironment and protection from native immune cells, emphasizing the need for scaffolds to aid cell therapy efficacy (243). Transcriptional space-time mapping has identified intricate coordination of immune and stromal cell patterns and gene programs, demonstrating the complex interaction of different cell types in healing (244). The deregulated recruitment of neutrophils and macrophages orchestrated by factors like FOXM1 and STAT3 in diabetic foot ulcers highlights a key role of these elements in immune dysfunction affecting wound healing (245). Single-cell analysis revealed the influence of Angptl4 on the immune cell response during wound healing. This includes coordinating neutrophil removal and resolving inflammation through an Angptl4-ifi202b axis, thus contributing to immune regulation during wound repair (246). Stem cell-derived exosomes have emerged as a promising option for wound healing by shaping the activity of immune cells and other skin cells during the healing process (247). Monocytes and lymphocytes are crucial in both physiologic wound healing and the impaired mechanisms observed in diabetic wound repair, impacting macrophage transition and ultimately influencing wound outcomes (248). STING activation is noted to inhibit diabetic wound healing by inducing M1 polarization of macrophages, thereby exacerbating inflammation and impeding healing (249). Mesenchymal stem cell-derived extracellular vesicles are advantageous over parental stem cells due to their high safety profile. They offer immune regulation, reduced embolism risk, and decreased risk of carcinogenesis, which is particularly beneficial for diabetic wound healing (250). These studies offer insight into the diverse roles and mechanisms through which immune cell therapies can modulate the wound healing process, providing innovative strategies to combat immune dysfunction in wounds.

5.5 CLINICAL APPLICATIONS AND FUTURE DIRECTIONS

One of the main challenges in chronic wound healing is the persistent inflammation that impairs tissue regeneration and increases the risk of infection. Chronic wounds are often associated with underlying conditions such as diabetes, obesity, or vascular diseases that affect the immune system and impair blood flow. To overcome these challenges, novel therapeutic targets have been identified that aim to reduce inflammation, enhance angiogenesis, stimulate growth factors, and recruit stem cells to the wound site. Some of these targets include cytokines, chemokines, toll-like receptors, and extracellular matrix components that regulate the function of immune cells such as macrophages, neutrophils, and lymphocytes. Acute wounds, on the other hand, usually heal without complications but often result in scar formation that can impair the function and appearance of the skin. Scarless healing is a desirable outcome for acute wounds, especially for those affecting visible areas such as the face. Scarless healing involves the restoration of the original skin structure and function without fibrosis or contraction. To achieve this goal, strategies to modulate immune responses to promote skin regeneration have been explored. These strategies include manipulating the balance between pro-inflammatory and anti-inflammatory cytokines, enhancing the resolution of inflammation, promoting the differentiation of stem cells into skin cells, and modulating the extracellular matrix remodeling.

The clinical applications and future directions of these strategies depend on several factors, such as their safety, efficacy, cost-effectiveness, and delivery methods. Some of these strategies have shown promising results in animal models but have not yet been translated into human trials or clinical practice. Moreover, some of these strategies may have adverse effects on other aspects of wound healing or immune function, such as infection resistance or autoimmune reactions. Therefore, more research is needed to understand the molecular mechanisms and interactions between the immune system and the skin during wound healing and to design novel regenerative therapies that can modulate immune responses in a specific and controlled manner (251–256).

5.5.1 Current Approaches in Wound Care

However, chronic wounds, such as diabetic ulcers, venous leg ulcers, and pressure ulcers, are characterized by persistent inflammation and impaired healing, which can lead to tissue loss, infection, and reduced quality of life (257). Therefore, strategies to modulate immune responses to promote wound healing are of great interest in the treatment of chronic wounds. Some of the current approaches in wound care include the use of biomaterials, growth factors, stem cells, gene therapy, and immunotherapy to modulate the immune system and enhance tissue regeneration (258). For example, biomaterials can be designed to mimic the extracellular matrix and provide signals that recruit and activate immune cells, such as macrophages, that can facilitate wound closure and tissue repair (259). Growth factors, such as platelet-derived growth factor (PDGF), vascular endothelial growth factor (VEGF), and transforming growth factor-beta (TGF-β), can be delivered to the wound site to stimulate angiogenesis, cell proliferation, and matrix synthesis (260). Stem cells, such as mesenchymal stem cells (MSCs), can be transplanted into the wound to modulate inflammation, promote angiogenesis, and differentiate into various cell types (261). Gene therapy can be used to introduce genes that encode for beneficial proteins, such as growth factors, cytokines, or anti-inflammatory molecules, into the wound cells or stem cells (262). Immunotherapy can be used to target specific immune cells or molecules that are involved in wound healing, such as macrophages, neutrophils, T cells, or toll-like receptors (TLRs). These approaches aim to restore the balance between pro-inflammatory and anti-inflammatory signals in the wound microenvironment and enhance the healing outcome (263).

5.5.2 Challenges and Opportunities in Immune Modulation

We will briefly review some of the strategies to modulate immune responses to promote wound healing and the current research gaps and directions. One of the main strategies to modulate immune responses to promote wound healing is to use biomaterials that can interact with immune cells and factors, such as cytokines, chemokines, and growth factors. Biomaterials can be designed to mimic the extracellular matrix, provide structural support, deliver drugs or cells, or induce specific immune responses. For example, biomaterials can be used to recruit or activate macrophages, which are key orchestrators of wound healing, and modulate their polarization from pro-inflammatory (M1) to anti-inflammatory (M2) phenotypes. Biomaterials can also be used to manipulate other immune cells, such as neutrophils, T cells, natural killer cells, and dendritic cells, that are involved in wound healing. Another strategy to modulate immune responses to promote wound healing is to use stem cells or their derivatives, such as exosomes or conditioned media. Stem cells have the potential to differentiate into various cell types and secrete paracrine factors that can modulate inflammation, angiogenesis, and tissue regeneration. Stem cells can also interact with immune cells and modulate their functions. For example, mesenchymal stem cells (MSCs) have been shown to suppress neutrophil activity, induce macrophage polarization, inhibit T cell proliferation, and modulate dendritic cell maturation. Stem cells can be delivered alone or in combination with biomaterials to enhance their retention and efficacy. Despite the promising results of these strategies in preclinical and clinical studies, there are still many challenges and opportunities in immune modulation for wound healing. Some of the challenges include understanding the complex interactions between different immune cell populations and factors in different wound types and stages, optimizing the delivery methods and dosages of biomaterials and stem cells, ensuring the safety and biocompatibility of these interventions, and developing standardized protocols and outcome measures for evaluation. Some of the opportunities include exploring novel biomaterials and stem cell sources that can modulate specific immune pathways or targets, developing smart or responsive biomaterials that can sense and adapt to the wound environment, combining different strategies to achieve synergistic effects, and personalizing the interventions based on the patient's immune status and wound characteristics (264–276).

5.5.3 Future Directions for Research and Development

Here are some of the latest future directions for research and development in this field.

5.5.3.1 Targeting Immune Cells in Wound Healing

Future research aims to better understand the role of specific immune cell types, such as macrophages, neutrophils, and regulatory T cells, in wound healing. Strategies for targeting these cells through immunomodulatory therapies could help promote optimal healing outcomes.

5.5.3.2 Biomaterials for Immune Modulation

The development of biomaterials that can modulate immune responses at the wound site is a promising direction for research. These materials could be designed to promote a pro-regenerative immune environment and enhance wound healing.

5.5.3.3 Precision Medicine Approaches

Personalized medicine approaches could tailor wound healing interventions based on individual immune profiles and wound characteristics. Utilizing technologies such as genomics, proteomics, and machine learning may help predict patient responses to immunomodulatory therapies.

5.5.3.4 Combination Therapies

Future research may explore the use of combination therapies that target multiple aspects of the immune response in wound healing. Combinations of immunomodulatory agents, growth factors, and biomaterials could provide synergistic effects and enhance therapeutic outcomes (277–280). These future directions for research and development in wound healing highlight the importance of understanding immune responses and leveraging immune modulation strategies to promote optimal healing outcomes. Further studies in these areas may lead to innovative therapies that improve wound healing processes and outcomes.

5.6 CONCLUSION

The immune system and how it interacts with the tissue environment are acknowledged as key factors in regulating wound healing in a variety of skin conditions, including diabetes, ulcers, burn injuries, skin cancer, and autoimmune skin abnormalities. ECM is crucial in controlling the immune system and how it functions during the wound healing process. The interplay between cells and their surroundings renders the extracellular matrix (ECM) a dynamic mediator for the optimal operation of immune cells. Additionally, the inherent immunomodulatory domains of ECM compositions interact with immune cell receptors to regulate immune cell function. The best approach to treat autoimmune illnesses is to use immune-regulating factors because of the significance of immune system components and how they interact with ECM. Biomaterial-based structures are one of the things that can stimulate or inhibit the immune system. Different biomaterials, both natural and artificial, can interact with immune cells in a controlled way and elicit distinct immunological responses depending on their chemical, mechanical, side group, and morphological characteristics. Coating carboxylic groups on the surface of biomaterials, for instance, has the ability to modify immune cell responses and aid in the healing of chronic wounds. Furthermore, polymeric skin patches and MN patches have the ability to precisely release bioactive molecules like cytokines and immunomodulators, which can significantly change how the immune system functions in autoimmune illnesses and skin ulcers. The technique of cell administration is one of the novel approaches to suppress or modulate the immune system. By injecting stem cells, such as MSCs, or loading cells into biomaterials, immune cells can modulate the skin's immune system and signal regulation. Recent research has shown the critical functions that the immune system plays in the healing process of wounds. Nonetheless, a number of widely used therapies, such as growth factors and cytokines, cell-based therapies, immune cell therapies, controlled release bioactive molecules, scaffolds for immune cell recruitment, and immunomodulation strategies, have successfully repaired skin defects while also hastening the healing of autoimmune skin diseases, diabetic wounds, and skin cancer. Using immune system-based tactics, certain strategies that have failed in clinical trials in the past might succeed.

REFERENCES

1. Roth J, Harré EM, Rummel C, Gerstberger R, Hübschle T. Signaling the brain in systemic inflammation: Role of sensory circumventricular organs. *Front Biosci.* 2004;9:290–300.
2. Nurden AT, Nurden P, Sanchez M, Andia I, Anitua E. Platelets and wound healing. *Front Biosci.* 2008;13(9):3532–48.
3. Messadi D V., Le A, Berg S, Huang G, Zhuang W, Bertolami CN. Effect of TGF-beta 1 on PDGF receptors expression in human scar fibroblasts. *Front Biosci.* 1998;3.
4. Barcia C, Emborg ME, Hirsch EC, Herrero MT. Blood vessels and parkinsonism. *Front Biosci.* 2004;9:277–82.
5. Castellone MD, Laukkanen MO. TGF-beta1, WNT, and SHH signaling in tumor progression and in fibrotic diseases. *Front Biosci—Scholar.* 2017 Jan 1;9(1):31–45.
6. Diegelmann RF, Evans MC. Wound healing: An overview of acute, fibrotic and delayed healing. *Front Biosci* [Internet]. 2004 Jan 1 [cited 2024 Feb 23];9(1):283–9. Available from: www.imrpress.com/journal/FBL/9/1/10.2741/1184
7. Wight TN. Arterial remodeling in vascular disease: A key role for hyaluronan and versican. *Front Biosci.* 2008;13(13):4933–7.
8. Sen CK, Gordillo GM, Roy S, Kirsner R, Lambert L, Hunt TK, et al. Human skin wounds: A major and snowballing threat to public health and the economy: Perspective article. *Wound Repair Regen.* 2009 Nov;17(6):763–71.
9. Stojadinovic O, Pastar I, Vukelic S, Mahoney MG, Brennan D, Krzyzanowska A, et al. Deregulation of keratinocyte differentiation and activation: A hallmark of venous ulcers. *J Cell Mol Med.* 2008 Dec;12(6B):2675–90.
10. Eming SA, Krieg T, Davidson JM. Inflammation in wound repair: Molecular and cellular mechanisms. *J Invest Dermatol.* 2007;127(3):514–25.
11. Eming SA, Martin P, Tomic-Canic M. Wound repair and regeneration: Mechanisms, signaling, and translation. *Sci Transl Med.* 2014 Dec 3 [cited 2024 Feb 23];6(265). Available from: www.science.org/doi/10.1126/scitranslmed.3009337
12. Brem H, Tomic-Canic M. Cellular and molecular basis of wound healing in diabetes. *J Clin Investig.* 2007 May 1;117(5):1219–22.
13. Wilgus TA, Roy S, McDaniel JC. Neutrophils and wound repair: Positive actions and negative reactions. *Adv Wound Care (New Rochelle).* 2013 Sep [cited 2024 Feb 23];2(7):379. Available from: /pmc/articles/PMC3763227/
14. Hendrijantini N, Kuntjoro M, Agustono B, Maya Sitalaksmi R, Dimas Aditya Ari M, Theodora M, et al. Human umbilical cord mesenchymal stem cells induction in peri-implantitis Rattus norvegicus accelerates and enhances osteogenesis activity and implant osseointegration. *Saudi Dent J.* 2023 Feb 1 [cited 2024 Feb 23];35(2):147–53. Available from: www.ncbi.nlm.nih.gov/pubmed/36942204
15. Liu W, Zhang X. Receptor activator of nuclear factor-κB ligand (RANKL)/RANK/osteoprotegerin system in bone and other tissues (Review). *Mol Med Rep.* 2015 May 1;11(5):3212–8.
16. Leibbrandt A, Penninger JM. Novel functions of RANK(L) signaling in the immune system. *Adv Exp Med Biol.* 2010 [cited 2024 Feb 23];658:77–94. Available from: https://pubmed.ncbi.nlm.nih.gov/19950018/
17. Serhan CN. Pro-resolving lipid mediators are leads for resolution physiology. *Nature.* 2014 [cited 2024 Feb 23];510(7503):92–101. Available from: https://pubmed.ncbi.nlm.nih.gov/24899309/
18. Islamuddin M, Mustfa SA, Ullah SNMN, Omer U, Kato K, Parveen S. Innate immune response and inflammasome activation during SARS-CoV-2 infection. *Inflammation.* 2022 Oct 1 [cited 2024 Feb 23];45(5):1849. Available from: /pmc/articles/PMC9371632/
19. Göschl L, Scheinecker C, Bonelli M. Treg cells in autoimmunity: From identification to Treg-based therapies. *Semin Immunopathol.* 2019 May 1;41(3):301–14.
20. Brown CY, Sadlon T, Hope CM, Wong YY, Wong S, Liu N, et al. Molecular Insights Into Regulatory T-Cell Adaptation to Self, Environment, and Host Tissues: Plasticity or Loss of Function in Autoimmune Disease. *Front Immunol.* 2020 Sep 15;11.

21. Qiu R, Zhou L, Ma Y, Zhou L, Liang T, Shi L, et al. Regulatory T Cell Plasticity and Stability and Autoimmune Diseases. *Clin Rev Allergy Immunol.* 2020 Feb 1;58(1):52–70.
22. Zhang X, Olsen N, Zheng SG. The progress and prospect of regulatory T cells in autoimmune diseases. *J Autoimmun.* 2020 Jul 1 [cited 2024 Feb 23];111. Available from: https://pubmed.ncbi.nlm.nih.gov/32305296/
23. Zickert A, Janković MR, Malmström V, Chemin K, Gunnarsson I. Occurrence and localization of FOXP3 + cells in kidney biopsies in lupus nephritis and ANCA-associated vasculitis. *Clin Rheumatol.* 2023 Oct 1;42(10):2889–95.
24. Sadlik J, Kosińska E, Słota D, Niziołek K, Tomala A, Włodarczyk M, et al. Bioactive Hydrogel Based on Collagen and Hyaluronic Acid Enriched with Freeze-Dried Sheep Placenta for Wound Healing Support. *Int J Mol Sci.* 2024 Feb 1;25(3).
25. Alkhilani MA, Hammoodi OT, Emran HA, Alhayani WA. Impact of using processed urinary bladder submucosa and hydrogel fabricated from tendon on skin healing process in rabbits. Al-Karagoly H, editor. *Vet Med Int.* 2024 Feb 6 [cited 2024 Feb 23];2024:6641975. Available from: www.ncbi.nlm.nih.gov/pubmed/38352052
26. Reinke JM, Sorg H. Wound repair and regeneration. *Eur Surg Res.* 2012 Aug [cited 2024 Feb 23];49(1):35–43. Available from: https://pubmed.ncbi.nlm.nih.gov/22797712/
27. Delavary BM, van der Veer WM, van Egmond M, Niessen FB, Beelen RHJ. Macrophages in skin injury and repair. *Immunobiology.* 2011 Jul;216(7):753–62.
28. Eming SA, Martin P, Tomic-Canic M. Wound repair and regeneration: Mechanisms, signaling, and translation. *Sci Transl Med.* 2014 Dec 3 [cited 2024 Feb 23];6(265). Available from: https://pubmed.ncbi.nlm.nih.gov/25473038/
29. Kaplani K, Koutsi S, Armenis V, Skondra FG, Karantzelis N, Champeris Tsaniras S, et al. Wound healing related agents: Ongoing research and perspectives. *Adv Drug Deliv Rev.* 2018 Apr 1;129:242–53.
30. Wohlsein JC, Meurer M, Mörgelin M, Nessler JN, Flegel T, Schenk HC, et al. Neutrophil extracellular traps in CSF and serum of dogs with steroid-responsive meningitis-arteritis. Corridon PR, editor. *PLoS One.* 2024 [cited 2024 Feb 23];19(1):e0295268. Available from: www.ncbi.nlm.nih.gov/pubmed/38241272
31. Yang L, He C, Wang W. Association between neutrophil to high-density lipoprotein cholesterol ratio and disease severity in patients with acute biliary pancreatitis. *Ann Med.* 2024 Dec [cited 2024 Feb 23];56(1):2315225. Available from: www.ncbi.nlm.nih.gov/pubmed/38335727
32. Medhasi S, Sriwarom A, Permpalung N, Torvorapanit P, Plongla R, Chindamporn A, et al. Ex vivo observation of Pythium insidiosum-antigen treated neutrophils on three Pythium insidiosum strains isolated from vascular pythiosis patients. *Hum Vaccin Immunother.* 2024 Dec 31 [cited 2024 Feb 23];20(1):2304372. Available from: www.ncbi.nlm.nih.gov/pubmed/38314761
33. Nathan C. Neutrophils and immunity: Challenges and opportunities. *Nat Rev Immunol.* 2006 Mar [cited 2024 Feb 23];6(3):173–82. Available from: https://pubmed.ncbi.nlm.nih.gov/16498448/
34. Ng MTH, Borst R, Gacaferi H, Davidson S, Ackerman JE, Johnson PA, et al. A single cell atlas of frozen shoulder capsule identifies features associated with inflammatory fibrosis resolution. *Nat Commun.* 2024 Feb 19 [cited 2024 Feb 23];15(1):1394. Available from: www.ncbi.nlm.nih.gov/pubmed/38374174
35. Oishi Y, Manabe I. Macrophages in inflammation, repair and regeneration. *Int Immunol.* 2018 Oct 29;30(11):511–28.
36. Wynn TA, Vannella KM. Macrophages in tissue repair, regeneration, and fibrosis. *Immunity.* 2016 Mar 15 [cited 2024 Feb 23];44(3):450–62. Available from: https://pubmed.ncbi.nlm.nih.gov/26982353/
37. Lafuse WP, Wozniak DJ, Rajaram MVS. Role of cardiac macrophages on cardiac inflammation, fibrosis and tissue repair. *Cells.* 2021 Dec 1;10(1):1–27.
38. Yuan X, Xin T, Yu H, Huang J, Xu Y, Ou C, et al. Transcription factor IRF7 is involved in psoriasis development and response to guselkumab treatment. *J Inflamm Res.* 2024 Feb [cited 2024 Feb 23];17:1039–55. Available from: www.ncbi.nlm.nih.gov/pubmed/38375022
39. Gupta A, Jamal T, Rajbhar P, Gaur AS, Chauhan SS, Parthasarathi R. Cytokines inhibitory mechanism of Prunus domestica L. (Plum) peptides as potential immunomodulators against systemic lupus erythematosus: An in-silico screening. *Silico Pharmacol.* 2024 Feb 14 [cited 2024 Feb 23];12(1):12. Available from: www.ncbi.nlm.nih.gov/pubmed/38370860
40. Gaffen SL, Jain R, Garg A V., Cua DJ. The IL-23-IL-17 immune axis: From mechanisms to therapeutic testing. *Nat Rev Immunol.* 2014 [cited 2024 Feb 23];14(9):585–600. Available from: https://pubmed.ncbi.nlm.nih.gov/25145755/
41. Nies JF, Panzer U. IL-17C/IL-17RE: Emergence of a unique axis in TH17 biology. *Front Immunol.* 2020 Feb 26;11.

42. Schilrreff P, Alexiev U. Chronic inflammation in non-healing skin wounds and promising natural bioactive compounds treatment. *Int J Mol Sci*. 2022 Apr 28 [cited 2024 Feb 23];23(9):4928. Available from: www.mdpi.com/1422-0067/23/9/4928/htm
43. Virtanen MI, Brinchmann MF, Patel DM, Iversen MH. Chronic stress negatively impacts wound healing, welfare, and stress regulation in internally tagged Atlantic salmon (Salmo salar). *Front Physiol*. 2023;14.
44. Jiang Y, Xu X, Xiao L, Wang L, Qiang S. The role of microRNA in the inflammatory response of wound healing. *Front Immunol*. 2022 Mar 21 [cited 2024 Feb 23];13:852419. Available from: www.frontiersin.org
45. Li D, Wu N. Mechanism and application of exosomes in the wound healing process in diabetes mellitus. *Diabetes Res Clin Pract*. 2022 May 1;187:109882.
46. Ogle ME, Segar CE, Sridhar S, Botchwey EA. Monocytes and macrophages in tissue repair: Implications for immunoregenerative biomaterial design. *Exp Biol Med (Maywood)*. 2016 May 1 [cited 2024 Feb 24];241(10):1084–97. Available from: https://pubmed.ncbi.nlm.nih.gov/27229903/
47. Martin KE, García AJ. Macrophage phenotypes in tissue repair and the foreign body response: Implications for biomaterial-based regenerative medicine strategies. *Acta Biomater*. 2021 Oct 1;133:4–16.
48. Shrestha S, McFadden MJ, Gramolini AO, Santerre JP. Proteome analysis of secretions from human monocyte-derived macrophages post-exposure to biomaterials and the effect of secretions on cardiac fibroblast fibrotic character. *Acta Biomater*. 2020 Jul 15;111:80–90.
49. Chen XY, Wang TT, Shen Q, Ma H, Li ZH, Yu XN, et al. Preclinical investigations on anti-fibrotic potential of long-term oral therapy of sodium astragalosidate in animal models of cardiac and renal fibrosis. *ACS Pharmacol Transl Sci*. 2024 Feb 9 [cited 2024 Feb 24];7(2):421–31. Available from: www.ncbi.nlm.nih.gov/pubmed/38357273
50. Diwan R, Bhatt HN, Beaven E, Nurunnabi M. Emerging delivery approaches for targeted pulmonary fibrosis treatment. *Adv Drug Deliv Rev*. 2024 Jan 1;204.
51. Hinz B, Phan SH, Thannickal VJ, Galli A, Bochaton-Piallat ML, Gabbiani G. The myofibroblast: One function, multiple origins. *Am J Pathol*. 2007 [cited 2024 Feb 24];170(6):1807–16. Available from: https://pubmed.ncbi.nlm.nih.gov/17525249/
52. Rosenstein RK, Rose JJ, Brooks SR, Tsai WL, Gadina M, Pavletic SZ, et al. Identification of fibroinflammatory and fibrotic transcriptomic subsets of human cutaneous sclerotic chronic graft-versus-host disease. JID Innovations: Skin science from molecules to population health. 2024 Mar [cited 2024 Feb 24];4(2):100246. Available from: www.ncbi.nlm.nih.gov/pubmed/38357212
53. Huang E, Peng N, Xiao F, Hu D, Wang X, Lu L. The roles of immune cells in the pathogenesis of fibrosis. *Int J Mol Sci*. 2020 Aug 1;21(15):1–27.
54. Antar SA, Ashour NA, Marawan ME, Al-Karmalawy AA. Fibrosis: Types, effects, markers, mechanisms for disease progression, and its relation with oxidative stress, immunity, and inflammation. *Int J Mol Sci*. 2023 Feb 1;24(4).
55. Cao Y, Harvey BP, Jin L, Westmoreland S, Wang J, Puri M, et al. Therapeutic TNF Inhibitors exhibit differential levels of efficacy in accelerating cutaneous wound healing. JID innovations: Skin science from molecules to population health. 2024 Jan [cited 2024 Feb 24];4(1):100250. Available from: www.ncbi.nlm.nih.gov/pubmed/38226320
56. Larouche J, Sheoran S, Maruyama K, Martino MM. Immune regulation of skin wound healing: Mechanisms and novel therapeutic targets. *Adv Wound Care (New Rochelle)*. 2018 Jul 1 [cited 2024 Feb 24];7(7):209–31. Available from: https://pubmed.ncbi.nlm.nih.gov/29984112/
57. Arenas Gómez CM, Sabin KZ, Echeverri K. Wound healing across the animal kingdom: Crosstalk between the immune system and the extracellular matrix. *Dev Dynam*. 2020 Jul 1;249(7):834–46.
58. Shashank MPVB, Chandana KKSSSMT, Prithvi SS et al. Sustainable development of plant tissue culture industry: The Indian scenario. *J Appl Biol Biotechnol*. 2021 Mar 10.
59. Raffa P, Easler M, Urciuolo A. Three-dimensional in vitro models of neuromuscular tissue. *Neural Regen Res*. 2022 Apr 1 [cited 2024 Feb 24];17(4):759. Available from: /pmc/articles/PMC8530117/
60. Jimi N, Hookabe N, Moritaki T, Kimura T, Imura S. First evidence of male dwarfism in scale worms: A new species of Polynoidae (Annelida) from hermit crab and molluscan shells. *J Zool Syst Evol Res*. 2021 May 1;59(4):801–18.
61. Renò F, Sabbatini M. Breaking a vicious circle: Lymphangiogenesis as a new therapeutic target in wound healing. *Biomedicines*. 2023 Mar 1 [cited 2024 Feb 24];11(3). Available from: https://pubmed.ncbi.nlm.nih.gov/36979635/
62. Swoboda L, Held J. Impaired wound healing in diabetes. *J Wound Care*. 2022 Oct 2;31(10):882–5.

63. He Y, Cen Y, Tian M. Immunomodulatory hydrogels for skin wound healing: Cellular targets and design strategy. *J Mater Chem B*. 2024 [cited 2024 Feb 27]. Available from: https://pubs.rsc.org/en/content/articlehtml/2024/tb/d3tb02626d
64. Nitzsche A, Poittevin M, Benarab A, Bonnin P, Faraco G, Uchida H, et al. Endothelial S1P1 signaling counteracts infarct expansion in ischemic stroke. *Circ Res*. 2021 Feb 2 [cited 2024 Feb 27];128(3):363. Available from: /pmc/articles/PMC7874503/
65. Zhang Z, Ji C, Wang D, Wang M, Song D, Xu X, et al. The burden of diabetes on the soft tissue seal surrounding the dental implants. *Front Physiol*. 2023 Feb 16;14:1136973.
66. Shafqat A, Abdul Rab S, Ammar O, Al Salameh S, Alkhudairi A, Kashir J, et al. Emerging role of neutrophil extracellular traps in the complications of diabetes mellitus. *Front Med (Lausanne)*. 2022 Aug 23;9:995993.
67. Sałata P, Szczuraszek P, Szczuraszek H, Bętkowska P, Tomkiewicz M, Licak G, et al. Nutrition of neonates and infants with severe congenital heart disease—the review of the literature. *Quality in Sport*. 2023 Mar 30 [cited 2024 Feb 27];10(1):52–61. Available from: https://apcz.umk.pl/QS/article/view/42897
68 .Leal EC, Carvalho E. Heme oxygenase-1 as therapeutic target for diabetic foot ulcers. *Int J Mol Sci*. 2022 Oct 10 [cited 2024 Feb 27];23(19):12043. Available from: www.mdpi.com/1422-0067/23/19/12043/htm
69. Hackl C, Schlitt HJ, Renner P, Lang SA. Liver surgery in cirrhosis and portal hypertension. *World J Gastroenterol*. 2016 Mar 7;22(9):2725–35.
70. Balikji J, Koyun AH, Hendriksen PA, Kiani P, Stock AK, Garssen J, et al. The impact of COVID-19 lockdowns in Germany on mood, attention control, immune fitness, and quality of life of young adults with self-reported impaired wound healing. *J Clin Med*. 2023 Apr 29 [cited 2024 Feb 27];12(9):3205. Available from: www.mdpi.com/2077-0383/12/9/3205/htm
71. Balikji J, Hoogbergen MM, Garssen J, Roth T, Verster JC. Insomnia complaints and perceived immune fitness in young adults with and without self-reported impaired wound healing. *Medicina*. 2022 Aug 4 [cited 2024 Feb 27];58(8):1049. Available from: www.mdpi.com/1648-9144/58/8/1049/htm
72. Zhang XT, Mao ZY, Jin XY, Wang YG, Dong YQ, Zhang C. Identification of a tsRNA contributor to impaired diabetic wound healing via high glucose-induced endothelial dysfunction. *Diabetes, Metabolic Syndrome and Obesity*. 2023 Jan 31 [cited 2024 Feb 27];16:285–98. Available from: www.dovepress.com/identification-of-a-tsrna-contributor-to-impaired-diabetic-wound-heali-peer-reviewed-fulltext-article-DMSO
73. Bu Y, Shih KC, Wong HL, Kwok SS, Lo ACY, Chan JYK, et al. The association between altered intestinal microbiome, impaired systemic and ocular surface immunity, and impaired wound healing response after corneal alkaline-chemical injury in diabetic mice. *Front Immunol*. 2023 Jan 31;14:1063069.
74. Park DJ, Duggan E, Ho K, Dorschner RA, Dobke M, Nolan JP, et al. Serpin-loaded extracellular vesicles promote tissue repair in a mouse model of impaired wound healing. *J Nanobiotechnology*. 2022 Dec 1 [cited 2024 Feb 27];20(1):1–17. Available from: https://jnanobiotechnology.biomedcentral.com/articles/10.1186/s12951-022-01656-7
75. Zhang Y, Lu Q. Immune cells in skin inflammation, wound healing, and skin cancer. *J Leukoc Biol*. 2023 Sep 18 [cited 2024 Feb 27]. Available from: https://dx.doi.org/10.1093/jleuko/qiad107
76. Rehak L, Giurato L, Meloni M, Panunzi A, Manti GM, Uccioli L. The immune-centric revolution in the diabetic foot: Monocytes and lymphocytes role in wound healing and tissue regeneration—A narrative review. *J Clin Med*. 2022 Feb 8 [cited 2024 Feb 27];11(3):889. Available from: www.mdpi.com/2077-0383/11/3/889/htm
77. Liu Z, Zhang Y, Youn JY, Zhang Y, Makino A, Yuan JXJ, et al. Flavored and nicotine-containing e-cigarettes induce impaired angiogenesis and diabetic wound healing via increased endothelial oxidative stress and reduced NO bioavailability. *Antioxidants*. 2022 May 5 [cited 2024 Feb 27];11(5):904. Available from: www.mdpi.com/2076-3921/11/5/904/htm
78. Selig D, Caridha D, Evans M, Kress A, Lanteri C, Ressner R, et al. Animal models in regulatory breakpoint determination: Review of new drug applications of approved antibiotics from 2014–2022. *J Pers Med*. 2024 Jan 19 [cited 2024 Feb 27];14(1). Available from: www.ncbi.nlm.nih.gov/pubmed/38276233
79. Aksamitiene E, Heffelfinger RN, Hoek JB, Pribitkin E deAzevedo. Standardized pre-clinical surgical animal model protocol to investigate the cellular and molecular mechanisms of ischemic flap healing. *Biol Proced Online*. 2024 Jan 17 [cited 2024 Feb 27];26(1):2. Available from: www.ncbi.nlm.nih.gov/pubmed/38229030
80. Mukherjee P, Roy S, Ghosh D, Nandi SK. Role of animal models in biomedical research: A review. *Lab Anim Res*. 2022 Dec 1 [cited 2024 Feb 27];38(1). Available from: https://pubmed.ncbi.nlm.nih.gov/35778730/

81. Da Silva-Álvarez S, Guerra-Varela J, Sobrido-Cameán D, Quelle A, Barreiro-Iglesias A, Sánchez L, et al. Cell senescence contributes to tissue regeneration in zebrafish. *Aging Cell.* 2020 Jan 1;19(1).
82. Albanesi C, Madonna S, Gisondi P, Girolomoni G. The interplay between keratinocytes and immune cells in the pathogenesis of psoriasis. *Front Immunol.* 2018 Jul 6;9(JUL).
83. Freund A, Orjalo A V., Desprez PY, Campisi J. Inflammatory networks during cellular senescence: Causes and consequences. *Trends Mol Med.* 2010 May;16(5):238–46.
84. Wilkinson HN, Hardman MJ. Senescence in wound repair: Emerging strategies to target chronic healing wounds. *Front Cell Dev Biol.* 2020 Aug 11 [cited 2024 Feb 27];8. Available from: /pmc/articles/PMC7431694/
85. Gualandris A, Noghero A, Cora' D, Astanina E, Arese M, Bussolino F. Role of TGFbeta1 and WNT6 in FGF2 and BMP4-driven endothelial differentiation of murine embryonic stem cells. *Angiogenesis.* 2022 Feb 1;25(1):113–28.
86. Dudley AC, Griffioen AW. Pathological angiogenesis: Mechanisms and therapeutic strategies. *Angiogenesis.* 2023 Apr 15 [cited 2024 Feb 27];26(3):313–47. Available from: https://link.springer.com/article/10.1007/s10456-023-09876-7
87. Francis CR, Kushner EJ. Trafficking in blood vessel development. *Angiogenesis.* 2022 Aug 1;25(3):291–305.
88. Resnik SR, Egger A, Abdo Abujamra B, Jozic I. Clinical implications of cellular senescence on wound healing. *Current Dermatology Reports.* 2020 Nov 7 [cited 2024 Feb 27];9(4):286–97. Available from: https://link.springer.com/article/10.1007/s13671-020-00320-3
89. Andrade AM, Sun M, Gasek NS, Hargis GR, Sharafieh R, Xu M. Role of senescent cells in cutaneous wound healing. *Biology.* 2022 Nov 29 [cited 2024 Feb 27];11(12):1731. Available from: www.mdpi.com/2079-7737/11/12/1731/htm
90. Wu YK, Cheng NC, Cheng CM. Biofilms in chronic wounds: Pathogenesis and diagnosis. *Trends Biotechnol.* 2019 May 1 [cited 2024 Feb 27];37(5):505–17. Available from: https://pubmed.ncbi.nlm.nih.gov/30497871/
91. Versey Z, da Cruz Nizer WS, Russell E, Zigic S, DeZeeuw KG, Marek JE, et al. Biofilm-innate immune interface: Contribution to chronic wound formation. *Front Immunol.* 2021 Apr 9 [cited 2024 Feb 27];12. Available from: /pmc/articles/PMC8062706/
92. Dight J, Zhao J, Styke C, Khosrotehrani K, Patel J. Resident vascular endothelial progenitor definition and function: The age of reckoning. *Angiogenesis.* 2022 Feb 1;25(1):15–33.
93. Graham K, Unger E. Overcoming tumor hypoxia as a barrier to radiotherapy, chemotherapy and immunotherapy in cancer treatment. *Int J Nanomedicine.* 2018;13:6049–58.
94. Pasut A, Becker LM, Cuypers A, Carmeliet P. Endothelial cell plasticity at the single-cell level. *Angiogenesis.* 2021 May 1;24(2):311–26.
95. Aplin AC, Nicosia RF. Tissue oxygenation stabilizes neovessels and mitigates hemorrhages in human atherosclerosis-induced angiogenesis. *Angiogenesis.* 2022 Feb 1;26(1):63.
96. Abiko Y, Selimovic D. The mechanism of protracted wound healing on oral mucosa in diabetes. *Rev Biomol Biomed.* 2010 Aug 20 [cited 2024 Feb 28];10(3):186–91. Available from: www.bjbms.org/ojs/index.php/bjbms/article/view/2683
97. Catrina SB, Zheng X. Disturbed hypoxic responses as a pathogenic mechanism of diabetic foot ulcers. *Diabetes Metab Res Rev.* 2016 Jan 1 [cited 2024 Feb 28];32:179–85. Available from: https://onlinelibrary.wiley.com/doi/full/10.1002/dmrr.2742
98. Prakashan D, Roberts A, Gandhi S. Recent advancement of nanotherapeutics in accelerating chronic wound healing process for surgical wounds and diabetic ulcers. *Biotechnol Genet Eng Rev.* 2023 [cited 2024 Feb 28]. Available from: https://pubmed.ncbi.nlm.nih.gov/36641600/
99. Huang Y, Ding Y, Wang B, Ji Q, Peng C, Tan Q. Neutrophils extracellular traps and ferroptosis in diabetic wounds. *Int Wound J.* 2023 Nov 1 [cited 2024 Feb 28];20(9):3840–54. Available from: https://onlinelibrary.wiley.com/doi/full/10.1111/iwj.14231
100. Wan X, Chen Y, Geng F, Sheng Y, Wang F, Guo J. Narrative review of the mechanism of natural products and scar formation in wound repair. *Ann Transl Med.* 2022 Feb [cited 2024 Feb 28];10(4):236–236. Available from: https://atm.amegroups.org/article/view/90324/html
101. Kim SA, Ryu HW, Lee KS, Cho JW. Application of platelet-rich plasma accelerates the wound healing process in acute and chronic ulcers through rapid migration and upregulation of cyclin A and CDK4 in HaCaT cells. *Mol Med Rep.* 2013 Feb;7(2):476–80.
102. Kita A, Saito Y, Miura N, Miyajima M, Yamamoto S, Sato T, et al. Altered regulation of mesenchymal cell senescence in adipose tissue promotes pathological changes associated with diabetic wound healing. *Communications Biology.* 2022 Apr 5 [cited 2024 Feb 28];5(1):1–16. Available from: www.nature.com/articles/s42003-022-03266-3

103. Yang P, Pei Q, Yu T, Chang Q, Wang D, Gao M, et al. Compromised wound healing in ischemic type 2 diabetic rats. *PLoS One.* 2016 Mar 1 [cited 2024 Feb 28];11(3):e0152068. Available from: https://journals.plos.org/plosone/article?id=10.1371/journal.pone.0152068
104. Bonnici L, Suleiman S, Schembri-Wismayer P, Cassar A. Targeting signalling pathways in chronic wound healing. *Int J Mol Sci.* 2023 Dec 19;25(1):50.
105. Mitchell K, Szekeres C, Milano V, Svenson KB, Nilsen-Hamilton M, Kreidberg JA, et al. α3β1 integrin in epidermis promotes wound angiogenesis and keratinocyte-to-endothelial-cell crosstalk through the induction of MRP3. *J Cell Sci.* 2009 Jun 1 [cited 2024 Feb 28];122(11):1778–87. Available from: https://dx.doi.org/10.1242/jcs.040956
106. Jin P, Pan Q, Lin Y, Dong Y, Zhu J, Liu T, et al. Platelets facilitate wound healing by mitochondrial transfer and reducing oxidative stress in endothelial cells. *Oxid Med Cell Longev.* 2023;2023.
107. Boykin J V. Wound nitric oxide bioactivity: A promising diagnostic indicator for diabetic foot ulcer management. *J Wound Ostomy Continence Nurs.* 2010 Jan [cited 2024 Feb 28];37(1):25–32. Available from: https://pubmed.ncbi.nlm.nih.gov/20075688/
108. Ma J, Yong L, Lei P, Li H, Fang Y, Wang L, et al. Advances in microRNA from adipose-derived mesenchymal stem cell-derived exosome: Focusing on wound healing. *J Mater Chem B.* 2022 Oct 24 [cited 2024 Feb 28];10(46):9565–77. Available from: https://pubmed.ncbi.nlm.nih.gov/36398750/
109. Tu J, Wan C, Zhang F, Cao L, Law PWN, Tian Y, et al. Genetic correction of Werner syndrome gene reveals impaired pro-angiogenic function and HGF insufficiency in mesenchymal stem cells. *Aging Cell.* 2020 May 1 [cited 2024 Feb 28];19(5):e13116. Available from: https://onlinelibrary.wiley.com/doi/full/10.1111/acel.13116
110. Zbinden JC, Mirhaidari GJM, Blum KM, Musgrave AJ, Reinhardt JW, Breuer CK, et al. The lysosomal trafficking regulator is necessary for normal wound healing. *Wound Repair and Regen.* 2022 Jan 1;30(1):82–99.
111. Zhang Z, Chen T, Liu W, Xiong J, Jiang L, Liu M. Paeonol accelerates skin wound healing by regulating macrophage polarization and inflammation in diabetic rats. *Korean J Physiol Pharmacol.* 2023 Sep 1 [cited 2024 Feb 28];27(5):437–48. Available from: www.kjpp.net/journal/view.html?doi=10.4196/kjpp.2023.27.5.437
112. Zhang Z, Chen T, Liu W, Xiong J, Jiang L, Liu M. Paeonol accelerates skin wound healing by regulating macrophage polarization and inflammation in diabetic rats. *Korean J Physiol Pharmacol.* 2023 Sep 1 [cited 2024 Feb 28];27(5):437–48. Available from: www.kjpp.net/journal/view.html?doi=10.4196/kjpp.2023.27.5.437
113. Chen X, Tong G, Fan J, Shen Y, Wang N, Gong W, et al. FGF21 promotes migration and differentiation of epidermal cells during wound healing via SIRT1-dependent autophagy. *Br J Pharmacol.* 2022 Mar 1;179(5):1102–21.
114. Cheng L, Shi H, Du L, Liu Q, Yue H, Zhang H, et al. Hemodynamic force dictates endothelial angiogenesis through MIEN1-ERK/MAPK-signaling axis. *J Cell Physiol.* 2024 [cited 2024 Feb 28]. Available from: https://onlinelibrary.wiley.com/doi/full/10.1002/jcp.31177
115. Chawla V, Sharma S, Singh Y. Yttrium oxide nanoparticle-loaded, self-assembled peptide gel with antibacterial, anti-inflammatory, and proangiogenic properties for wound healing. *ACS Biomater Sci Eng.* 2023 May 8 [cited 2024 Feb 28];9(5):2647–62. Available from: https://pubs.acs.org/doi/abs/10.1021/acsbiomaterials.3c00134
116. Coquenlorge S, Van Landeghem L, Jaulin J, Cenac N, Vergnolle N, Duchalais E, et al. The arachidonic acid metabolite 11β-ProstaglandinF2α controls intestinal epithelial healing: Deficiency in patients with Crohn's disease. *Sci Reports.* 2016 May 3 [cited 2024 Feb 28];6(1):1–14. Available from: www.nature.com/articles/srep25203
117. Dorschner RA, Lee J, Cohen O, Costantini T, Baird A, Eliceiri BP. ECRG4 regulates neutrophil recruitment and CD44 expression during the inflammatory response to injury. *Sci Adv.* 2020 [cited 2024 Feb 28];6(11). Available from: www.science.org/doi/10.1126/sciadv.aay0518
118. Zhang Z, Chen T, Liu W, Xiong J, Jiang L, Liu M. Paeonol accelerates skin wound healing by regulating macrophage polarization and inflammation in diabetic rats. *Korean J Physiol Pharmacol.* 2023 Sep 1 [cited 2024 Feb 28];27(5):437–48. Available from: www.kjpp.net/journal/view.html?doi=10.4196/kjpp.2023.27.5.437
119. Kwiatkowska A, Drabik M, Lipko A, Grzeczkowicz A, Stachowiak R, Marszalik A, et al. Composite membrane dressings system with metallic nanoparticles as an antibacterial factor in wound healing. *Membranes.* 2022 Feb 13 [cited 2024 Feb 28];12(2):215. Available from: www.mdpi.com/2077-0375/12/2/215/htm

120. Bhat MJ, Ayed HY, Alrasheed AM, Alghamdi MA, Alsaleh SS, Alrashid YM, et al. Public knowledge about how common chronic diseases affect wound healing postoperatively in aseer region. *Cureus*. 2022 Sep 30 [cited 2024 Feb 28];14(9). Available from: www.cureus.com/articles/105037-public-knowledge-about-how-common-chronic-diseases-affect-wound-healing-postoperatively-in-aseer-region
121. Shu Y, Liu X, Huang H, Wen Q, Shu J. Research progress of natural compounds in anti-liver fibrosis by affecting autophagy of hepatic stellate cells. *Mol Biol Rep*. 2021 Feb 1 [cited 2024 Feb 28];48(2):1915–24. Available from: https://link.springer.com/article/10.1007/s11033-021-06171-w
122. Goracci C, Balzano E, Cunto G De, Bartalesi B, Cavarra E, Lungarella G, et al. Cellular senescence as fibrogenic mechanism in smoking lung. *The FASEB Journal*. 2022 May [cited 2024 Feb 28];36(S1). Available from: https://onlinelibrary.wiley.com/doi/full/10.1096/fasebj.2022.36.S1.L7685
123. Oluwole DO, Coleman L, Buchanan W, Chen T, La Ragione RM, Liu LX. Antibiotics-free compounds for chronic wound healing. *Pharmaceutics*. 2022 May 9 [cited 2024 Feb 28];14(5):1021. Available from: www.mdpi.com/1999-4923/14/5/1021/htm
124. Churko JM, Shao Q, Gong X qun, Swoboda KJ, Bai D, Sampson J, et al. Human dermal fibroblasts derived from oculodentodigital dysplasia patients suggest that patients may have wound-healing defects. *Hum Mutat*. 2011 Apr;32(4):456–66.
125. Qi M. Human and mouse models suggest that subclinical defects of wound healing contribute to oculodentodigital dysplasia (ODDD). *Hum Mutat*. 2011 Apr 1 [cited 2024 Feb 28];32(4):v–v. Available from: https://onlinelibrary.wiley.com/doi/full/10.1002/humu.21496
126. Kharaziha M, Baidya A, Annabi N. Rational design of immunomodulatory hydrogels for chronic wound healing. *Adv Mater*. 2021 Oct 1 [cited 2024 Feb 28];33(39):2100176. Available from: https://onlinelibrary.wiley.com/doi/full/10.1002/adma.202100176
127. Batool F, Özçelik H, Stutz C, Gegout PY, Benkirane-Jessel N, Petit C, et al. Modulation of immune-inflammatory responses through surface modifications of biomaterials to promote bone healing and regeneration. *J Tissue Eng*. 2021 Oct 26 [cited 2024 Feb 28];12. Available from: https://journals.sagepub.com/doi/10.1177/20417314211041428
128. Whitaker R, Hernaez-Estrada B, Hernandez RM, Santos-Vizcaino E, Spiller KL. Immunomodulatory biomaterials for tissue repair. *Chem Rev*. 2021 Sep 22 [cited 2024 Feb 28];121(18):11305–35. Available from: https://pubs.acs.org/doi/abs/10.1021/acs.chemrev.0c00895
129. Whitaker R, Hernaez-Estrada B, Hernandez RM, Santos-Vizcaino E, Spiller KL. Immunomodulatory biomaterials for tissue repair. *Chem Rev*. 2021 Sep 22 [cited 2024 Feb 28];121(18):11305–35. Available from: https://pubs.acs.org/doi/abs/10.1021/acs.chemrev.0c00895
130. Alhahri G, Badran A, Hijazi A, Daou A, Baydoun E, Nasser M, et al. The therapeutic wound healing bioactivities of various medicinal plants. *Life*. 2023 Jan 23 [cited 2024 Feb 28];13(2):317. Available from: www.mdpi.com/2075-1729/13/2/317/htm
131. Joorabloo A, Liu T. Recent advances in nanomedicines for regulation of macrophages in wound healing. *J Nanobiotechnology*. 2022 Sep 9 [cited 2024 Feb 28];20(1):1–19. Available from: https://jnanobiotechnology.biomedcentral.com/articles/10.1186/s12951-022-01616 1
132. Shi G, Chen W, Zhang Y, Dai X, Zhang X, Wu Z. An antifouling hydrogel containing silver nanoparticles for modulating the therapeutic immune response in chronic wound healing. *Langmuir*. 2019 Feb 5 [cited 2024 Feb 28];35(5):1837–45. Available from: https://pubs.acs.org/doi/abs/10.1021/acs.langmuir.8b01834
133. Lantonio B, Rosano M, Brown BN, Wells A, Yates CC. Multipotent stromal cell and fibroblast co-transplantation alter wound microenvironment to normalize aged-deficient wound healing. *The FASEB Journal*. 2017 Apr [cited 2024 Feb 28];31(S1):333.3–333.3. Available from: https://onlinelibrary.wiley.com/doi/full/10.1096/fasebj.31.1_supplement.333.3
134. Xiong Y, Lin Z, Bu P, Yu T, Endo Y, Zhou W, et al. A whole-course-repair system based on neurogenesis-angiogenesis crosstalk and macrophage reprogramming promotes diabetic wound healing. *Adv Mater*. 2023 May 1 [cited 2024 Feb 28];35(19):2212300. Available from: https://onlinelibrary.wiley.com/doi/full/10.1002/adma.202212300
135. Li W, Jiang L, Wu S, Yang S, Ren L, Cheng B, et al. A shape-programmable hierarchical fibrous membrane composite system to promote wound healing in diabetic patients. *Small*. 2022 Mar 1 [cited 2024 Feb 28];18(11):2107544. Available from: https://onlinelibrary.wiley.com/doi/full/10.1002/smll.202107544
136. Griffin DR, Archang MM, Kuan CH, Weaver WM, Weinstein JS, Feng AC, et al. Activating an adaptive immune response from a hydrogel scaffold imparts regenerative wound healing. *Nat Mater*. 2020 Nov 9 [cited 2024 Feb 28];20(4):560–9. Available from: www.nature.com/articles/s41563-020-00844-w

137. Renò F, Sabbatini M. Breaking a vicious circle: Lymphangiogenesis as a new therapeutic target in wound healing. *Biomedicines*. 2023 Feb 21 [cited 2024 Feb 28];11(3):656. Available from: www.mdpi.com/2227-9059/11/3/656/htm
138. Falanga A, Galdiero S. Emerging therapeutic agents on the basis of naturally occurring antimicrobial peptides. *Amino Acids, Peptides and Proteins*. 2017;42:190–227.
139. Holland R, Gomes Filho JE, Cintra LTA, Queiroz IODA, Estrela C. Factors affecting the periapical healing process of endodontically treated teeth. *J Appl Oral Sci*. 2017 Sep 1 [cited 2024 Feb 28];25(5):465–76. Available from: www.scielo.br/j/jaos/a/dRgMC4hj3SXWRWBh4NVqNCw/?lang=en
140. Li Q, Xia S, Yin Y, Guo Y, Chen F, Jin P. miR-5591–5p regulates the effect of ADSCs in repairing diabetic wound via targeting AGEs/AGER/JNK signaling axis. *Cell Death Dis*. 2018 May 1 [cited 2024 Mar 2];9(5). Available from: /pmc/articles/PMC5948214/
141. Li H, Cheng F, Wu G, Tyavambiza C, Meyer M, Meyer S. Cellular and molecular events of wound healing and the potential of silver based nanoformulations as wound healing agents. *Bioengineering*. 2022 Nov 19 [cited 2024 Mar 6];9(11):712. Available from: www.mdpi.com/2306-5354/9/11/712/htm
142. Ohadi M, Forootanfar H, Dehghannoudeh N, Banat IM, Dehghannoudeh G. The role of surfactants and biosurfactants in the wound healing process: A review. 2023 Apr 8 [cited 2024 Mar 6];32:XXXIX–XLVI. Available from: www.magonlinelibrary.com/doi/10.12968/jowc.2023.32.Sup4a.xxxix
143. Nasseri S, Sharifi M. Therapeutic Potential of Antimicrobial Peptides for Wound Healing. *Int J Pept Res Ther*. 2022 Jan 1 [cited 2024 Mar 6];28(1):1–15. Available from: https://link.springer.com/article/10.1007/s10989-021-10350-5
144. Bacci S, Mandal D, Gupta J. Burgeoning nanotechnology for diabetic wound healing: A novel approach to the future. *Med Sci Forum*. 2023 Mar 29 [cited 2024 Mar 6];21(1):33. Available from: www.mdpi.com/2673-9992/21/1/33/htm
145. Golkar N, Ashoori Y, Heidari R, Omidifar N, Abootalebi SN, Mohkam M, et al. A novel effective formulation of bioactive compounds for wound healing: Preparation, in vivo characterization, and comparison of various postbiotics cold creams in a rat model. *Evid Based Complementary Altern Med*. 2021;2021.
146. Yan S, Li J, Gao Y, You J, Xu S, Wang C, et al. Encapsulation of sericin-decorated efficient agents in silk hydrogels for wound dressings. *ACS Appl Mater Interfaces*. 2023 Oct 25 [cited 2024 Mar 6];15(42):48952–62. Available from: https://pubs.acs.org/doi/abs/10.1021/acsami.3c10044
147. Panjaitan J, Girsang E, Chiuman L. Effectiveness of temulawak (Curcuma xanthorrhiza) ointments as wound healing agents in wistar rats. *Jurnal Teknologi Laboratorium*. 2022 Nov 14 [cited 2024 Mar 6];11(2):95–104. Available from: www.teknolabjournal.com/index.php/Jtl/article/view/376
148. Protzman NM, Mao Y, Long D, Sivalenka R, Gosiewska A, Hariri RJ, et al. Placental-derived biomaterials and their application to wound healing: A review. *Bioengineering*. 2023 Jul 12 [cited 2024 Mar 6];10(7):829. Available from: www.mdpi.com/2306-5354/10/7/829/htm
149. Ellatif SA, Abdel Razik ES, Abu-Serie MM, Mahfouz A, Shater AF, Saleh FM, et al. Immunomodulatory efficacy-mediated anti-HCV and Anti-HBV potential of kefir grains; unveiling the in vitro antibacterial, antifungal, and wound healing activities. *Molecules*. 2022 Mar 21 [cited 2024 Mar 6];27(6):2016. Available from: www.mdpi.com/1420-3049/27/6/2016/htm
150. Wijayadi LJ, Kelvin K. Effectiveness of helichrysum italicum essential oil on wound healing. *J Food Pharm Sci*. 2022 Sep 7 [cited 2024 Mar 6];681–97. Available from: https://jurnal.ugm.ac.id/v3/JFPS/article/view/5324
151. Li W, Chen Y, Cai Z, He X, Yang L, Zhu J, et al. Traditional Chinese medicine Qingre Huoxue decoction enhances wound healing in through modulation of angiogenic and inflammatory pathways. *Int Wound J*. 2024 Mar 1 [cited 2024 Mar 6];21(3):e14724. Available from: https://onlinelibrary.wiley.com/doi/full/10.1111/iwj.14724
152. Kim SY, Lee JO, Lee S, Heo J, Cho KH, Bahuguna A, et al. Ozonated Sunflower Oil (OSO) alleviates inflammatory responses in oxazolone-induced atopic dermatitis (AD)-like mice and LPS-treated RAW 264.7 cells. *J Microbiol Biotechnol*. 2024 Apr 28 [cited 2024 Mar 6];34(4):1–10. Available from: www.jmb.or.kr/journal/view.html?doi=10.4014/jmb.2310.10037
153. . Z, Sudheer P. Chitosan a natural anti-inflammatory and wound healing agent: A brief update. *Nat Resour Human Health*. 2024 Feb 28 [cited 2024 Mar 6];4(2):115–27. Available from: www.nrfhh.com/Chitosan-a-natural-anti-inflammatory-and-wound-healing-agent-A-brief-update,177602,0,2.html
154. Zanetta P, Ballacchino C, Squarzanti DF, Amoruso A, Pane M, Azzimonti B. Lactobacillus johnsonii LJO02 (DSM 33828) cell-free supernatant and Vitamin D improve wound healing and reduce interleukin-6 production in staphylococcus aureus-infected human keratinocytes. *Pharmaceutics*. 2023 Dec 21 [cited 2024 Mar 6];16(1):18. Available from: www.mdpi.com/1999-4923/16/1/18/htm

155. Zhang J, Wang L, Xu C, Cao Y, Liu S, Reis RL, et al. Transparent silk fibroin film-facilitated infected-wound healing through antibacterial, improved fibroblast adhesion and immune modulation. *J Mater Chem B.* 2024 Jan 3 [cited 2024 Mar 7];12(2):475–88. Available from: https://pubs.rsc.org/en/content/articlehtml/2024/tb/d3tb02146g
156. Kapate N, Liao R, Sodemann RL, Stinson T, Prakash S, Kumbhojkar N, et al. Backpack-mediated anti-inflammatory macrophage cell therapy for the treatment of traumatic brain injury. *PNAS Nexus.* 2023 Dec 21 [cited 2024 Mar 7];3(1). Available from: https://dx.doi.org/10.1093/pnasnexus/pgad434
157. He T, Xiao Y, Guo Z, Shi Y, Tan Q, Huang Y, et al. Modulation of macrophage function by bioactive wound dressings with an emphasis on extracellular matrix-based scaffolds and nanofibrous composites. *Pharmaceutics.* 2023 Feb 28 [cited 2024 Mar 7];15(3):794. Available from: www.mdpi.com/1999-4923/15/3/794/htm
158. Zhang Y, Kang J, Chen X, Zhang W, Zhang X, Yu W, et al. Ag nanocomposite hydrogels with immune and regenerative microenvironment regulation promote scarless healing of infected wounds. *J Nanobiotechnology.* 2023 Dec 1 [cited 2024 Mar 7];21(1):1–17. Available from: https://jnanobiotechnology.biomedcentral.com/articles/10.1186/s12951-023-02209-2
159. Joorabloo A, Liu T. Recent advances in nanomedicines for regulation of macrophages in wound healing. *J Nanobiotechnology.* 2022 Sep 9 [cited 2024 Mar 7];20(1):1–19. Available from: https://jnanobiotechnology.biomedcentral.com/articles/10.1186/s12951-022-01616-1
160. Zhang X, Feng J, Feng W, Xu B, Zhang K, Ma G, et al. Glycosaminoglycan-based hydrogel delivery system regulates the wound microenvironment to rescue chronic wound healing. *ACS Appl Mater Interfaces.* 2022 Jul 20 [cited 2024 Mar 7];14(28):31737–50. Available from: https://pubs.acs.org/doi/abs/10.1021/acsami.2c08593
161. Marinelli L, Cacciatore I, Costantini E, Dimmito MP, Serra F, Di Stefano A, et al. Wound-healing promotion and anti-inflammatory properties of carvacrol prodrugs/hyaluronic acid formulations. *Pharmaceutics.* 2022 Jul 1 [cited 2024 Mar 7];14(7):1468. Available from: www.mdpi.com/1999-4923/14/7/1468/htm
162. Yang S, Jiang H, Qian M, Ji G, Wei Y, He J, et al. MSC-derived sEV-loaded hyaluronan hydrogel promotes scarless skin healing by immunomodulation in a large skin wound model. *Biomed Mater.* 2022 Apr 29 [cited 2024 Mar 7];17(3):034104. Available from: https://iopscience.iop.org/article/10.1088/1748-605X/ac68bc
163. Kharaziha M, Baidya A, Annabi N. Rational design of immunomodulatory hydrogels for chronic wound healing. *Adv Mater.* 2021 Oct 1 [cited 2024 Mar 7];33(39):2100176. Available from: https://onlinelibrary.wiley.com/doi/full/10.1002/adma.202100176
164. Merecz-Sadowska A, Sitarek P, Śliwiński T, Zajdel R. Anti-inflammatory activity of extracts and pure compounds derived from plants via modulation of signaling pathways, especially PI3K/AKT in macrophages. *Int J Mol Sci.* 2020 Dec 16 [cited 2024 Mar 7];21(24):9605. Available from: www.mdpi.com/1422-0067/21/24/9605/htm
165. Zielińska M, Pawłowska A, Orzeł A, Sulej L, Muzyka-Placzyńska K, Baran A, et al. Wound microbiota and its impact on wound healing. *Int J Mol Sci.* 2023 Dec 10 [cited 2024 Mar 7];24(24):17318. Available from: www.mdpi.com/1422-0067/24/24/17318/htm
166. Morio KA, Sternowski RH, Zeng E, Brogden KA. Antimicrobial peptides and biomarkers induced by ultraviolet irradiation have the potential to reduce endodontic inflammation and facilitate tissue healing. *Pharmaceutics.* 2022 Sep 1 [cited 2024 Mar 7];14(9):1979. Available from: www.mdpi.com/1999-4923/14/9/1979/htm
167. Amiri N, Golin AP, Jalili RB, Ghahary A. Roles of cutaneous cell-cell communication in wound healing outcome: An emphasis on keratinocyte-fibroblast crosstalk. *Exp Dermatol.* 2022 Apr 1 [cited 2024 Mar 7];31(4):475–84. Available from: https://onlinelibrary.wiley.com/doi/full/10.1111/exd.14516
168. Ramírez-Martínez G, Jiménez-Álvarez LA, Cruz-Lagunas A, Ignacio-Cortés S, Gómez-García IA, Rodríguez-Reyna TS, et al. Possible role of matrix metalloproteinases and TGF-β in COVID-19 severity and sequelae. *J Interferon Cytokine Res.* 2022 Aug 1 [cited 2024 Mar 7];42(8):352–68. Available from: www.liebertpub.com/doi/10.1089/jir.2021.0222
169. Tsai CY, Hsieh SC, Liu CW, Lu CS, Wu CH, Liao HT, et al. Cross-talk among polymorphonuclear neutrophils, immune, and non-immune cells via released cytokines, granule proteins, microvesicles, and neutrophil extracellular trap formation: A novel concept of biology and pathobiology for neutrophils. *Int J Mol Sci.* 2021 Mar 18 [cited 2024 Mar 7];22(6):3119. Available from: www.mdpi.com/1422-0067/22/6/3119/htm
170. Villalobos V, Garrido M, Reyes A, Fernández C, Diaz C, Torres VA, et al. Aging envisage imbalance of the periodontium: A keystone in oral disease and systemic health. *Front Immunol.* 2022 Oct 20;13:1044334.

171. Liu Y, Segura T. Biomaterials-mediated regulation of macrophage cell fate. *Front Bioeng Biotechnol.* 2020 Dec 11;8:609297.
172. Zhang W, Pajulas A, Kaplan MH. γδ T Cells in skin inflammation. *Crit Rev™ in Immunol.* 2022 [cited 2024 Mar 7];42(5):43–56. Available from: www.dl.begellhouse.com/journals/2ff21abf44b19838,732aaf de4bb9be0d,262dfd8824e9f2b5.html
173. Ramey-Ward AN, Walthall HP, Smith S, Barrows TH. Human keratin matrices promote wound healing by modulating skin cell expression of cytokines and growth factors. *Wound Repair Regen.* 2023 [cited 2024 Mar 7]. Available from: https://onlinelibrary.wiley.com/doi/full/10.1111/wrr.13137
174. Abedin-Do A, Zhang Z, Douville Y, Méthot M, Rouabhia M. Effect of electrical stimulation on diabetic human skin fibroblast growth and the secretion of cytokines and growth factors involved in wound healing. *Biology.* 2021 Jul 9 [cited 2024 Mar 7];10(7):641. Available from: www.mdpi.com/2079-7737/10/7/641/htm
175. Yang L, Zhang YP, Peng LH. Research progress on role of Chinese medicinal materials and their topical preparations in wound healing by regulating cytokines and growth factors. *Zhongguo Zhongyao Zazhi.* 2021 Oct 15;46(20):5173–84.
176. Viaña-Mendieta P, Mirna l, Anchez LS, Benavides J, Lorena M, Anchez S. Rational selection of bioactive principles for wound healing applications: Growth factors and antioxidants. *Int Wound J.* 2022 Jan 1 [cited 2024 Mar 7];19(1):100–13. Available from: https://onlinelibrary.wiley.com/doi/full/10.1111/iwj.13602
177. Kant V, Jangir BL, Kumar V, Nigam A, Sharma V. Quercetin accelerated cutaneous wound healing in rats by modulation of different cytokines and growth factors. *Growth Factors.* 2020 Mar 3 [cited 2024 Mar 7];38(2):105–19. Available from: www.tandfonline.com/doi/abs/10.1080/08977194.2020.1822830
178. Jimi S, Jaguparov A, Nurkesh A, Sultankulov B, Saparov A. Sequential delivery of cryogel released growth factors and cytokines accelerates wound healing and improves tissue regeneration. *Front Bioeng Biotechnol.* 2020 Apr 17;8:513103.
179. Nurkesh A, Jaguparov A, Jimi S, Saparov A. Recent advances in the controlled release of growth factors and cytokines for improving cutaneous wound healing. *Front Cell Dev Biol.* 2020 Jul 14;8:551685.
180. Smagul S, Kim Y, Smagulova A, Raziyeva K, Nurkesh A, Saparov A. Biomaterials loaded with growth factors/cytokines and stem cells for cardiac tissue regeneration. *Int J Mol Sci.* 2020 Aug 19 [cited 2024 Mar 7];21(17):5952. Available from: www.mdpi.com/1422-0067/21/17/5952/htm
181. Pereira RVS, EzEldeen M, Ugarte-Berzal E, Martens E, Malengier-Devlies B, Vandooren J, et al. Physiological fibrin hydrogel modulates immune cells and molecules and accelerates mouse skin wound healing. *Front Immunol.* 2023 Apr 24;14:1170153.
182. Miricescu D, Badoiu SC, Stanescu-Spinu II, Ripszky Totan A, Stefani C, Greabu M. Growth factors, reactive oxygen species, and metformin—promoters of the wound healing process in burns? *Int J Mol Sci.* 2021 Sep 1 [cited 2024 Mar 7];22(17):9512. Available from: www.mdpi.com/1422-0067/22/17/9512/htm
183. Canchy L, Kerob D, Demessant A, Amici JM. Wound healing and microbiome, an unexpected relationship. *J Eur Acad Dermatol Venereol.* 2023 Mar 1 [cited 2024 Mar 7];37(S3):7–15. Available from: https://onlinelibrary.wiley.com/doi/full/10.1111/jdv.18854
184. Liu C, Xu Y, Lu Y, Du P, Li X, Wang C, et al. Mesenchymal stromal cells pretreated with proinflammatory cytokines enhance skin wound healing via IL-6-dependent M2 polarization. *Stem Cell Res Ther.* 2022 Dec 1 [cited 2024 Mar 7];13(1):1–17. Available from: https://stemcellres.biomedcentral.com/articles/10.1186/s13287-022-02934-9
185. Barrientos S, Stojadinovic O, Golinko MS, Brem H, Tomic-Canic M. Growth factors and cytokines in wound healing. *Wound Repair Regen.* 2008 Sep [cited 2024 Mar 19];16(5):585–601. Available from: https://pubmed.ncbi.nlm.nih.gov/19128254/
186. Yang B, Zhou L, Peng B, Sun Z, Dai Y, Zheng J. In vitro comparative evaluation of recombinant growth factors for tissue engineering of bladder in patients with neurogenic bladder. *J Surg Res.* 2014 Jan;186(1):63–72.
187. Shen C, Sun L, Zhu N, Qi F. Kindlin-1 contributes to EGF-induced re-epithelialization in skin wound healing. *Int J Mol Med.* 2017 [cited 2024 Mar 19];39(4):949–59. Available from: https://pubmed.ncbi.nlm.nih.gov/28290610/
188. Meyer M, Müller AK, Yang J, Moik D, Ponzio G, Ornitz DM, et al. FGF receptors 1 and 2 are key regulators of keratinocyte migration in vitro and in wounded skin. *J Cell Sci.* 2012 Dec [cited 2024 Mar 19];125(Pt 23):5690–701. Available from: https://pubmed.ncbi.nlm.nih.gov/22992463/
189. Zhang C, Tan CK, McFarlane C, Sharma M, Tan NS, Kambadur R. Myostatin-null mice exhibit delayed skin wound healing through the blockade of transforming growth factor-β signaling by decorin. *Am J*

Physiol Cell Physiol. 2012 [cited 2024 Mar 19];302(8). Available from: https://pubmed.ncbi.nlm.nih.gov/22277753/

190. Ashcroft GS, Jeong MJ, Ashworth JJ, Hardman M, Jin W, Moutsopoulos N, et al. Tumor necrosis factor-alpha (TNF-α) is a therapeutic target for impaired cutaneous wound healing. *Wound Repair Regen*. 2012 Jan [cited 2024 Mar 19];20(1):38–49. Available from: https://pubmed.ncbi.nlm.nih.gov/22151742/
191. Ishida Y, Kondo T, Kimura A, Matsushima K, Mukaida N. Absence of IL-1 receptor antagonist impaired wound healing along with aberrant NF-kappaB activation and a reciprocal suppression of TGF-beta signal pathway. *J Immunol*. 2006 May 1 [cited 2024 Mar 19];176(9):5598–606. Available from: https://pubmed.ncbi.nlm.nih.gov/16622029/
192. Gallucci RM, Berryann K, Simeonova PP, Matheson JM, Luster MI, Sugawara T, et al. Interleukin-6 treatment augments cutaneous wound healing in immunosuppressed mice. *J Interferon Cytokine Res*. 2001 [cited 2024 Mar 19];21(8):603–9. Available from: https://pubmed.ncbi.nlm.nih.gov/11559438/
193. Kondo T, Ohshima T, Mori R, Guan DW, Ohshima K, Eisenmenger W. Immunohistochemical detection of chemokines in human skin wounds and its application to wound age determination. *Int J Legal Med*. 2002 [cited 2024 Mar 19];116(2):87–91. Available from: https://pubmed.ncbi.nlm.nih.gov/12056526/
194. Liu W, Zu L, Wang S, Li J, Fei X, Geng M, et al. Tailored biomedical materials for wound healing. *Burns Trauma*. 2023 Jan 1 [cited 2024 Mar 9];11. Available from: https://dx.doi.org/10.1093/burnst/tkad040
195. Nour S, Baheiraei N, Imani R, Khodaei M, Alizadeh A, Rabiee N, et al. A review of accelerated wound healing approaches: biomaterial-assisted tissue remodeling. *J Mater Sci Mater Med*. 2019 Oct 1 [cited 2024 Mar 9];30(10):1–15. Available from: https://link.springer.com/article/10.1007/s10856-019-6319-6
196. Yang Y, Wang L, Zhou Y, He Y, Lin S, Zeng Y, et al. Antioxidant-enriched autologous biogel promoted diabetic wound healing by remodeling inherent posttraumatic inflammatory patterning and restoring compromised microenvironment homeostasis. *Regen Biomater*. 2022 Apr 8 [cited 2024 Mar 9];9. Available from: https://dx.doi.org/10.1093/rb/rbac023
197. Bloise N, Rountree I, Polucha C, Montagna G, Visai L, Coulombe KLK, et al. Engineering immunomodulatory biomaterials for regenerating the infarcted myocardium. *Front Bioeng Biotechnol*. 2020 Apr 7;8:521672.
198. Wang L, Zhong Y, Wu Q, Wang Y, Tang R, Zhou S, et al. Spermidine-functionalized biomaterials to modulate implant-induced immune response and enhance wound healing. *Chem Eng J*. 2023 Nov 15;476:146416.
199. Protzman NM, Mao Y, Long D, Sivalenka R, Gosiewska A, Hariri RJ, et al. Placental-derived biomaterials and their application to wound healing: A review. *Bioengineering*. 2023 Jul 12 [cited 2024 Mar 9];10(7):829. Available from: www.mdpi.com/2306-5354/10/7/829/htm
200. Apte A, Liechty KW, Zgheib C. Immunomodulatory biomaterials on chemokine signaling in wound healing. *Front Pharmacol*. 2023 Apr 21;14:1084948.
201. Mao J, Chen L, Cai Z, Qian S, Liu Z, Zhao B, et al. Advanced biomaterials for regulating polarization of macrophages in wound healing. *Adv Funct Mater*. 2022 Mar 1 [cited 2024 Mar 9];32(12):2111003. Available from: https://onlinelibrary.wiley.com/doi/full/10.1002/adfm.202111003
202. Batool F, Özçelik H, Stutz C, Gegout PY, Benkirane-Jessel N, Petit C, et al. Modulation of immune-inflammatory responses through surface modifications of biomaterials to promote bone healing and regeneration. *J Tissue Eng*. 2021 Oct 26 [cited 2024 Mar 9];12. Available from: https://journals.sagepub.com/doi/10.1177/20417314211041428
203. Wang J, Liu X, Wang Y, Zhang Y, Gao R, Gao Y, et al. Pro-regenerative glycopeptide hydrogel activates type 2 immune response for wound healing via macrophage-T cell crosstalk. *Adv Funct Mater*. 2024 [cited 2024 Mar 9];2307711. Available from: https://onlinelibrary.wiley.com/doi/full/10.1002/adfm.202307711
204. Wang C, Chu C, Zhao; Xiwen, Hu C, Liu; Li, Li J, et al. The diameter factor of aligned membranes facilitates cutaneous wound healing by promoting epithelialization in an immune way. *bioRxiv*. 2021 Jun 3 [cited 2024 Mar 9];2021.02.25.432929. Available from: www.biorxiv.org/content/10.1101/2021.02.25.432929v3
205. Swoboda L, Held J. Impaired wound healing in diabetes. 2022 Oct 14 [cited 2024 Mar 9];31(10):882–5. Available from: www.magonlinelibrary.com/doi/10.12968/jowc.2022.31.10.882
206. Ahmad T, McGrath S, Sirafim C, Do Amaral RJFC, Soong SL, Sitram R, et al. Development of wound healing scaffolds with precisely-triggered sequential release of therapeutic nanoparticles. *Biomater Sci*. 2021 Jun 15 [cited 2024 Mar 9];9(12):4278–88. Available from: https://pubs.rsc.org/en/content/articlehtml/2021/bm/d0bm01277g

207. Qin J, Chen F, Wu P, Sun G. Recent advances in bioengineered scaffolds for cutaneous wound healing. *Front Bioeng Biotechnol.* 2022 Mar 1;10:841583.
208. Griffin DR, Archang MM, Kuan CH, Weaver WM, Weinstein JS, Feng AC, et al. Activating an adaptive immune response from a hydrogel scaffold imparts regenerative wound healing. *bioRxiv.* 2020 May 30 [cited 2024 Mar 9];2020.05.27.117317. Available from: www.biorxiv.org/content/10.1101/2020.05.27.117317v1
209. Fan R, Zhang C, Li F, Li B, McCarthy A, Zhang Y, et al. Hierarchically assembled nanofiber scaffolds with dual growth factor gradients promote skin wound healing through rapid cell recruitment. *Adv Sci.* 2024 [cited 2024 Mar 9];2309993. Available from: https://onlinelibrary.wiley.com/doi/full/10.1002/advs.202309993
210. Liu Y, Suarez-Arnedo A, Caston ELP, Riley L, Schneider M, Segura T. Exploring the role of spatial confinement in immune cell recruitment and regeneration of skin wounds. *Adv Mater.* 2023 Dec 1 [cited 2024 Mar 9];35(49):2304049. Available from: https://onlinelibrary.wiley.com/doi/full/10.1002/adma.202304049
211. Sawaya AP, Stone RC, Brooks SR, Pastar I, Jozic I, Hasneen K, et al. Deregulated immune cell recruitment orchestrated by FOXM1 impairs human diabetic wound healing. *Nat Commun.* 2020 Sep 16 [cited 2024 Mar 9];11(1):1–14. Available from: www.nature.com/articles/s41467-020-18276-0
212. Wee WKJ, Low ZS, Ooi CK, Henategala BP, Lim ZGR, Yip YS, et al. Single-cell analysis of skin immune cells reveals an Angptl4-ifi20b axis that regulates monocyte differentiation during wound healing. *Cell Death & Disease.* 2022 Feb 24 [cited 2024 Mar 9];13(2):1–9. Available from: www.nature.com/articles/s41419-022-04638-7
213. Sivaraj D, Chen K, Chattopadhyay A, Henn D, Wu W, Noishiki C, et al. Hydrogel scaffolds to deliver cell therapies for wound healing. *Front Bioeng Biotechnol.* 2021 May 3;9:660145.
214. Rehak L, Giurato L, Meloni M, Panunzi A, Manti GM, Uccioli L. The immune-centric revolution in the diabetic foot: monocytes and lymphocytes role in wound healing and tissue regeneration—a narrative review. *J Clin Med.* 2022 Feb 8 [cited 2024 Mar 9];11(3):889. Available from: www.mdpi.com/2077-0383/11/3/889/htm
215. Tian DM, Wan HH, Chen JR, Ye Y Bin, He Y, Liu Y, et al. In-situ formed elastin-based hydrogels enhance wound healing via promoting innate immune cells recruitment and angiogenesis. *Mater Today Bio.* 2022 Jun 1;15:100300.
216. Castillo-Henríquez L, Castro-Alpízar J, Lopretti-Correa M, Vega-Baudrit J. Exploration of bioengineered scaffolds composed of thermo-responsive polymers for drug delivery in wound healing. *Int J Mol Sci.* 2021 Jan 30 [cited 2024 Mar 9];22(3):1408. Available from: www.mdpi.com/1422-0067/22/3/1408/htm
217. Palaniyandi S, Hildebrandt G. S242: Mast cells are mediators of inflammatory effector cell recruitment and fibrosis in dermal chronic graft-versus-host disease. *Hemasphere.* 2023 Aug 8 [cited 2024 Mar 9];7(S3):e22000bd. Available from: https://journals.lww.com/hemasphere/fulltext/2023/08003/s242__mast_cells_are_mediators_of_inflammatory.144.aspx
218. Sharifi E, Sadati SA, Yousefiasl S, Sartorius R, Zafari M, Rezakhani L, et al. Cell loaded hydrogel containing Ag-doped bioactive glass–ceramic nanoparticles as skin substitute: Antibacterial properties, immune response, and scarless cutaneous wound regeneration. *Bioeng Transl Med.* 2022 Sep 1 [cited 2024 Mar 9];7(3):e10386. Available from: https://onlinelibrary.wiley.com/doi/full/10.1002/btm2.10386
219. Bernardes BG, Veiga A, Barros J, García-González CA, Oliveira AL. Sustainable silk-based particulate systems for the controlled release of pharmaceuticals and bioactive agents in wound healing and skin regeneration. *Int J Mol Sci.* 2024 Mar 8 [cited 2024 Mar 9];25(6):3133. Available from: www.mdpi.com/1422-0067/25/6/3133/htm
220. Scull G, Brown AC. Development of novel microenvironments for promoting enhanced wound healing. *Curr Tissue Microenv Rep.* 2020 Jul 29 [cited 2024 Mar 9];1(3):73–87. Available from: https://link.springer.com/article/10.1007/s43152-020-00009-6
221. Zhu S, Mbugua J, Chase M, Nicholson III TR, Jung JH, Maskrod J, et al. Electronically-controlled drug release system to promote wound healing. [cited 2024 Mar 9]. Available from: www.nsti.org,
222. Yuan J, Hou Q, Zhong L, Dai X, Lu Q, Li M, et al. Sustained release of inhibitor from bionic scaffolds for wound healing and functional regeneration. *Biomater Sci.* 2020 Oct 13 [cited 2024 Mar 9];8(20):5647–55. Available from: https://pubs.rsc.org/en/content/articlehtml/2020/bm/d0bm00929f
223. Heydari P, Kharaziha M, Varshosaz J, Javanmard SH. Current knowledge of immunomodulation strategies for chronic skin wound repair. *J Biomed Mater Res B Appl Biomater.* 2022 Feb 1 [cited 2024 Mar 9];110(2):265–88. Available from: https://onlinelibrary.wiley.com/doi/full/10.1002/jbm.b.34921

224. Chung CW, Liao BW, Huang SW, Chiou SJ, Chang CH, Lin SJ, et al. Magnetic responsive release of nitric oxide from an MOF-derived Fe3O4@PLGA microsphere for the treatment of bacteria-infected cutaneous wound. *ACS Appl Mater Interfaces*. 2022 Feb 9 [cited 2024 Mar 9];14(5):6343–57. Available from: https://pubs.acs.org/doi/abs/10.1021/acsami.1c20802
225. Johnson MD, Lloyd J, Tekkam S, Crooke SN, Witherden DA, Havran WL, et al. Degradable hydrogels for the delivery of immune-modulatory proteins in the wound environment. *ACS Appl Bio Mater*. 2020 Aug 17 [cited 2024 Mar 9];3(8):4779–88. Available from: https://pubs.acs.org/doi/abs/10.1021/acsabm.0c00301
226. Araújo RG, Zavala NR, Castillo-Zacarías C, Barocio ME, Hidalgo-Vázquez E, Parra-Arroyo L, et al. Recent advances in prodigiosin as a bioactive compound in nanocomposite applications. *Molecules*. 2022 Aug 5 [cited 2024 Mar 9];27(15):4982. Available from: www.mdpi.com/1420-3049/27/15/4982/htm
227. Chung CW, Liao BW, Huang SW, Chiou SJ, Chang CH, Lin SJ, et al. Magnetic responsive release of nitric oxide from an MOF-derived Fe3O4@PLGA microsphere for the treatment of bacteria-infected cutaneous wound. *ACS Appl Mater Interfaces*. 2022 Feb 9 [cited 2024 Mar 9];14(5):6343–57. Available from: https://pubs.acs.org/doi/abs/10.1021/acsami.1c20802
228. Borbora A, Xu Y, Dey S, Wang X, Yao Y, Mandal BB, et al. Lubricated interfaces enabling simultaneous pulsatile and continuous chemical release modes. *Adv Mater*. 2023 Aug 1 [cited 2024 Mar 9];35(31):2302264. Available from: https://onlinelibrary.wiley.com/doi/full/10.1002/adma.202302264
229. Bloise N, Rountree I, Polucha C, Montagna G, Visai L, Coulombe KLK, et al. Engineering immunomodulatory biomaterials for regenerating the infarcted myocardium. *Front Bioeng Biotechnol*. 2020 Apr 7;8:521672.
230. Li Z, Maitz P. Cell therapy for severe burn wound healing. *Burns Trauma*. 2018 Dec 1 [cited 2024 Mar 9];6. Available from: https://dx.doi.org/10.1186/s41038-018-0117-0
231. Han SK. Cell therapy. Innovations and advances in wound healing. 2023 [cited 2024 Mar 9];289–344. Available from: https://link.springer.com/chapter/10.1007/978-981-19-9805-8_12
232. Golchin A, Shaikhnia F, Heidari F, Mahdi D, Hemmati Y, Tayebi L. Cell-derived materials for wound healing. *Handbook of the Extracellular Matrix*. 2023 [cited 2024 Mar 9];1–22. Available from: https://link.springer.com/referenceworkentry/10.1007/978-3-030-92090-6_36-1
233. Kou Z, Li B, Aierken A, Tan N, Li C, Han M, et al. Mesenchymal stem cells pretreated with collagen promote skin wound-healing. *Int J Mol Sci*. 2023 May 12 [cited 2024 Mar 9];24(10):8688. Available from: www.mdpi.com/1422-0067/24/10/8688/htm
234. Wang M, Xu X, Lei X, Tan J, Xie H. Mesenchymal stem cell-based therapy for burn wound healing. *Burns Trauma*. 2021 Jan 1 [cited 2024 Mar 9];9:2. Available from: https://dx.doi.org/10.1093/burnst/tkab002
235. Saadh MJ, Ramírez-Coronel AA, Saini RS, Arias-Gonzáles JL, Amin AH, Gavilán JCO, et al. Advances in mesenchymal stem/stromal cell-based therapy and their extracellular vesicles for skin wound healing. *Human Cell*. 2023 Apr 17 [cited 2024 Mar 9];36(4):1253–64. Available from: https://link.springer.com/article/10.1007/s13577-023-00904-8
236. Zhu M, Cao L, Melino S, Candi E, Wang Y, Shao C, et al. Orchestration of mesenchymal stem/stromal cells and inflammation during wound healing. *Stem Cells Transl Med*. 2023 Sep 15 [cited 2024 Mar 9];12(9):576–87. Available from: https://dx.doi.org/10.1093/stcltm/szad043
237. Zhao W, Zhang H, Liu R, Cui R. Advances in immunomodulatory mechanisms of mesenchymal stem cells-derived exosome on immune cells in scar formation. *Int J Nanomedicine*. 2023 Jul 3 [cited 2024 Mar 9];18:3643–62. Available from: www.dovepress.com/advances-in-immunomodulatory-mechanisms-of-mesenchymal-stem-cells-deri-peer-reviewed-fulltext-article-IJN
238. Yang J, Li H, Ran M, Yang S, Ma K, Zhang C, et al. Transplantation of umbilical cord-derived mesenchymal stem cells attenuates surgical wound-induced blood-brain barrier dysfunction in mice. *Stem Cells Int*. 2023;2023.
239. Almalki SG. Adipose-derived mesenchymal stem cells and wound healing. *Saudi Med J*. 2022 Oct 1 [cited 2024 Mar 9];43(10):1075–86. Available from: https://smj.org.sa/content/43/10/1075
240. Wang Z, Feng C, Liu H, Meng T, Huang WQ, Song KX, et al. Exosomes from circ-Astn1-modified adipose-derived mesenchymal stem cells enhance wound healing through miR-138–5p/SIRT1/FOXO1 axis regulation. *World J Stem Cells*. 2023 May 26 [cited 2024 Mar 9];15(5):476–89. Available from: www.ncbi.nlm.nih.gov/pubmed/37342222
241. Ouyang L, Qiu D, Fu X, Wu A, Yang P, Yang Z, et al. Overexpressing HPGDS in adipose-derived mesenchymal stem cells reduces inflammatory state and improves wound healing in type 2 diabetic

mice. *Stem Cell Res Ther.* 2022 Dec 1 [cited 2024 Mar 9];13(1):1–15. Available from: https://stemcellres.biomedcentral.com/articles/10.1186/s13287-022-03082-w

242. Li Q, Qi G, Lutter D, Beard W, Souza CRS, Highland MA, et al. Injectable peptide hydrogel encapsulation of mesenchymal stem cells improved viability, stemness, anti-inflammatory effects, and early stage wound healing. *Biomolecules.* 2022 Sep 1 [cited 2024 Mar 9];12(9):1317. Available from: www.mdpi.com/2218-273X/12/9/1317/htm
243. Sivaraj D, Chen K, Chattopadhyay A, Henn D, Wu W, Noishiki C, et al. Hydrogel scaffolds to deliver cell therapies for wound healing. *Front Bioeng Biotechnol.* 2021 May 3;9:660145.
244. Hu KH, Kuhn NF, Courau T, Tsui J, Samad B, Ha P, et al. Transcriptional space-time mapping identifies concerted immune and stromal cell patterns and gene programs in wound healing and cancer. *Cell Stem Cell.* 2023 Jun 1 [cited 2024 Mar 9];30(6):885–903.e10. Available from: www.cell.com/article/S1934590923001716/fulltext
245. Sawaya AP, Stone RC, Brooks SR, Pastar I, Jozic I, Hasneen K, et al. Deregulated immune cell recruitment orchestrated by FOXM1 impairs human diabetic wound healing. *Nat Commun.* 2020 Sep 16 [cited 2024 Mar 9];11(1):1–14. Available from: www.nature.com/articles/s41467-020-18276-0
246. Wee WKJ, Low ZS, Ooi CK, Henategala BP, Lim ZGR, Yip YS, et al. Single-cell analysis of skin immune cells reveals an Angptl4-ifi20b axis that regulates monocyte differentiation during wound healing. *Cell Death & Disease.* 2022 Feb 24 [cited 2024 Mar 9];13(2):1–9. Available from: www.nature.com/articles/s41419-022-04638-7
247. Zhou C, Zhang B, Yang Y, Jiang Q, Li T, Gong J, et al. Stem cell-derived exosomes: emerging therapeutic opportunities for wound healing. *Stem Cell Research & Therapy.* 2023 Apr 26 [cited 2024 Mar 9];14(1):1–18. Available from: https://stemcellres.biomedcentral.com/articles/10.1186/s13287-023-03345-0
248. Rehak L, Giurato L, Meloni M, Panunzi A, Manti GM, Uccioli L. The immune-centric revolution in the diabetic foot: Monocytes and lymphocytes role in wound healing and tissue regeneration—a narrative review. *J Clin Med.* 2022 Feb 8 [cited 2024 Mar 9];11(3):889. Available from: www.mdpi.com/2077-0383/11/3/889/htm
249. Geng K, Ma X, Jiang Z, Huang W, Gu J, Wang P, et al. High glucose-induced STING activation inhibits diabetic wound healing through promoting M1 polarization of macrophages. *Cell Death Discovery.* 2023 Apr 26 [cited 2024 Mar 9];9(1):1–16. Available from: www.nature.com/articles/s41420-023-01425-x
250. Jiang M, Jiang X, Li H, Zhang C, Zhang Z, Wu C, et al. The role of mesenchymal stem cell-derived EVs in diabetic wound healing. *Front Immunol.* 2023 Feb 28;14:1136098.
251. Petkovic M, Vangmouritzen M, Mojsoska B, Jenssen H. Immunomodulatory properties of host defence peptides in skin wound healing. *Biomolecules.* 2021 Jun 28 [cited 2024 Mar 11];11(7):952. Available from: www.mdpi.com/2218-273X/11/7/952/htm
252. Adib Y, Bensussan A, Michel L. Cutaneous wound healing: A review about innate immune response and current therapeutic applications. *Mediators Inflamm.* 2022;2022.
253. Krizanova O, Penesova A, Sokol J, Hokynkova A, Samadian A, Babula P. Signaling pathways in cutaneous wound healing. *Front Physiol.* 2022 Nov 25;13:1030851.
254. Notari L, Nardini P, Grandi V, Corsi A, Pimpinelli N, Bacci S. Neuroimmunomodulation in chronic wound healing after treatment with photodynamic therapy: The role of iNOs. *Med Sci Forum.* 2023 Mar 8 [cited 2024 Mar 11];21(1):44. Available from: www.mdpi.com/2673-9992/21/1/44/htm
255. Chen RR, Silva EA, Yuen WW, Mooney DJ. Spatio–temporal VEGF and PDGF delivery patterns blood vessel formation and maturation. *Pharm Res.* 2007 Feb;24(2):258–64.
256. Masoomikarimi M, Salehi M. Modulation of the immune system promotes tissue regeneration. *Mol Biotechnol.* 2022 Jan 13 [cited 2024 Mar 11];64(6):599–610. Available from: https://link.springer.com/article/10.1007/s12033-021-00430-8
257. Singer AJ, Clark RAF. Cutaneous wound healing. Epstein FH, editor. *N Engl J Med.* 1999 Sep 2 [cited 2024 Mar 11];341(10):738–46. Available from: https://pubmed.ncbi.nlm.nih.gov/10471461/
258. Michel T, Luft D, Abraham MK, Reinhardt S, Salinas Medina ML, Kurz J, et al. Cationic nanoliposomes meet mRNA: Efficient delivery of modified mRNA using hemocompatible and stable vectors for therapeutic applications. *Mol Ther Nucleic Acids.* 2017 Jul 25 [cited 2024 Mar 11];8:459–68. Available from: https://europepmc.org/articles/PMC5545769
259. Sahana TG, Rekha PD. Biopolymers: Applications in wound healing and skin tissue engineering. *Mol Biol Rep.* 2018 Dec 1 [cited 2024 Mar 11];45(6):2857–67. Available from: https://pubmed.ncbi.nlm.nih.gov/30094529/
260. Fadilah NIM, Phang SJ, Kamaruzaman N, Salleh A, Zawani M, Sanyal A, et al. Antioxidant biomaterials in cutaneous wound healing and tissue regeneration: A critical review. *Antioxidants.* 2023 Apr 1 [cited 2024 Mar 11];12(4). Available from: /pmc/articles/PMC10198372/

261. Bharadvaja N, Gautam S, Singh H. Natural polyphenols: A promising bioactive compounds for skin care and cosmetics. *Mol Biol Rep*. 2023 Feb 1 [cited 2024 Mar 11];50(2):1817–28. Available from: https://pubmed.ncbi.nlm.nih.gov/36494596/
262. Gallo RL. Human skin is the largest epithelial surface for interaction with microbes. *J Investig Dermatol*. 2017 Jun 1;137(6):1213–4.
263. Quintero-Rincón P, Mesa-Arango AC, Flórez-Acosta OA, Zapata-Zapata C, Stashenko EE, Pino-Benítez N. Exploring the potential of extracts from Sloanea medusula and S. calva: Formulating two skincare gels with antioxidant, sun protective factor, and anti-candida albicans activities. *Pharmaceuticals (Basel)*. 2023 Jul 11 [cited 2024 Mar 11];16(7). Available from: www.ncbi.nlm.nih.gov/pubmed/37513902
264. Nqoro X, Taziwa R. Polymer-based functional materials loaded with metal-based nanoparticles as potential scaffolds for the management of infected wounds. *Pharmaceutics*. 2024 Jan 23 [cited 2024 Mar 11];16(2):155. Available from: www.ncbi.nlm.nih.gov/pubmed/38399218
265. Heydari P, Kharaziha M, Varshosaz J, Javanmard SH. Current knowledge of immunomodulation strategies for chronic skin wound repair. *J Biomed Mater Res B Appl Biomater*. 2022 Feb 1;110(2):265–88.
266. Larouche J, Sheoran S, Maruyama K, Martino MM. Immune regulation of skin wound healing: mechanisms and novel therapeutic targets. *Adv Wound Care* (New Rochelle). 2018 Jul 1 [cited 2024 Mar 11];7(7):209–31. Available from: https://pubmed.ncbi.nlm.nih.gov/29984112/
267. San Emeterio CL, Hymel LA, Turner TC, Ogle ME, Pendleton EG, York WY, et al. Nanofiber-based delivery of bioactive lipids promotes pro-regenerative inflammation and enhances muscle fiber growth after volumetric muscle loss. *Front Bioeng Biotechnol*. 2021 Mar 19;9.
268. Aamodt JM, Grainger DW. Extracellular matrix-based biomaterial scaffolds and the host response. *Biomaterials*. 2016 Apr 1;86:68–82.
269. Karkanitsa M, Fathi P, Ngo T, Sadtler K. Mobilizing endogenous repair through understanding immune reaction with biomaterials. *Front Bioeng Biotechnol*. 2021 Nov 30;9:730938.
270. Cui D, Wu B, He D, Wang Y, Jiao Y, Zhang B. 3D-Printed cold preservation device in renal autotransplantation for the treatment of a patient with renal artery stenosis. *Front Bioeng Biotechnol*. 2022 Jan 3;9.
271. Alcazar CA, Hu C, Rando TA, Huang NF, Nakayama KH. Transplantation of insulin-like growth factor-1 laden scaffolds combined with exercise promotes neuroregeneration and angiogenesis in a preclinical muscle injury model. *Biomater Sci*. 2020 Oct 7;8(19):5376–89.
272. García-García A, Martin I. Extracellular matrices to modulate the innate immune response and enhance bone healing. *Front Immunol*. 2019 Sep 20;10.
273. Larouche J, Sheoran S, Maruyama K, Martino MM. Immune regulation of skin wound healing: Mechanisms and novel therapeutic targets. [cited 2024 Mar 11]. Available from: www.liebertpub.com
274. Ellis S, Lin EJ, Tartar D. Immunology of wound healing. *Curr Dermatol Rep*. 2018 Dec 1;7(4):350–8.
275. Masoomikarimi M, Salehi M. Modulation of the immune system promotes tissue regeneration. *Mol Biotechnol*. 2022 Jun 1 [cited 2024 Mar 11];64(6):599–610. Available from: https://pubmed.ncbi.nlm.nih.gov/35022994/
276. Frantz S, Vincent KA, Feron O, Kelly RA. Innate immunity and angiogenesis. *Circ Res*. 2005 Jan 21;96(1):15–26.
277. Chopra H, Kumar S, Singh I. Strategies and therapies for wound healing: A review. *Curr Drug Targets*. 2022 Apr 16 [cited 2024 Mar 11];23(1):87–98. Available from: https://pubmed.ncbi.nlm.nih.gov/33858310/
278. Balta MG, Papathanasiou E, Blix IJ, Van Dyke TE. Host modulation and treatment of periodontal disease. *J Dent Res*. 2021 Jul 1 [cited 2024 Mar 11];100(8):798–809. Available from: https://pubmed.ncbi.nlm.nih.gov/33655803/
279. Sadtler K, Singh A, Wolf MT, Wang X, Pardoll DM, Elisseeff JH. Design, clinical translation and immunological response of biomaterials in regenerative medicine. *Nat Rev Mater*. 2016 Jun 7;1(7).
280. Adams S, Wuescher LM, Worth R, Yildirim-Ayan E. Mechano-immunomodulation: Mechanoresponsive changes in macrophage activity and polarization. *Ann Biomed Eng*. 2019 Nov 1 [cited 2024 Mar 11];47(11):2213–31. Available from: https://pubmed.ncbi.nlm.nih.gov/31218484/

6 The Role of the Extracellular Matrix (ECM) in Wound Healing (I.E., Matrix Metalloproteinases (MMPs) and Growth Factors)

Maya Magdy Abdelwahab,[1*] *Dina Mostafa Abohashema,*[1] *Mahmoud Samy Abdelwahab,*[1] *Mahmoud Talaat Fayed,*[1] *Abdelrahman Ghanem,*[1] *and Osama A. A. Mohamed*[2,3,4]

[1] Faculty of Medicine, Helwan University, Cairo, Egypt.

[2] Universitat Autònoma de Barcelona, Barcelona, Spain.

[3] University of Naples Federico II, Campania, Italy.

[4] Aristotle University of Thessaloniki, Thessaloniki, Greece.

* Corresponding author: Maya Magdy Abdelwahab

6.1 INTRODUCTION

Wound healing is a normal physiological reaction to tissue injury in the human body. It refers to replacing damaged tissue with living tissue, accomplished through tissue regeneration or scar formation (Wallace et al., 2023) The wound healing process encompasses four intertwined phases: hemostasis, inflammation, proliferation, and tissue remodeling or resolution. These phases must occur in the proper order, at a precise time, and persist for optimal durations with the right intensity (Guo, 2010) Initially, Hemostasis, occurring immediately after injury, prevents excessive bleeding through vasoconstriction and platelet activation, adhesion, and aggregation, which halt bleeding at the injury site (Schultz et al., 2011) Following this, the inflammatory phase involves various inflammatory cells, starting with neutrophils, infiltrating the site and releasing cytokines and mediators to stimulate processes like angiogenesis, thrombosis, and re-epithelialization. This limits further damage, closes the wound, removes cellular debris and bacteria, and promotes cellular migration. These stages usually last several days. The proliferative phase includes forming granulation tissue, re-epithelialization, and neovascularization, lasting several weeks. Fibroblasts deposit extracellular components to provide structural support. Finally, the remodeling phase, during which the wound achieves peak strength, signifies maturation (Wallace et al., 2023).

The extracellular matrix (ECM) is a complex network consisting of various multidomain macromolecules arranged in a manner specific to each cell or tissue type. ECM components link together to create a structurally stable framework, influencing the mechanical characteristics of tissues (Yue, 2014)m Additionally, the ECM serves as a reservoir for growth factors and other bioactive molecules. It is a dynamic entity crucial for regulating essential cellular processes such as proliferation, adhesion, migration, polarity, differentiation, and apoptosis (Mecham, 2012). Approximately 300 proteins comprise the "core matrisome," including major components like collagens, proteoglycans, elastin, and cell-binding glycoproteins, each with unique physical and biochemical properties

DOI: 10.1201/9781003605966-6

(Hynes & Naba, 2012). Enzymes such as matrix metalloproteinases and specific glycosidases like heparanase and hyaluronidases play roles in matrix remodeling, influencing human health. Specific ECM components impact cellular processes and functions, including cell proliferation and survival, migration, differentiation, autophagy, angiogenesis, and immune regulation. Structural alterations in the ECM are closely associated with the progression of diseases (Karamanos et al., 2021).

The wound healing process relies on the synthesis, localization, and organization of extracellular matrix (ECM) molecules. ECM molecules are involved in each stage of wound healing, penetrating the entire process (Lin, 2023). Following tissue injury, wound healing begins with hemostasis, where a clot forms with the assistance of ECM proteins like collagens, providing a scaffold for cell migration (Farndale et al., 2004). Epithelialization follows, involving the migration of epidermal cells over the wound site, facilitated by interactions with ECM proteins. (Clark, 1996). Angiogenesis, driven by growth factors and ECM interactions, forms new blood vessels to supply nutrients (Tonnesen et al., 2000). Proliferation entails fibroblast proliferation and ECM synthesis, leading to granulation tissue formation. Lastly, remodeling involves ECM degradation by matrix metalloproteinases (MMPs), leading to scar formation characterized by collagen accumulation (Velnar, 2009).

This chapter reviews the ECM's critical role in wound healing, emphasizing the functions of matrix metalloproteinases (MMPs) and growth factors. It overviews wound healing and its stages, highlighting the importance of the ECM in each stage. It further discusses the ECM's structure and functions, including its role in providing structural support, regulating cell behavior and signaling pathways, modulating angiogenesis, and influencing tissue remodeling. Additionally, it discusses the significant role of MMPs in modulating the ECM during wound healing and their regulation. Lastly, the chapter covers therapeutic approaches targeting the ECM for wound healing.

6.2 STRUCTURE AND COMPOSITION OF ECM

6.2.1 Organization in Tissues

The ECM is an essential, active, organized meshwork of numerous secreted molecules and enzymes that regulate cellular metabolism, control morphogenesis, and provide structural support for cells and tissues (Xue & Jackson, 2015). Two forms of ECM have been identified: the basement membrane and the interstitial ECM, as shown in **Figure 6.1**. The basement membrane can also be described as the pericellular matrix, which provides cellular attachment. On the other hand, the interstitial matrix plays a role in enhancing tissue integrity (Dzobo & Dandara, 2023).

6.2.1.1 Basement Membrane

Basement membranes (BMs) are specialized structures within the extracellular matrix that compartmentalize tissues, offer structural reinforcement, and influence cell behavior and signaling. They underlie epithelial and endothelial cells, encase smooth muscle, adipose tissue, and Schwann cells, and are found in synaptic clefts (Randles et al., 2017). BMs connect cells with the interstitial matrix containing fibrillar collagens like collagen I. BMs vary in composition to provide diverse biomechanical and biochemical properties suited to their specific functions (Gatseva et al., 2019).

6.2.1.2 Interstitial Matrix

The interstitial extracellular matrix (ECM) comprises different components that assemble to generate a complex 3D network of molecules underlying the basement membrane (Dzobo & Dandara, 2023). Variations in ECM gene expression, posttranscriptional splicing, and posttranslational modifications result in the production of specific types of ECM for numerous tissues and organs (Yin et al., 2021). The molecules of ECM are of different sizes, shapes, and spatial arrangements. The ECM is dynamic and continuously changes during tissue development and disease progression (W. Chen et al., 2022).

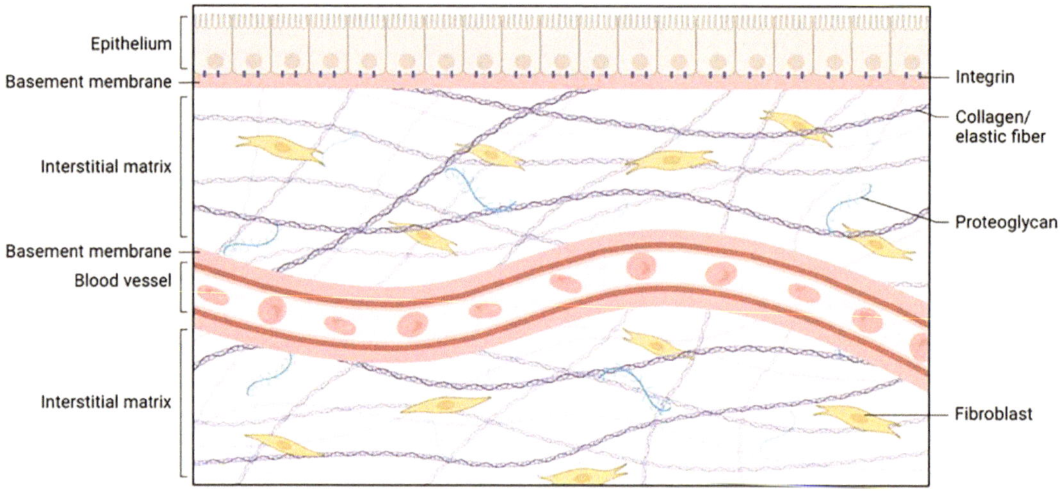

FIGURE 6.1 Schematic representation of ECM structural organization.

This figure illustrates the two primary extracellular matrix (ECM) forms—the basement membrane and the interstitial matrix. It highlights the distinct structural and compositional differences between these ECM forms (Huang E., 2023).

6.2.2 Components

6.2.2.1 Collagen

Collagens are significant ECM proteins, all of which share the same structure: the triple helix. They are considered the most abundant proteins in the body. Twenty-eight types of collagen have been identified and assigned Roman numbers to differentiate them (I–XXVIII) (Karamanos et al., 2021). Collagen is important for providing a substrate for cellular differentiation and proliferation. It also plays a role in the elasticity and mechanical strength of tissues. Fibroblasts are the source of collagen synthesis in a healing wound (Sorushanova et al., 2019). Several factors influence skin healing, including collagen organization, type, and amount. Collagen type I is the most abundant collagen in the skin. However, the primary collagen synthesized during the early stages of wound healing is type III, which is later replaced by type I (Mathew-Steiner et al., 2021). After wound closure, collagen remains in a dynamic state of remodeling, which is necessary to improve the tensile strength of the repaired tissues months after the initial insult (Mathew-Steiner et al., 2021)

6.2.2.2 Elastin

Elastic fibers are responsible for the stretch and recoil properties of the skin. Elastin, a key component of the ECM, specifically in the dermis, is composed of tropoelastin that attaches to microfibrils and forms elastic fibers (Baumann et al., 2021) Elastin formation occurs in a developing fetus's lungs, skin, and blood vessels during the early neonatal period, a process known as elastogenesis (Baumann et al., 2021). The primary role of elastin in wound healing is to provide mechanical elasticity and improve dermis regeneration by acting on cells that reduce wound contraction (Almine et al., 2012). Various studies have mentioned tropoelastin's significant role in the different stages of wound healing; for example, it induces transient expression of chemokines, which are crucial for tissue repair and recovery (Zhang et al., 2022). Since tropoelastin expression stops after the embryonic and neonatal periods, no elastin production exists in adults. However, its synthesis is induced by injury or disease, and the expression of elastic fiber proteins is influenced by various signaling molecules such as transforming growth factor β (TGF-β) and tumor necrosis factor-α (Almine et al., 2012). Scar tissue has reduced elasticity compared to the original healthy tissue due to limited and insufficient production of elastic fibers, which blocks the reformation of an elastic fiber network in the scar tissue. (Rodríguez-Cabello et al., 2018)

6.2.2.3 Proteoglycans

Proteoglycans are a multifunctional family of cell-associated proteins and matrix components, each with a specific localization in the skin (Smith & Melrose, 2015). The primary structure of proteoglycans (PGs) is made up of core proteins or glycoproteins attached to side chains of glycosaminoglycans (GAGs). PGs play essential roles in cellular interactions, signaling, migration, and cell proliferation associated with wound healing (Ghatak et al., 2015). Numerous PG types have been recognized. They are further classified based on their GAG side chain and their location in tissues, with an explicit distinction between those that reside in the ECM and those that are cell-associated (Smith & Melrose, 2015). PGs help hydrate the ECM, allow the permeability of low molecular weight solutes, and organize the matrix by interacting with other ECM components (Couchman & Pataki, 2012)

Composition of Extracellular Matrix (ECM)

6.3 FUNCTIONS OF ECM IN WOUND HEALING

6.3.1 Provision of Structural Support

The ECM plays an important role in providing cells with structural support. It provides a structural skeleton that holds cells within connective tissue together, which is fundamental for tissue integrity and function (Discher et al., 2005; Hynes, 2009). Furthermore, integrins help the ECM serve as an anchoring platform for cells, allowing cell adhesion and signaling (Larsen et al., 2006).

The anatomy of the ECM is organized into three-dimensional (3D) patterns, mostly made up of the interstitial matrix and the basement membrane (BM). The BM provides tissues with the structural support and tensile strength required due to its richness in fibronectin, laminins, and collagen type IV, the BM. Additionally, the BM is vital for maintaining endothelial and epithelial tissue integrity and blood vessel formation (Kim et al., 2011; Lu et al., 2012). Integrins control cell shape, motility, and signaling through mediating contacts between the BM and cells (Tanzer, 2006).

Tissue architecture and function also depend on the interstitial matrix between connective tissue cells and the BM. Comprising fibronectin, elastin, and collagen, the interstitial matrix creates a three-dimensional support network. Fibronectin, in particular, helps to organize the matrix structure. (Frantz et al., 2010; Kumar et al., 2013).

Functionally, the ECM's structural support offers anchoring sites and directs cell division and migratory tracking, which promotes cell migration. Besides, the ECM influences cell communication pathways and serves as a repository for growth factor (GF) signaling molecules, further controlling cell behavior (Lu et al., 2012). Furthermore, the ECM's biomechanical characteristics, such as its elasticity, significantly influence cell behavior and differentiation processes (Engler et al., 2006).

6.3.2 Regulation of Cell Behavior

ECM regulates cell behavior at different stages of wound healing. During the inflammatory phase, the ECM promotes cellular migration and proliferation, aiding tissue repair (Midwood et al., 2004). Factors including platelet-derived growth factor (PDGF) and transforming growth factor β (TGF-β) promote wound healing by facilitating the migration of macrophages, neutrophils, fibroblasts, and smooth muscle cells to the wound site (Anitua et al., 2004; Diegelmann & Evans, 2004) Moreover, ECM helps in the formation of granulation tissue and the proliferation of fibroblasts through the action of growth factors released from macrophages, which developed from circulating monocytes (Diegelmann & Evans, 2004). ECM influences the resolution of inflammation by regulating the migration of fibroblasts, endothelial cells, and mesenchymal stem cells, which is essential for angiogenesis (Tonnesen et al., 2000).

The ECM facilitates the creation of new cells during the proliferative phase. New blood vessels and lactate synthesized by macrophages are two stimulants that cause fibroblasts to create a new ECM (Ghani et al., 2004). One of the most important steps is the restoration of the permeability barrier achieved by the action of growth factors produced by macrophages, activated platelets, and keratinocytes, which facilitates epithelialization ((Mascharak et al., 2021; R. A. Yates et al., 1991). Later, myofibroblasts formed from fibroblasts promote wound constriction (R. A. Clark et al., 1996; Noskovicova et al., 2021). Matrix production and remodeling occur simultaneously during the proliferative phase, where the ECM is the basic support for tissue remodeling ((Coelho & McCulloch, 2016; Tonnesen et al., 2000).

ECM is also essential during tissue remodeling. Collagen in the ECM is synthesized and degraded by macrophages and fibroblasts to restore tissue strength (Diegelmann & Evans, 2004; Mascharak et al., 2021). Enzymes like transglutaminases and lysyl oxidases mediate the cross-linking and structural alterations of the collagenous matrix, which results in the replacement of the fibronectin-rich matrix with a stronger collagenous matrix (Coelho & McCulloch, 2016; Gurtner et al., 2008).

6.3.3 Modulation of Cell Signaling Pathways

During wound healing, ECM regulates cell signaling by controlling matrix metalloproteinases (MMPs) such as collagenases and gelatinases. These MMPs target collagen and elastin, important ECM proteins, and release bioactive fragments known as matrikines, which promote tissue regeneration by facilitating fibroblast proliferation and collagen synthesis (Rose et al., 2021; Wells et al., 2016).

Furthermore, fibronectin is an essential ECM protein with a significant role in cell signaling. It interacts with integrins and TGF-β to regulate lysyl oxidase activity, promote collagen deposition, and stimulate fibroblast migration and proliferation (Weisel & Litvinov, 2017). In addition, vitronectin and tenascin-C, other ECM constituents, influence wound healing by regulating fibroblast migration, adhesion, and proliferation, which is important for cell signaling as well (Luo et al., 2017; C. C. Yates et al., 2011). Besides, during wound healing, matrix-cell proteins such as osteopontin, osteonectin, thrombospondins, fibulins, and CCN family proteins function as signaling molecules inside the ECM, controlling fibroblast activity, ECM formation, and tissue remodeling (Abdelaziz Mohamed et al., 2019; Kaasbøll et al., 2018; C. C. Yates et al., 2011).

6.3.4 Influence on Tissue Remodeling

(Diegelmann & Evans, 2004; Mascharak et al., 2021). This stage, which usually starts two to three weeks following the injury, may continue for a year or more (Gurtner et al., 2008). The length of remodeling is primarily determined by the production and breakdown of collagen within the ECM (Harper et al., 2014). As wound healing advances, TGF-β1 stimulates collagen production from fibroblasts, forming a stronger collagenous matrix that replaces the fibronectin-rich matrix (Gurtner et al., 2008; Li et al., 2007). This matrix is further strengthened by transglutaminases and lysyl oxidase enzymes through cross-linking and modifications (Coelho & McCulloch, 2016).

Proper wound healing requires a balance between collagen synthesis and breakdown during remodeling (R. A. Clark et al., 1996; Lee et al., 2021). MMPs break down ECM proteins, gradually replacing type III collagen with type I collagen, resulting in a type I collagenous scar (Overall & López-Otín, 2002; Visse & Nagase, 2003). This scar has a lower tensile strength than normal tissue due to the absence of vascularization and anatomical organization (Abercrombie et al., 1956; Kumar et al., 2009). A maximum of 70–80% of the tensile strength of normal tissue can be achieved despite comprehensive wound healing (Carr, 2022; R. A. Clark et al., 1996; Diegelmann & Evans, 2004).

6.4 ROLE OF ECM IN WOUND HEALING PHASES

6.4.1 Hemostasis

The initial phase of wound healing begins as soon as an injury is noticed (Shah et al., 2012). The process starts with the narrowing of the injured vasculature, which is brought on by the action of vasoconstriction hormones. These include adrenaline, thromboxane A2, and serotonin (Wu & Chen, 2014). The platelets are early regulators of the healing process (Diegelmann & Evans, 2004). When they come into contact with the collagen of the injured vessels, they go through adhesion, aggregation, and activation. This causes the release of adhesion glycoprotein and ADP from the damaged vessels, which in turn promotes more platelet aggregation (Wu & Chen, 2014). The primary glycoproteins extruded from platelet granules include von Willebrand's factor, fibrinogen, fibronectin, vitronectin, and thrombospondin (Brass, 2003). The surface of activated platelets serves as a site for prothrombin activation concurrently, resulting in the production of active thrombin, the critical component of the coagulation process that catalyzes the conversion of fibrinogen into fibrin and, ultimately, the formation of a blood clot (Hoffman et al., 2006). Many growth factors are released from α granules by the aggregated platelets that are "trapped" in the provisional matrix (Olczyk et al., 2014). Therefore, PDGF and TGF-β recruit neutrophils and monocytes into the wound area to start the inflammatory response (Diegelmann & Evans, 2004). The by-products of the breakdown of bacteria and C5a complement are additional chemotactic stimuli that increase the recruitment of neutrophils. TGF-α, Bfgf, and VEGF stimulate endothelial cells to start angiogenesis. PDGF and IGF-1, on the other hand, activate and recruit fibroblasts, causing them to migrate into the wound area, proliferate, and begin the manufacture of collagen and glycosaminoglycans (Yang et al., 2011).

6.4.2 Inflammatory Stage

The healing process's inflammatory phase begins 24 hours after an injury occurs and can last up to 48 hours on average (Busti et al., 2005). The first vessel contraction in the early inflammatory phase of the wound healing cascade is followed by a widening of the lumen and an increase in the vascular permeability of the walls, which encourages plasma to "leak" into the injured tissue area (Sinno & Prakash, 2013). Prostaglandins, histamine, kinins, leukotrienes, and acid hydrolases all contribute to the alterations (Wu & Chen, 2014). The latter plays a significant role in the immunological defense of the wound and is a primary inducer of VEGF synthesis (Olczyk et al., 2014). Neutrophils and monocytes/macrophages are the main inflammatory phase cells (Koh & DiPietro, 2011). These cells not only actively phagocytose and debride the wound to maintain its aseptic state, but they also produce a significant amount of active mediators, such as growth factors and cytokines, whose release is essential for starting the subsequent stage of the healing process (Gurtner et al., 2008). The first inflammatory cells to show up in the wound region are neutrophils. Neutrophils are "attracted" to the site of damage by chemotactic agents. These cells are the first line of defense against infections, phagocytizing and destroying the bacteria by producing reactive oxygen and nitrogen species and breaking down the injured connective tissue components by releasing proteases (Taylor et al., 2004). After two or three days in the wound area, monocytes replace the reduced neutrophils due to apoptosis (Shah et al., 2012). Monocytes go from the capillary to the extracellular matrix (ECM), where they become macrophages in response to inflammatory mediators like TGF-N and byproducts of fibrin and fibronectin breakdown originating from the "temporary" wound matrix (Guo & DiPietro, 2010). Macrophages, like neutrophils, have two roles in the healing process (Eming et al., 2007). By releasing the plasminogen activator, they induce the fibrin fabric to be removed. In addition to managing inflammation, the aforementioned mediators also affect angiogenesis, collagen synthesis, and epithelialization (Reinke & Sorg, 2012). During the advanced stages of inflammation, lymphocytes penetrate the wound environment and impact the proliferation of fibroblasts and the production of collagen (Guo & DiPietro, 2010).

6.4.3 Proliferation Phase

Following two to three days of hemostasis and inflammation, the injured tissue regeneration process picks up speed (Guo & DiPietro, 2010). The rise in cell density in the wound bed during this period is linked to the migration and proliferation of keratinocytes, endothelial cells, and fibroblasts. The aforementioned mediators induce and regulate angiogenesis, epithelialization, and ECM production (Olczyk et al., 2014). Collagen matrix, rich in proteoglycans, glycosaminoglycans, and noncollagenous glycoproteins, replaces a temporary matrix mainly composed of a fibrin and fibronectin network. This process ultimately restores the correct tissue structure and function (Midwood et al., 2004). Fibroblasts are the main cells in the phase under discussion. They are mostly made of nondifferentiated mesenchymal cells in the dermis that change into fibroblasts when exposed to growth factors and cytokines generated by neutrophils, macrophages, and blood platelets (Sorrell & Caplan, 2004). Within 48 to 72 hours of the injury first appearing, these cells move to the damaged area (Busti et al., 2005). According to the chemotactic PDGF, EGF, IGF-1, and TGF-β gradient, the affected cells are "attracted" to the wound location, where growth factors boost their proliferation. Subsequently, the synthesis of ECM components and the developing "granulation" tissue commence (Barrientos et al., 2008). The newly formed connective tissue framework that is "intertwined" by numerous capillaries has a distinct, granulated appearance, which is the source of the word "granulation" (Olczyk et al., 2014). This tissue becomes visible about the fourth day following the injury (Nauta et al., 2011). The early granulation tissue matrix (up to the third-day postinjury) is heavily concentrated in hyaluronic acid and fibronectin (Olczyk et al., 2014). On the other hand, fibronectin produces "scaffolding" that promotes collagen fibrogenesis (Busti et al., 2005). The hyaluronic acid content in the wound area rapidly drops beginning on the third day following the injury, and collagen replaces the glycosaminoglycan. From the time the wound first formed until the third week, the granulation tissue's collagen content rises. This is followed by a gradual reduction in fibroblasts until their disappearance during the apoptotic phase (Hinz, 2007). Types I and III collagen predominate in the dermis and are found in a 4:1 ratio. However, collagen type III is more common in the early stages of healing. The freshly formed tissue is "toughened" by this protein, which gives it tensile strength (Hoffman et al., 2006). The granulation tissue eventually becomes a scar during remodeling, temporarily replacing the dermis (Olczyk et al., 2014). Then, epithelialization occurs; the multiphase process involves rebuilding the epithelium following an injury (Diegelmann & Evans, 2004). As they contribute to the closure of the wound surface, epithelial cells come from both the borders of the wound and epithelial appendages such as sweat glands, sebaceous glands, and hair follicles. Growth factors, including EGF, KGF, and TGF-α, are the mediators that boost the migration and proliferation of the aforementioned cells (Reinke & Sorg, 2012).

6.4.4 Remodeling Phase

The final stage of the healing process is remodeling. The injured surface contracts as it heals. The phenotypic development of preexisting fibroblasts into myofibroblasts is the primary process responsible for wound contracture. The latter comprises microfilaments called alpha-smooth muscle actin (α-SMA), which provide the cells with their ability to contract (Olczyk et al., 2014). The integrin receptors $\alpha 1\ \beta 1$ and $\alpha 2\ \beta 1$ then interact with particular locations on collagen to mediate the granulation tissue's contraction. Because of the aforementioned change, myofibroblasts become the most prevalent cell group in the granulation tissue during the second week of healing (Gurtner et al., 2008). The granulation tissue "matures" into a scar during this stage of the healing process, and the produced tissue also gains more mechanical strength. The granulation tissue's maturation process includes a decrease in capillaries by clumping together to form larger blood vessels and a decrease in the water content associated with glycosaminoglycans and proteoglycans (Diegelmann & Evans, 2004). The tissue's tensile strength is enhanced by changes in the mutual proportion of collagen types (type I collagen content increases in favor of type III collagen content), total collagen content, spatial organization, and the number of covalent cross-links. After many months of rebuilding, the tensile strength equals 80% of that of the unmodified tissue. In contrast, in the event of a wound newly covered with epithelium, it equals 25% compared to the dermis (Vedrenne et al., 2012). Fibroblast count and vascular density both decline during the remodeling period. The original scar tissue, which was initially composed of finely arranged, inadvertent collagen fibers typical of the proliferation phase, is replaced by a dermis-like matrix in which mature, cross-linked collagen fibers with the appropriate diameter form the framework of the newly created tissue (Olczyk et al., 2014). Growth factors and cytokines control the cells' activities in the healing process. Integrin receptors and sticky molecules facilitate contact between these cells and other extracellular matrix (ECM) components. Endothelial and fibroblast cells release matrix metalloproteinases, which allow them to migrate; neutrophil and macrophage proteases, on the other hand, remove damaged matrix components, aiding in the remodeling of the original scar tissue. Extracellular matrix elements are crucial in healing (Reinke & Sorg, 2012) (**Figure 6.2**).

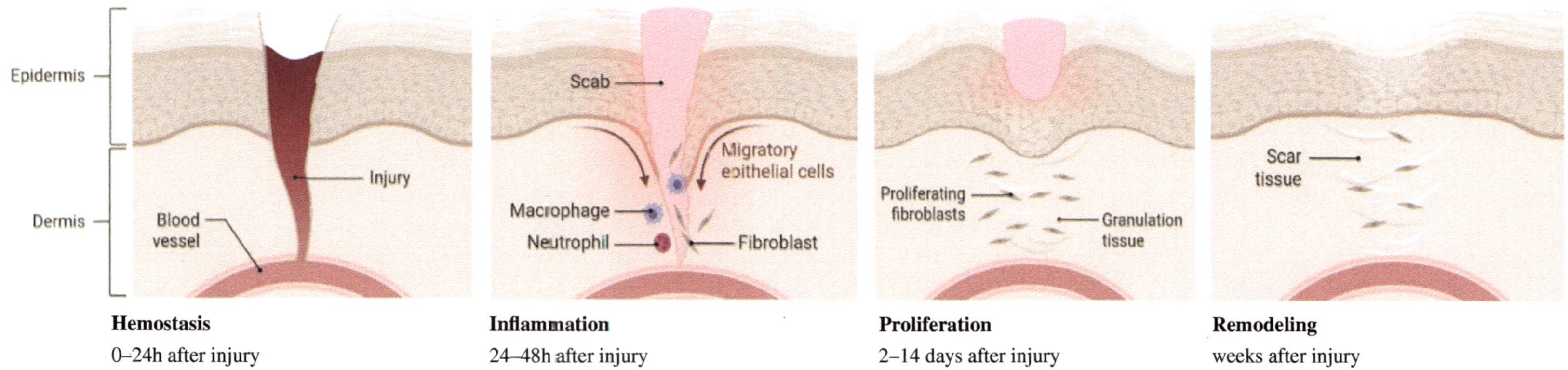

FIGURE 6.2 Stages of Wound Healing.

This figure illustrates the chronological sequence and key activities across the four major phases of wound healing: Hemostasis, Inflammatory, Proliferation, and Remodeling. (BioRender.com. 2024).

6.5 ECM ALTERATIONS ASSOCIATED WITH WOUND HEALING IMPAIRMENT

6.5.1 Fibrosis and Chronic Wounds

Chronic wounds, such as hypertrophic and keloid ones, are likely associated with impaired wound healing as they exhibit ECM composition alterations crucial to the wound healing process (Profyris et al., 2012; Sidgwick & Bayat, 2012). These chronic wounds develop abnormal scars that raise significant challenges due to their functional limitations and disfiguring appearance (Castoldi & Rosing, 2010).

Hypertrophic scars, resulting from thermal or traumatic injuries, are characterized by excessive dermal fibrosis and altered collagen synthesis and degradation during wound healing (Armour et al., 2007). The primary cause of these changes is aberrant ECM remodeling, which results in elevated, red, and stiff scars. However, the more severe type of scars, known as keloid scars, exhibit abnormal collagen production and deposition, creating enormous fibrous masses that extend beyond the initial incision area. Keloid fibroblasts are more active in the synthesis of collagen and fibronectin, which is why keloid scars can grow indefinitely and are aggressive in nature (Baisch & Riedel, 2006; Fujiwara et al., 2005; Gauglitz et al., 2011).

Hypertrophic and keloid scars can be further distinguished from normal tissue by variations in the composition and structure of collagen (Gauglitz et al., 2011; Shih et al., 2010; Verhaegen et al., 2009). Comparing keloids to normotrophic scars, the former shows large, thick collagen fibers with less organization and fewer cross-links. Hypertrophic scars, conversely, have comparatively smaller collagen bundles, which is one of their unique morphological features (Linares et al., 1972).

Proteoglycans, including versican and decorin, are essential for modifying the composition of the extracellular matrix and skin properties (Armour et al., 2007). Expression variations of proteoglycans, specifically decorin levels, decrease in hypertrophic scars, resulting in abnormal collagen organization and accelerated extracellular matrix formation, aggravating the impairment of wound healing (Sayani et al., 2000; Hayat et al., 2023; Zhang et al., 2004).

The ECM changes within wound healing are further exacerbated through the dysregulated activity of matrix metalloproteinases (MMPs) and transforming growth factor-β (TGF-β). Stimulating fibroblasts to produce more collagen and aggravating the scarring process, TGF-β1 expression is upregulated in these scars. In contrast, TGF-β3 levels are reduced, interrupting the balance between the various TGF-β isoforms and encouraging fibrosis (Eto et al., 2012; Ulrich et al., 2010). However, low MMP levels and high tissue inhibitors of metalloproteinases (TIMPs) distort normal ECM turnover and remodeling, further impairing wound healing in chronic wounds (Abdou et al., 2011; Bertina et al., 1994).

As fibrosis's hallmark, extracellular matrix accumulation can significantly slow wound healing (Eto et al., 2012; Ulrich et al., 2010). This is apparent where there are shifts in the composition of the ECM seen in scar formation. For example, the behavior of fibroblasts in hypertrophic scars markedly differs from that of normal fibroblasts (Blackburn & Cosman, 1966; Ehrlich et al., 1994). Moreover, the irregular arrangement of collagen and the increased production of ECM are attributed to the decrease in decorin levels within hypertrophic scar tissue (Sayani et al., 2000; Zhang et al., 2004). Along with the changes in the composition of proteoglycans, these alterations play an essential role in hypertrophic scarring development.

Furthermore, compared to fibroblasts in normal scars or skin, keloid fibroblasts show raised activity in collagen and fibronectin synthesis, which contributes to the aggressive nature of keloid scars (Babu et al., 1989; Chua et al., 2011).

6.5.2 Aging-Related Alterations

A factor that significantly contributes to the deterioration of wound healing, especially chronic wounds, is the extracellular matrix's (ECM) age-related changes (**Table 6.1**). As the ECM structure

TABLE 6.1
Comparison of ECM Components and Their Role in Normal, Hypertrophic, and Keloid Scars

ECM Component	Normal Tissue	Hypertrophic Scars	Keloid Scars	Role in Wound Healing
Collagen	Well-organized, balanced	Excessive, smaller bundles	Excessive, large, disorganized fibers	Provides structural integrity and strength
Proteoglycans (decorin)	Normal levels	Decreased levels	Decreased levels	Regulates collagen organization and ECM structure
TGF-β1	Normal expression	Increased expression	Increased expression	Promotes collagen deposition and fibrosis
TGF-β3	Normal expression	Decreased expression	Decreased expression	Balances TGF-β1 activity, reduces fibrosis
MMPs	Normal levels	Decreased levels	Decreased levels	Degrades ECM components, enabling remodeling
TIMPs	Normal levels	Increased levels	Increased levels	Inhibits MMPs, preventing ECM degradation
Fibroblast activity	Normal collagen synthesis	Increased collagen synthesis	Highly active collagen synthesis	Synthesizes ECM components, crucial for wound healing
Fibronectin	Normal levels	Normal or elevated levels	Elevated levels	Facilitates cell adhesion and ECM assembly

and composition change with age, it affects the healing of scars, which may lead to aberrant scar formation (Castoldi & Rosing, 2010; Profyris et al., 2012).

Prominent features of age-related ECM changes are collagen production and breakdown dysregulation. Normally, the skin's elasticity and integrity are maintained by the balance between collagen I and III levels. However, there is an imbalance in the synthesis and breakdown of collagen in aging skin and chronic wounds, such as keloid and hypertrophic scars. This results in excessive collagen deposition and morphological alterations, leading to aberrant scars (Armour et al., 2007; Sayani et al., 2000; Zhang et al., 2004).

Moreover, age-related modifications in the expression of proteoglycans, such as decorin, also lead to changes in the extracellular matrix that are linked to poor wound healing. Decorin is a vital proteoglycan in healthy skin that controls the production of collagen and the proliferation of fibroblasts. These decorin levels are markedly lowered in hypertrophic scars; thus, the collagen organization becomes aberrant, and ECM levels are elevated. The former changes worsen wound healing and play a role in developing chronic wounds in older people (Sayani et al., 2000; Zhang et al., 2004).

ECM remodeling during wound healing is also influenced by age-related changes in growth factor signaling, specifically TGF-β. For example, TGF-β1 is a highly expressed factor in aged skin and chronic wounds, whereas TGF-β3 expression is lower, resulting in excessive collagen deposition and fibrosis (Abdou et al., 2011; Honardoust et al., 2012).

Additionally, age-related alterations in MMPs and TIMPs contribute to impaired ECM turnover and remodeling in chronic wounds. Decreased MMP levels and increased TIMP levels lead to tissue fibrosis and excessive scar formation (Eto et al., 2012; Ulrich et al., 2010).

6.6 MATRIX METALLOPROTEINASES IN WOUND HEALING

6.6.1 Overview of Matrix Metalloproteinases (MMP)

The extracellular matrix is degraded and reorganized in various physiological and pathological settings through a family of zinc-dependent endopeptidases known as matrix metalloproteinases (Page-McCaw et al., 2007). Their crucial roles impact numerous biological processes, such as tissue remodeling, wound healing, angiogenesis, and immune response modulation (Nagase et al., 2006). Dysregulation of MMP activity has been linked to several illnesses, such as neurological disorders, cancer, and rheumatoid arthritis (Sternlicht & Werb, 2001).

When matrix metalloproteinases are first synthesized, they are inactive, and proteolytic activation is essential for their catalytic potential to be released (Brew & Nagase, 2010). After being activated, MMPs coordinate the breakdown of different components of the extracellular matrix, such as glycoproteins, collagens, and proteoglycans, affecting the tissues' structural integrity (Nagase et al., 2006).

Furthermore, MMPs perform nonproteolytic functions such as engaging in cell signaling, migration, and immune modulation, extending their functional spectrum beyond ECM degradation. These various involvements underline the different contributions of MMPs to physiological and pathological processes (Overall & Kleifeld, 2006).

MMPs, such as collagenases, gelatinases, stromelysins, and membrane-type MMPs (MT-MMPs), are divided into many subgroups according to their structural properties and substrate specificity. Hence, these subgroups have distinct cellular localization and substrate preferences, allowing them to engage in particular physiological processes (Nagase et al., 2006).

6.6.2 Role of Matrix Metalloproteinases in Extracellular Remodeling

Matrix metalloproteinases (MMPs) are key factors in the complex process of extracellular remodeling. They control a range of physiological and pathological processes, specifically wound healing, as they precisely regulate the dynamic equilibrium, determining the extracellular matrix's composition and structure (Nagase & Woessner, 1999). Moreover, to provide room for repair activities, MMPs primarily break down and process a variety of ECM components, which is necessary to deposit new ECM (Nagase & Woessner, 1999).

Additionally, MMPs have many roles in regulating cell behavior during tissue remodeling, such as migration, proliferation, and differentiation, exerting a profound influence over the cellular dynamics integral to wound repair (Murphy & Nagase, 2008). By controlling the bioavailability of cytokines, growth factors, and cell adhesion molecules, MMPs can dictate the cellular responses necessary for successful tissue regeneration. Moreover, MMPs play a role in the structural reorganization of the extracellular matrix (ECM) essential for restoring tissue architecture and function (Visse & Nagase, 2003).

6.6.3 Regulation of Matrix Metalloproteinases Activity during Wound Healing

Normally, MMPs are expressed at basal levels, but during tissue remodeling—such as wound healing—their levels can quickly increase (Page-McCaw et al., 2007). They are expressed by various skin cell types, including keratinocytes, fibroblasts, endothelial cells, and inflammatory cells, highlighting their varied functions in tissue repair (Page-McCaw et al., 2007).

Many signals, including cytokines, hormones, and cell-ECM interactions, induce the expression of MMPs (Sternlicht & Werb, 2001). These signals result in MMP transcriptional activation through the activation of many signaling pathways, such as NF-κB, MAP kinase, Smad-dependent pathways, focal adhesion kinase (FAK), and Wnt signaling (Sternlicht & Werb, 2001).

MMP activity is strictly regulated as it must be enzymatically activated to become catalytically active, as they are first expressed as inactive proenzymes (pro-MMP) (Nagase et al., 2006). By breaking the link between the prodomain and the zinc ion in the active site, serine proteinases or other MMPs can activate pro-MMPs (Nagase et al., 2006). Although they selectively block the enzymatic activity of MMPs, tissue inhibitors of metalloproteinases (TIMPs) also control MMP activity (Nagase et al., 2006).

Alongside their traditional role in ECM degradation and remodeling, MMPs have various functions in regulating cell-cell and cell-matrix interactions (Sternlicht & Werb, 2001). They can also change cell surface receptors and junctional proteins, liberate cytokines and growth factors that have been trapped in the extracellular matrix, and control biological processes, including inflammation and cell death (Sternlicht & Werb, 2001) Keratinocyte migration during wound healing is promoted through the release of physiologically active fragments by MMPs from extracellular matrix proteins like laminin-332 (Sternlicht & Werb, 2001).

Some MMPs have unique functions in cell migration and ECM remodeling during wound healing. For instance, MMP-1, MMP-8, and MMP-13 facilitate collagen degradation, promoting keratinocyte migration (Page-McCaw et al., 2007). MMP-2 and MMP-9 stimulate angiogenesis by activating proangiogenic cytokines such as VEGF and control keratinocyte migration by cleaving laminin-332 (Page-McCaw et al., 2007). Dysregulated MMP expression or activity can impair wound healing, as observed in chronic wounds, underscoring the importance of MMP regulation in tissue repair (Page-McCaw et al., 2007).

6.7 GROWTH FACTORS IN WOUND HEALING

The intricate chain of events, including cell migration, division, differentiation, and protein expression during wound healing, is mediated by intercellular signaling, embodied in growth factors. Wound healing involves several growth factors, such as EGF, TGF-β, FGF, VEGF, and PDGF, the expression of which changes during the process, as does the expression of their receptors (Vaidyanathan, 2021).

6.7.1 Epidermal Growth Factor (EGF)

Growth factors such as epidermal growth factor (EGF) and epidermal growth factor receptor (EGFR) bind to promote cell growth and proliferation (Okabe et al., 2013). Other family members include betacellulin, amphiregulin, epiregulin, neuregulin, heparin-binding EGF (HB-EGF), and transforming growth factors (Werner & Grose, 2003). EGF, a crucial controller of epithelial cell motility, influences the rate of re-epithelialization by interacting with EGFR in the cells at the wound site. It promotes dermal maturation and helps to aid in wound contraction by inducing fibroblast migration and proliferation (Bodnar, 2013) (**Table 6.2**).

6.7.2 Transforming Growth Factor (TGF)

About 33 distinct proteins, members of the human TGF-b superfamily, are important players in the control of tissue homeostasis in multicellular animals through cellular signaling (Clark & Coker, 1998). Neutrophils, fibroblasts, platelets, and other cell types abundantly produce TGF-b. This major cytokine binds to the type II serine-threonine kinase receptor to start intracellular signaling pathways. This process affects gene transcription, affecting all stages of wound healing (Jagadeesan & Bayat, 2007). By encouraging endothelial progenitor cells to ease blood flow to the wound site, TGF-b1 promotes angiogenesis (W. Chen & Ten Dijke, 2016). Myofibroblast differentiation is a primary function of TGF-b1, making it a promising target for treating hypertrophic scars and keloids.

TABLE 6.2
Types and Functions of MMPs

MMP Subtype	Examples	Primary Substrates	Physiological Roles	Pathological Implications
Collagenases	MMP-1, MMP-8, MMP-13	Collagens (I, II, III)	Tissue remodeling, wound healing	Osteoarthritis, rheumatoid arthritis
Gelatinases	MMP-2, MMP-9	Gelatins, collagen IV, elastin	Angiogenesis, wound healing	Cancer metastasis, chronic wounds
Stromelysins	MMP-3, MMP-10	Proteoglycans, fibronectin, laminin	ECM remodeling, cell migration	Arthritis, cancer
Membrane-type MMPs	MT1-MMP, MT2-MMP	Collagen, laminin, proteoglycans	Cell migration, invasion	Cancer invasion, chronic inflammation
Other MMPs	MMP-7 (matrilysin), MMP-12 (metalloelastase)	Various ECM components, elastin	ECM degradation, inflammation regulation	Pulmonary diseases, cancer

6.7.3 Platelet-Derived Growth Factor (PDGF)

Platelet-derived growth factor (PDGF) is key in controlling numerous cellular processes associated with wound healing, such as chemotaxis, angiogenesis, and the mitogenesis of fibroblasts. The protein has three isoforms: AA, AB, and BB. It comprises A and B polypeptide chains (Vaidyanathan, 2021). Certain cell types exhibit chemotaxis when their PDGF receptors are activated. Yet, some cell types, such as smooth muscle cells and fibroblasts, have been shown to exhibit chemotaxis inhibition in response to PDGF receptor activation (Heldin & Westermark, 1999). The intracellular Ca2+ concentrations rise in response to both receptor types. Accordingly, it is known that the receptors have an anti-apoptotic impact and prevent cell-to-cell communication via gap junctions. Incisional and excisional wounds were treated with PDGF at concentrations of 20 – 200 picomoles, which accelerated the infiltration of fibroblasts and inflammatory cells, extracellular matrix deposition, and collagen formation. The effect was exaggerated, leading to an accelerated healing process compared to the untreated models (Pierce et al., 1991). In a previous study, Lynch et al. (1987) found that the platelet-derived growth factor was more effective at inducing fibroblast migration, collagen and glycosaminoglycan formation, and DNA synthesis when partially purified instead of from a pure fraction.

6.7.4 Vascular Endothelial Growth Factor (VEGF)

An interesting proangiogenic molecule known as vascular endothelial growth factor (VEGF) induces angiogenesis, a crucial occurrence in several natural cellular processes, including wound healing. Each of the seven VEGF family members has a homology domain in common, and it is believed that they link to particular vascular endothelial growth factor receptors to transduce signals. At the injury site, activated platelets release VEGF-A, which binds to VEGFR-1 on inflammatory cells to draw monocytes and neutrophils from the bloodstream. Neutrophils that have been recruited additionally introduce proinflammatory cytokines such as TnF-a and IL-1â into the system (Vaidyanathan, 2021).

Furthermore, it is being researched if VEGF-A plays a role in controlling the synthesis of plasminogen activator during the re-epithelialization phase44 (Hoeben et al., 2004). Research indicates that it is crucial in encouraging vascularity and closure in chronic hypoxic wounds. Decreased angiogenesis and VEGF production are thought to be the causes of the poor healing of diabetic wounds. Hypoxia, a state brought on by traumatic wounds or ischemic injury, is hypothesized to control the expression of VEGF and its receptors (Vaidyanathan, 2021).

6.7.5 Granulocyte Macrophage Colony-Stimulating Factor (GM-CSF)

GM-CSF, which functions in an autocrine manner to promote epidermal proliferation, is secreted by the keratinocytes in response to damage. When GM-CSF antagonist-overexpressing mice were studied in an experimental setting, it became clear how important GM-CSF is for healthy wound healing because the mice displayed delayed epithelialization and neovascularization (Mann et al., 2006). By improving keratinocyte migration, GM-CSF facilitates their entry into the wound site, an amazing process that creates granulation tissue and epithelialization (Mann et al., 2001). The synthesis of the vascular collagenous matrix was faulty in the GM-CSF deficit, indicating its importance in preserving vascular integrity. Exogenous GM-CSF improved wound healing in diabetic mice by upregulating the synthesis of IL-6 and monocyte chemoattractant protein-1. At a concentration of 200μg/mL, it was discovered that topical application of GM-CSF might alleviate decubital ulcers in cancer patients (Raderer et al., 1997). Gulcelik et al. (2006) demonstrated that a local injection of GM-CSF improved the healing of an incisional lesion in rats treated with Adriamycin, albeit at a lower concentration (Gulcelik et al., 2006). Furthermore, GM-CSF can chemotact a variety of inflammatory cells and mediators, which indirectly aids in the inflammation and healing of wounds. Thus, GM-CSF fulfilled all the conditions needed to be an active agent, speeding up wound healing mechanisms and satisfying scientists' desire to create it in the lab using recombinant technology (Castrogiovanni et al., 2010). In an observational case study, a patient with bilateral hemiplegia had a sacral pressure ulcer cured by local GM-CSF injections. Within a few days, a hard granulation tissue developed; a sample of the granulation tissue revealed fibroblasts and inflammatory cells, with increased infiltration at the injection site. Randomized controlled studies involving individuals with diabetic foot ulcers demonstrated the importance of GM-CSF in infiltrating the inflammatory cells by promoting the release of neutrophil endothelial progenitor cells from bone marrow. In a pilot experimental research, systemic treatment with GM-CSF succeeded in healing lesions caused by dystrophic epidermolysis bullosa (Vaidyanathan, 2021).

6.7.6 Connective Tissue Growth Factor (CTGF)

A protein called CTGF controls several biological processes, including angiogenesis, adhesion, cell division, and proliferation. Exogenous CTGF has been thoroughly studied for its potential to heal in both human and animal experiments. Diabetes caused by streptozotocin, using full-thickness excision wounds generated in an experimental setting, utilized Sprague-Dawley rats as diabetes wound models. Although not statistically significant, the topical use of recombinant CTGF for seven days resulted in a good change in wound closure in the diabetic group (Ramazani et al., 2018)(Alfaro et al., 2013). Following long-term recombinant CTGF treatment, additional experimental excision wound models were found, revealing increased fibroblast proliferation and collagen deposition at the granulation tissue. In comparison to the controls, the treated model had improved vascularity. CTGF therapy influences the wound cellular microenvironment, particularly by increasing the number of macrophages. The number of macrophages will always play a role in the development and subsidence of wound inflammation (Henshaw et al., 2015). Diabetes wounds treated with CTGF demonstrated an increase in collagen IV, which is important to move the healing process into a remodeling phase (Vaidyanathan, 2021). In addition to acting as a chemotactic factor for fibroblasts, CTGF controls scarring. It is known that CTGF functions as a component of the positive feedback loop by triggering thrombospondin, which activates the latent TGF-â precursor into the active form (Gibson et al., 2014) (**Table 6.3**).

TABLE 6.3
Overview of Key Growth Factors Involved in Wound Healing

Growth Factor	Source Cells	Main Functions in Wound Healing	Notable Effects and Mechanisms
EGF	Keratinocytes, macrophages	Promotes cell growth, proliferation, and differentiation	Accelerates re-epithelialization, promotes dermal maturation, and induces fibroblast migration and proliferation
TGF-β	Neutrophils, fibroblasts, platelets, macrophages	Regulates gene transcription affecting all stages of wound healing	Promotes angiogenesis, myofibroblast differentiation, and collagen production
PDGF	Platelets, endothelial cells	Controls chemotaxis, angiogenesis, and mitogenesis	Enhances fibroblast and inflammatory cell infiltration, ECM deposition, and collagen formation
VEGF	Activated platelets, macrophages	Induces angiogenesis, attracts inflammatory cells	Enhances vascularity and wound closure, which is important in hypoxic conditions and chronic wounds
GM-CSF	Keratinocytes	Promotes epidermal proliferation and neovascularization	Facilitates keratinocyte migration, granulation, tissue formation, and epithelialization
CTGF	Fibroblasts, endothelial cells	Regulates angiogenesis, adhesion, and cell division	Increases fibroblast proliferation, collagen deposition, and macrophage activity

6.8 THERAPEUTIC APPROACHES TARGETING ECM IN WOUND HEALING

6.8.1 ECM-Based Biomaterials

Therapeutic wound healing techniques have shown promising evolution throughout the past decades. ECM-based biomaterials exhibit unique characteristics and the ability to imitate the natural tissue environment; thus, they are useful in tissue engineering and regeneration (Yu et al., 2021). One of the most important types of ECM-based biomaterials is the decellularized extracellular matrix (dECM). dECM materials have abundant bioactive elements and are made from allogeneic, autologous, or xenogeneic tissues; besides, they have remarkable biological activity and natural scaffold characteristics (Zhang et al., 2021). Moreover, these qualities allow dECM to eliminate biological components while keeping the original tissue structure unaffected (R. Yu et al., 2021; L. Chen et al., 2021).

The acellular dermal matrix (ADM), one of the dECM biomaterials, has demonstrated promise in accelerating wound healing, especially with its allogenic and xenogenic use (Gowda et al., 2016; Y. H. Kim et al., 2019). ADM has a wealth of collagen fibers, elastin, and other natural elements essential for tissue formation and cell adhesion derived from human or animal dermal tissues (Sobti & Liao, 2016). Furthermore, ADM has undergone some modifications, such as adding growth factors, which alleviated its ability to promote tissue regeneration (Chu et al., 2018).

Comparably, decellularized adipose tissue (DAT) scaffolds present a promising avenue for soft tissue regeneration and wound healing (C. Yu et al., 2013). DAT scaffolds support angiogenesis and cell proliferation, both essential for wound healing, by maintaining the structure of the extracellular matrix and bioactive chemicals (Pati et al., 2014). Novel techniques such as DAT-loaded thermosensitive hydrogels have been shown to increase vascularization and speed up wound closure (E. J. Kim et al., 2016).

Moreover, decellularized small intestine submucosa (DSIS) has become a useful extracellular matrix material for treating wounds (Fan et al., 2014). DSIS is a promising option for treating chronic wounds because of its immunosuppressive qualities and capacity to stimulate angiogenesis (M. S. Kim et al., 2005). Besides, the therapeutic capabilities can be improved by combining them with stem cells or using nanoparticles to alter their physical qualities (Mondalek et al., 2008).

6.8.2 Growth Factors and Cytokines

Growth factors and cytokines have drawn interest as therapeutic approaches targeting the ECM in wound healing due to their importance in regulating the complex cellular processes needed for tissue regeneration and wound repair (Yamakawa & Hayashida, 2019). Growth factors include vascular endothelial growth factor (VEGF), platelet-derived growth factor (PDGF), fibroblast growth factor (FGF), and epidermal growth factor (EGF), all of which are necessary for aiding tissue repair by stimulation of cell proliferation, angiogenesis, and collagen deposition (Yamakawa & Hayashida, 2019).

Biomaterials have been designed to provide growth factors in a controlled way to overcome issues such as short bioactivity and limited skin penetration (Zhang et al., 2018). Hydrogels are a great example with high potential for acting as a stage for long-term drug administration. They protect GFs from rapid degradation and provide slow-releasing rates at the wound site, thus extending the therapeutic action period of the drug (Zhang et al., 2018). According to the recent knowledge available in the literature, combining growth factors with biomaterials reduces inflammation, facilitates tissue regeneration, and improves wound closure (Zhang et al., 2018). In addition, the impact on wound repair can be augmented by administering a combination of various growth factors, which have been shown to create synergistic effects (Yoo et al., 2018) (**Table 6.4**).

TABLE 6.4
Overview of ECM-Based Biomaterials and Their Applications in Wound Healing

ECM-Based Biomaterial	Source	Key Components	Applications	Enhancements	References
Decellularized ECM (dECM)	Autologous, allogeneic, xenogenic tissues	Bioactive elements, natural scaffolds	Tissue restoration, wound healing	–	Zhang et al., 2021; R. Yu et al., 2021; L. Chen et al., 2021
Acellular dermal matrix (ADM)	Human or animal dermal tissues	Collagen fibers, elastin, natural elements	Skin wound healing, tissue formation, cell adhesion	Addition of growth factors	Gowda et al., 2016; Sobti & Liao, 2016; Y. H. Kim et al., 2019; Chu et al., 2018
Decellularized adipose tissue (DAT)	Adipose tissue	Bioactive chemicals, extracellular matrix	Soft tissue regeneration, wound healing	DAT-loaded thermosensitive hydrogels	C. Yu et al., 2013; Pati et al., 2014; E. J. Kim et al., 2016
Decellularized small intestine submucosa (DSIS)	Small intestine submucosa	Bioactive elements, extracellular matrix	Treatment of chronic wounds, angiogenesis	Combination with stem cells, nanoparticle modifications	Fan et al., 2014; M. S. Kim et al., 2005; Mondalek et al., 2008

6.8.3 Stem Cell-Based Therapies

Many factors, such as aging, a sedentary lifestyle, and chronic diseases such as diabetes, can impair wound healing. However, stem cell-based therapies targeting the ECM present promising pathways for overcoming these problems by reducing scarring and improving tissue regeneration (Gurtner et al., 2008; Sen et al., 2009). Mesenchymal stem cells (MSCs) are a potent stem cell therapy that can be extracted from different sources, such as adipose tissue and bone marrow (Crovetti et al., 2004). MSCs have demonstrated high efficacy in reducing inflammation, promoting wound closure, and improving wound healing (Aarabi et al., 2007). Furthermore, other types, including embryonic stem cells (ESCs) and induced pluripotent stem cells (iPSCs), can develop into multiple cell types, thus providing reasonable alternatives (Pastar et al., 2014).

On the contrary, the deficit of MSCs and the uncertainty of their lasting efficacy pose great obstacles (Harvey, 2005). Additionally, ESCs and iPSCs show risks of developing tumors, and their unresolved ethical concerns must be closely evaluated before any therapeutic application (Grose & Werner, 2004). Despite these issues, stem cell-based therapies target the wound's underlying pathophysiological changes and promote competent tissue repair, thus improving wound care (Rhett et al., 2008). These treatments need to be introduced into standard clinical practice and undergo more investigations for further improvements and scientific satisfaction (Pastar et al., 2014).

6.9 CONCLUSION

The extracellular matrix (ECM), a dynamic and complex scaffold that affects cellular behavior and tissue regeneration, is essential to wound healing. The extracellular matrix (ECM) changes significantly during wound healing, going through several phases, including hemostasis, inflammation, proliferation, and remodeling. First, the extracellular matrix (ECM) aids cell adhesion, migration, and differentiation by acting as a reservoir for growth factors and structural support. In addition to providing mechanical strength, this complex matrix—made up of proteins including collagen, elastin, fibronectin, and glycoproteins—also controls the biochemical processes necessary for tissue repair. Coordinating the cellular responses required for efficient tissue repair is one of the ECM's most essential roles in wound healing. Integrin-mediated signaling pathways drive the interaction of fibroblasts, endothelial cells, and keratinocytes with the extracellular matrix (ECM), promoting angiogenesis, re-epithelialization, and matrix deposition. Matrix metalloproteinases (MMPs) and their inhibitors mediate the destruction and subsequent production of extracellular matrix (ECM) components, essential for the transition between the various stages of healing. This dynamic process can become unbalanced and result in fibrosis or chronic wounds, emphasizing the significance of well-controlled ECM remodeling. There are significant implications for both clinical practice and future research. Comprehending the molecular mechanisms underlying ECM dynamics provides opportunities for innovative therapeutic approaches. Promising outcomes in acute and chronic wound care are observed when biomaterials that resemble the natural extracellular matrix (ECM) are used to improve wound healing. Targeting particular ECM elements or signaling pathways may also result in novel therapeutic approaches for pathological disorders like fibrosis linked to abnormal ECM remodeling. Wound treatment is revolutionizing clinical practice with advances in ECM-based therapeutics, such as bioengineered skin substitutes and ECM-derived scaffolds. These developments highlight the necessity of personalized treatment strategies that consider patient-specific ECM traits to maximize healing results. Further investigation into the ECM's function in wound healing can significantly enhance patients' quality of life, highlighting the matrix's significance in regenerative medicine and therapeutic settings.

REFERENCES

Aarabi, S., Longaker, M. T., & Gurtner, G. C. (2007). Hypertrophic scar formation following burns and trauma: New approaches to treatment. *PLoS Medicine*, *4*(9), e234. https://doi.org/10.1371/journal.pmed.0040234

Abdelaziz Mohamed, I., Gadeau, A.-P., Hasan, A., Abdulrahman, N., & Mraiche, F. (2019). Osteopontin: A promising therapeutic target in cardiac fibrosis. *Cells*, *8*(12), 1558. https://doi.org/10.3390/cells8121558

Abdou, A. G., Maraee, A. H., Al-Bara, A. M., & Diab, W. M. (2011). Immunohistochemical expression of TGF-β1 in keloids and hypertrophic scars. *The American Journal of Dermatopathology*, *33*(1), 84–91. https://doi.org/10.1097/DAD.0b013e3181d0c3ad

Abercrombie, M., Flint, M. H., & James, D. W. (1956). Wound contraction in relation to collagen formation in scorbutic Guinea-pigs. *Development*, *4*(2), 167–175. https://doi.org/10.1242/DEV.4.2.167

Adapted from "Wound Healing", by BioRender.com (2024). Available from https://app.biorender.com/biorender-templates/figures/all/t-5fa1b1622a60ac00a3d858ec-wound-healing

Alfaro, M. P., Deskins, D. L., Wallus, M., DasGupta, J., Davidson, J. M., Nanney, L. B., A Guney, M., Gannon, M., & Young, P. P. (2013). A physiological role for connective tissue growth factor in early wound healing. *Laboratory Investigation; a Journal of Technical Methods and Pathology*, *93*(1), 81–95. https://doi.org/10.1038/labinvest.2012.162

Almine, J. F., Wise, S. G., & Weiss, A. S. (2012). Elastin signaling in wound repair. *Birth Defects Research Part C—Embryo Today: Reviews*, *96*(3), 248–257. https://doi.org/10.1002/bdrc.21016.

Anitua, E., Andia, I., Ardanza, B., Nurden, P., & Nurden, A. (2004). Autologous platelets as a source of proteins for healing and tissue regeneration. *Thrombosis and Haemostasis*, *91*(01), 4–15. https://doi.org/10.1160/TH03-07-0440

Armour, A., Scott, P. G., & Tredget, E. E. (2007). Cellular and molecular pathology of HTS: basis for treatment. *Wound Repair and Regeneration : Official Publication of the Wound Healing Society [and] the European Tissue Repair Society*, *15*(Suppl 1), S6–17. https://doi.org/10.1111/j.1524-475X.2007.00219.x

Babu, M., Diegelmann, R., & Oliver, N. (1989). Fibronectin is overproduced by keloid fibroblasts during abnormal wound healing. *Molecular and Cellular Biology*, *9*(4), 1642–1650. https://doi.org/10.1128/mcb.9.4.1642-1650.1989

Baisch, A., & Riedel, F. (2006). [Hyperplastic scars and keloids. Part I: basics and prevention]. *HNO*, *54*(11), 893–904; quiz 905. https://doi.org/10.1007/s00106-006-1462-z

Barrientos, S., Stojadinovic, O., Golinko, M. S., Brem, H., & Tomic-Canic, M. (2008). Growth factors and cytokines in wound healing. *Wound Repair and Regeneration : Official Publication of the Wound Healing Society [and] the European Tissue Repair Society*, *16*(5), 585–601. https://doi.org/10.1111/j.1524-475X.2008.00410.x

Baumann, L., Bernstein, E. F., Weiss, A. S., Bates, D., Humphrey, S., Silberberg, M., & Daniels, R. (2021). Clinical relevance of elastin in the structure and function of skin. *Aesthetic Surgery Journal Open Forum*, *3*(3). https://doi.org/10.1093/asjof/ojab019.

Bertina, R. M., Koeleman, B. P., Koster, T., Rosendaal, F. R., Dirven, R. J., de Ronde, H., van der Velden, P. A., & Reitsma, P. H. (1994). Mutation in blood coagulation factor V associated with resistance to activated protein C. *Nature*, *369*(6475), 64–67. https://doi.org/10.1038/369064a0

Blackburn, W. R., & Cosman, B. (1966). Histologic basis of keloid and hypertrophic scar differentiation. Clinicopathologic correlation. *Archives of Pathology*, *82*(1), 65–71.

Bodnar, R. J. (2013). Epidermal growth factor and epidermal growth factor receptor: The yin and yang in the treatment of cutaneous wounds and cancer. *Advances in Wound Care*, *2*(1), 24–29. https://doi.org/10.1089/wound.2011.0326

Brass, L. F. (2003). Thrombin and platelet activation. *Chest*, *124*(3 Suppl), 18S–25S. https://doi.org/10.1378/chest.124.3_suppl.18s

Brew, K., & Nagase, H. (2010). The tissue inhibitors of metalloproteinases (TIMPs): an ancient family with structural and functional diversity. *Biochimica et Biophysica Acta*, *1803*(1), 55–71. https://doi.org/10.1016/j.bbamcr.2010.01.003

Busti, A. J., Hooper, J. S., Amaya, C. J., & Kazi, S. (2005). Effects of perioperative antiinflammatory and immunomodulating therapy on surgical wound healing. *Pharmacotherapy*, *25*(11), 1566–1591. https://doi.org/10.1592/phco.2005.25.11.1566

Carr, N. J. (2022). The pathology of healing and repair. *Surgery (United Kingdom)*, *40*(1), 13–19. https://doi.org/10.1016/j.mpsur.2021.11.003

Castoldi, E., & Rosing, J. (2010). APC resistance: Biological basis and acquired influences. *Journal of Thrombosis and Haemostasis: JTH*, *8*(3), 445–453. https://doi.org/10.1111/j.1538-7836.2009.03711.x

Castrogiovanni, P., Ventimiglia, P., & Imbesi, R. (2010). Wound healing: Experience with rHuGM-CSF. *Wounds : A Compendium of Clinical Research and Practice*, *22*(10), 256–260.

Chen, L., Li, Z., Zheng, Y., Zhou, F., Zhao, J., Zhai, Q., Zhang, Z., Liu, T., Chen, Y., & Qi, S. (2021). 3D-printed dermis-specific extracellular matrix mitigates scar contraction via inducing early angiogenesis and macrophage M2 polarization. *Bioactive Materials*, *10*, 236–246. https://doi.org/10.1016/J.BIOACTMAT.2021.09.008

Chen, W., & Ten Dijke, P. (2016). Immunoregulation by members of the TGFβ superfamily. *Nature Reviews. Immunology*, *16*(12), 723–740. https://doi.org/10.1038/nri.2016.112

Chen, W., Yuan, Y., Li, C., Mao, H., Liu, B., & Jiang, X. (2022). Modulating tumor extracellular matrix by simultaneous inhibition of two cancer cell receptors. *Advanced Materials*, *34*(10). https://doi.org/10.1002/adma.202109376.

Chu, J., Shi, P., Deng, X., Jin, Y., Liu, H., Chen, M., Han, X., & Liu, H. (2018). Dynamic multiphoton imaging of acellular dermal matrix scaffolds seeded with mesenchymal stem cells in diabetic wound healing. *Journal of Biophotonics*, *11*(7). https://doi.org/10.1002/jbio.201700336

Chua, A. W. C., Gan, S. U., Ting, Y., Fu, Z., Lim, C. K., Song, C., Sabapathy, K., & Phan, T. T. (2011). Keloid fibroblasts are more sensitive to Wnt3a treatment in terms of elevated cellular growth and fibronectin expression. *Journal of Dermatological Science*, *64*(3), 199–209. https://doi.org/10.1016/j.jdermsci.2011.09.008

Clark, D. A., & Coker, R. (1998). Transforming growth factor-beta (TGF-beta). *The International Journal of Biochemistry & Cell Biology*, *30*(3), 293–298. https://doi.org/10.1016/s1357-2725(97)00128-3

Clark, R. A. F., Ashcroft, G. S., Spencer, M. J., Larjava, H., & Ferguson, M. W. J. (1996). Re-epithelialization of normal human excisional wounds is associated with a switch from αvβ5 to αvβ6 integrins. *British Journal of Dermatology*, *135*(1), 46–51.

Coelho, N. M., & McCulloch, C. A. (2016). Contribution of collagen adhesion receptors to tissue fibrosis. *Cell and Tissue Research*, *365*(3), 521–538. https://doi.org/10.1007/s00441-016-2440-8

Couchman, J. R., & Pataki, C. A. (2012). An introduction to proteoglycans and their localization. *Journal of Histochemistry and Cytochemistry*, *60*(12), 885–897. https://doi.org/10.1369/0022155412464638.

Crovetti, G., Martinelli, G., Issi, M., Barone, M., Guizzardi, M., Campanati, B., Moroni, M., & Carabelli, A. (2004). Platelet gel for healing cutaneous chronic wounds. *Transfusion and Apheresis Science*, *30*(2), 145–151. https://doi.org/10.1016/j.transci.2004.01.004

Diegelmann, R. F., & Evans, M. C. (2004). Wound healing: An overview of acute, fibrotic and delayed healing. *Frontiers in Bioscience: A Journal and Virtual Library*, *9*, 283–289. https://doi.org/10.2741/1184

Discher, D. E., Janmey, P., & Wang, Y. L. (2005). Tissue cells feel and respond to the stiffness of their substrate. *Science (New York, N.Y.)*, *310*(5751), 1139–1143. https://doi.org/10.1126/SCIENCE.1116995

Dzobo, K., & Dandara, C. (2023). The extracellular matrix: Its composition, function, remodeling, and role in tumorigenesis. In *Biomimetics* (Vol. 8, Issue 2). MDPI. https://doi.org/10.3390/biomimetics8020146.

Ehrlich, H. P., Desmoulière, A., Diegelmann, R. F., Cohen, I. K., Compton, C. C., Garner, W. L., Kapanci, Y., & Gabbiani, G. (1994). Morphological and immunochemical differences between keloid and hypertrophic scar. *The American Journal of Pathology*, *145*(1), 105–113.

Eming, S. A., Krieg, T., & Davidson, J. M. (2007). Inflammation in wound repair: molecular and cellular mechanisms. *The Journal of Investigative Dermatology*, *127*(3), 514–525. https://doi.org/10.1038/sj.jid.5700701

Engler, A. J., Sen, S., Sweeney, H. L., & Discher, D. E. (2006). Matrix elasticity directs stem cell lineage specification. *Cell*, *126*(4), 677–689. https://doi.org/10.1016/J.CELL.2006.06.044

Eto, H., Suga, H., Aoi, N., Kato, H., Doi, K., Kuno, S., Tabata, Y., & Yoshimura, K. (2012). Therapeutic potential of fibroblast growth factor-2 for hypertrophic scars: upregulation of MMP-1 and HGF expression. *Laboratory Investigation; a Journal of Technical Methods and Pathology*, *92*(2), 214–223. https://doi.org/10.1038/labinvest.2011.127

Fan, M.-R., Gong, M., Da, L.-C., Bai, L., Li, X.-Q., Chen, K.-F., Li-Ling, J., Yang, Z.-M., & Xie, H.-Q. (2014). Tissue engineered esophagus scaffold constructed with porcine small intestinal submucosa and synthetic polymers. *Biomedical Materials*, *9*(1), 015012. https://doi.org/10.1088/1748-6041/9/1/015012

Farndale, R. W., Sixma, J. J., Barnes, M. J., & De Groot, P. G. (2004). The role of collagen in thrombosis and hemostasis. *Journal of Thrombosis and Haemostasis*, *2*(4), 561–573.

Frantz, C., Stewart, K. M., & Weaver, V. M. (2010). The extracellular matrix at a glance. *Journal of Cell Science*, *123*(Pt 24), 4195–4200. https://doi.org/10.1242/JCS.023820

Fujiwara, M., Muragaki, Y., & Ooshima, A. (2005). Keloid-derived fibroblasts show increased secretion of factors involved in collagen turnover and depend on matrix metalloproteinase for migration. *The British Journal of Dermatology*, *153*(2), 295–300. https://doi.org/10.1111/j.1365-2133.2005.06698.x

Gatseva, A., Sin, Y. Y., Brezzo, G., & van Agtmael, T. (2019). Basement membrane collagens and disease mechanisms. *Essays in Biochemistry*, *63*(3), 297–312. Portland Press Ltd. https://doi.org/10.1042/EBC20180071.

Gauglitz, G. G., Korting, H. C., Pavicic, T., Ruzicka, T., & Jeschke, M. G. (2011). Hypertrophic scarring and keloids: pathomechanisms and current and emerging treatment strategies. *Molecular Medicine (Cambridge, Mass.)*, *17*(1–2), 113–125. https://doi.org/10.2119/molmed.2009.00153

Ghani, Q. P., Wagner, S., Becker, H. D., Hunt, T. K., & Hussain, M. Z. (2004). *Regulatory Role of Lactate in Wound Repair* (pp. 565–575). https://doi.org/10.1016/S0076-6879(04)81036-X

Ghatak, S., Maytin, E. v., MacK, J. A., Hascall, V. C., Atanelishvili, I., Moreno Rodriguez, R., Markwald, R. R., & Misra, S. (2015). Roles of proteoglycans and glycosaminoglycans in wound healing and fibrosis. In *International Journal of Cell Biology* (Vol. 2015). Hindawi Publishing Corporation. https://doi.org/10.1155/2015/834893.

Gibson, D. J., Pi, L., Sriram, S., Mao, C., Petersen, B. E., Scott, E. W., Leask, A., & Schultz, G. S. (2014). Conditional knockout of CTGF affects corneal wound healing. *Investigative Ophthalmology & Visual Science*, *55*(4), 2062–2070. https://doi.org/10.1167/iovs.13-12735

Gowda, A. U., McNichols, C. H. L., Asokan, I., Matthews, J. A., Buckingham, E. B., Sabino, J., Maddox, J. S., Slezak, S., Rasko, Y., & Singh, D. P. (2016). Porcine acellular dermal matrix for hernia repair in transplant patients. *Annals of Plastic Surgery*, *77*(6), 674–677. https://doi.org/10.1097/SAP.0000000000000929

Grose, R., & Werner, S. (2004). Wound-healing studies in transgenic and knockout mice. *Molecular Biotechnology*, *28*(2), 147–166. https://doi.org/10.1385/MB:28:2:147

Gulcelik, M. A., Dinc, S., Dinc, M., Yenidogan, E., Ustun, H., Renda, N., & Alagol, H. (2006). Local granulocyte-macrophage colony-stimulating factor improves incisional wound healing in adriamycin-treated rats. *Surgery Today*, *36*(1), 47–51. https://doi.org/10.1007/s00595-005-3097-1

Guo, S., & DiPietro, L. A. (2010). Factors affecting wound healing. *Journal of Dental Research*, *89*(3), 219–229. https://doi.org/10.1177/0022034509359125

Gurtner, G. C., Werner, S., Barrandon, Y., & Longaker, M. T. (2008). Wound repair and regeneration. *Nature*, *453*(7193), 314–321. https://doi.org/10.1038/nature07039

Hayat, M., Nawaz, A., Chinnam, S., Muzammal, M., Latif, M. S., Yasin, M., Ashique, S., Zengin, G., & Farid, A. (2023). Formulation development and optimization of herbo synthetic gel: In vitro biological evaluation and in vivo wound healing studies. *Process Biochemistry*, *130*, 116–126.

Harper, D., Young, A., & McNaught, C. E. (2014). The physiology of wound healing. *Surgery (United Kingdom)*, *32*(9), 445–450. https://doi.org/10.1016/j.mpsur.2014.06.010

Harvey, C. (2005). Wound healing. *Orthopaedic Nursing*, *24*(2), 143???157. https://doi.org/10.1097/00006416-200503000-00012

Heldin, C. H., & Westermark, B. (1999). Mechanism of action and in vivo role of platelet-derived growth factor. *Physiological Reviews*, *79*(4), 1283–1316. https://doi.org/10.1152/physrev.1999.79.4.1283

Henshaw, F. R., Boughton, P., Lo, L., McLennan, S. V, & Twigg, S. M. (2015). Topically applied connective tissue growth factor/CCN2 improves diabetic preclinical cutaneous wound healing: potential role for CTGF in human diabetic foot ulcer healing. *Journal of Diabetes Research*, *2015*, 236238. https://doi.org/10.1155/2015/236238

Hinz, B. (2007). Formation and function of the myofibroblast during tissue repair. *Journal of Investigative Dermatology*, *127*(3), 526–537. https://doi.org/10.1038/sj.jid.5700613

Hoeben, A., Landuyt, B., Highley, M. S., Wildiers, H., Van Oosterom, A. T., & De Bruijn, E. A. (2004). Vascular endothelial growth factor and angiogenesis. *Pharmacological Reviews*, *56*(4), 549–580. https://doi.org/10.1124/pr.56.4.3

Hoffman, M., Harger, A., Lenkowski, A., Hedner, U., Roberts, H. R., & Monroe, D. M. (2006). Cutaneous wound healing is impaired in hemophilia B. *Blood*, *108*(9), 3053–3060. https://doi.org/10.1182/blood-2006-05-020495

Honardoust, D., Varkey, M., Marcoux, Y., Shankowsky, H. A., & Tredget, E. E. (2012). Reduced decorin, fibromodulin, and transforming growth factor-β3 in deep dermis leads to hypertrophic scarring. *Journal of Burn Care & Research : Official Publication of the American Burn Association*, *33*(2), 218–227. https://doi.org/10.1097/BCR.0b013e3182335980

Huang, E. (2023). The Extracellular Matrix (ECM). Available from https://app.biorender.com/biorender-templates/figures/all/t-64945f1b3957a0de690f12d3-the-extracellular-matrix-ecm

Hynes, R. O. (2009). The extracellular matrix: not just pretty fibrils. *Science (New York, N.Y.)*, *326*(5957), 1216–1219. https://doi.org/10.1126/SCIENCE.1176009

Hynes, R. O., & Naba, A. (2012). Overview of the matrisome-an inventory of extracellular matrix constituents and functions. *Cold Spring Harbor Perspectives in Biology*, 4(1), a004903. https://doi.org/10.1101/cshperspect.a004903

Jagadeesan, J., & Bayat, A. (2007). Transforming growth factor beta (TGFbeta) and keloid disease. *International Journal of Surgery (London, England)*, *5*(4), 278–285. https://doi.org/10.1016/j.ijsu.2006.04.007

Kaasbøll, O. J., Gadicherla, A. K., Wang, J.-H., Monsen, V. T., Hagelin, E. M. V., Dong, M.-Q., & Attramadal, H. (2018). Connective tissue growth factor (CCN2) is a matricellular preproprotein controlled by proteolytic activation. *Journal of Biological Chemistry*, *293*(46), 17953–17970. https://doi.org/10.1074/jbc.RA118.004559

Karamanos, N. K., Theocharis, A. D., Piperigkou, Z., Manou, D., Passi, A., Skandalis, S. S., Vynios, D. H., Orian-Rousseau, V., Ricard-Blum, S., Schmelzer, C. E. H., Duca, L., Durbeej, M., Afratis, N. A., Troeberg, L., Franchi, M., Masola, V., & Onisto, M. (2021). A guide to the composition and functions of the extracellular matrix. *The FEBS Journal*, *288*(24), 6850–6912. https://doi.org/10.1111/febs.15776

Kim, E. J., Choi, J. S., Kim, J. S., Choi, Y. C., & Cho, Y. W. (2016). Injectable and thermosensitive soluble extracellular matrix and methylcellulose hydrogels for stem cell delivery in skin wounds. *Biomacromolecules*, *17*(1), 4–11. https://doi.org/10.1021/acs.biomac.5b01566

Kim, M. S., Hong, K. D., Shin, H. W., Kim, S. H., Kim, S. H., Lee, M. S., Jang, W. Y., Khang, G., & Lee, H. B. (2005). Preparation of porcine small intestinal submucosa sponge and their application as a wound dressing in full-thickness skin defect of rat. *International Journal of Biological Macromolecules*, *36*(1–2), 54–60. https://doi.org/10.1016/j.ijbiomac.2005.03.013

Kim, S. H., Turnbull, J., & Guimond, S. (2011). Extracellular matrix and cell signalling: the dynamic cooperation of integrin, proteoglycan and growth factor receptor. *The Journal of Endocrinology*, *209*(2), 139–151. https://doi.org/10.1530/JOE-10-0377

Kim, Y. H., Hwang, K. T., Kim, K. H., Sung, I. H., & Kim, S. W. (2019). Application of acellular human dermis and skin grafts for lower extremity reconstruction. *Journal of Wound Care*, *28*(Sup4), S12–S17. https://doi.org/10.12968/JOWC.2019.28.SUP4.S12

Koh, T. J., & DiPietro, L. A. (2011). Inflammation and wound healing: the role of the macrophage. *Expert Reviews in Molecular Medicine*, *13*, e23. https://doi.org/10.1017/S1462399411001943

Kumar, I., Staton, C. A., Cross, S. S., Reed, M. W. R., & Brown, N. J. (2009). Angiogenesis, vascular endothelial growth factor and its receptors in human surgical wounds. *The British Journal of Surgery*, *96*(12), 1484–1491. https://doi.org/10.1002/BJS.6778

Kumar, V., Abbas, A. K., Aster, J. C., & Robbins, S. L. (2013). Robins basic pathology. In *Robbins Basic Pathology*. https://books.google.com/books/about/Robbins_Basic_Pathology.html?id=aLV9nU_7X6UC

Larsen, M., Artym, V. V., Green, J. A., & Yamada, K. M. (2006). The matrix reorganized: extracellular matrix remodeling and integrin signaling. *Current Opinion in Cell Biology*, *18*(5), 463–471. https://doi.org/10.1016/J.CEB.2006.08.009

Lee, Y. I., Lee, S. G., Kim, J., Choi, S., Jung, I., & Lee, J. H. (2021). Proteoglycan combined with hyaluronic acid and hydrolyzed collagen restores the skin barrier in mild atopic dermatitis and dry, eczema-prone skin: A pilot study. *International Journal of Molecular Sciences*, *22*(19), 10189. https://doi.org/10.3390/ijms221910189

Li, J., Chen, J., & Kirsner, R. (2007). Pathophysiology of acute wound healing. *Clinics in Dermatology*, *25*(1), 9–18. https://doi.org/10.1016/J.CLINDERMATOL.2006.09.007

Lin, P., Zhang, G., Li, H., (2023) The role of extracellular matrix in wound healing. *Dermatologic Surgery*, *49*(5S), S41–S48. https://doi.org/10.1097/dss.0000000000003779

Linares, H. A., Kischer, C. W., Dobrkovsky, M., & Larson, D. L. (1972). The histiotypic organization of the hypertrophic scar in humans. *The Journal of Investigative Dermatology*, *59*(4), 323–331. https://doi.org/10.1111/1523-1747.ep12627386

Lu, P., Weaver, V. M., & Werb, Z. (2012). The extracellular matrix: a dynamic niche in cancer progression. *The Journal of Cell Biology*, *196*(4), 395–406. https://doi.org/10.1083/JCB.201102147

Luo, M., Ji, Y., Luo, Y., Li, R., Fay, W. P., & Wu, J. (2017). Plasminogen activator inhibitor-1 regulates the vascular expression of vitronectin. *Journal of Thrombosis and Haemostasis*, *15*(12), 2451–2460. https://doi.org/10.1111/jth.13869

Lynch, S. E., Nixon, J. C., Colvin, R. B., & Antoniades, H. N. (1987). Role of platelet-derived growth factor in wound healing: Synergistic effects with other growth factors. *Proceedings of the National Academy of Sciences of the United States of America*, *84*(21), 7696–7700. https://doi.org/10.1073/pnas.84.21.7696

Mann, A., Breuhahn, K., Schirmacher, P., & Blessing, M. (2001). Keratinocyte-derived granulocyte-macrophage colony stimulating factor accelerates wound healing: Stimulation of keratinocyte proliferation, granulation tissue formation, and vascularization. *The Journal of Investigative Dermatology*, *117*(6), 1382–1390. https://doi.org/10.1046/j.0022-202x.2001.01600.x

Mann, A., Niekisch, K., Schirmacher, P., & Blessing, M. (2006). Granulocyte-macrophage colony-stimulating factor is essential for normal wound healing. *The Journal of Investigative Dermatology. Symposium Proceedings*, *11*(1), 87–92. https://doi.org/10.1038/sj.jidsymp.5650013

Mascharak, S., desJardins-Park, H. E., Davitt, M. F., Griffin, M., Borrelli, M. R., Moore, A. L., Chen, K., Duoto, B., Chinta, M., Foster, D. S., Shen, A. H., Januszyk, M., Kwon, S. H., Wernig, G., Wan, D. C., Lorenz, H. P., Gurtner, G. C., & Longaker, M. T. (2021). Preventing *Engrailed-1* activation in fibroblasts yields wound regeneration without scarring. *Science*, *372*(6540). https://doi.org/10.1126/science.aba2374

Mathew-Steiner, S. S., Roy, S., & Sen, C. K. (2021). Collagen in wound healing. In *Bioengineering* (Vol. 8, Issue 5). MDPI AG. https://doi.org/10.3390/bioengineering8050063.

Mecham, R. P. (2012). Overview of extracellular matrix. *Current Protocols in Cell Biology*, Chapter 10, 10.1.1–10.1.16. https://doi.org/10.1002/0471143030.cb1001s57

Midwood, K. S., Williams, L. V., & Schwarzbauer, J. E. (2004). Tissue repair and the dynamics of the extracellular matrix. *The International Journal of Biochemistry & Cell Biology*, *36*(6), 1031–1037. https://doi.org/10.1016/j.biocel.2003.12.003

Mondalek, F. G., Lawrence, B. J., Kropp, B. P., Grady, B. P., Fung, K. M., Madihally, S. V., & Lin, H. K. (2008). The incorporation of poly(lactic-co-glycolic) acid nanoparticles into porcine small intestinal submucosa biomaterials. *Biomaterials*, *29*(9), 1159–1166. https://doi.org/10.1016/j.biomaterials.2007.11.020

Murphy, G., & Nagase, H. (2008). Progress in matrix metalloproteinase research. *Molecular Aspects of Medicine*, *29*(5), 290–308. https://doi.org/10.1016/j.mam.2008.05.002

Nagase, H., & Woessner, J. F. (1999). Matrix metalloproteinases. *The Journal of Biological Chemistry*, *274*(31), 21491–21494. https://doi.org/10.1074/jbc.274.31.21491

Nagase, H., Visse, R., & Murphy, G. (2006). Structure and function of matrix metalloproteinases and TIMPs. *Cardiovascular Research*, *69*(3), 562–573. https://doi.org/10.1016/j.cardiores.2005.12.002

Nauta, A., Gurtner, G., & Longaker, M. (2011). Wound healing and regenerative strategies. *Oral Diseases*, *17*(6), 541–549. https://doi.org/10.1111/j.1601-0825.2011.01787.x

Noskovicova, N., Hinz, B., & Pakshir, P. (2021). Implant fibrosis and the underappreciated role of myofibroblasts in the foreign body reaction. *Cells*, *10*(7), 1794. https://doi.org/10.3390/cells10071794

Okabe, K., Hayashi, R., Aramaki-Hattori, N., Sakamoto, Y., & Kishi, K. (2013). Wound treatment using growth factors. *Modern Plastic Surgery*, *03*(03), 108–112. https://doi.org/10.4236/mps.2013.33022

Olczyk, P., Mencner, Ł., & Komosinska-Vassev, K. (2014). The role of the extracellular matrix components in cutaneous wound healing. *BioMed Research International*, *2014*, 747584. https://doi.org/10.1155/2014/747584

Overall, C. M., & Kleifeld, O. (2006). Tumour microenvironment—opinion: validating matrix metalloproteinases as drug targets and anti-targets for cancer therapy. *Nature Reviews. Cancer*, *6*(3), 227–239. https://doi.org/10.1038/nrc1821

Overall, C. M., & López-Otín, C. (2002). Strategies for MMP inhibition in cancer: innovations for the post-trial era. *Nature Reviews Cancer*, *2*(9), 657–672. https://doi.org/10.1038/nrc884

Page-McCaw, A., Ewald, A. J., & Werb, Z. (2007). Matrix metalloproteinases and the regulation of tissue remodelling. *Nature Reviews. Molecular Cell Biology*, *8*(3), 221–233. https://doi.org/10.1038/nrm2125

Pastar, I., Stojadinovic, O., Yin, N. C., Ramirez, H., Nusbaum, A. G., Sawaya, A., Patel, S. B., Khalid, L., Isseroff, R. R., & Tomic-Canic, M. (2014). Epithelialization in wound healing: A comprehensive review. *Advances in Wound Care*, *3*(7), 445–464. https://doi.org/10.1089/wound.2013.0473

Pati, F., Jang, J., Ha, D.-H., Won Kim, S., Rhie, J.-W., Shim, J.-H., Kim, D.-H., & Cho, D.-W. (2014). Printing three-dimensional tissue analogues with decellularized extracellular matrix bioink. *Nature Communications*, *5*(1), 3935. https://doi.org/10.1038/ncomms4935

Pierce, G. F., Mustoe, T. A., Altrock, B. W., Deuel, T. F., & Thomason, A. (1991). Role of platelet-derived growth factor in wound healing. *Journal of Cellular Biochemistry*, *45*(4), 319–326. https://doi.org/10.1002/jcb.240450403

Profyris, C., Tziotzios, C., & Do Vale, I. (2012). Cutaneous scarring: Pathophysiology, molecular mechanisms, and scar reduction therapeutics Part I. The molecular basis of scar formation. *Journal of the American Academy of Dermatology*, *66*(1), 1–10; quiz 11–12. https://doi.org/10.1016/j.jaad.2011.05.055

Raderer, M., Kornek, G., Hejna, M., Koperna, K., Scheithauer, W., & Base, W. (1997). Topical granulocyte-macrophage colony-stimulating factor in patients with cancer and impaired wound healing. *Journal of the National Cancer Institute*, *89*(3), 263. https://doi.org/10.1093/jnci/89.3.263

Ramazani, Y., Knops, N., Elmonem, M. A., Nguyen, T. Q., Arcolino, F. O., van den Heuvel, L., Levtchenko, E., Kuypers, D., & Goldschmeding, R. (2018). Connective tissue growth factor (CTGF) from basics to clinics. *Matrix Biology: Journal of the International Society for Matrix Biology*, *68–69*, 44–66. https://doi.org/10.1016/j.matbio.2018.03.007

Randles, M. J., Humphries, M. J., & Lennon, R. (2017). Proteomic definitions of basement membrane composition in health and disease. In *Matrix Biology* (Vols. 57–58, pp. 12–28). Elsevier B.V. https://doi.org/10.1016/j.matbio.2016.08.006.

Reinke, J. M., & Sorg, H. (2012). Wound repair and regeneration. *European Surgical Research. Europaische Chirurgische Forschung. Recherches Chirurgicales Europeennes*, *49*(1), 35–43. https://doi.org/10.1159/000339613

Rhett, J. M., Ghatnekar, G. S., Palatinus, J. A., O'Quinn, M., Yost, M. J., & Gourdie, R. G. (2008). Novel therapies for scar reduction and regenerative healing of skin wounds. *Trends in Biotechnology*, *26*(4), 173–180. https://doi.org/10.1016/j.tibtech.2007.12.007

Rodríguez-Cabello, J. C., González de Torre, I., Ibañez-Fonseca, A., & Alonso, M. (2018). Bioactive scaffolds based on elastin-like materials for wound healing. In *Advanced Drug Delivery Reviews* (Vol. 129, pp. 118–133). Elsevier B.V. https://doi.org/10.1016/j.addr.2018.03.003.

Rose, K. W. J., Taye, N., Karoulias, S. Z., & Hubmacher, D. (2021). Regulation of ADAMTS Proteases. *Frontiers in Molecular Biosciences*, *8*. https://doi.org/10.3389/fmolb.2021.701959

Sayani, K., Dodd, C. M., Nedelec, B., Shen, Y. J., Ghahary, A., Tredget, E. E., & Scott, P. G. (2000). Delayed appearance of decorin in healing burn scars. *Histopathology*, *36*(3), 262–272. https://doi.org/10.1046/j.1365-2559.2000.00824.x

Schultz, G. S., Chin, G. A., Moldawer, L., Fitridge, R., & Thompson, M. (2011). Principles of wound healing. In R. Fitridge & M. Thompson (Eds.), *Mechanisms of Vascular Disease: A Reference Book for Vascular Specialists*. University of Adelaide Press. Available from: www.ncbi.nlm.nih.gov/books/NBK534261/

Sen, C. K., Gordillo, G. M., Roy, S., Kirsner, R., Lambert, L., Hunt, T. K., Gottrup, F., Gurtner, G. C., & Longaker, M. T. (2009). Human skin wounds: A major and snowballing threat to public health and the economy. *Wound Repair and Regeneration*, *17*(6), 763–771. https://doi.org/10.1111/j.1524-475X.2009.00543.x

Shah, J. M. Y., Omar, E., Pai, D. R., & Sood, S. (2012). Cellular events and biomarkers of wound healing. *Indian Journal of Plastic Surgery: Official Publication of the Association of Plastic Surgeons of India*, *45*(2), 220–228. https://doi.org/10.4103/0970-0358.101282

Shih, B., Garside, E., McGrouther, D. A., & Bayat, A. (2010). Molecular dissection of abnormal wound healing processes resulting in keloid disease. *Wound Repair and Regeneration: Official Publication of the Wound Healing Society [and] the European Tissue Repair Society*, *18*(2), 139–153. https://doi.org/10.1111/j.1524-475X.2009.00553.x

Sidgwick, G. P., & Bayat, A. (2012). Extracellular matrix molecules implicated in hypertrophic and keloid scarring. *Journal of the European Academy of Dermatology and Venereology: JEADV*, *26*(2), 141–152. https://doi.org/10.1111/j.1468-3083.2011.04200.x

Sinno, H., & Prakash, S. (2013). Complements and the wound healing cascade: An updated review. *Plastic Surgery International*, *2013*, 146764. https://doi.org/10.1155/2013/146764

Smith, M. M., & Melrose, J. (2015). Proteoglycans in normal and healing skin. *Advances in Wound Care*, *4*(3), 152–173. https://doi.org/10.1089/wound.2013.0464.

Sobti, N., & Liao, E. C. (2016). Surgeon-controlled study and meta-analysis comparing FlexHD and AlloDerm in immediate breast reconstruction outcomes. *Plastic and Reconstructive Surgery*, *138*(5), 959–967. https://doi.org/10.1097/PRS.0000000000002616

Sorrell, J. M., & Caplan, A. I. (2004). Fibroblast heterogeneity: More than skin deep. *Journal of Cell Science*, *117*(5), 667–675. https://doi.org/10.1242/jcs.01005

Sorushanova, A., Delgado, L. M., Wu, Z., Shologu, N., Kshirsagar, A., Raghunath, R., Mullen, A. M., Bayon, Y., Pandit, A., Raghunath, M., & Zeugolis, D. I. (2019). The collagen suprafamily: From biosynthesis to advanced biomaterial development. In *Advanced Materials* (Vol. 31, Issue 1). Wiley-VCH Verlag. https://doi.org/10.1002/adma.201801651

Sternlicht, M. D., & Werb, Z. (2001). How matrix metalloproteinases regulate cell behavior. *Annual Review of Cell and Developmental Biology*, *17*, 463–516. https://doi.org/10.1146/annurev.cellbio.17.1.463

Tanzer, M. L. (2006). Current concepts of extracellular matrix. *Journal of Orthopaedic Science: Official Journal of the Japanese Orthopaedic Association*, *11*(3), 326–331. https://doi.org/10.1007/S00776-006-1012-2

Taylor, K. R., Trowbridge, J. M., Rudisill, J. A., Termeer, C. C., Simon, J. C., & Gallo, R. L. (2004). Hyaluronan fragments stimulate endothelial recognition of injury through TLR4*. *Journal of Biological Chemistry*, *279*(17), 17079–17084. https://doi.org/10.1074/jbc.M310859200

Tonnesen, M. G., Feng, X., & Clark, R. A. F. (2000). Angiogenesis in wound healing. *Journal of Investigative Dermatology Symposium Proceedings*, *5*(1), 40–46. https://doi.org/10.1046/j.1087-0024.2000.00014.x

Ulrich, D., Ulrich, F., Unglaub, F., Piatkowski, A., & Pallua, N. (2010). Matrix metalloproteinases and tissue inhibitors of metalloproteinases in patients with different types of scars and keloids. *Journal of Plastic, Reconstructive & Aesthetic Surgery : JPRAS*, *63*(6), 1015–1021. https://doi.org/10.1016/j.bjps.2009.04.021

Vaidyanathan, L. (2021). Growth factors in wound healing—A review. *Biomedical and Pharmacology Journal, 14*(3), 1469–1480. https://doi.org/10.13005/bpj/2249

Vedrenne, N., Coulomb, B., Danigo, A., Bonté, F., & Desmoulière, A. (2012). The complex dialogue between (myo)fibroblasts and the extracellular matrix during skin repair processes and ageing. *Pathologie-Biologie, 60*(1), 20–27. https://doi.org/10.1016/j.patbio.2011.10.002

Velnar, T., Bailey, T., & Smrkolj, V. (2009). The wound healing process: an overview of the cellular and molecular mechanisms. *Journal of International Medical Research, 37*(5), 1528–1542.

Verhaegen, P. D. H. M., van Zuijlen, P. P. M., Pennings, N. M., van Marle, J., Niessen, F. B., van der Horst, C. M. A. M., & Middelkoop, E. (2009). Differences in collagen architecture between keloid, hypertrophic scar, normotrophic scar, and normal skin: An objective histopathological analysis. *Wound Repair and Regeneration : Official Publication of the Wound Healing Society [and] the European Tissue Repair Society, 17*(5), 649–656. https://doi.org/10.1111/j.1524-475X.2009.00533.x

Visse, R., & Nagase, H. (2003). Matrix metalloproteinases and tissue inhibitors of metalloproteinases: structure, function, and biochemistry. *Circulation Research, 92*(8), 827–839. https://doi.org/10.1161/01.RES.0000070112.80711.3D

Wallace, H. A., Basehore, B. M., & Zito, P. M. (2023, June 12). Wound healing phases. In StatPearls [Internet]. StatPearls Publishing. www.ncbi.nlm.nih.gov/books/NBK470443/

Weisel, J. W., & Litvinov, R. I. (2017). *Fibrin Formation, Structure and Properties* (pp. 405–456). https://doi.org/10.1007/978-3-319-49674-0_13

Wells, A., Nuschke, A., & Yates, C. C. (2016). Skin tissue repair: Matrix microenvironmental influences. *Matrix Biology, 49*, 25–36. https://doi.org/10.1016/j.matbio.2015.08.001

Werner, S., & Grose, R. (2003). Regulation of wound healing by growth factors and cytokines. *Physiological Reviews, 83*(3), 835–870. https://doi.org/10.1152/physrev.2003.83.3.835

Wu, Y.-S., & Chen, S.-N. (2014). Apoptotic cell: linkage of inflammation and wound healing. *Frontiers in Pharmacology, 5*, 1. https://doi.org/10.3389/fphar.2014.00001

Xue, M., & Jackson, C. J. (2015). Extracellular matrix reorganization during wound healing and its impact on abnormal scarring. *Advances in Wound Care, 4*(3), 119–136. https://doi.org/10.1089/wound.2013.0485.

Yamakawa, S., & Hayashida, K. (2019). Advances in surgical applications of growth factors for wound healing. In *Burns and Trauma* (Vol. 7). Oxford University Press. https://doi.org/10.1186/s41038-019-0148-1

Yang, H. S., Shin, J., Bhang, S. H., Shin, J. Y., Park, J., Im, G. Il, Kim, C. S., & Kim, B. S. (2011). Enhanced skin wound healing by a sustained release of growth factors contained in platelet-rich plasma. *Experimental & Molecular Medicine, 43*(11), 622–629. https://doi.org/10.3858/emm.2011.43.11.070

Yates, C. C., Bodnar, R., & Wells, A. (2011). Matrix control of scarring. *Cellular and Molecular Life Sciences, 68*(11), 1871–1881. https://doi.org/10.1007/s00018-011-0663-0

Yates, R. A., Nanney, L. B., Gates, R. E., & King, L. E. (1991). Epidermal growth factor and related growth factors. *International Journal of Dermatology, 30*(10), 687–694. https://doi.org/10.1111/j.1365-4362.1991.tb02609.x

Yin, H., Wang, J., Li, H., Yu, Y., Wang, X., Lu, L., Lv, C., Chang, B., Jin, W., Guo, W., Ren, C., & Yang, G. (2021). Extracellular matrix protein-1 secretory isoform promotes ovarian cancer through increasing alternative mRNA splicing and stemness. *Nature Communications, 12*(1). https://doi.org/10.1038/s41467-021-24315-1.

Yoo, Y., Hyun, H., Yoon, S.-J., Kim, S. Y., Lee, D.-W., Um, S., Hong, S. O., & Yang, D. H. (2018). Visible light-cured glycol chitosan hydrogel dressing containing endothelial growth factor and basic fibroblast growth factor accelerates wound healing in vivo. Journal of Industrial and Engineering Chemistry, *67*, 365–372.

Yu, C., Bianco, J., Brown, C., Fuetterer, L., Watkins, J. F., Samani, A., & Flynn, L. E. (2013). Porous decellularized adipose tissue foams for soft tissue regeneration. *Biomaterials, 34*(13), 3290–3302. https://doi.org/10.1016/j.biomaterials.2013.01.056

Yu, R., Zhang, H., & Guo, B. (2021). Conductive biomaterials as bioactive wound dressing for wound healing and skin tissue engineering. *Nano-Micro Letters, 14*(1). https://doi.org/10.1007/S40820-021-00751-Y

Yue, B. (2014). Biology of the extracellular matrix: An overview. *Journal of Glaucoma, 23*(8 Suppl 1), S20–S23. https://doi.org/10.1097/IJG.0000000000000108

Zhang, X., Alanazi, Y. F., Jowitt, T. A., Roseman, A. M., & Baldock, C. (2022). Elastic fibre proteins in elastogenesis and wound healing. *International Journal of Molecular Sciences, 23*(8). MDPI. https://doi.org/10.3390/ijms23084087.

Zhang, X., Chen, X., Hong, H., Hu, R., Liu, J., & Liu, C. (2021). Decellularized extracellular matrix scaffolds: Recent trends and emerging strategies in tissue engineering. *Bioactive Materials*, *10*, 15–31. https://doi.org/10.1016/J.BIOACTMAT.2021.09.014

Zhang, X., Kang, X., Ji, L., Bai, J., Liu, W., & Wang, Z. (2018). Stimulation of wound healing using bioinspired hydrogels with basic fibroblast growth factor (bFGF). *International Journal of Nanomedicine*, *13*, 3897–3906. https://doi.org/10.2147/IJN.S168998

Zhang, Z., Liu, Y., Zhang, X., & Xu, W.-S. (2004). [The content of decorin and its mRNA expression in normal human skin and hyperplastic scars]. *Zhonghua Shao Shang Za Zhi = Zhonghua Shaoshang Zazhi = Chinese Journal of Burns*, *20*(2), 76–78.

7 Tissue Engineering for Wound Healing

Recent Advancements and Opportunities

Khalid Nawaz,[1] Muneeb Ullah,[2] Jin-Wook Yoo,[2] Abdul Wahab,[3] Nurhasni Hasan,[4] and Muhammad Naeem[5]*

[1] Department of Molecular and Cell Biology, Far Eastern Federal University, Vladivostok, Russia

[2] College of Pharmacy, Pusan National University, Busandaehak-ro 63 beon-gil 2, Geoumjeong-gu, Busan 46241, Republic of Korea

[3] Department of Pharmacy, Kohat University of Science, and Technology (KUST), Kohat, 26000, Khyber Pakhtunkhwa, Pakistan

[4] Faculty of Pharmacy, Universitas Hasanuddin, Jl. Perintis Kemerdekaan Km 10, Makassar 90245, Republic of Indonesia

[5] Department of Biological Sciences, National University of Medical Sciences (NUMS) Rawalpindi 46000, Punjab, Pakistan.

* Corresponding author: Muhammad Naeem

7.1 INTRODUCTION

The wound healing process is characterized by the intricate coordination of multiple overlapping events, including blood clotting, inflammation, cellular proliferation, and extracellular remodeling regulated by the secretion of various growth factors, chemokines, and cytokines [1]. Consequently, any disruption in skin integrity can precipitate tissue breakdown, resulting in the formation of acute or chronic wounds [2]. Acute wounds typically arise from traumatic injuries such as burns or surgical incisions and typically heal within an expected timeframe, whereas chronic wounds, which encompass conditions such as ulcers (diabetic, venous, pressure) or postsurgical wounds, exhibit a prolonged healing process, often failing to progress through the normal stages of wound healing repair, resulting in a lack of significant improvement over an extended period [3]. Abnormalities in the aforementioned process can impede proper wound healing and thwart tissue regeneration, thereby hindering treatment and compromising tissue integrity restoration [4]. Numerous conventional and regenerative studies have been dedicated to developing effective wound therapies, aiming to ease healthcare costs, promote successful scar healing, and provide long-term relief [5].

Nevertheless, regenerative therapy is preferred over conventional methods, as it tries to restore skin function by re-establishing damaged cells and skin tissues without scarring [6]. This therapeutic approach encompasses a range of strategies, including but not limited to tissue engineering, biomaterials, growth factors therapy, and stem cell transplantation [7]. For example, in a study involving mice with full-thickness wounds, the administration of adipose-derived stem cells via intravenous, intramuscular, and topical routes accelerated the wound healing process and improved wound closure [8].

DOI: 10.1201/9781003605966-7

Tissue engineering aims at producing functional tissue constructs for use in the reconstruction or regeneration of damaged or lost tissues and organs, such as skin, spinal cord, and other organs. In addition, secondary benefits of tissue engineering have been the development of models for studying function, disease, and drug testing and development. Tissue engineering can be accomplished either ex vivo or in situ through the use of various molecules, materials, or cells to stimulate the local regenerative capacity of tissue [9]. It is crucial to recognize that this effort in tissue engineering marked a fundamental shift in the approach to treating tissue loss [10]. With the ultimate aim of generating functional organs with a complex interplay of diverse cell signals and scaffolds, the effort necessitated a "systemic" approach guided by engineering principles instead of the traditional reductionist methodology of experimentation [11]. The primary objective extended beyond the knowledge of a fundamental mechanism to rather produce a product to mitigate or cure a disease state [12].

In this book chapter, we aim to explore the recent advancements in tissue engineering for wound healing while also examining various challenges and opportunities in wound management. The field of tissue engineering is continuously progressing, taking advantage of recent developments in areas such as polymeric nanomaterials, both natural and synthetic. Cells, scaffolds, stem cell-based, gene therapy, and vascularization-based strategies are being developed to design three-dimensional (3D) structural units aimed at the functional restoration of cutaneous wound tissue.

7.2 WOUND HEALING AND TISSUE ENGINEERING

Classical wound healing follows a continuum of three distinct but continuous phases: inflammation, proliferation, and remodeling [13]. The inflammatory stage commences with hemostasis and the formation of platelet embolism, followed by the consolidation of the fibrin matrix into scaffolds to facilitate cell homing [14]. Inflammatory cells, such as neutrophils and macrophages, play an essential role in cleaning the dead cells, bacteria, and contagious organisms [15]. However, excessive inflammation often leads to the formation of scar tissue, underscoring the significance of balanced inflammatory responses in wound healing outcomes [16]. During the proliferation phase, angiogenic cells proliferate and migrate to the scaffolds, facilitating the formation of vascularized new tissues [17]. Meanwhile, macrophages also contribute to and accelerate angiogenesis in this phase. The degeneration, transformation, and regeneration of extracellular matrix (ECM) take place simultaneously and last for months during the tissue remodeling stage [18]. Cutaneous wounds can be treated through either repair therapy or regenerative therapy. Repair therapy represents the conventional wound healing process, typically accompanied by scar formation, while regenerative therapy aims to restore the full dermal layers with skin appendages (**Figure 7.1**). Recent advancements in wound healing treatments offer patients newfound hope in restoring skin integrity and wound healing through tissue engineering methodologies.

Tissue engineering can be considered a multidisciplinary and emerging field in technology, aimed at regenerating damaged organs, generating complex tissues, and maintaining normal cell homeostasis [19]. A primary objective of tissue engineering is the design of new tissue and organ replacements that closely mimic the physiological environment for cells [20]. In the context of wound healing, tissue engineering products can be classified into two main categories: cell-based skin constructs and acellular products [21]. These products encompass a diverse array of materials, categorized as epidermal and dermal substitutes. The dermis comprises dense irregular connective tissue containing collagen fibers, elastic fibers, and an extrafibrillar matrix, which serves to anchor the dermis to the epidermis [22]. The papillary region, located in the superficial area (upper layer) of the dermis, is a cell-rich layer housing macrophages and fibroblasts, facilitating interactions between the dermis and epidermis. This interaction induces the synthesis of ECM components, which in turn stimulates the differentiation and growth of keratinocytes. While some allogeneic and xenogenic tissue-engineered skin substitutes have been developed, they often carry the risk of graft rejection [23]. The need for artificial skin substitutes to address large skin defects became apparent in the 1970s, prompting plastic surgeons and researchers to explore dermal tissue fabrication in the labs using tissue engineering techniques [24]. Subsequently, numerous studies have been conducted

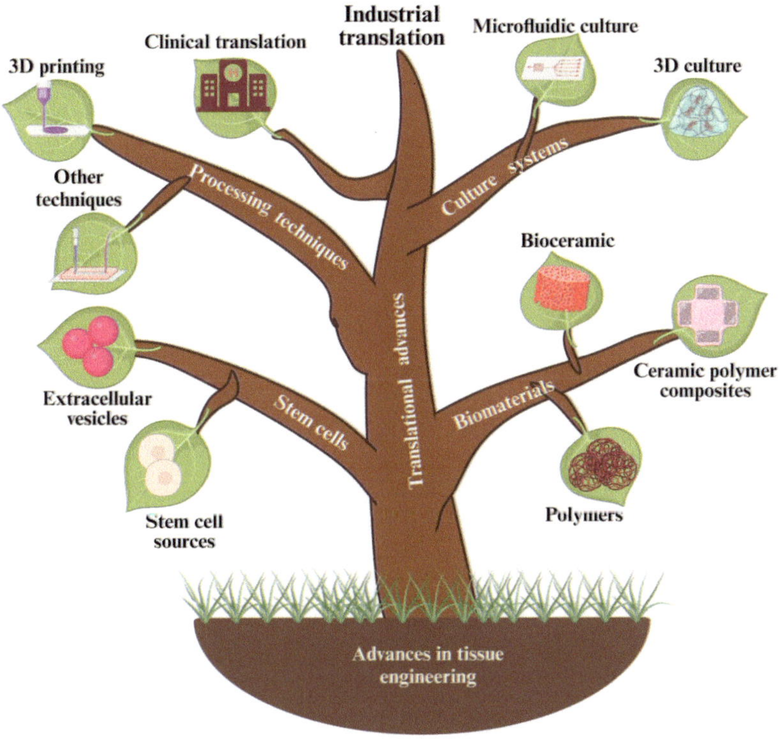

FIGURE 7.1 Schematic Illustrated tress of the advances in tissue engineering in wound healing.

on the production of dermal tissue substitutes, leading to the development of commercially available products approved by the FDA for the treatment of large burn wounds [25].

Over the past decades, a number of cell-based skin constructs have been developed to address full-thickness skin defects [26]. In a pioneering achievement in 1975, keratinocyte colonies were successfully cultured from single-cell suspensions of human epidermal cells [27]. This method emerged as a viable option for closing large burn wounds, particularly when autologous skin grafts were limited, offering a potential lifesaving approach [28]. Since then, several clinical case reports have been published describing the use of cultured epidermal autografts (CEA) as confluent sheets for burn wound coverage [29]. However, the poor survival rate of keratinocytes in these cell sheets has posed a significant concern. The introduction of novel biocompatible carriers for CEA has permitted grafting at earlier culturing stages of keratinocytes, enhancing their viability [30]. Additionally, keratinocyte colonies have been utilized for grafting onto burn wound sites before forming a single-cell layer sheet. Despite these advancements, only a few products are commercially available as cell-based skin substitutes, primarily consisting of human allogeneic skin cells [31]. One such example is Apligraf ® (Organogenesis Inc., MA, USA), a Food and Drug Administration (FDA)-approved skin substitute consisting of human allogeneic keratinocytes and fibroblasts derived from neonatal foreskins seeded onto a bovine type I collagen gel [32].

7.3 ADVANCEMENTS IN TISSUE ENGINEERING IN WOUND HEALING

7.3.1 Advancements in Biomaterial

In tissue engineering, biomaterials are used to provide micro- and nano-structural characteristics, morphology, and surface properties that support cells and can be loaded with appropriate growth factors [33]. These biomaterials, typically in the form of matrices or scaffolds, are engineered to guide cell growth through specific designs. Although biomaterials have been integral to tissue engineering since the early 1990s, their utilization and applications continue to evolve. Recent advancements in the use of biomaterials for tissue engineering are summarized in **Table 7.1**. It is noteworthy that the field of tissue engineering has often an eye on commercialization, such that

TABLE 7.1
Advantages and Disadvantages of Various Biomaterials Used for Tissue Engineering

Biomaterial	Advantages	Limitations	Types of Tissue Engineering Products	References
ECM	Mimicking native tissue		Skin, meniscus, kidney, and bone	[41, 42]
Synthetic polymers	Excellent mechanical characteristics Exhibit bioresorbability and undergo controlled processing	Lack of cell adhesion molecules Inflammation	Brain, nerve, bone, and cartilage	[42]
Natural polymers	Biocompatibility High processability Elasticity Degradability	Limited mechanical properties	Several tissues such as bone, cartilage, heart, and liver	[43]
Smart and functional polymer composites	Biological properties Physical properties, e.g., stimuli, responsiveness, self-healing, and shape memory	Environment may affect the responsiveness and controllability	Injectable therapeutics for regenerative bone disorders	[44]
Hydrogels	Delivery of cells, drugs, and biomolecule	Cell adhesion Adhesive strength Mechanical properties	3D bioprinting Minimally invasive regenerative therapies Injectable materials and drug delivery vehicles for regeneration Cartilage regeneration	[41, 45]
Metals	Biocompatibility Enhanced mechanical properties Degradable metal alloys	Uncontrolled carrion	3D pours scaffolds Absorbable implants for bone repair	[46]
Bioceramics	Biocompatibility Bioactive 3D printed scaffolds with mechanical characteristics comparable to human cortical bone High compression strength	Brittleness Low tensile strength Weak under cyclic or high loads	Hard tissue engineering such as bone, tooth, and cartilage	[41, 45]

FDA approval for a new biomaterial may be a hurdle that shapes the evolution of tissue engineering constructs [34]. This regulatory barrier can, at times, constrain the innovation within the field, because it is often more expedient to adapt a natural biomaterial or utilize one with an established FDA track record rather than develop a novel material that will require extensive vetting before clinical use [35].

7.3.2 Advancements in Polymeric Biomaterials

7.3.2.1 Natural Polymers

Natural polymers that are widely used in tissue engineering because of their availability and biocompatibility include gelatin, collagen, chitosan, hyaluronic acid, alginate, and polyhydroxyalkanoates (PHAs) [36]. Peptides present on some of them, such as collagen, aid in cell attachment, migration, and function. Silk, another natural polymer, has gained popularity in tissue engineering applications due to its high processability, strength, and elasticity. PHAs represent a special group of natural polymers due to their degradation through surface erosion, which helps maintain their overall structure [37]. In vitro studies have demonstrated guided growth of neuronal cells using highly aligned electrospun fibers of a blend of poly(3-hydroxybutyrate) [p(3HB)] and poly(3-hydroxy octanoate) [p(3HO)] [38]. In another application, p(3HO) was used in the production of cardiac patches, which exhibited favorable mechanical properties closely resembling native cardiac muscle and surface topography that enabled efficient cell adhesion and proliferation [39]. To produce new constructs with improved properties, polymers can also be used in combination; for example, electrospun fibers made from a combination of chitosan and gelatin have been shown to enhance bone regeneration capability. Furthermore, blending PHAs with the synthetic polymer polycaprolactone (PCL) to produce a scaffold for delivering seeded cardiac progenitor cells has demonstrated the adhesion of cardiac progenitor cells, stem cell proliferation, and retention when implanted in the postmortem murine hearts [40].

7.3.2.2 Synthetic Polymers

Synthetic polymers are extensively used in tissue engineering due to their widespread availability, cost-effectiveness, bioresorbability, and versatility in processing, making them suitable for various applications [47]. Commonly used synthetic polymers include polyglycolide (PGA), polylactide (PLA), poly(lactide-co-glycolide) (PLGA), polycaprolactone (PCL), polydimethylsiloxane (PDMS), and poly(glycerol sebacate) (PGS) [48]. Among synthetic polymers, PDMS holds a unique application in tissue engineering because of its high oxygen (O2) diffusivity, biocompatibility, flexibility, and ease of fabrication [49]. It has been explored for the engineering of cell sheets and is widely utilized in the development of 3D organ-on-a-chip (OoC) cultures that significantly advance the field of tissue model engineering [50]. However, PDMS is a nonbiodegradable polymer, limiting its use primarily to ex vivo rather than implantable tissue engineering constructs.

Blends of synthetic polymers were also explored to combine the properties of different materials. Moreover, synthetic polymers have been used in combination to create distinct phases within the resulting structure [51]. For example, in a study by Fang et al., core electrospun nanofibers using PGA and PLA were produced to create a shell, resulting in materials that accelerated wound healing in vivo [52]. Synthetic polymers can also be combined with natural biomaterials to create semisynthetic polymers. For example, Jiao et al. melt-blended PCL and hyaluronic acid to 3D print scaffolds for bone tissue engineering, resulting in constructs with improved mechanical characteristics compared to those made from PCL alone [35]. The bioresorbability of many synthetic polymers offers advantages in various tissue engineering applications. Nevertheless, regulating the rate of degradation can present clinical challenges. Xu et al. conducted an experiment involving the addition of magnesium to PLGA to form composite films with low ranges of magnesium weight percentages, and they found that the addition of magnesium extended the degradation duration while also improving the tensile strength of the films [53].

7.3.2.3 Hydrogel/Polymer-Based Scaffolds

Hydrogels represent a rapidly advancing frontier in tissue engineering, particularly for use in 3D bioprinting, which has witnessed active research growth in recent years [54]. These versatile materials can be derived from various natural or synthetic polymers and have been used for the engineering of diverse tissues. One of their key advantages is their ability to encapsulate cells while maintaining the permeability necessary for the diffusion of O2 and nutrients across the material. However, a longstanding challenge with hydrogels has been their mechanical properties and lack of adhesiveness [55]. Recent advancements have addressed these shortcomings. For instance, Shirzaei Sani et al. developed an adhesive hybrid hydrogel composed of HA and elastin-like polypeptide, characterized by remarkable adhesive and tunable physical properties and antimicrobial activity [56]. This breakthrough enhances the clinical translation of hydrogels, which was previously limited by their low adhesive strength, poor mechanical characteristics, and their weakness to inhibit bacterial colonization [57]. Hydrogel reinforcement can be achieved through interpenetrating secondary networks, a strategy that expands the use of conventional hydrogels and facilitates their utilization in emerging biofabrication techniques aimed at achieving hierarchical architectures and personalized medicine [58].

These interpenetrating hydrogels hold immense potential across various domains, including tissue engineering, drug delivery systems, and the development of in vitro disease models for drug discovery and screening [59]. Hybrid hydrogels are reported to have superior adhesive strength to engineered tissues compared to commercially available tissue adhesives [60]. Moreover, a significant advantage of hydrogels is their use as injectable materials to deliver cells, biomolecules, and drugs for regenerative purposes, often through minimally invasive techniques [61]. For instance, recent research into cartilage repair revealed that HA hydrogels could encapsulate chondrocytes, supporting cell survival and the regeneration of cartilaginous tissue [62]. Beyond HA, hydrogels derived from collagen, alginate, and ECM have also been used in tissue engineering and cell encapsulation. In addition, microencapsulation of cells to produce microgels was also explored. These hydrogels can be blended and processed using a 3D bioprinting method, wherein cells are encapsulated and printed into designed structures before being cross-linked to provide appropriate mechanical properties. For example, 3D-printed scaffolds of collagen/alginate hydrogel have been used for cartilage tissue engineering and ECM hydrogel for cardiac patches [63].

The regeneration of damaged tissue can be pursued either via ex vivo or in situ methods. In ex vivo tissue engineering, scaffolds are combined with cells and biomolecules outside the body to generate cell-laden tissue constructs for implantation (**Figure 7.2**). However, ex vivo tissue regeneration presents certain limitations, such as tissue morbidity and the lack of reliable cell sources [64]. On the other hand, in situ tissue engineering requires precise control over the biochemical and biophysical cues to stimulate resident host cells and attract cells to the site of injury requiring regeneration [9]. This approach can be achieved by either stimulating endogenous cells using extracellular signals or through cell reprogramming. In the former approach, cells are primed via extracellular factors, such as by modulating the biochemical and biophysical characteristics of the biomaterial. In the latter approach, direct manipulation of the cellular gene expression program is accomplished through cellular reprogramming. Given its relation to biomaterials, we focus on reviewing the first approach in this section.

7.3.2.4 Functional and Smart Polymers

Smart polymers such as those with self-healing, stimuli-responsiveness, or shape memory properties are used in tissue engineering [65]. The ability to change the shape of 3D-printed objects in response to environmental stimuli, such as light, pH, heat, moisture, or water as a function of

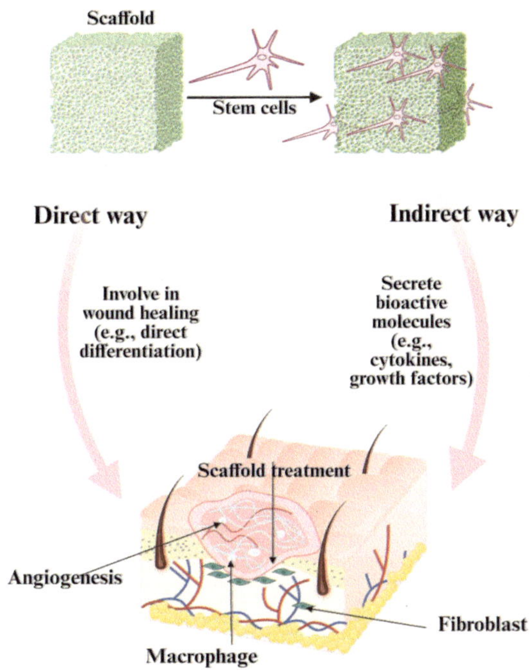

FIGURE 7.2 Stem cell-encapsulated scaffold-based delivery for wound healing and tissue regeneration.

time, is known as four-dimensional (4D) printing, and it has gained significant interest in recent years. For example, Invernizzi et al. introduced a novel 4D-printable smart material comprising PCL and 2 ureido-4[1H]-pyrimidinone (UPy), a thermally activated memory polymer with self-repairing capabilities [66]. The incorporation of methacrylate-bearing UPy (UPyMA) monomers has provided the 4D-printed structures with self-healing properties, showcasing the possibility to fabricate actuators for soft robotics for the first time [67]. The synthesis of smart hydrogels offering both self-healing and shape memory properties concurrently is anticipated to be a subject of further investigation in the coming years.

Although for many reasons smart and functional polymers have garnered attention for their promising applications in tissue engineering, it is essential to acknowledge their inherent drawbacks. For example, the complex environment of the body, pH, temperature, or triggering thresholds of changes may impact the responsiveness of these polymers, which in turn could affect the controllability of the construct and the behavior of the included cells [68, 69]. Furthermore, the placement of shape memory materials within the body carries the risk of injury to neighboring tissues or functional impairment after returning to their original shape. Mitigating these challenges is imperative to gain the full benefits of smart polymers in tissue engineering.

Stimuli-responsive and self-healing hydrogels have emerged as promising pharmaceutical carriers for tissue engineering applications. One notable example is their role in bone regeneration, where they can create an optimal microenvironment for new bone formation and facilitate therapeutic drug delivery [70]. Unlike conventional hydrogels, these constructs show resistance against mechanical stress, safeguarding their therapeutic cargos from degradation and enabling sustained release for the extended periods necessary for effective bone tissue healing. In pursuit of this goal, researchers have developed a hydrogel composed of chondroitin sulfate (ChS), renowned for its regenerative properties, especially developed for bone tissue repair, with the material being cross-linked to mimic cranial bone characteristics [71].

The ChS-based hydrogel demonstrates remarkable attributes of excellent self-healing, injectability, and strong adhesion to in vivo tissues. When encapsulated with rat-derived mesenchymal stem cells (MSCs), this hydrogel exhibits favorable cytocompatibility [72]. Importantly, in a murine bone defect model, the injection of bone morphogenetic protein (BMP)-4-loaded hydrogel resulted in the repair of the defect, evidenced by the formation of new cranial bone tissue and a significant reduction in defect size after 12 weeks, compared to phosphate-buffered saline-loaded hydrogel [73]. In recent years, there has been active progress in the development of advanced functional polymers with tunable chemical and biological properties. This progress has smoothed the way for novel applications in 3D and 4D bioprinting and drug delivery, as evidenced by Zhang et al.'s recent study on the development of biocompatible hydrogel ink, which contains self-healing PR cross-linked hydrogel microparticles of chitosan methacrylate and polyvinyl alcohol hybrid hydrogels [74]. Their findings proved the rapid production of advanced structures with high aspect ratios and excellent shape accuracy at organ-proper scales. Several other drug delivery systems have been devised based on advanced polymers. Consequently, functional hydrogels capable of delivering the required dosage with precise chronology and localization have the potential to mitigate complications and enhance the success of clinical applications.

7.4 STEM CELL APPLICATION

One of the foremost challenges in advancing stem cell-based treatments in tissue engineering lies in discerning the biochemical and biophysical environments specific to different tissues. In addition to describing the differentiation and growth factors that replicate the stem cell environment, studies have determined the significance of characterizing the physical properties and mechanical forces of the stem cell matrix, including its morphology and stiffness. Investigating the impact of ECM biological, chemical, and physical effects on stem cells is important to develop methodologies capable of influencing cell differentiation [75]. Research has also revealed that aligning the surface architecture and stiffness properties of biomaterials with those of specific native tissues favors the differentiation of stem cells into cell types characteristic of these tissues [76].

Controlling stem cell fate is a crucial aspect of stem cell research and its application. In a recent study, magnetic nanoparticles were used to direct stem cell differentiation [77]. Through the application of an externally applied magnetic field, iron oxide particle-laden embryonic stem cells (ESCs) were aggregated to form spheroids [78]. Then, opposing magnetic fields were used to stretch these spheroids, leading to differentiation toward the cardiac lineage. Combining both chemical and physical factors resulted in a significantly higher degree of myogenic differentiation. The differentiation of stem cells into the desired cell type is possible through the identification of factors such as the matrix microenvironment and epigenetic mechanisms that regulate stem cell fate. For example, an injectable hydrogel composed of HA, galactose oxidase, horseradish peroxidase, and tyramine was developed for cross-linking purposes [79]. Experiments in mice demonstrated the material's biocompatibility, which makes it a good candidate for use in biomedical applications such as tissue engineering.

A recent study showed how the porosity of hydrogel biomaterial can influence MSCs and their response to insulin-like growth factor-1 (IGF-1) [80]. Unlike nanoporous alginate hydrogel, microporous variants were found to sensitize MSCs to the growth factor. Additionally, the addition of a cell–cell adhesion mediating molecule (N-cadherin) mimicking peptide to nanoporous alginate replicated the effect observed with macroporous materials, enhancing MSCs' paracrine activity in response to IGF-1 [81]. This further underscores the role of the physical properties of the biomaterials and also the possibility to modulate them through chemical means. Together, these approaches will further advance our abilities to control stem cell behavior in the future and tailor it toward desired activities and fate for the regeneration of specific tissues.

7.5 D BIOPRINTED SKIN GRAFTING IN WOUND HEALING

Traditional tissue-engineered [3] skin constructs have provided significant benefits to patients with chronic wounds or burns; however, they still face challenges such as inadequate vascularization, lack of hair follicles and appendages, and limited pigmentation properties, etc. In recent years, 3D bioprinting has emerged as a promising approach to address these challenges by aiming to recapitulate fully functional native skin, despite the immense challenges, including the restoration of multiple cell types and the complex micro-architecture of ECM components with sufficient resolution [82]. Skin bioprinting considers several parameters such as the precise positioning of cell layers, cell-cell interactions, and cell-matrix interactions, which conventional tissue engineering strategies lack [83]. 3D bioprinting represents a flexible, fully automated, and advanced manufacturing platform for designing and engineering human tissues and organs. It involves the layer-by-layer precise positioning of living cells, biochemical/growth factors, and biological materials using computer-aided design [26], offering a high degree of flexibility, spatial control of functional components placement, and repeatability, which are otherwise lacking in conventional skin tissue engineering-based approaches. Therefore, bioprinting enables concurrent engineering design of tissues and organs, spanning from micro to macro scales to restore the natural niche for skin cells [84]. Many bioprinting approaches, such as autonomous self-assembly, biomimicry, and mini tissue building blocks, have been developed with appropriate biological and mechanical properties to recapitulate the clinical function of various tissues and organs.

A number of bioprinting approaches, including laser, inkjet, extrusion, and microvalve-based systems, are being utilized for the deposition and positioning of biological components [85]. In skin tissue bioprinting, laser-based, inkjet-based, and microvalve-based techniques have been primarily employed to fabricate multilayered skin constructs, often using collagen-based bio-ink. For example, a bilayered skin graft was printed using a modified inkjet printer to deposit human microvascular endothelial cells onto a collagen gel containing human dermal fibroblast cells (neonatal) and human epidermal keratinocytes (neonatal) [86]. This bilayer bioprinted construct was utilized for the healing of full-thickness wounds in athymic nude mice and showed neoskin-like properties with improved wound contraction (by up to 17%) compared to the commercially available skin graft, Apligraf [87]. Koch et al. demonstrated the bioprinting of 20 layers each of fibroblasts and keratinocytes onto a sheet of decellularized dermal matrix (Matriderm^R) using collagen-based bio-ink [88]. In this method, 20 cell-collagen sublayers of HaCaT keratinocytes create 3D bilayer cell constructs containing fibroblasts and keratinocytes. The printed skin constructs showed epidermal markers such as cadherins and connexin 43 (Cx43) after 10 days of cultivation, which signifies tissue morphogenesis and cohesion [21].

In another study, a printed skin construct demonstrated good integration of the graft with the surrounding tissue and improved vascularization after 11 days of *in vivo* transplantation in nude mice [21]. Furthermore, collagen and gelatin constructs containing multilayered keratinocytes and fibroblasts were bioprinted and showed notably elevated cell viability with fluidic channels (85% viability) [89]. In a recent assessment, bioprinted collagen-based skin constructs were utilized to mimic the native cellular density of different skin cells [90]. Moreover, skin bioprinting can be conducted *in situ* by directly printing onto the wound area. For instance, in one study, a microvalve-based bioprinting method was employed to print amniotic fluid-derived stem cells (AFSCs) onto full-thickness skin wounds (2 cm x 2 cm) of nu/nu mice [21]. This approach resulted in heightened wound closure rates and angiogenesis due to the secretion of trophic factors from AFSCs. Furthermore, the *in situ* microvalve-based printing technique was investigated on full-thickness large wounds (10 cm x 10 cm) of nude mice [91]. In this study, a fibrinogen/collagen hydrogel precursor containing fibroblasts (1.0 x 105 cells/cm2) was bioprinted as a first layer, followed by keratinocytes (1.0 x 107 cells/cm2) above the fibroblast layer as the second layer [21]. Despite difficulties associated with cellular deposition, complete re-epithelialization of the large wound was observed after 8 weeks. In a recent study, a polyelectrolyte gelatin-chitosan hydrogel-based skin construct was bioprinted for skin tissue engineering [39]. The bioprinted constructs were three-layered (~400 μm thickness) and represented the outer epidermal layer and a part of the dermal region. Skin bioprinting presents comparatively fewer challenges in printing skin constructs when compared to thicker tissues or organs, and it closely mimics the characteristics of mature native skin tissue [92]. Therefore, continued advancements in skin bioprinting and in situ skin bioprinting have the potential to enable the on-demand fabrication of autologous skin constructs wound treatment. Skin 3D bioprinting offers relatively less difficulty in printing skin constructs as compared to other thicker tissues and/or organs and shows a close resemblance to matured native skin tissue [83]. Therefore, further advancement in skin bioprinting and *in situ* skin bioprinting might lead to the on-demand fabrication of autologous skin constructs for wound treatment [93]. Although the complete replication of a fully functional skin equivalent through bioprinting remains a goal yet to be achieved, skin bioprinting holds promising potential in generating pigmented skin models, hair follicles, and vasculature networks in skin tissue engineering. Hence, further refinement of skin bioprinting techniques is required before they are applied for immediate wound repair in humans.

7.6 GENE-BASED THERAPEUTIC APPROACHES IN TISSUE ENGINEERING IN WOUND HEALING

The skin possesses significant potential for effective and functional regeneration. In this regard, tissue-engineered skin substitutes have shown promising results in skin repair [94]. However, the current bioengineered skin grafts are challenged by hypertrophic scarring, insufficient vascularization, appendages, and lack of hair follicles, thereby creating an inhospitable environment for optimal skin repair. Moreover, chronic and infected wounds are biologically and biochemically complex wound sites and impose considerable problems in wound healing [95]. Therefore, interventions aimed at postnatal wound healing repair, such as therapeutic applications involving growth factors, could enhance the healing capacity in wounds that are pathologically altered, minimizing scar formation and inflammation [96]. Additionally, the skin's high epidermal turnover rate and the transient upregulation and downregulation of numerous growth factors and cytokines make it an ideal candidate for genetic manipulations and gene therapy research [97].

Skin tissue is an attractive target for in vitro and in vivo gene transfer therapeutic strategies due to its easy accessibility and susceptibility to a variety of skin diseases. The feasibility of gene transfer research in the skin is facilitated by the ease of harvesting and culturing skin cells (fibroblasts and keratinocytes), which can be genetically manipulated [97]. These genetically modified skin cells serve as vehicles for gene transfer in skin diseases, particularly in wound healing [98]. Through genetic molecular approaches, genetically modified cells have been utilized for the synthesis and controlled delivery of specific growth factors to wound sites in a time-dependent manner. In a previous study, particle bombardment-based gene delivery of epidermal growth factor (EGF) resulted in increased EGF concentration and accelerated wound healing in a porcine wound healing model [21]. Similarly, in another study, accelerated wound healing and re-epithelialization were observed through the transplantation of fibroblast suspensions transiently transfected with EGF plasmid [99]. Moreover, polyglycolic acid (PGA) scaffolds seeded with genetically modified fibroblasts encoding the human PDGF-B gene demonstrated 40% higher re-epithelialization acceleration in full-thickness skin wounds in diabetic mice [100]. Initially, genetic manipulation of skin cells relied on viral vectors, which suffer from limitations, including causing inflammation and the risk of infection. To overcome these drawbacks associated with viral vector-based gene delivery, various nonviral gene transfer techniques have been employed in recent years, such as high-pressure injection, gene gun, naked DNA application, microseeding, and liposomal vectors.

Recently, gene-activated matrix (GAM) has been explored as a novel approach for wound healing that combines gene therapy and tissue engineering principles [101]. In the GAM system, gene delivery from polymeric biomaterials can occur through direct incorporation of DNA/RNA/small interfering RNA (siRNA) into the scaffold or by seeding genetically modified cells (obtained via viral or nonviral gene transfer) onto the matrix to induce the desired response [21]. The method employed for incorporating genes onto the matrix is also considered a key factor in the GAM system. In this approach, genetically modified skin cells/nucleic acids are incorporated into a biomaterial scaffold to create a regenerative wound environment by promoting cellular adhesion, migration, proliferation, and differentiation. This alternative strategy of gene delivery from polymeric biomaterials addresses the limitations of viral and nonviral gene delivery methods, offering better protection against DNA degradation and allowing for greater control at the protein level while minimizing the inflammatory response. Guo *et al.* demonstrated enhanced angiogenesis of chitosan/collagen-based bilayer equivalent (BDE) using DNA encoding VEGF [21, 102]. In this study, a complex consisting of a cationic gene delivery vector N, N, N trimethyl chitosan chloride (TMC), and plasmid DNA encoding VEGF-165 was utilized to create a gene-activated BDE. The released plasmid DNA from BDE retained its structural integrity and showed efficient transfection, leading to enhanced angiogenesis in both *in vitro* and *in vivo* conditions [103]. Following transplantation of an ultrathin graft for 112 days, the healed skin had approximately 80% of the tensile strength of native skin and displayed a similar structural composition. Similarly, treatment with the gene-activated BDE enhanced vascularization with the formation of a significantly higher number of mature blood vessels in a burn wound pig model.

Currently, scarring remains a considerable problem associated with conventional skin grafts and bioengineered skin substitutes. Scarring not only leads to the loss of dermal functionality but also causes local ulceration and disfigurement. Therefore, achieving scarless repair is of utmost need not only for aesthetic purposes but also for functional requirements. In efforts to address this, a combination of TMC/siRNA complexed with BDE (RNAi-BDE) has shown inhibition of scarring by downregulating the TGF-β1 pathway. This suppression of TGF-β1 has been found to inhibit the expression of various scar-related factors such as α-smooth muscle actin, collagen type I, and collagen type III [104]. Hence, the GAM might be a promising approach for the regeneration of skin appendages and ultimately recapitulation of fully functional skin in the future [21]. Despite successful applications in enhancing vascularization and wound healing, there have been few clinical trials using a GAM-based approach. However, a recent report highlighted a significantly reduced amputation rate of diabetic feet using an injection of genetically modified endothelial cells [105]. Therefore, further research is required to optimize the selection of matrices, gene delivery vehicles, and gene manipulation methods for wound healing.

7.7 VASCULARIZATION-BASED STRATEGIES

In the past decade, the majority of the available skin substitutes have consisted primarily of dermo-epidermal constructs, which combine dermal and epidermal layers within an acellular matrix using keratinocytes and fibroblasts, respectively [106]. However, the inclusion of multiple layers in these constructs can trigger the drifting of cell layers away from the wound surface, leading to subsequent cell death due to hypoxia. Even though acute hypoxia during the healing process has been reported to stimulate fibroblasts and keratinocytes, prolonged hypoxia may result in tissue loss, fibroblast dysfunction, and reduced migration and proliferation of keratinocytes [107]. Consequently, partial necrosis caused by insufficient vascularization poses a serious threat to their clinical use. Vascularization plays a key role in regenerative medicine and represents a critical bottleneck in skin repair and regeneration. Enhanced vascularization promotes better integration and survival of cells in the thick tissue-engineered skin substitutes [108]. Therefore, several strategies have been developed to improve vascularization, such as growth factor delivery systems, cell-based vascularization strategies using vessel-forming cells, and modulation of structural and physicochemical properties of dermal scaffolds. Additionally, other emerging techniques include miRNA modulation, cell sheet engineering, microvascular fragments derived from adipose tissue, and advanced methodologies like 3D bioprinting and microfluidics, which have also been introduced in recent years [109].

In tissue engineering of skin, various cell sources contribute to facilitating vascularization through processes like angiogenesis, vasculogenesis, or both. When selecting cell sources for the vascularization of tissue-engineered skin substitutes, certain desirable properties are sought, including ease of harvesting, rapid expansion rate, nonimmunogenicity, and no risk of malignant transformation. Over time, different cell sources have been employed to enhance vascularization in engineered skin. Among these, endothelial cells (ECs) and progenitor ECs, such as outgrowth endothelial cells (OECs), are mostly utilized [110]. Endothelial origin cells characteristically express endothelial markers such as cadherin, von Willebrand factor (vWF), vascular endothelial (VE), and platelet endothelial cell adhesion molecule-1 (PECAM-1 or CD31), and their delivery to promote neovascularization offers the most appealing option.

The delivery of ECs is usually carried out by two main approaches; one involves culturing cells into the skin substitutes followed by the implantation of the system without cellular organization, while the second approach entails revascularization of the system before implantation [111]. In a study by Hendrickx *et al.*, enhanced vascularization was demonstrated within thick human dermal fibroblast sheets when combined with late outgrowth endothelial progenitor cells, leading to the organization of the epithelial matrix and prevention of excessive wound contraction after implantation [112]. Similarly, in a recent study, co-culturing juvenile dermal fibroblasts with both mature human umbilical vein endothelial cells (HUVECs) and progenitor ECs resulted in enhanced vascularization along with increased expression of papillary fibroblast markers (podoplanin) and ECM secretion [113]. The acceleration in vascularization may be attributed to angiogenic, vasculogenic, or arteriogenic mechanisms (*via* smooth muscle cells coating support). Although the precise mechanism remains unclear, it is assumed that cells possessing a combination of these three properties, such as blood outgrowth endothelial cells (BOEC) and multipotent adult progenitor cells (MAPC), might exert a more functional and robust effect on vascularization [21]. Recent research has also demonstrated the superior vascular potential of BOEC compared to HUVECs.

In the past decade, many researchers have directed their attention toward exploring the potential of fibroblasts in improving vascularization. Guerreiro *et al.* demonstrated the induction of blood vessel ingrowth following the implantation of Matrigel plugs impregnated with neonatal human dermal fibroblasts in mice, indicating their role in recruiting ECs in vivo [21]. Additionally, researchers have shown the improvement of highly anastomosed capillary-like structure formation through the culture of human dermal fibroblasts with HUVECs. In a co-culture system, angiogenesis is regulated by the interaction between ECs and other cell types, which depends on the intricate communication pathways mediated by the diffusible signaling molecules and cell-cell contacts [114]. Kim *et al.* also demonstrated the *in vitro* generation of perusable micro vessel networks by co-culturing normal human lung fibroblasts with HUVECs through microfluidic technology [115]. Another study reported ECs sprouting when they utilized a combination of fibroblast-derived proteins, such as angiogenin, tumor necrosis factor (TNF), and hepatocyte growth factor (HGF) using 3D fibrin gels with normal human lung fibroblasts [116]. Furthermore, ECM proteins like collagen 1 and insulin-like growth factor-binding protein 7 (IGFBP7) supported lumen formation by HUVECs [116]. Besides HUVECs, fibroblasts have also been co-cultured with OECs using 3D tissue constructs to enhance vascularization and wound healing.

Mesenchymal stem cells (MSCs) are also frequently used to improve artificial skin vascularization. Similar to other cell types, MSCs experience hypoxia within scaffolds and release hypoxia-inducible factor [117]-1α-regulated growth factors, which promote angiogenesis in the surrounding host tissue following implantation during the initial phase [118]. In addition, MSCs possess the ability to differentiate into ECs, allowing the direct integration and formation of newly developed microvascular networks. Human sweat gland-derived stem cells have been employed previously to promote vascularization with the formation of capillary-like structures during dermal regeneration[119]. Laschke *et al.*, recently demonstrated enhancement in the vasculogenic process when

MSCs were seeded on scaffolds as multicellular spheroids rather than single cells, using a green fluorescent protein (GFP)+/GFP-based crossover approach [109]. Therefore, MSC spheroids may promote the simultaneous development of new microvascular networks at different locations in skin substitutes and serve as an effective vascularization unit. In another study, aggregates of bone marrow-derived mesenchymal stem cells (BMMSCs) were utilized as optimal for vascularization and regeneration of full-layer cutaneous wounds [120]. In their study to confirm the pro-regenerative ability of BMMSCs and to compare their regenerative efficacy with available stem cell administration strategies (subcutaneous and intravenous), cell aggregates constructed with green fluorescent protein-expressing (GFP+) BMMSCs were applied to a full-thickness cutaneous wound model [117]. *In vitro*, cultured cell-aggregate transplanted models indicated higher expression of several pro-healing genes, better vascularization with the formation of capillary structures, and enhanced ECM deposition. Several other recent emerging strategies have also been employed in skin tissue engineering for the improvement of vascularization. These strategies include cell-sheet technology, 3D bioprinting, microfluidics, and miRNA modulation. Cell-sheet technology involves the generation of scaffold-free artificial skin composed of multilayered cell sheets, which alleviate complications related to cell seeding efficiency and/or biocompatibility associated with conventional scaffold-based approaches [121]. Yu *et al.* demonstrated stimulation of vascularization in bioglass-activated fibroblast cell sheets, as evidenced by elevated secretion of multiple proangiogenic growth factors and ECM deposition [122]. However, MSC-based cell sheets already exhibit a high vessel-forming capacity and endogenous proangiogenic activity.

Cerqueira *et al.*, in their study on recent skin healing, reported improved re-epithelialization, vascularization, and ECM modulation using combined cell sheets composed of fibroblasts, keratinocytes, and human dermal microvascular endothelial cells (HDMECs) [123]. Moreover, to create highly organized microvascular networks with hierarchical and branching patterns, emerging technologies like 3D bioprinting and microfluidics have received significant attention in recent years [124]. The bioprinting technique is known for its reproducibility and high throughput, which enables the assembly of phase-changing hydrogels, endothelial and parenchymal cells, and other soluble factors in a structured manner. In a mouse wound model, improved vascularization was evident using amniotic fluid-derived stem cells in bioprinted fibrin collagen hydrogels. More recently, improved vascularization was reported in a synthetic injectable material for the first time using a microfluidic water-in-oil emulsion approach [125]. This method utilized microfluidic water-in-oil droplet segmentation to form self-assembled microporous annealed particle (MAP) gels with interconnected pores. These MAP gels possess a unique combination of microporosity, modular assembly, and injectability, underscoring their potential to revolutionize the landscape of tissue regeneration and tissue creation *in vivo* and de novo, respectively [126]. Furthermore, the injectable nature of these modular scaffolds allows for their application to wounds of any shape and size [127].

Along with the aforementioned advanced tissue engineering techniques, miRNA has also emerged as a promising technique for regulating angiogenesis and enhancing the engraftment of skin substitutes through pretreatment of wounds. miR-126, a candidate miRNA and marker for skin endothelial cells, has shown promise in enhancing wound angiogenesis [128]. Moreover, transfection with specific miRNAs has also been employed for the improvement of vascularization. For instance, a study reported subcutaneous implantation of collagen–fibronectin gels containing miR-132-transfected HUVECs and demonstrated higher microvessel density compared to non-transfected samples [21].

7.8 CONCLUSION AND FUTURE PERSPECTIVES

Skin disease or damage accounts for about half of the world's annual healthcare expenditure. Skin tissue engineering has recently garnered considerable interest as an alternative to conventional treatment methods, revolutionizing the treatment of damaged or wounded skin. The regeneration of engineered skin relies on three main components: cells, growth factors, and biomaterials. Despite significant progress in this field, several challenges remain, such as limited effusiveness, high cost, and the inability to fully reconstitute skin appendages in tissue-engineered substitutes, which require solutions. The goal of achieving fully native skin has not yet become a reality, emphasizing the critical need for exploring novel biomaterials with better physicochemical properties to develop in vitro skin models (full thickness, wound healing model) for skin regeneration and wound healing. Furthermore, the incorporation of single or multiple factors in tissue-engineered skin has shown promise in producing more natural substitutes. In recent years, innovations such as stem cell-based strategies, nanotechnology, gene therapy, 3D bioprinting, microfluidics, and others have emerged for clinical translation to bring approved tissue-engineered skin products to the market. The ongoing progress and development in skin tissue engineering and wound healing will find tremendous applications in populations suffering from chronic wounds and burn problems in the near future.

The development of advanced skin tissue engineering strategies in the past few decades has magnificently led to the success of skin graft implantation. A notable example is the facial transplant surgery of Patrick Hardison in 2015, which highlights the remarkable achievements in tissue engineering and transplantation methodologies, greatly improving patients' quality of life. Currently, available artificial skin grafts in the market not only serve the purpose of skin grafting but are also used in the treatment of chronic wounds. Extensive research has explored the understanding of cellular interactions and crosstalk within the wound microenvironment to discover therapeutics for different types of chronic wounds. While natural biomaterials like collagen, fibrin, HA, and GAGs have shown effectiveness in skin tissue engineering applications, they suffer from drawbacks such as cost-effectiveness and batch-to-batch variation. Synthetic polymers offer the advantage of being molded into desired constructs and can be further functionalized to efficiently target cell signaling and biochemical pathways in the future. Targeting the interplay between cells, bioactive factors, and ECM may help uncover hidden problems related to the grafting and healing mechanisms. Modeling approaches using the aforementioned engineering techniques could potentially solve the current problems associated with skin grafts. Moreover, constructs that mimic natural architecture to provide ideal physical and mechanical cues, incorporate a cocktail of bioactive molecules to offer chemical cues, and cellular therapy to provide biological cues may represent an ideal approach for designing skin grafts.

Bioengineered skin products currently available in the market predominantly consist of decellularized skin or xenogenic collagen. Though these are well-established skin grafts, they are highly expensive and carry the inherent risk of disease transmission due to viruses and prions. Advancements in constructing cell-laden artificial grafts have successfully taken over some limitations of acellular grafts; however, the lack of off-the-shelf availability of cellular grafts remains a major issue. Hospitals require sophisticated laboratory and cell expansion facilities to utilize such constructs effectively. Additionally, cell-based constructs suffer from drawbacks such as expensive storage conditions, limited shelf life, difficulty in transportation, friable nature, infections, and high prices. The current facilities are not reaching all the victims due to high-cost treatments. Dramatic improvement in patient survival rates could be achieved by developing affordable technologies and scaling up graft production. However, government regulations, lengthy clinical trials, and hurdles present in translating research to the market are major drawbacks that hinder technological advancement.

The current products available in the market can only partially restore some functions of natural skin; they have not yet achieved full replacement of the entire structural and functional attributes of skin. Restoration of other skin functions like sensation, thermoregulation, perspiration, and pigmentation persists as a major challenge in the way of skin tissue engineering. Although attempts to develop prevascularized skin grafts have been successful in various studies, these products have not yet reached the market. Similarly, regeneration of skin appendages like hair follicles, sebaceous glands, arrector pili muscles, and neurons has been achieved in animal models, but translating these successes to human patients remains challenging. Complications such as uncontrolled scarring, contractions, skin malfunction, and the formation of keloids are other problems that harm skin aesthetics after improper healing. The mechanism behind scar formation involves complex pathways encompassing various cells, cytokines, growth factors, miRNA, and ECM components. Research efforts aimed at regulating collagen fibers using poly-L-arginine as instructive substrates to modulate fibroblasts for collagen production and fiber orientation have been explored for scar healing. Further investigation into the molecular interplay and crosstalk between cells, growth factors, ECM, miRNA, and other bioactive factors is necessary to mitigate complications during later stages of grafting. Stem cell therapy, gene therapy, and the use of growth factors have emerged as advanced approaches for treating chronic cutaneous wounds and promoting skin regeneration, giving hope to many through the development of skin tissue engineering. However, these strategies have not yet been fully optimized. Further exploration of the signaling pathways of stem cells and growth factors is required to avoid complications arising from tumor growth and metastasis. Therefore, smart strategies should be applied in skin graft engineering, which may help circumvent the long-term adverse effects of growth factor signaling and undifferentiated stem cells.

REFERENCES

1. Strang, H., et al., Role of cytokines and chemokines in wound healing, in *Wound healing, tissue repair, and regeneration in diabetes*. 2020, Elsevier. p. 197–235.
2. Flanagan, M., *Wound healing and skin integrity: Principles and practice*. 26 Feb 2013, John Wiley & Sons.
3. Atiyeh, B.S., et al., Management of acute and chronic open wounds: The importance of moist environment in optimal wound healing. 2002. **3**(3): Elsevier. p. 179–195.
4. Kaplani, K., et al., Wound healing related agents: Ongoing research and perspectives. 2018. **129**: Elsevier. p. 242–253.
5. Man, E. and C.J.E.J.O.P.S. Hoskins, Towards advanced wound regeneration. 2020. **149**: Elsevier. p. 105360.
6. Riha, S.M., M. Maarof, and M.B.J.P. Fauzi, Synergistic effect of biomaterial and stem cell for skin tissue engineering in cutaneous wound healing: A concise review. 2021. **13**(10): p. 1546.
7. Kwon, S.G., et al., Recent advances in stem cell therapeutics and tissue engineering strategies. 2018. **22**: Springer. p. 1–8.
8. Álvaro-Afonso, F.J., et al., Adipose-derived mesenchymal stem cells in the treatment of diabetic foot ulcers: A review of preclinical and clinical studies. *"Angiology"*, 2020. **71**(9): p. 853–863.
9. Gaharwar, A.K., I. Singh, and A.J.N.R.M. Khademhosseini, Engineered biomaterials for in situ tissue regeneration. 2020. **5**(9): p. 686–705.
10. Henkel, J., et al., Bone regeneration based on tissue engineering conceptions—a 21st century perspective. 2013. **1**(1): p. 216–248.
11. Ogle, M.E., et al., Monocytes and macrophages in tissue repair: Implications for immunoregenerative biomaterial design. 2016. **241**(10): p. 1084–1097.
12. Eming, S.A., P. Martin, and M.J.S.T.M. Tomic-Canic, Wound repair and regeneration: Mechanisms, signaling, and translation. 2014. **6**(265): p. 265sr6.
13. Cañedo-Dorantes, L. and M.J.I.J.O.I. Cañedo-Ayala, Skin acute wound healing: A comprehensive review. 2019(1): Wiley Online Library. p. 3706315.
14. Gardikiotis, I., et al., Borrowing the features of biopolymers for emerging Wound Healing Dressings: A review. 2022. **23**(15): p. 8778.
15. Burton, J.L. and R.J.J.V.C.F.A.P. Erskine, Immunity and mastitis some new ideas for an old disease. 2003. **19**(1): p. 1–45.
16. El Ayadi, A., J.W. Jay, and A.J.I.J.O.M.S. Prasai, Current approaches targeting the wound healing phases to attenuate fibrosis and scarring. 2020. **21**(3): p. 1105.
17. Serbo, J.V., S.J.S.C.R. Gerecht, and therapy, Vascular tissue engineering: Biodegradable scaffold platforms to promote angiogenesis. 2013. **4**: p. 1–8.
18. Mata, R., et al., The dynamic inflammatory tissue microenvironment: Signality and disease therapy by biomaterials. 2021. **2021**(4189516): p. 31.
19. Gomes, M.E., et al., Tissue engineering and regenerative medicine: New trends and directions—a year in review. 2017. **23**(3): p. 211–224.
20. Shieh, S.-J. and J.P.J.S. Vacanti, State-of-the-art tissue engineering: From tissue engineering to organ building. 2005. **137**(1): p. 1–7.

21. Bhardwaj, N., D. Chouhan, and B.J.C.P.D. B Mandal, Tissue engineered skin and wound healing: Current strategies and future directions. 2017. **23**(24): p. 3455–3482.
22. Nyström, A. and L. Bruckner-Tuderman. Matrix molecules and skin biology, in *Seminars in cell & developmental biology*, 2019. **89**, pp. 136–146. Academic Press.
23. Oualla-Bachiri, W., et al., From grafts to human bioengineered vascularized skin substitutes. 2020. **21**(21): p. 8197.
24. Yu, J.R., et al., Current and future perspectives on skin tissue engineering: Key features of biomedical research, translational assessment, and clinical application. 2019. **8**(5): p. 1801471.
25. Kelangi, S.S., et al., On skin substitutes for wound healing: Current products, limitations, and future perspectives. 2020. **8**(01n02): p. 8–14.
26. Chocarro-Wrona, C., et al., Therapeutic strategies for skin regeneration based on biomedical substitutes. 2019. **33**(3): p. 484–496.
27. Lian, W., et al., Single-cell sequencing reveals increased LAMB3-positive basal keratinocytes and ZNF90-positive fibroblasts in autologous cultured epithelium. 2024. **7**(1): p. 79.
28. Ghio, S.C., et al., The role of cultured autologous bilayered skin substitutes as epithelial stem cell niches after grafting: A systematic review of clinical studies. 2021. **5**(2): p. 56–66.
29. Atiyeh, B.S. and M.J.B. Costagliola, Cultured epithelial autograft (CEA) in burn treatment: Three decades later. 2007. **33**(4): p. 405–413.
30. Hynds, R.E., P. Bonfanti, and S.M.J.E.M.M. Janes, Regenerating human epithelia with cultured stem cells: Feeder cells, organoids and beyond. 2018. **10**(2): p. 139–150.
31. van Uelft, M., *Characterization of Ceramide synthase 2 dysfunctions caused by targeted mutations in transgenic mice*. 2020, Dissertation, Bonn, Rheinische Friedrich-Wilhelms-Universität Bonn, 2020.
32. Dai, C., S. Shih, and A.J.J.O.D.T. Khachemoune, Skin substitutes for acute and chronic wound healing: An updated review. 2020. **31**(6): p. 639–648.
33. Padmanabhan, J., T.R.J.W.I.R.N. Kyriakides, and Nanobiotechnology, Nanomaterials, inflammation, and tissue engineering. 2015. **7**(3): p. 355–370.
34. Chen, F.-M. and X.J.P. I.P.S. Liu, Advancing biomaterials of human origin for tissue engineering. 2016. **53**: p. 86–168.
35. Ashammakhi, N., et al., Highlights on advancing frontiers in tissue engineering. 2022. **28**(3): p. 633–664.
36. Biswal, T.J.M.T.P., Biopolymers for tissue engineering applications: A review. 2021. **41**: p. 397–402.
37. Meereboer, K.W., M. Misra, and A.K.J.G.C. Mohanty, Review of recent advances in the biodegradability of polyhydroxyalkanoate (PHA) bioplastics and their composites. 2020. **22**(17): p. 5519–5558.
38. Mukherjee, A. and M.J.B. Koller, Microbial polyHydroxyAlkanoate (PHA) biopolymers—Intrinsically natural. 2023. **10**(7): p. 855.
39. Chai, J.M., et al., Surface-modified highly biocompatible bacterial-poly (3-hydroxybutyrate-co-4-hydroxybutyrate): a review on the promising next-generation biomaterial. 2020. **13**(1): p. 51.
40. Constantinides, C., et al., In vivo tracking and 1H/19F magnetic resonance imaging of biodegradable polyhydroxyalkanoate/polycaprolactone blend scaffolds seeded with labeled cardiac stem cells. 2018. **10**(30): p. 25056–25068.
41. Khoshnood, N. and A. Zamanian, Decellularized extracellular matrix bioinks and their application in skin tissue engineering. *Bioprinting*, 2020. **20**: p. e00095.
42. Yi, S., et al., Extracellular matrix scaffolds for tissue engineering and regenerative medicine. *Current Stem Cell Research & Therapy*, 2017. **12**(3): p. 233–246.
43. Asti, A. and L. Gioglio, Natural and synthetic biodegradable polymers: Different scaffolds for cell expansion and tissue formation. *The International Journal of Artificial Organs*, 2014. **37**(3): p. 187–205.
44. Zhang, F. and M.W. King, Biodegradable polymers as the pivotal player in the design of tissue engineering scaffolds. *Advanced Healthcare Materials*, 2020. **9**(13): p. 1901358.
45. Gater, R., et al., Scaffolds mimicking the native structure of tissues, in *Handbook of tissue engineering scaffolds*, **1**, pp. 51–71. Woodhead Publishing.
46. Venditti, I., Nanostructured materials based on noble metals for advanced biological applications. *Nanomaterials*, **9**(11), p. 1593.
47. Socci, M.C., et al., Polymeric materials, advances and applications in tissue engineering: A review. 2023. **10**(2): p. 218.
48. Ciolacu, D.E., R. Nicu, and F.J.B. Ciolacu, Natural polymers in heart valve tissue engineering: Strategies, advances and challenges. 2022. **10**(5): p. 1095.
49. Khan, F., M. Tanaka, and S.R.J.J.O.M.C.B. Ahmad, Fabrication of polymeric biomaterials: A strategy for tissue engineering and medical devices. 2015. **3**(42): p. 8224–8249.

50. Ahadian, S., et al., Organ-on-a-chip platforms: A convergence of advanced materials, cells, and microscale technologies. 2018. **7**(2): p. 1700506.
51. Sionkowska, A.J.P. i.p.s., Current research on the blends of natural and synthetic polymers as new biomaterials. 2011. **36**(9): p. 1254–1276.
52. Fang, Y., et al., Biodegradable core-shell electrospun nanofibers based on PLA and γ-PGA for wound healing. 2019. **116**: p. 30–37.
53. Xu, T.O., et al., Self-neutralizing PLGA/magnesium composites as novel biomaterials for tissue engineering. *Biomedical Materials*, 2018. **13**(3): p. 035013.
54. Askari, M., et al., Recent progress in extrusion 3D bioprinting of hydrogel biomaterials for tissue regeneration: A comprehensive review with focus on advanced fabrication techniques. 2021. **9**(3): p. 535–573.
55. Zhang, W., et al., Adhesive and tough hydrogels: From structural design to applications. 2021. **9**(30): p. 5954–5966.
56. Shirzaei Sani, E., et al., Engineering adhesive and antimicrobial hyaluronic acid/elastin-like polypeptide hybrid hydrogels for tissue engineering applications. 2018. **4**(7): p. 2528–2540.
57. Shirzaei Sani, E., *Hydrogel-based bioadhesives for tissue engineering and surgical applications.* University of California, Los Angeles; 2020.
58. Dhand, A.P., J.H. Galarraga, and J.A.J.T.I.B. Burdick, Enhancing biopolymer hydrogel functionality through interpenetrating networks. 2021. **39**(5): p. 519–538.
59. Jacob, S., et al., Emerging role of hydrogels in drug delivery systems, tissue engineering and wound management. 2021. **13**(3): p. 357.
60. Crosby, C.O., et al., Interpenetrating polymer network hydrogels as bioactive scaffolds for tissue engineering. 2022. **38**(3): p. 347–361.
61. Vega, S.L., et al., Recent advances in hydrogels for cartilage tissue engineering. 2017. **33**: p. 59.
62. Tsanaktsidou, E., O. Kammona, and C.J.P. Kiparissides, Recent developments in hyaluronic acid-based hydrogels for cartilage tissue engineering applications. 2022. **14**(4): p. 839.
63. Naghieh, S., et al., Printability of 3D printed hydrogel scaffolds: Influence of hydrogel composition and printing parameters. 2019. **10**(1): p. 292.
64. Liu, X., et al., Utilization of ex vivo tissue model to study skin regeneration following microneedle stimuli. *Scientific Reports*, 2022. **12**(1): p. 18115.
65. Hunter, A.C. and S.M.J.P.C. Moghimi, Smart polymers in drug delivery: A biological perspective. 2017. **8**(1): p. 41–51.
66. Antony, G.J.M., et al., Stimuli-responsive self-healable materials. 2021: p. 361–377.
67. Suriano, R., et al., 4D printing of smart stimuli-responsive polymers. 2019. **166**(9): p. B3274–B3281.
68. Mano, J.F.J.A.E.M., Stimuli-responsive polymeric systems for biomedical applications. 2008. **10**(6): p. 515–527.
69. Schmaljohann, D.J.A.d.d.r., Thermo-and pH-responsive polymers in drug delivery. 2006. **58**(15): p. 1655–1670.
70. Zhu, G., et al., Bone physiological microenvironment and healing mechanism: Basis for future bone-tissue engineering scaffolds. 2021. **6**(11): p. 4110–4140.
71. Alizadeh, S., et al., Sulfated polysaccharide as biomimetic biopolymers for tissue engineering scaffolds fabrication: Challenges and opportunities. 2024: p. 122124.
72. Azevedo, S., et al., Bioinspired ultratough hydrogel with fast recovery, self-healing, injectability and cytocompatibility. 2017. **29**(28): p. 1700759.
73. Johnson, C.T., et al., Lysostaphin and BMP-2 co-delivery reduces S. aureus infection and regenerates critical-sized segmental bone defects. *Science Advances*, 2019. **5**(5): p. eaaw1228.
74. Devi VK, A., et al., Self-healing hydrogels: Preparation, mechanism and advancement in biomedical applications. 2021. **13**(21): p. 3782.
75. Akhmanova, M., et al., Physical, spatial, and molecular aspects of extracellular matrix of in vivo niches and artificial scaffolds relevant to stem cells research. *Stem Cells International*, 2015. (1): p. 167025.
76. Kong, Y., et al., Regulation of stem cell fate using nanostructure-mediated physical signals. 2021. **50**(22): p. 12828–12872.
77. Shah, B., P.T. Yin, and K.-B.J.A.C. Lee, Multimodal magnetic core-shell nanoparticles for effective stem cell differentiation and imaging. 2013. **52**(24): p. 6190.
78. Khademhosseini, A., et al., Embryonic stem cells as a cell source for tissue engineering, in *Principles of tissue engineering*. 2020, Elsevier. p. 467–490.
79. El-Sayed, N.S., S.J.C. Kamel, and Interfaces, Polysaccharides-based injectable hydrogels: Preparation, characteristics, and biomedical applications. 2022. **6**(4): p. 78.
80. Min, Q., et al., Controlled delivery of insulin-like growth factor-1 from bioactive glass-incorporated alginate-poloxamer/silk fibroin hydrogels. 2020. **12**(6): p. 574.

81. Qazi, T.H., et al., Biomaterials that promote cell-cell interactions enhance the paracrine function of MSCs. 2017. **140**: p. 103–114.
82. Hayat M, Nawaz A, Chinnam S, Muzammal M, Latif MS, Yasin M, Ashique S, Zengin G, Farid A. Formulation development and optimization of herbo synthetic gel: In vitro biological evaluation and in vivo wound healing studies. *Process Biochemistry*, **2023**.130: p. 116–126.
83. Wei, Q., et al., Three-dimensional bioprinting of tissue-engineered skin: Biomaterials, fabrication techniques, challenging difficulties, and future directions: A review. *International Journal of Biological Macromolecules*, 2024: 131281.
84. Mota, C., et al., Bioprinting: From tissue and organ development to in vitro models. 2020. **120**(19): p. 10547–10607.
85. Ng, W.L., et al., Microvalve-based bioprinting–process, bio-inks and applications. 2017. **5**(4): p. 632–647.
86. Baltazar, T., et al., Three dimensional bioprinting of a vascularized and perfusable skin graft using human keratinocytes, fibroblasts, pericytes, and endothelial cells. 2020. **26**(5–6): p. 227–238.
87. Tan, S.H., et al., Recent advances in the design of three-dimensional and bioprinted scaffolds for full-thickness wound healing. 2022. **28**(1): p. 160–181.
88. He, P., et al., Bioprinting of skin constructs for wound healing. *Burns Trauma*, 2018. **6**: p. 5.
89. Daikuara, L.Y., et al., 3D bioprinting constructs to facilitate skin regeneration. 2022. **32**(3): p. 2105080.
90. Xu, J., et al., Advances in the research of bioinks based on natural collagen, polysaccharide and their derivatives for skin 3D bioprinting. 2020. **12**(6): p. 1237.
91. Mehrotra, S., et al., 3D printing/bioprinting based tailoring of in vitro tissue models: Recent advances and challenges. 2019. **2**(4): p. 1385–1405.
92. Vijayavenkataraman, S., et al., 3D bioprinting of tissues and organs for regenerative medicine. 2018. **132**: p. 296–332.
93. Albanna, M., et al., In situ bioprinting of autologous skin cells accelerates wound healing of extensive excisional full-thickness wounds. 2019. **9**(1): p. 1856.
94. Horch, R.E., et al., Tissue engineering of cultured skin substitutes. 2005. **9**(3): p. 592–608.
95. Liu, Y.-F., et al., Therapeutic strategies for chronic wound infection. *Chinese Journal of Traumatology*, **25**(01), pp. 11–16.
96. Singer, A.J.J.T.E.P.B.R., Healing mechanisms in cutaneous wounds: Tipping the balance. 2022. **28**(5): p. 1151–1167.
97. Amjadian, S., et al., The emerging therapeutic targets for scar management: Genetic and epigenetic landscapes. 2022. **35**(5): p. 247–265.
98. Eming, S.A., T. Krieg, and J.M.J.C.I.D. Davidson, Retracted: Gene therapy and wound healing. 2007. **25**(1): p. 79–92.
99. Rousselle, P., F. Braye, and G.J.A.D.D.R. Dayan, Re-epithelialization of adult skin wounds: Cellular mechanisms and therapeutic strategies. 2019. **146**: p. 344–365.
100. Las Heras, K., et al., Chronic wounds: Current status, available strategies and emerging therapeutic solutions. 2020. **328**: p. 532–550.
101. Bonadio, J.J.A.D.D.R., Tissue engineering via local gene delivery: Update and future prospects for enhancing the technology. 2000. **44**(2–3): p. 185–194.
102. Mao, Z., et al., Enhanced angiogenesis of porous collagen scaffolds by incorporation of TMC/DNA complexes encoding vascular endothelial growth factor. *Acta Biomater*, 2009. **5**(8): p. 2983–94.
103. Guo, R., et al., The healing of full-thickness burns treated by using plasmid DNA encoding VEGF-165 activated collagen–chitosan dermal equivalents. 2011. **32**(4): p. 1019–1031.
104. Urciuolo, F., et al., Bioengineered skin substitutes: The role of extracellular matrix and vascularization in the healing of deep wounds. *Journal of Clinical Medicine*, 2019. **8**(12): p. 2083.
105. Moreira, H.R. and A.P. Marques, Vascularization in skin wound healing: Where do we stand and where do we go? *Current Opinion in Biotechnology*, 2022. **73**: p. 253–262.
106. Varkey, M., J. Ding, and E.E.J.J.O.F.B. Tredget, Advances in skin substitutes—potential of tissue engineered skin for facilitating anti-fibrotic healing. 2015. **6**(3): p. 547–563.
107. Lokmic, Z., et al., Hypoxia and hypoxia signaling in tissue repair and fibrosis. 2012. **296**: p. 139–185.
108. Gibot, L., et al., A preexisting microvascular network benefits in vivo revascularization of a microvascularized tissue-engineered skin substitute. 2010. **16**(10): p. 3199–3206.
109. Frueh, F.S., et al., Current and emerging vascularization strategies in skin tissue engineering. 2017. **37**(5): p. 613–625.
110. Peters, E.B.J.T.E.P.B.R., Endothelial progenitor cells for the vascularization of engineered tissues. 2018. **24**(1): p. 1–24.

111. Nguyen, J., Y.-Y. Lin, and S. Gerecht, The next generation of endothelial differentiation: Tissue-specific ECs. *Cell Stem Cell*, 2021. **28**(7): p. 1188–1204.
112. Shahin, H., et al., Vascularization is the next challenge for skin tissue engineering as a solution for burn management. *Burns Trauma*, 2020. **8**: p. tkaa022.
113. Shah Mohammadi, M., et al., Critical considerations for regeneration of vascularized composite tissues. 2021. **27**(4): p. 366–381.
114. Piard, C., et al., 3D printed HUVECs/MSCs cocultures impact cellular interactions and angiogenesis depending on cell-cell distance. 2019. **222**: p. 119423.
115. Hasan, A., et al., Microfluidic techniques for development of 3D vascularized tissue. 2014. **35**(26): p. 7308–7325.
116. Newman, A.C., et al., The requirement for fibroblasts in angiogenesis: Fibroblast-derived matrix proteins are essential for endothelial cell lumen formation. 2011. **22**(20): p. 3791–3800.
117. Shang, F., et al., Advancing application of mesenchymal stem cell-based bone tissue regeneration. *Bioactive Materials*, 2021. **6**(3): p. 666–683.
118. Della Rocca, Y., et al., Hypoxia: Molecular pathophysiological mechanisms in human diseases. *Journal of Physiology and Biochemistry*, 2022. **78**(4): p. 739–752.
119. Yuan, X., et al., Reciprocal interaction between vascular niche and sweat gland promotes sweat gland regeneration. 2023. **21**: p. 340–357.
120. An, Y., et al., Bone marrow mesenchymal stem cell aggregate: An optimal cell therapy for full-layer cutaneous wound vascularization and regeneration. 2015. **5**(1): p. 17036.
121. Kaur, A., et al., Functional skin grafts: Where biomaterials meet stem cells. 2019. **2019**.
122. Yu, H., et al., Bioglass activated skin tissue engineering constructs for wound healing. 2016. **8**(1): p. 703–715.
123. Cerqueira, M., et al., Cell sheet technology-driven re-epithelialization and neovascularization of skin wounds. 2014. **10**(7): p. 3145–3155.
124. Li, H., et al., Technology for the formation of engineered microvascular network models and their biomedical applications. *Nano Convergence*, 2024. **11**(1): p. 10.
125. Skardal, A., et al., Bioprinted amniotic fluid-derived stem cells accelerate healing of large skin wounds. *Stem Cells Translational Medicine*, 2012. **1**(11): p. 792–802.
126. Smith, B.T., *Development of synthetic polymeric and ceramic composite constructs for bone tissue engineering*. 2019, Rice University.
127. Griffin, D.R., et al., Accelerated wound healing by injectable microporous gel scaffolds assembled from annealed building blocks. *Nature Materials*, 2015. **14**(7): p. 737–744.
128. Beavers, K.R., C.E. Nelson, and C.L. Duvall, MiRNA inhibition in tissue engineering and regenerative medicine. *Advanced Drug Delivery Reviews*, 2015. **88**: p. 123–137.

8 Nanomaterials in Wound Healing

Mechanism, Recent Advancements, Limitations, and Future Perspectives

Sandhya Vasanth,[1] *Alima Misiriya,*[2] *Fathima Thahsin,*[3] *Ayesha Sultana,*[4] *and Mohammed Gulzar Ahmed*[5*]

[1,2,3,4,5] Department of Pharmaceutics, Yenepoya Pharmacy College & Research Centre, Yenepoya (Deemed to Be University), Mangalore, 575018, Karnataka, India

* Corresponding author: Mohammed Gulzar Ahmed

8.1 INTRODUCTION

Wound healing is a dynamic and complex process driven by biological events that restore tissue integrity and function. Whether through trauma or surgery or from any other number of medical conditions, wounds represent a significant challenge to healthcare systems globally. Great strides have been made in medical science. However, there remains a pressing need to develop innovative strategies for wound management. In recent years, promising research on the application of nanotechnology has revolutionized wound healing by leveraging the unique properties of NMs. NMs are materials whose at least one dimension falls within the nanometer range. These properties stem from their large surface area-to-volume ratio, quantum confinement effects, and surface reactivity, making them unique and appropriate for biomedical applications (Pormohammad et al., 2021). There is enormous potential in wound healing. NMs provide an unmatched possibility for targeted and controlled delivery of therapeutic agents, modulation of cellular responses, and creating biomimetic scaffolds for tissue regeneration. NMs offer promising approaches for the revolutionary treatment of various wounds, from acute injuries to chronic ulcers. Researchers and clinicians hope to bring innovative, novel strategies to life to stimulate faster healing, reduce scarring, and prevent complications related to infections by exploiting the unique properties of materials at the nanoscale. With almost complete precision, it facilitates the targeted delivery of nanoparticles carrying bioactive molecules, such as growth factors, cytokines, and antimicrobial agents, to the wound site. Targeted delivery of such therapeutic payloads minimizes systemic side effects and accelerates wound healing. One of the primary uses of NMs in wound healing is through drug delivery (Nandhini et al., 2024; M. Wang et al., 2021a).

Nanoparticles can be engineered to carry therapeutic agents, such as growth factors, cytokines, antibiotics, and anti-inflammatory drugs, to target and time-release such drugs at the wound site. This localization reduces the systemic side effects while increasing the therapeutic effect of the drugs to accelerate the healing time, with a concomitant reduction in the risks associated with infections (Wang et al., 2021a). Furthermore, NMs can act as artificial scaffolds in tissue engineering and regeneration. Nanofibrous scaffolds, hydrogels, and other nanostructured matrices are

 DOI: 10.1201/9781003605966-8

biomimetic matrices through which cells can attach (Zhao et al., 2021a), proliferate, and differentiate (Blanco-Fernandez et al., 2021; Kushwaha et al., 2022). By biomimicking the extracellular matrix, these nanomaterial-based scaffolds provide a supportive environment for the regeneration of tissues and the closure of massive tissue loss or chronic wounds. In addition to drug delivery and tissue engineering, NMs can modulate the wound microenvironment for optimal healing conditions. Inflammation, angiogenesis, and remodeling of the extracellular matrix can be regulated by transitioning from an inflammatory stage into the proliferative and remodeling phases of healing. In this way, NMs modulate these biological processes to accelerate healing and improve patient outcomes.

Recent advances in nanotechnology are rapidly developing NMs with sophisticated applications in wound healing. Stimuli-responsive nanocarriers capable of controlled drug release have been designed. NMs integrated with advanced scaffolds and matrices for tissue engineering, coupled with nanotechnology-based imaging modalities, enable real-time monitoring of wound healing processes (Naskar & Kim, 2020). However, there are a few challenges and limitations regarding the enormous potential of NMs for wound healing. These include apprehensions about their biocompatibility, toxicity, and long-term safety. In addition, there is still a lack of scalability, cost-effectiveness, and even regulatory approval regarding clinical use. The mechanisms involved in the complex interplay between NMs and biological systems still need to be further explored and optimized for their therapeutic efficacy (Niculescu & Grumezescu, 2022; Pormohammad et al., 2021).

The future of NMs in healing is exciting; interdisciplinary collaborations for nanomaterial-based approaches can overcome existing limitations and embrace emerging opportunities while fostering future research in nanotechnology, materials science, and biomedical engineering. These collaborations will continue to support the development of precise, personalized, real-time monitoring of the healing process between physicians and patients. These changes will alter how wounds are managed globally for better patient outcomes (Bellu et al., 2021).

8.2 ADVANTAGES

Enhanced Surface Area: Recent studies revealed that NMs have a high surface area-to-volume ratio, allowing enhanced interaction with biological systems, such as cells and tissues. This property can encourage cell adhesion, proliferation, and migration, thereby accelerating wound healing.

Controlled Drug Delivery: NMs can be loaded with drugs or growth factors, facilitating controlled release at the wound site. This enhances therapeutic efficacy while muting systemic side effects (Wang et al., 2019, 2021b).

Antimicrobial Properties: Some NMs possess an inherent antimicrobial feature or can be surface functionalized with antimicrobial agents. This capability accelerates the elimination of infections, a frequent complication of the healing process for wounds, thus increasing the efficiency of healing. It also provides a sterile environment for the wound (Wang et al., 2019, 2021b).Click or tap here to enter text.

Inflammation Modulation: NMs may modulate the inflammatory response at the wound site, promoting a balanced immune reaction that supports healing while minimizing excessive inflammation that may hinder the whole process (Kushwaha et al., 2022).

Scaffolding: NMs can be used as scaffolds for tissue regeneration and provide a support structure for cell attachment, proliferation, and differentiation. This function is beneficial in the healing of large or complicated wounds (Girija AR, 2022a).

Biocompatibility and Biodegradability: Most NMs used in wound healing are biocompatible and biodegradable, thus reducing the risk of adverse reactions and integrating into the host tissue after a certain period without long-term complications (Wang et al., 2019, 2021c).

Diagnostic Abilities: Some NMs can be engineered to contain diagnostic agents that can track the healing course of a wound and detect complications, like infection or impaired healing (Naskar & Kim, 2020).

FIGURE 8.1 Advantages of NMs in wound healing.

Enhanced Wound Healing Potential: Healing of wounds and regeneration of tissues are expedited, resulting in better results than conventional treatments, which may even go as far as healing chronic wounds or injuries (Barroso et al., 2020).

Targeted delivery: Delivering therapeutic drugs to wound sites reduces systemic adverse effects and customizes treatment according to the unique features of the wound (Singh et al., 2019).

Improved patient compliance: Advanced wound dressings improve patient comfort and quality of life throughout the healing process by offering greater comfort and flexibility and lowering the need for invasive treatments or frequent dressing changes (Vijayan et al., 2021a).

Adaptability and Flexibility capacity to modify the properties of NMs to fit particular types of wounds provides flexibility in design and manufacture to suit a range of applications and enables integration with other therapeutic modalities for synergistic benefits (Tavakoli & Klar, 2020a) (**Figure 8.1**).

8.3 MECHANISM OF NMS IN WOUND HEALING

8.3.1 Enhanced Drug Delivery

Wound healing is a complex process that involves various cellular and molecular interactions. A critical factor in wound healing is the enhanced surface area of NMs. These NMs have a high surface area-to-volume ratio, allowing increased interaction with cells and tissues. This enhanced interaction can stimulate cell adhesion, proliferation, and migration, leading to accelerated wound healing. Furthermore, the use of NMs in wound dressings has shown promising results. These dressings, made from chitosan blends and electrospun nanofibers, have demonstrated favorable properties such as absorption of exudates and prevention of bacterial penetration (Osanloo et al., 2024). The increased surface area provided by nanofibrous scaffolds promotes cell attachment and proliferation, making them ideal for wound healing applications. In addition, the use of hybrid nanofibers has also shown potential for wound healing. The combination of NMs and wound healing is promising for developing advanced wound dressings and therapies. Using NMs with enhanced surface area has shown promising results in promoting wound healing. The hybrid nanofibers, composed of chitosan and silk fibroin, are especially promising for wound healing applications (Osanloo et al., 2024). Electrospun nanofibrous scaffolds have also been explored in various biomedical applications, including wound healing. These nanofiber dressings have been found to have favorable structural, mechanical, and biological properties that enhance wound healing. The enhanced surface area of NMs plays a crucial role in promoting wound healing. The increased interaction with cells and tissues due to their high surface area-to-volume ratio encourages cell adhesion, proliferation, and migration, thereby accelerating wound healing (Juncos Bombin et al., 2020a).

Targeted Delivery: Nanoparticles can be designed to target specific cells or tissues within a wound, ensuring that therapeutic agents are delivered directly to the site where they are most needed. This targeted approach can help to minimize systemic side effects and maximize the efficacy of the treatment. Nanoparticles can be coated with molecules that specifically bind to receptors or proteins in the wound area. For instance, a nanocarrier could be functionalized with antibodies that target fibroblast growth factor receptors, active during wound healing. Specific cells in the wound environment may be involved in the uptake and transport of the nanoparticles. For example, macrophages, which are abundant in wounds, can engulf nanoparticles that can then be engineered to improve healing or modulate inflammation. Due to inflammation and increased vascular permeability, nanoparticles can passively accumulate in the wound site. For example, nanoparticles might be small enough to penetrate these gaps in the blood vessels to localize in the wound tissue (Singh et al., 2019).

i. **Polymeric nanoparticles**: These can encapsulate and sustain the release of growth factors, cytokines, and other bioactive molecules that promote tissue regeneration and minimize inflammation (Zielinska et al., 2020).
ii. **Silver nanoparticles**: While primarily used for their antimicrobial properties, they can also be loaded with anti-inflammatory agents, allowing localized action at the wound site (M. Singh et al., 2022).
iii. **Lipid nanoparticles**: These biocompatible carriers can deliver various drugs, including anti-inflammatory compounds, directly to the wound, improving healing outcomes (Motsoene et al., 2023).
iv. **Nanofibrous scaffolds**: These are designed to mimic the natural extracellular matrix and can be loaded with therapeutic agents released over time to modulate the wound healing process (Vijayan et al., 2021b; Gupta et al., 2014).
v. **Nanogels**: These hydrogel-based nanoparticles can deliver a controlled range of therapeutic agents, including anti-inflammatories and antibiotics, to the wound area (Tavakoli & Klar, 2020b; Liu Y et al., 2022).

Controlled Release: Nanocarriers can encapsulate drugs and release them in a controlled manner over an extended period, which is particularly beneficial in wound healing. This approach involves loading NMs with drugs or growth factors and releasing them at the wound site in a controlled manner, enhancing the therapeutic efficacy of the drugs or growth factors while minimizing systemic side effects. Nanotechnology has been developed to improve the delivery of drugs, including anticancer drugs, to specific tissue sites while minimizing distribution and toxicity in healthy tissues. Particles within the nanoscale range have shown great potential as carriers for drugs in controlled drug delivery systems. These nanocarriers can alter drugs' reactivity, strength, and electrical properties, ultimately influencing their behavior in vivo. Controlled drug delivery using NMs in wound healing has shown promising results, enhancing the therapeutic efficacy of drugs or growth factors while minimizing systemic side effects. NMs can be engineered to respond to specific stimuli in the wound microenvironment, such as pH-responsive nanocarriers that release their payload in response to the acidic environment of a wound and biodegradable nanoparticles that slowly release therapeutic agents as they degrade at the wound site. This innovative approach has great potential for optimizing wound healing outcomes and enhancing therapeutic efficacy while minimizing systemic side effects (Naskar & Kim, 2020).

i. **Polymeric Nanoparticles**: These can be loaded with various drugs, including growth factors and antibiotics, released over time to promote tissue repair and control infection (M. Hayat et al., 2023).
ii. **Lipid Nanoparticles**: These are used to control the release of drugs, such as anti-inflammatory agents, to the wound site, which could reduce swelling and support healing processes (M. Hayat et al., 2023).
iii. **Silver Nanoparticles**: In addition to their inherent antimicrobial properties, silver nanoparticles can deliver therapeutic agents sustainably, helping to manage infection and inflammation within the wound bed (Hamdan et al., 2017; Tain J et al., 2007).
iv. **Nanofibrous Scaffolds**: These are materials engineered to support tissue growth and can be functionalized with bioactive agents that are gradually released to aid in healing and modulate the immune response (Jiang et al., 2023).
v. **Nanogels**: Hydrogel-based nanoparticles can encapsulate drugs and protect them until they are released at the wound site. They provide hydration and promote a conducive environment for wound healing while also delivering therapeutics in a controlled manner (M. Hayat et al., 2023).

8.3.2 Cell Adhesion and Proliferation

NMs can significantly influence cell adhesion and proliferation, crucial wound healing processes. The mechanism involves interaction at the cellular level, where NMs provide a conducive environment and cues for cells to adhere to the wound site and multiply to repair tissue. They improve cell adhesion through surface topography and chemistry by altering the physical and chemical properties of NMs, which can be tailored to closely mimic the natural extracellular matrix, enhancing the adhesion of cells necessary for wound healing, such as fibroblasts and keratinocytes. For example, nanofibrous scaffolds with a fibrous structure similar to collagen in the extracellular matrix can promote better cell adhesion. Nanoparticles can be loaded with growth factors released locally to stimulate cell proliferation (Jiang et al., 2023).

Nanoparticles are encapsulated with epidermal growth factor, which can be released in a controlled manner to promote keratinocyte proliferation. NMs' size, shape, and surface properties can send positive signals to cells for proliferation. Electrospun nanofibrous scaffolds, for instance, provide a high surface area that offers numerous binding sites for cell receptors to interact with, thereby increasing cellular activities, including proliferation. NMs can be functionalized with molecules such as integrins or fibronectin that bind selectively to cellular receptors, differentiating between types of cells that populate the wound. For instance, a scaffold might promote the preferential adhesion and proliferation of endothelial cells over fibroblasts, thus encouraging angiogenesis in the early stages of wound healing.

Nanofibers can mimic the structure of the natural extracellular matrix, providing a scaffold that supports cell adhesion and proliferation. They help create an environment conducive to tissue regeneration. NMs can be functionalized with specific biochemical cues that promote cell adhesion. For example, RGD peptide sequences on NMs have enhanced cell adhesion by activating integrins, which mediate cell attachment to the ECM (Kahraman et al., 2016; Bombin A D J et al., 2020). Biocompatible NMs do not elicit a significant immune response and are conducive to cellular growth. Studies involving biocompatible nanofibrous scaffolds have shown good cellular compatibility, facilitating cell adhesion and proliferation (Zhou et al., 2017). NMs can be engineered to release growth factors or other bioactive substances that stimulate cell.

1. **Nanofibrous Scaffolds**: Mimicking the natural extracellular matrix, these scaffolds support cell attachment and proliferation. They can be made from materials like polycaprolactone, polyethylene oxide, or collagen (Jiang et al., 2023).
2. **Carbon Nanotubes**: These can be incorporated into scaffolds to enhance mechanical strength and electrical conductivity, which can promote cell attachment and stimulate cell growth (Blanco-Fernandez et al., 2021).
3. **Nanocomposite Materials**: Combinations of polymers with inorganic nanoparticles can improve scaffold properties, like bioactivity and stiffness, thus aiding in cell adhesion and proliferation necessary for tissue regeneration (Kushwaha et al., 2022).
4. **Hydrogel Nanoparticles**: These water-containing NMs can create a moist and nutrient-rich environment that supports cell migration and proliferation in the wound healing process (Tavakoli & Klar, 2020c; Liu Y et al., 2022).
5. **Graphene**: This material's unique properties, including its high surface area and biocompatibility, make it beneficial for supporting cell growth on its surface (Zhao et al., 2021b; Fan Z et al., 2014).

8.3.3 Antimicrobial Properties

Some NMs possess an inherent feature of antimicrobial activity or can be surface-functionalized with antimicrobial agents. This capability accelerates the elimination of infections, a frequent complication of the healing process for wounds, thus accelerating and increasing the efficiency of healing. It also provides a sterile environment for the wound. NMs, such as silver nanoparticles, are known for their antimicrobial properties, which can help eliminate infections more swiftly and effectively than traditional treatments. By preventing and combating microbial contamination, they contribute to a more sterile environment, which is essential for proper wound healing (Maťátková et al., 2022).

These antimicrobial properties can also reduce the risk of infection-related complications and promote faster recovery (Chakrabarti et al., 2018; Aramwit P et al., 2014). Nanoparticles have a high surface area relative to their volume, which allows them to interact more effectively with microorganisms. This extensive contact can disrupt the cell walls of bacteria or other pathogens, leading to their inactivation or death (Abdeen et al., 2013; Arafa M G et al., 2018) Silver nanoparticles are known to release metal ions that are toxic to microorganisms. These ions can damage cellular components such as DNA and proteins, thereby inhibiting the growth of or killing the bacteria (Abdeen et al., 2013) (**Figure 8.2**). Nanoparticles can physically disrupt the cell membranes of microbes, causing structural damage that leads to cellular content leakage and, ultimately, cell death. Nanoparticles can generate reactive oxygen species that damage bacterial cells by oxidizing lipids, proteins, and DNA. Nanoparticles may interfere with the ability of bacteria to uptake necessary nutrients, effectively starving the cells and preventing their proliferation. Some NMs can prevent the formation of biofilms, which are protective communal structures that bacteria can form on wound surfaces to shield themselves from antimicrobial agents. NMs with specific topographies or surface charges can interfere with the initial attachment of bacteria to the wound surface, thus preventing biofilm establishment. Nanoparticles with a positive charge can attract and disrupt negatively charged bacterial cell membranes, inhibiting their ability to form biofilms. NMs can release ions or other chemical agents that are toxic to bacteria. Silver nanoparticles are a typical example; they release silver ions that interfere with bacterial DNA and cell wall synthesis, effectively killing the bacteria and preventing biofilm formation. Quorum sensing is the communication method that bacteria use to coordinate biofilm formation. Nanoparticles can be engineered to release agents that block these signaling pathways, preventing the bacteria from initiating biofilm formation. For example, zinc

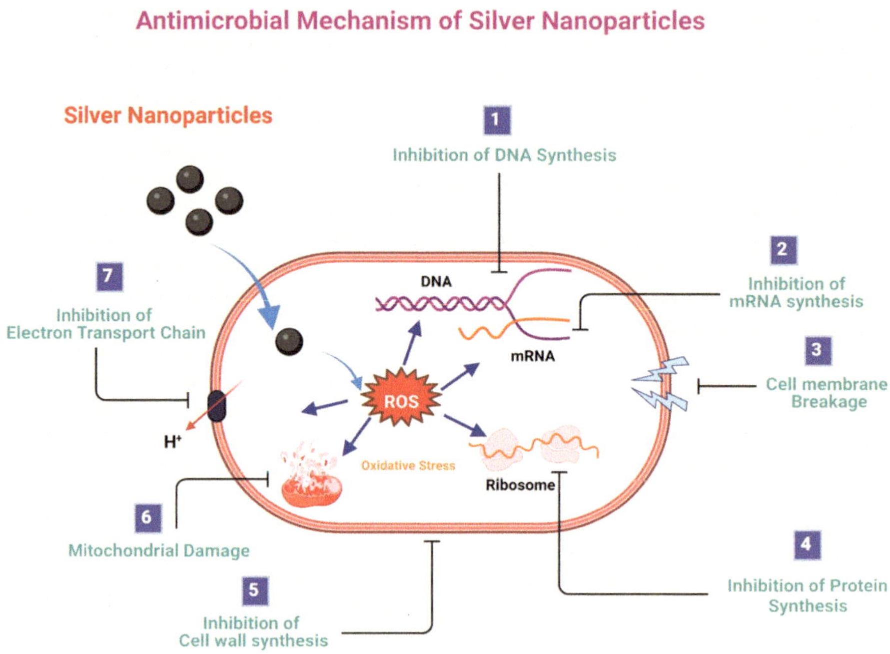

FIGURE 8.2 Mechanism of silver nanoparticles in wound healing (Jain et al., 2021).

oxide nanoparticles have been shown to interfere with quorum sensing. NMs can mimic the action of enzymes that degrade biofilm matrix components (Han et al., 2020). For example, nanoparticles can be designed to produce catalase-like activities that degrade the protective hydrogen peroxide barrier formed by the biofilms, which leads to their disruption.

8.3.4 Anti-inflammatory Effects

NMs can be engineered to deliver anti-inflammatory drugs directly to the wound site, reducing the production or activity of pro-inflammatory cytokines. For example, nanoparticles containing anti-inflammatory agents like corticosteroids can target inflamed tissues and release their payload in a controlled manner, thereby decreasing local inflammation. Gold and silver nanoparticles have been shown to possess anti-inflammatory activities. Through their interaction with inflammatory cells, they can modulate the release of pro-inflammatory cytokines, such as tumor necrosis factor-alpha and various interleukins (Niculescu & Grumezescu, 2022).

Nanoparticles made from biocompatible and biodegradable polymers like polylactic acid or polylactic-co-glycolic acid can be loaded with anti-inflammatory drugs (like NSAIDs or corticosteroids) that suppress the release of pro-inflammatory cytokines at the wound site. Liposomes and solid lipid nanoparticles can encapsulate and deliver hydrophilic and lipophilic anti-inflammatory compounds to the targeted inflammatory cells within the wound microenvironment, thereby reducing the levels of cytokines involved in the inflammation process. Cerium oxide nanoparticles act as ROS scavengers and indirectly inhibit the release of pro-inflammatory cytokines by reducing oxidative stress in the wound area. Excessive ROS contributes to inflammation and impedes healing. Specific NMs have antioxidant properties, allowing them to act as scavengers for ROS, thus reducing oxidative stress and consequent inflammation. For instance, cerium oxide nanoparticles act as ROS scavengers, protecting the tissue from oxidative damage. Gold nanoparticles have been studied for their ability to inhibit NF-κB, a transcription factor that controls the expression of various inflammatory genes. Some NMs can affect the behavior of immune cells, such as macrophages, which play a significant role in the inflammatory phase of wound healing. By altering the surface chemistry or delivering specific signals, nanoparticles can encourage macrophages to transition from a pro-inflammatory (M1) to an anti-inflammatory (M2) phenotype, which is more conducive to tissue repair and regeneration. Nanoparticles made from natural biopolymers like chitosan have been shown to modulate immune responses. They can interact with macrophages and other immune cells, influencing cytokine production and favoring a healing phenotype (Barroso et al., 2020). NMS

8.3.4.1 Modulate the Immune Cell Activity: Biopolymer-Based Nanoparticles

Due to their inert nature and ease of functionalization, these have been studied for their ability to influence immune cell activity. They can be tailored to carry specific signals that interact with immune cells, thus helping to reduce inflammation and promote wound healing.

i. **Cerium Oxide Nanoparticles**: These nanoparticles have enzyme-mimetic activities, specifically catalase and superoxide dismutase activities, which enable them to neutralize ROS such as hydrogen peroxide and superoxide radicals. Their regenerative property allows them to cycle through oxidation states, continually scavenging ROS without being consumed (Xue et al., 2024).
ii. **Silver Nanoparticles**: While primarily known for their antimicrobial effects, silver nanoparticles also possess anti-inflammatory properties partly due to their ability to modulate ROS levels in the wound environment (M. Singh et al., 2022). They also modulate inflammatory responses within the wound microenvironment (Singh et al., 2019).
iii. **Fullerenes**: These carbon-based NMs can act as free radical sponges, trapping and neutralizing ROS due to their large surface area and the ability to accept or donate electrons (Shershakova et al., 2023).
iv. **Lipid-Based Nanocarriers**: Liposomes, for example, can be loaded with drugs that induce a switch from M1 to M2 macrophages, thereby facilitating the transition from inflammation to healing in the wound (Motsoene et al., 2023).
v. **Metal Oxide Nanoparticles**: Other metal oxides such as zinc oxide and manganese dioxide nanoparticles have also shown ROS scavenging abilities, reducing oxidative stress in wound sites (Maťátková et al., 2022). They have been reported to have immunomodulatory effects that can support wound healing.
vi. **Gold Nanoparticles**: These can interfere with NF-κB, a key regulator of inflammation, reducing the expression of pro-inflammatory cytokines and adhesion molecules (Wang et al., 2021d; Ovais M et al., 2018).

8.3.5 Angiogenesis Promotion

NMs can be engineered to release angiogenic growth factors such as vascular endothelial growth factor, angiopoietin, and fibroblast growth factor that directly stimulate the growth of new blood vessels. The structure of nanofiber scaffolds can mimic the extracellular matrix, providing physical cues that promote endothelial cell migration and blood vessel formation. NMs can be functionalized with specific peptides or proteins that promote angiogenesis. These biochemical signals can enhance endothelial cell activation and tubule formation. By maintaining an environment that is nontoxic and supports cellular interaction, biocompatible NMs can facilitate the natural process of angiogenesis required for successful tissue repair. Nanocarriers can provide localized, controlled delivery of angiogenic cytokines, ensuring they are released at the wound site at therapeutic concentrations to stimulate angiogenesis effectively. The inflammatory response contributes to angiogenesis in the early stages of wound healing. NMs can interact with inflammatory cells, promoting a response encouraging angiogenesis.

8.3.6 Matrix Remodeling

NMs contribute to matrix remodeling in wound healing through the following mechanisms: Nanofibers can replicate the ECM's structure, supporting cellular activities crucial in the remodeling phase, such as migration, differentiation, and deposition of new tissue. Nanocarriers can deliver MMPs or their inhibitors to regulate the breakdown and remodeling of the extracellular matrix, which is essential for proper wound healing. Nanomaterial scaffolds designed to degrade over time can progressively release space for growing tissues, allowing the new matrix to replace the scaffold and effectively reform the skin's architecture. Some NMs can influence fibroblast function, enhancing the production and organization of collagen fibers, a primary component of the repair matrix. The mechanical properties of nanomaterial scaffolds, like elasticity and strength, can be tuned to match those of the native ECM, aiding in the remodeling phase by providing the appropriate mechanical cues for cell and tissue differentiation. By controlling the release of cytokines and growth factors, NMs can modulate cell signaling pathways that orchestrate matrix remodeling. This includes attracting fibroblasts and promoting their activity in ECM synthesis.

8.3.7 Scavenging Reactive Oxygen Species

NMs can contribute to wound healing by scavenging reactive oxygen species, typically overproduced during the inflammatory phase of wound repair. Excessive ROS can lead to additional tissue damage and impede the healing process. Some NMs possess antioxidant properties, allowing them to neutralize ROS and reduce oxidative stress at the wound site. This action helps to protect cells and extracellular matrix components, promoting a more conducive environment for tissue repair and regeneration. The reduced oxidative stress also aids in progressing through the wound healing phases more effectively, from inflammation to proliferation and remodeling. However, it is essential to note that NMs have shown potential in this area, so further research is necessary to understand their mechanisms fully and ensure their clinical safety and effectiveness (Wan & Ying, 2010). Furthermore, using NMs in wound healing has shown promise in promoting angiogenesis or the formation of new blood vessels. These new blood vessels provide improved oxygen and nutrient delivery to the wound site, aiding tissue regeneration. Additionally, NMs can enhance cell migration and proliferation at the wound site, facilitating faster tissue regeneration. Overall, NMs have the potential to revolutionize wound healing by addressing key factors such as oxidative stress, angiogenesis, and cell migration. Incorporating NMs in wound healing holds excellent potential for improving the overall outcome of tissue regeneration.

8.3.8 Immune Response Modulation

NMs can modulate the immune response during wound healing through several mechanisms: By scavenging reactive oxygen species, NMs can help reduce the inflammatory response that can otherwise slow the healing process (NMs for in situ cell delivery and tissue regeneration, 2010). Macrophages play a crucial role in the immune response to tissue injury. Some NMs have been shown to encourage macrophages to adopt an M2 phenotype, which supports tissue repair and regeneration. Nanofibrous structures, similar to the natural extracellular matrix, can promote cellular interactions and better integration with the surrounding tissue (Bombin A D J et al., 2020; Zhou et al., 2017). This can help to attenuate the immune response against the material and support a more regenerative rather than pro-inflammatory environment. The immune response during wound healing often has to deal with pathogens. Some NMs possess antimicrobial properties, which can prevent or reduce infection and thus support the immune system's function in wound repair (Robust et al. of Tungsten Oxide (WO3-x) Nanodots, 2019). NMs can deliver anti-inflammatory or regenerative drugs directly to the wound site. The controlled release of these agents can modulate the immune response to favor healing. However, further research is essential to optimize these materials for immune modulation and to ensure their compatibility and safety in clinical applications (Wan & Ying, 2010).

8.3.9 Photothermal Therapy

Photothermal therapy is a treatment that involves the use of photothermal agents, which absorb light (usually from a laser) and convert it into heat. In the context of wound healing, the mechanism of photothermal therapy can be described as follows: Upon exposure to a specific wavelength of light, the photothermal agents generate localized heat. This heat can kill bacteria and other pathogens in the wound, reducing the infection and eliminating one of the barriers to wound healing. The warmth generated by PTT can enhance blood flow to the wound area, accelerating the healing process. Increased blood flow brings more oxygen and nutrients to the tissue, aiding in regeneration and helping to remove cell debris and toxins. Heat can also stimulate the activity of cells involved in wound healing, such as fibroblasts, which are vital in the creation of new connective tissue, and endothelial cells, which are essential for angiogenesis, the formation of new blood vessels (Wang et al., 2020). Properly controlled heat can help reduce inflammation, a critical component of the initial phase of wound healing. By modulating the inflammatory response, PTT may contribute to a more balanced healing process. PTT can support remodeling, the last stage of wound healing, in which the initial, often weaker tissue is replaced by more muscular, more organized tissue. It may stimulate collagen rearrangement, improving tensile strength in the healed tissue. While there is potential for PTT in wound healing due to these mechanisms, careful control is required to ensure that the heat generated does not harm surrounding healthy tissue. Also, it is crucial to mention that more research is necessary to fully understand and optimize photothermal therapy's effects and safety (Zhou et al., 2017; Aslam et al., 2022) (**Figure 8.3**).

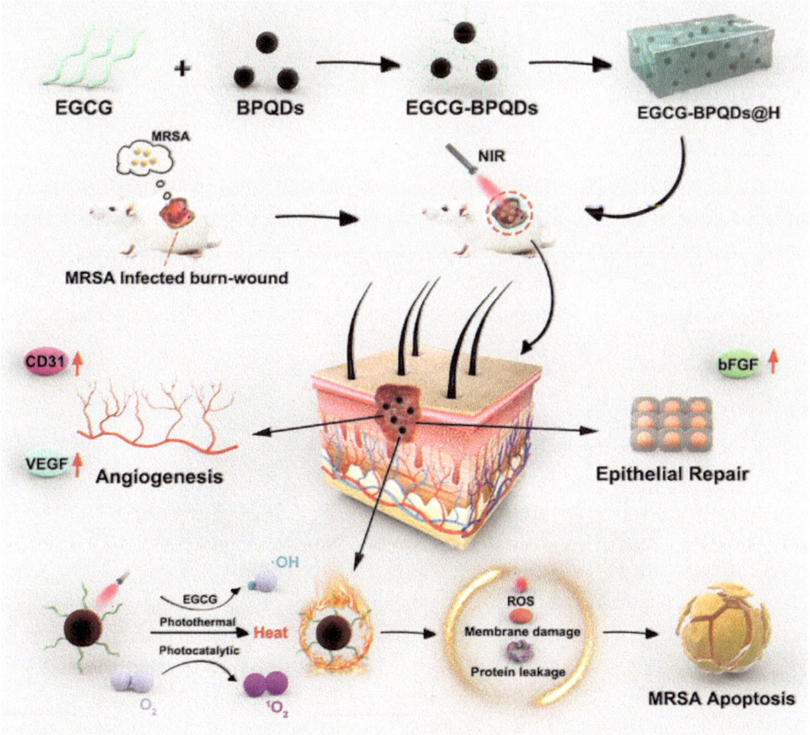

FIGURE 8.3 Graphical abstract of tea polyphenol-modified, photothermal-responsive, and ROS-generative black phosphorus quantum dots as nanoplatforms for promoting MRSA-infected wound healing in diabetic rats (Xu et al., 2021).

8.3.10 Gene Delivery

Gene delivery using NMs in wound healing is a sophisticated approach employing nanoscale carriers to deliver specific genes into cells at a wound site. Such genes are usually intended to code for therapeutic proteins that can enhance healing. Nanocarriers can be designed to target specific cells within the wound, such as skin cells, fibroblasts, or endothelial cells, which are crucial for tissue repair and regeneration. Once at the site, the NMs facilitate the genetic material's endocytosis or cellular uptake. This is due to their small size and ability to interact with cell membranes (Zhang et al., 2022). After endocytosis, the genetic material must escape the endosomes before it is directed to lysosomes and degraded. NMs can be engineered to promote endosomal escape. Once inside the cell and released from the endosome, the genetic material must reach the nucleus, where it can be transcribed. The resulting mRNA is then translated into proteins that can directly influence wound healing. The proteins produced can have various roles, such as promoting cell proliferation and migration, enhancing angiogenesis, or modulating the immune response to support the wound healing process. Ideally, the NMs will be biocompatible and biodegradable, meaning that they will eventually break down into nontoxic components that can be cleared from the body without causing adverse effects. NMs like DNA polyplexes and carbon nanotubes have been investigated for these purposes. However, further characterization in clinical safety is required before these materials can be commonly used for therapeutic applications in wound healing. The precise engineering of NMs for gene delivery holds significant potential for advancing wound care and regeneration (Kim et al., 2019)

8.4 RECENT ADVANCES IN NMS IN WOUND HEALING

Nanotechnology is contributing to improvements in wound healing by providing novel approaches to enhancing the healing process.

8.4.1 Antimicrobial NMs

8.4.1.1 Metal and Metal Oxide NMs

Employing metal and metal oxide nanoparticles in wound healing treatments has more benefits than using conventional materials. Because of their inherent characteristics, they also share similarities in how they cure wounds and have antibacterial properties. NMs' effectiveness in biological applications is generally influenced by size, shape, hydrolytic stability, zeta potential, surface functionalization, porosity, and metallic nanoparticles (Naskar & Kim, 2020). The most studied metallic nanoparticles are silver, gold, and zinc oxide.

8.4.1.2 Silver Nanoparticles

Silver's antibacterial properties have been used in wound dressings and artificial implants for the last few years to prevent infection and accelerate healing. Their increased surface area and small size at a nanoscale are why they can act against bacteria. These features of AgNPs allow them to break through bacterial cell walls and cause damage within cells, thereby preventing any further bacterial growth (**Figure 8.4**). Furthermore, silver nanoparticles and biopolymers are safe synergistic bioNMs for noncytotoxic wound dressings. Modern wound bandages have different types of silver coatings that aid drug delivery, especially in chronic wound care. AgNPs and collagen have antibacterial solid activity, making them suitable for dressing infected wounds. The green synthesis of silver nanoparticles using plant extracts was done by Batool et al.; these were then incorporated into a polymer blend (polyvinyl alcohol and starch) to form nanocomposite films. The film is up-and-coming for dressings because it has good physical properties and is highly antimicrobial. L-AgAPs-gel can be used widely as an effective, safe, antibacterial, and anti-inflammatory agent for treating wounds.

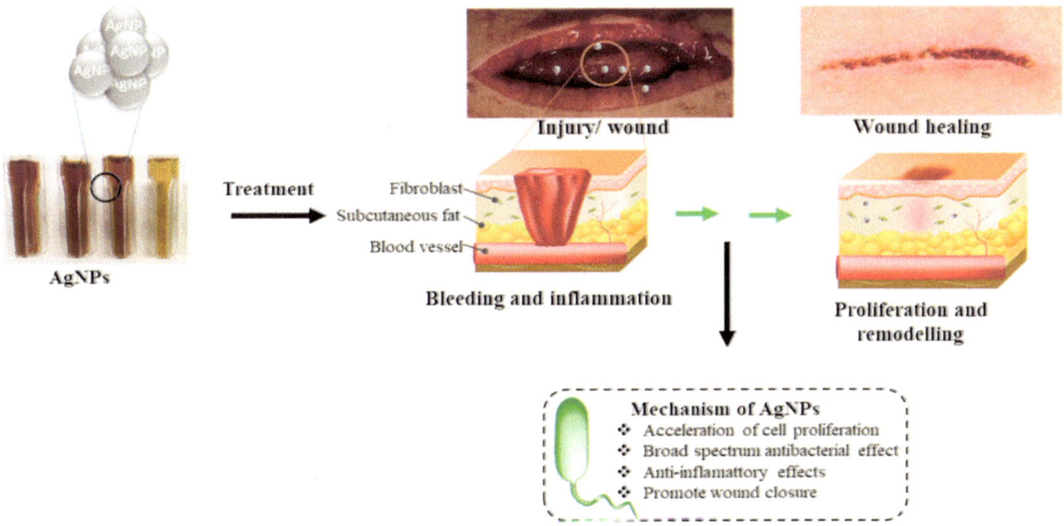

FIGURE 8.4 Mechanism of wound healing silver nanoparticles (Nqakala et al., 2021).

8.4.1.3 Gold Nanoparticles (Au NPs)

Their ability to absorb near IR light and ease of synthesis, besides the chemical stability, suggests excellent potential for the application of wound healing. Also, their surface plasmon resonance property can be adjusted with good effect for antibacterial and wound healing properties. Au NPs can inhibit the growth of multidrug-resistant (MDR) pathogens, such as *S. aureus*, by adhering to bacterial DNA and inhibiting the double helix from uncoiling during replication or transcription, thereby leading to the death of bacteria. They can also affect *P. aeruginosa*, by infiltrating into bacterial cells and impacting the bacterial membrane potential, suppressing the bacterial energy metabolism. This, in turn, blocks the function of the ATP synthase enzyme, thereby killing the bacterial cell. Hsu et al. reported that combining Au NPs and chitosan improves the healing activity of Au NPs severalfold with good biocompatibility. Au NP-chitosan increases the free radical scavenging activity of Au NPs severalfold with enhanced biocompatibility. In addition, the nanocomposite performed better than chitosan alone in the rat surgical wound model, enhancing epithelial tissue formation, hemostasis, and a quick healing rate. In burn wounds, the use of Au NPs along with cryopreserved human fibroblasts was seen to have shown an increased overall healing rate, with a lowered inflammatory phase and higher deposition of collagen, as reported by Volkova et al.

8.4.1.4 Zinc Oxide Nanoparticles (ZnO NPs)

ZnO NPs are widely used as an antibacterial agent with high biosafety and low toxicity. They are utilized in solar energy conversion, as antimicrobial agents, and for the photocatalytic degradation of environmental contaminants. Hence, ZnO NPs have unique characteristics that boost epithelial development, strengthen local defense systems, and minimize bacterial infection (killing bactericidal by the production of ROS) and inflammation, accelerating the wound's regeneration process. ZnO-containing hyaluronic acid-silk fibroin wound dressings have a high ZnO NPs content. The antibacterial activity of wound dressings against *E. coli* and *S. aureus* was enhanced, but a high concentration of ZnO (>3%, wt%) was toxic to cells.

ZnO nanoparticles were prepared by Wang et al. from the leaf extract of *Coleus amboinicus*, which displayed remarkable antibacterial action against a broad spectrum of Gram-positive and

Gram-negative bacteria, which would help in the healing of infected burns. ZnO composite scaffolds prepared by Thanusha et al. can reduce inflammation and facilitate tissue remodeling in second-degree burn healing. ZnO NP aqueous solutions produce significant ROS, indicating antibacterial activity. It was also found that ZnO NPs could effectively kill *Mycobacterium* by direct contact and high ROS production. In a dose-dependent manner, ZnO NPs could inhibit the biofilm formation of *P. aeruginosa* and *S. aureus*. ZnO NPs embedded in chitosan hydrogel collagen dressings or cellulose sheets showed antibacterial and tissue regeneration properties, making them ideal for countering the chances of infection during wound healing.

8.4.1.5 Hydrogels

Hydrogels are three-dimensional hydrophilic materials that hold much water within their structure. They are also highly biocompatible and sensitive to pH, temperature, and light changes. Based on their ability to hold water where it is required the most at the wound site, hydrogels have quickly become the dressing of choice for wound healing. This is because hydrogels create the perfect environment for repair by maintaining a suitable moisture content in the skin and promoting the removal of necrotic tissue. Hydrogels release water vapor and produce a humid environment that promotes healing while preventing the entry of bacteria, so they are preferable to other materials used in wound dressing. A hydrogel was made to treat wounds using adipose-derived stem cells from mice and pig models.

These in vivo models showed that wounds healed quite nicely. In another study that administered growth factors contained in hydrogels as a wound dressing, researchers noted that the hydrogels promoted angiogenesis and collagen production through the release of growth factors and an environment that provided moisture; the healing process in diabetic mice with chronic wounds was hastened.

Hydrogels are essential in wound care. First, the wound's hydrogels keep the wound environment wet and provide a nondiscomforting condition for patients. Hydrogel dressings have a high level of flexibility and durability and thus provide an optimal barrier around the wound during the whole process of healing. Thirdly, they can be used as vehicles for several drugs and bioactive substances that can speed up the healing of wounds. Many types of hydrogel dressings loaded with nanomedicines have now been created. Hydrogel-loaded nanomedicines can be divided into broad categories: oxygen delivery, antimicrobial, nucleic acid, and other nanodelivered medications that have different functions in wound healing.

Zhang et al. (2021) designed a hydrogel wound dressing that consists of two molecular layers. The upper layer is Ag@MOF-loaded chitosan nanoparticles, while a freeze-thaw process on the lower layer produces the PACS (polyvinyl alcohol/sodium alginate/chitosan) hydrogel. Although the lower layer, PACS, promotes epithelial growth, the upper layer, (Ag@MOF/CSNPs, has good antibacterial activity and can repress microbial invasion. It also has uniform pore distribution, good water absorptive capacity, swelling capacity, oxygen permeability, and biocompatibility. Being a bilayer dressing, it can be ideal in speeding up wound healing because of the ability of the dressing to not only promote re-epithelialization but also to decrease the inflammatory response, as these have been obtained from in vivo trials (Liu Y et al., 2022).

8.4.2 Nanofiber Scaffolds

In recent times, due to their unique characteristics and flexibility, NMs have been considered the best materials for making nanofiber scaffolds used in wound dressing. Nanofiber scaffolds have many benefits, including a large surface area-to-volume ratio, pores that can be adjusted accordingly, and their similarity to the extracellular matrix of tissues, hence being suitable for cell attachment support, proliferation promotion, and regeneration of new tissue. Many kinds of NMs, such as metals, ceramics, and polymers, are being investigated by scientists trying to produce nanofibrous scaffolding with specific properties aimed at different wound healing scenarios.

A critical breakthrough in nanomaterial-based nanofibrous scaffold design has been achieved by adding bioactive compounds or growth factors into the scaffold matrix. This means that various molecules, like anti-inflammatory drugs, antimicrobial agents, growth factors, etc., can be mixed within the fibers so that they will directly act on the wound, thereby accelerating its healing process while minimizing infections, inflammation, and other problems associated with them. Moreover, this integration also allows controlled release kinetics, which ensures sustained therapeutic effects at the site of injury without causing harm to other parts of the body through the general circulation system (Nandhini et al., 2024; Naskar & Kim, 2020).

Moreover, nanofiber scaffolds based on NMs have improved their wound healing performance due to recent developments. Electrospinning and self-assembly, among other fabrication methods, have been used to control these scaffolds' structure, composition, and mechanical properties with high precision. This has allowed researchers to modify any scaffold feature to make it suitable for a particular type or location of the injury. Additionally, multifunctional nanocomposite systems comprising several types of NMs exhibited better overall performance than those containing only one kind. Consequently, this means there is still more room for growth in this area, which can lead to significant changes in faster recovery rates and better outcomes for patients with wounds (Aminzai & Patan, 2024).

8.4.3 Smart NMs

The development of intelligent NMs has taken a great leap forward in recent years. These innovative materials could do wonders for treating wounds. Smart NMs react to certain stimulants in the wound area, such as pH levels, temperature changes, or specific biomarkers. This allows healthcare providers to deliver treatments where needed most in a patient's body. For instance, pH-sensitive nanoparticles may be used as therapeutic agents that release drugs only when exposed to an acidic microenvironment such as those found around chronic ulcers—thereby initiating healing without causing collateral damage elsewhere in the system. Alternatively, heating areas affected by inflammation until it reaches temperatures beyond what normal cells can tolerate is another way temperature-responsive nanoparticles could be utilized, hence providing localized heat treatment at sites with high concentrations of activated immune cells involved in chronic inflammation, promoting tissue repair processes during wound healing (Chopra et al., 2023)

Furthermore, scientists have combined NMs with more advanced sensing and imaging capabilities so that doctors can monitor wounds in real time as they heal over days or even weeks. This means we no longer need to wait until someone's condition worsens before intervening; instead, we can detect signs of infection early on and act accordingly—potentially saving lives. Furthermore, nanomaterial-based dressings equipped with sensors can monitor various wound parameters, such as pH, moisture level, and bacterial load, offering personalized wound care solutions and improving patient outcomes. Recent advancements in intelligent NMs hold tremendous potential for revolutionizing wound management by offering innovative strategies for tailored and responsive therapeutic interventions (R. Singh et al., 2022)

8.4.4 Gene Delivery

Recent developments in NMs have created exciting opportunities for exploiting gene delivery approaches in wound treatment. Gene therapy provides the ability to target the cellular and molecular components at the core of the wound healing cascades. Specifically, it provides the potential to improve cellular proliferation and differentiation, tissue regeneration, and wound closure. NMs are excellent gene carriers to the target cells in the wound site due to various challenges, such as poor transfection efficacy and the susceptibility of nucleic acids to degradation.

A noteworthy recent progress is the ability to shield and transport nucleic acids, such as plasmid DNA or small interfering RNA (siRNA), to specific cell populations in the wound bed through nanoparticles. These NMs may be fabricated with targeting moieties that bind to overexpressed cell surface receptors during healing, thereby ensuring that therapeutic genes are delivered selectively into relevant cell types like fibroblasts, keratinocytes, or endothelial cells. Additionally, multifunctional nanomaterial design allows for incorporating other features, including stimulus-responsive release mechanisms or imaging agents, thus increasing the accuracy and efficiency of gene delivery while enabling constant monitoring of treatment responses. Also, the design of nanomaterials has been improved to mimic natural extracellular matrix (ECM) architecture, leading to biomimetic nanocarriers that promote cellular adhesion, uptake, and intracellular trafficking of therapeutic genes. These biomimetic NMs improve biocompatibility and bioactivity within gene delivery systems, thus enhancing cellular uptake and transfection efficiency while reducing cytotoxicity and immunogenicity to minimum levels.

Furthermore, NMs as gene carriers achieve gene expression and may make it possible to spatially and temporally control the membrane and its critical signaling pathways, including angiogenesis, inflammation, and matrix degradation. Therefore, the developments in gene delivery systems through NMs have paved the way for a revolution in treating and managing wounds with prospective field-based and personalized therapeutic strategies to repair, heal, and rejuvenate tissues (Bai et al., 2022; Kushwaha et al., 2022)

8.4.5 D Printing with NMs

The development of NMs has substantially expanded the possibilities of 3D printing in wound healing. Researchers can use NMs in 3D printing protocols to produce intricate scaffolds while controlling architecture, physical properties, and bioactivity to simulate the targeted tissue's native microenvironment. Nanomaterial-reinforced bioinks allow highly biocompatible cell-laden structure deposition while facilitating cell connection, spreading, and development at the intended wound location. Nanoparticles and nanofibers or nanocomposites, which are embedded into 3D printed scaffolds, can offer additional qualities, such as antibacterial activity, controlled drug distribution, and enhanced mechanical durability, according to the specific requirements of tissue expansion.

A recent advance of particular significance is the emergence of nanocomposite bioinks for 3D printing, which involves using NMs to improve printed constructs' structural integrity and bioactivity. As a type of material, nanocomposite bio-inks combine the advantages of NMs and biopolymers, providing superior printability, mechanical properties, and functionalities for generating regenerative tissues. They can also facilitate real-time monitoring of the wound healing process when NMs with modern imaging or sensing functionality are incorporated into the 3D-printed scaffolds. For instance, clinicians might include nanoparticles featuring imaging contrast agents or biosensors in the scaffold matrix to monitor critical wound parameters, such as pH levels, oxygen tension, or inflammatory markers, without the need for perhaps invasive procedures. This information might allow clinicians to address concerns as they arise and enhance therapeutic outcomes.

In addition, patient-specific wound dressings produced with the application of 3D printing technology may be shaped and characterized (made unique) based on the patient's specific parameters due to progress in the field. Nanomaterial-enhanced bio-printed dressings are produced because they provide better healing by helping to cover and adhere correctly to an injury site, support it structurally during the recovery period due to their excellent mechanical strength, which does not easily fall off when worn for long periods, as well as by time-releasing therapeutic agents in prescribed doses, thus making them more efficient than traditional wound healing methods. Furthermore, combining NMs with either growth factors or stem cells in 3D-printed scaffolds also opens possibilities for improving wound dressings and quickening tissue renewal, which may result in faster wound healing. In general terms, modern developments of new technologies such as nano-enhanced 3D printing provide some original methods leading to targeted approaches toward wound treatment in medical practice through regenerative perspectives for the future and beyond in this new era known as regenerative medicine (Yayehrad et al., 2023).

8.4.6 Biodegradable NMs

Biodegradable NMs, with recent advances in NMs, hold remarkable promise in applications related to wound healing. Biodegradable NMs promise several advantages in and of themselves. Primarily, they can degrade over time into nontoxic byproducts, thus being compatible with the body's natural healing processes and reducing the likelihood of long-term adverse effects. These constructs could serve as temporary scaffolds for tissue regeneration as they are gradually absorbed by the body and, therefore, obviate the need for surgical removal, reducing patient morbidity. Another promising recent development is the fabrication of biodegradable nanofibrous scaffolds that mimic the structural and mechanical properties of the extracellular matrix to promote cell attachment, migration, and proliferation. These scaffolds can be designed by controlling fiber morphology, porosity, and degradation kinetics through electrospinning methods. As biodegradable polymers such as PLA, PGA, or copolymers are used in the fabrication process, the degradation rate of the scaffolds can be tuned to be proportional to the time required for healing, providing transient support for tissue regeneration and minimizing foreign body reactions (Arafa et al., 2018; Aminzai & Patan, 2024).

Most recently, advances in nanomaterial chemistry have led to the development of multifunctional biodegradable NMs showing increased therapeutic efficacy in wound healing. By loading biologically active molecules, such as growth factors, antimicrobial agents, or anti-inflammatory drugs, into biodegradable NMs, scientists can prepare composite systems that promote tissue regeneration while preventing infection and inflammation. In addition, integrating NMs with imaging and sensing functionalities provides real-time monitoring of wound healing, allowing for timely interventions and personalized treatment strategies. Recent advances in biodegradable NMs promise to revolutionize wound care by offering safe, effective, and patient-friendly solutions for tissue regeneration and wound healing (Nandhini et al., 2024; Aramwit et al., 2014) (**Figure 8.5**).

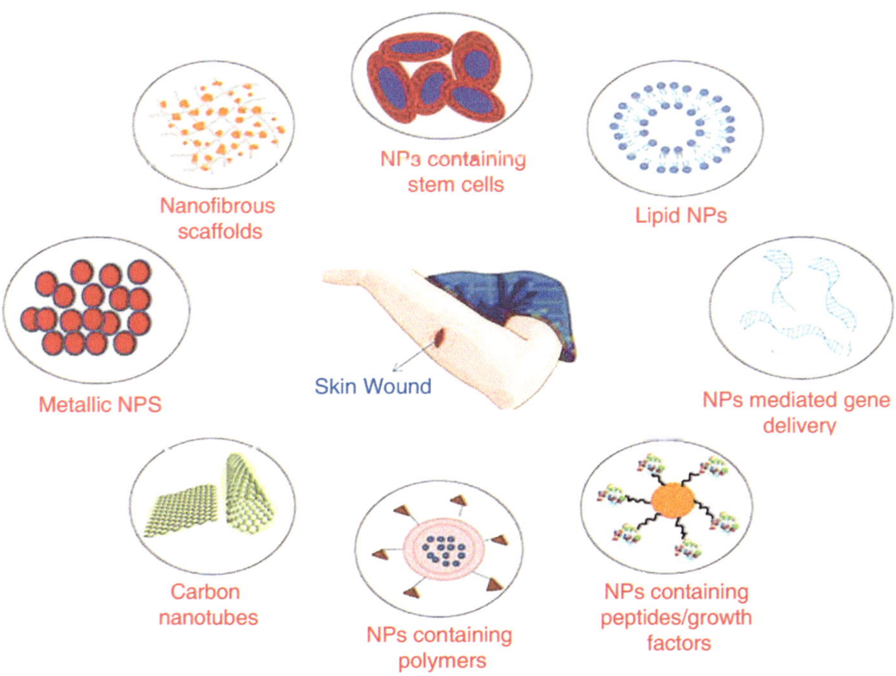

FIGURE 8.5 Recent advances in NMs for wound healing (Soni et al., 2023).

8.5 FUTURE PERSPECTIVES OF NMS IN WOUND HEALING

Future research could produce nanoparticles that release proteins, extracellular matrix components, and growth promoters, all of which accelerate wound healing by promoting angiogenesis, tissue regeneration, and cell proliferation. The functionalization of NMs has the potential to alter the immunologic response. Nanomaterial-based growth promoters as delivery tools hold significant potential for recuperative health care development and bettering patient outcomes. These frameworks' advantages over traditional methods include targeted ligands and regulated release kinetics. The ability of nanoparticles to be versatile is essential for wound healing (Nandhini et al., 2024).

8.5.1 Personalized Medicine

Personalized medicine substantially impacts wound healing, depending on the type of wound, medical history, and genetics. Customized bandages, foam, hydrogel, nano or microgels, film, nanofiber scaffolds, etc., can all be used as wound dressings. Utilizing NMs in wound healing therapies creates and explores potential enhancements in scar healing, targeted drug delivery, infection prevention, and tissue regeneration (Maťátková et al., 2022). To optimize wound healing, nanoparticles can offer versatile systems with longer drug release, antimicrobial capabilities, and bioactive molecule signaling delivery. In addition, they can duplicate the extracellular matrix, promote cell movement and adhesiveness, and enhance and alter signaling pathways linked to tissue regeneration.

Nanofabricated customized wound dressings hold much promise for personalized wound care. Natural polymeric nanoparticles (NPs), combined with biomolecules, growth hormones, and antibiotics, may be able to solve the shortcomings of the materials currently employed in wound healing and create more mechanically resistant scaffolds. A hybrid nanocomposite composed of natural and inorganic polymers and antibacterial nanoparticles may enhance wound healing (Nandhini et al., 2024) Functionalized wound dressings have been shown to promote angiogenesis, promote wound recovery, and regulate recently established collagen types. They reduce the amount of collagen surrounding the wound site by strengthening the preexisting structure of the collagen (Stoica et al., 2020; Coentro et al., 2019) Combining nanotechnology with personalized medicine to cut costs and shorten treatment times could make healthcare more individualized.

8.5.2 Combination Therapies

Using nanoparticles to co-encapsulate many medications or therapeutic ingredients into a single nanocarrier can simplify combination therapy. This methodology enables the progressive release of distinct pharmaceuticals, the simultaneous delivery of supplementary medications, or synergistic effects via various techniques. To allow for targeted, localized drug delivery to the area of the wound, drugs, growth promoters, or therapeutic agents can be encapsulated in NMs like liposomes, polymeric nanoparticles, or dendrimers. Packed with antibiotics, **xenobiotics** are transforming wound care by curing skin illnesses, treating bacteria resistant to traditional drugs, and providing customized treatment alternatives (Stoica et al., 2020).

Negative pressure wound therapy (NPWT) accelerates burn wound recuperation by encouraging blood flow and tissue restoration, improving the quality of life for burn patients and reducing the risk of infection and scarring.

Combining silver nanoparticles and ingredients like chitosan, starch, collagen, cellulose, and hyaluronic acid effectively promotes wound healing. Gold NPs and chitosan together can heal wounds. Chitosan and gold nanoparticles enhance their biocompatibility and provide free radical scavenging activity. Combining cryopreserved human fibroblasts with gold nanoparticles expedites collagen deposition, shortens the inflammatory phase, and heals burn wounds more quickly. When antibiotics and copper nanoparticles were applied to infected wounds, germs dramatically dropped, and the healing process accelerated. Furthermore, in addition to having potent antibacterial properties, titanium dioxide NP wound dressings with chitosan and pectin showed promising results in wound healing. Animal studies have shown that iron oxide-coated carbon nanotubes NMs have broad-spectrum antibacterial properties and effectively heal wounds.

Laser **technology** accelerated the healing process of fungal or bacterial wounds treated with graphene oxide. Researchers built a scaffold out of graphene and seeded it with mesenchymal stem cells and antimicrobial compounds.

Recent wound repair and renewal advancements have prompted the use of various bioresource-based nanocomposite dressings. Traditional bandages can cause tissue damage and are not very successful in preventing infection. Advanced healing dressings could address these issues by **combining electrospinning technologies** to create fiber scaffolds with controlled porosity (Singh et al., 2018). These scaffolds, resembling the extracellular matrix, promote in vivo tissue regeneration, prevent dehydration, promote cell adhesion, and provide mechanical strength (Han et al., 2020).

For chronic ulcers, **hydrogels containing platelet-derived growth factors (PDGF)** promote wound healing by releasing angiogenesis and collagen formation growth factors, limiting bacterial entrance, and establishing a humid environment. Because hydrogel-based scaffolds mimic the extracellular matrix found in the body and have a three-dimensional environment, they present a promising alternative for wound healing. Hydrogels are adaptable scaffolds for tissue repair and drug delivery. Researchers use advanced techniques to create antimicrobial hydrogels containing antimicrobial polymers such as antibiotics, peptides, biocides, and nanoparticles. Hydrogels are extensively utilized in tissue formation and medication delivery due to their elevated water content and tissue-like characteristics. They were created to manage cellular differentiation and regeneration and heal wounds. Composite NP-loaded gel dressing formulations hybridized with silver, zinc, and titanium dioxide nanoparticles are widely used to treat wounds. These hydrogels can decrease bacterial resistance and deliver metal NPs and antibiotics.

Adipose-derived stem cell (ASC) sheets are helpful for wound recovery and tissue building because they are simple to create and stimulate using L-ascorbate 2-phosphate. Recent research has shown that ASCs are an effective topical wound therapy alternative, may treat fibrosis and difficult scarring, and can enhance skin wound healing (Yu et al., 2018; Juncos Bombin et al., 2020b; Stoica et al., 2020).

8.5.3 Advanced Drug Delivery

Since nanoparticles have inherent qualities that facilitate wound healing and can be used as delivery systems for medicinal drugs, they are frequently used in wound therapy. Metallic NMs have demonstrated antimicrobial activity against a range of infections, assisting in the healing of wounds and preventing infections. To encourage tissue restoration and wound recovery, hydrogels and nanofibers, two types of polymeric NMs, are excellent for cellular adherence, movement, and multiplication. Liposomes and solid lipid nanoparticles are examples of lipid-based nanoparticles that enable focused and regulated drug delivery, guaranteeing the long-term release of drugs at the area of the wound. Inorganic nanoparticles have been demonstrated to improve neovascularization, connective tissue, and cell amplification, hence promoting tissue restoration and wound scarring. Solid lipids and nanolipid carriers are two varieties of lipid nanoparticles that have been used to create efficient carriers for the treatment of wounds (Nandhini et al., 2024; Stoica et al., 2020).

Recently, gene-activated matrix treatment has focused on skin, bone, and cartilage regeneration. DNA stability is the primary benefit of gene therapy for functions related to wound healing. Currently, methods like direct injection, gene gun transfer, and electroporation require numerous injections or are hindered by uneven gene expression. The issue can be addressed using NMs like electrospun nanofibrous meshes as a matrix encasing genes and wound bandage material. Scaffolds made of nanoparticles are often fabricated using electrospinning, phase separation, and self-assembly methods. The technique researchers most commonly use is electrospinning to create nanoscaffolds. (Maťátková et al., 2022).

In addition to having antimicrobial qualities, silver nanoparticles encourage angiogenesis, lessen inflammation, and aid in the stimulation of cell migration and proliferation in wounds. SilverIon®, CuraFoam®, DynaGinate (Ag foam dressing), Aquacel Ag®, and DynaGinate (Ag calcium alginate dressing) are examples of silver nanoparticles. Aside from being useful as antibiotic nanocarriers, gold nanoparticles can also be utilized to stimulate wounds electrically and are beneficial in photothermal therapy and gene monotherapy. The synthesis of collagen, angiogenesis, and wound healing are all improved by copper nanoparticles. Unlike traditional bandages, they treat ulcers, minimize wound size, and improve healing outcomes (Nandhini et al., 2024).

Zinc sulfide nanoparticles stimulate skin regeneration, collagen, and silica NP nanocomposites improve wound dressings' mechanical strength, while titanium dioxide nanoparticles offer antibacterial and anti-inflammatory properties. Iron oxide nanoparticles can modify their surface, combine easily, and possess a broad antimicrobial spectrum, reducing toxicity and increasing biodegradation. Ceramic nanoparticles promote angiogenesis and cell growth by slowly releasing silicon, and the presence of calcium ions and other substances in wound areas makes them an excellent material for restoring bone tissue (Nandhini et al., 2024).

When combined with other metallic composites, carbon nanotubes significantly enhance the antimicrobial properties of bacterial cell membranes. A nanofiber scaffold successfully healed a mimicked severe burn lesion by encouraging cellular transportation and proliferation and mimicking the skin's extracellular matrix. Graphene, combined with hyaluronic acid, PVA, and chitosan, exhibits potent antibacterial properties, accelerating the healing process of bacterial wounds. NMs offer promising antimicrobial therapy for wound infections, with black phosphorus (BP) being a superior nanomaterial for wound healing and other biological uses (Nandhini et al., 2024).

8.5.4 Smart Dressing and Sensors

Nanomaterial-based innovative scaffolds can exhibit a response to their surroundings around the wound, releasing chemicals or drugs in response to specific stimuli, hastening the healing process. These advances in nanotechnology provide personalized medicine with a solid basis on which to build, allowing for individualized patient therapies. One kind of scaffold that can help people with compromised immune systems is hydrogel, which can lower the risk of infection and bacterial growth in wounds. Its sensors monitor wound healing and offer prompt wound treatment.

Integrating nanosensors with nanomaterials can improve cancer medicine development and patient outcomes by monitoring treatment response in real-time, reducing toxicity, optimizing dosing regimens, and revolutionizing cancer diagnostics. This allows for early intervention, identification, and therapy success, leading to tailored treatment and reduced healthcare costs.

Researchers have created intelligent dressings for wound growth and healing that include biomimetic nanofiber membranes, microenvironment sensors, and gelatin methacryloyl (GelMA) mixed with β-cd UV-cross-linked hydrogel. Fluorescent nanosensors detect infection, monitor wound healing, and utilize magnesium hydroxide's antibacterial and anti-inflammatory properties in dressings, revolutionizing wound treatment with affordable, practical solutions. Magnesium-based innovative dressing technology is the future invention (Naskar & Kim, 2020).

8.5.5 Artificial Intelligence and Remote Monitoring Systems

NMs often signal the start of a new period of more inventive, personalized, and efficient wound healing techniques that can analyze wound data, predict healing outcomes, and optimize treatment plans. At the same time, remote monitoring systems enable healthcare professionals to monitor wound progress remotely. Intelligent wearable medical sensors enable accurate pathophysiological data sensing and real-time physiological status monitoring. Specifically, for the aim of monitoring health, wearable sensors identify the intended analytes in a range of biofluids (blood, tears, perspiration, tissue, and wound fluids, etc.) both in vivo and in vitro.

Wearable bandages and dressings aid in real-time wound healing assessment and treatment assessment. Advanced wearable sensors connect to the body, enabling live sensing, data recording, and computation. Real-time wireless communication enables *in situ*, diverse parameter wound assessment. (Cheng et al., 2021).

8.6 LIMITATIONS

a) **Biocompatibility and Toxicity**: Although NMs are a boon, they may induce adverse reactions in biological systems. Size, shape, surface chemistry, and composition can affect their biocompatibility and toxicity. Therefore, it is essential to ensure that the NMs used for wound healing are biocompatible and nontoxic to avoid harmful effects on cells, tissues, and organs (Girija AR, 2022b; Stoica et al., 2020).

b) **Regulatory Challenges**: With high regulatory scrutiny over the NMs used in wound healing owing to concerns over safety and efficacy, regulatory challenges are also part of the associated landscape for nanomedicine products. Regulatory approval for nanomaterial-based wound healing products is yet to be explored in depth, adding to the complexity and time-consuming nature of bringing these products to the market. Adequate safety, efficacy, and quality data are needed for NMs, adding to the challenges of getting these products to market (Wang et al., 2021e; Joorabloo & Liu, 2022).

c) **Scalability and Cost**: Nanomaterial manufacturing is relatively costly, thus impeding full-scale adoption. Scaling up large-scale NM manufacturing is expensive and complicated. The high costs associated with nanomaterial-based wound healing products may exclude the products from access and affordability, especially in low-resource settings (Wang et al., 2021f; Stoica et al., 2020).

d) **Stability and Durability**: NMs can also be unstable or degrade over time, and such factors might affect their performance and therapeutic efficacy. The performance of NMs used in wound healing could be affected by environmental and storage conditions, as well as the interaction of these materials with biological fluids. The long term stability and durability of nanomaterial-based products for wound healing are clinically relevant to their success (Kushwaha et al., 2022).

e) **Targeting Efficiency and Specificity**: It is vital to ensure the achievement of good targeting of NMs to wound sites without being taken up by nontarget sites. The NMs may cause nonspecific binding, fast bloodstream clearance, and poor tissue penetration. Innovative strategies to increase the efficiency and specificity of targeting NMs in the wound healing process should include surface modification, functionalization, and targeting using ligands (Singh et al., 2019)

f) **Complexity of Fabrication and Characterization**: The fabrication and characterization of NMs with desired properties for wound healing applications are technically challenging. Variability in nanomaterial synthesis, material properties, and formulation methods may lead to problems related to reproducibility and consistency in the fabricated nanomaterial-based wound healing products. Standardization of protocols in fabrication and characterization is essential for the quality and reliability of NMs in wound healing (Duan et al., 2019).

g) **Immunogenicity and Host Response**: The immunogenicity and host responses evoked by NMs may vary in different biological systems, sometimes in ways that critically affect the wound healing process. Size, shape, surface charge, and composition are critical parameters that dictate immunogenicity and host response to NMs. It is, therefore, crucial to establish and minimize immunological effects in NMs to minimize adverse reactions and optimize therapeutic potency in wound healing (Zhang et al., 2022).

h) **Standardization and Guidelines Lacking**: A lack of standardized protocols and guidelines on using NMs in wound healing impedes clinical translation and uptake. The variety of NMs, compositions, formulations, and routes of administration hinders comparisons of results across studies, thus impeding the development of a consensus set of guidelines. Developing standardized protocols and regulatory frameworks to evaluate and approve nanomaterial-based wound healing products is crucial for ensuring safety and efficacy (Wang et al., 2021e)

8.7 CONCLUSION

Integrating NMs into wound healing strategies holds immense promise for the future of regenerative medicine. Though there is so much left to be learned from their mechanisms of action, recent advancements, limitations, and future perspectives have already cast a paradigm shift toward more efficient and personalized approaches in wound care.

NMs have demonstrated versatility and efficiency in wound healing through mechanisms other than merely their enhanced surface area interaction, controlled drug delivery, antimicrobial properties, anti-inflammatory influence, and ability to promote tissue regeneration. This provides unprecedented opportunities to address the diverse challenges of wound management. By applying these unique properties, researchers and clinicians have made considerable strides in developing nanomaterial-based wound dressings, scaffolds, drug delivery systems, and even diagnostics that may accelerate healing, minimize complications, and improve patient outcomes.

However, the promise of such improved NMs in wound healing is tempered by some fundamental limitations that must be overcome. These include, but are not limited to, biocompatibility and toxicity issues, regulatory hurdles, issues related to cost and scalability, stability and durability limitations, inefficiencies in targeting, complexity of fabrication and characterization, immunogenicity risk, and a lack of standardization and guidelines. These challenges require concerted efforts by researchers, clinicians, industry partners, and regulatory agencies to ensure safe, effective, and standardized nanomaterial-based wound healing therapies.

Future advances in NMs for wound healing, fueled by nanotechnology, materials science, biomedical engineering, and translational research, could provide an exciting avenue. Next-generation NMs with enhanced functionalities, better biocompatibility, and optimized therapeutic outcomes will shape the future of wound care. Multifunctional NMs for targeted drug delivery, real-time monitoring, and personalized treatment approaches are poised to reshape the landscape of wound care. Finally, integrating nanotechnology with emerging technologies such as artificial intelligence and 3D bioprinting brings exciting prospects to the forefront of advancing strategies for wound healing and expediting the translation of nanomaterial-based therapies into clinical practice.

In conclusion, although NMs in wound healing hold opportunities and challenges, the potential for changing patient outcomes and quality of life is not to be underestimated. Now, to overcome the present limitations and groom a forward-looking perspective of NMs, we can herald a new era of wound care that is safe, effective, and fine-tuned to individual patient needs. As we explore nanotechnology's vast potential in wound healing, let us keep our eyes on advancing the field through novel research, collaboration, and a patient-centric approach. As a result, we can harness NMs to heal wounds, restore function, and enhance the well-being of millions of patients globally.

REFERENCES

Aminzai, M. T., & Patan, A. (2024). Recent applications and evaluation of metal nanoparticle-polymer hybrids as chronic wound dressings. *Journal of Nanomaterials*, *2024*. https://doi.org/10.1155/2024/3280349

Arafa, M. G., El-Kased, R. F., & Elmazar, M. M. (2018). Thermoresponsive gels containing gold nanoparticles as smart antibacterial and wound healing agents. *Scientific Reports*, *8*(1), 1–16. https://doi.org/10.1038/s41598-018-31895-4

Aramwit, P., Bang, N., Ratanavaraporn, J., & Ekgasit, S. (2014). Green synthesis of silk sericin-capped silver nanoparticles and their potent anti-bacterial activity. *Nanoscale Research Letters*, *9*(1), 1–7. https://doi.org/10.1186/1556-276X-9-79

Aslam, Z., Roome, T., Razzak, A., Aslam, S. M., Zaidi, M. B., Kanwal, T., Sikandar, B., Bertino, M. F., Rehman, K., & Shah, M. R. (2022). Investigation of wound healing potential of photo-active curcumin-ZnO-nanoconjugates in excisional wound model. *Photodiagnosis and Photodynamic Therapy*, *39*, 102956. https://doi.org/10.1016/j.pdpdt.2022.102956

Bai, Q., Zheng, C., Chen, W., Sun, N., Gao, Q., Liu, J., Hu, F., Pimpi, S. H., Yan, X., Zhang, Y., & Lu, T. (2022). Current challenges and future applications of antibacterial nanomaterials and chitosan hydrogel in burn wound healing. *Materials Advances*, *3*(17), 6707–6727. https://doi.org/10.1039/d2ma00695b

Barroso, A., Mestre, H., Ascenso, A., Simões, S., & Reis, C. (2020). Nanomaterials in wound healing: From material sciences to wound healing applications. *Nano Select*, *1*(5), 443–460. https://doi.org/10.1002/nano.202000055

Bellu, E., Medici, S., Coradduzza, D., Cruciani, S., Amler, E., & Maioli, M. (2021). Nanomaterials in skin regeneration and rejuvenation. *International Journal of Molecular Sciences*, *22*(13). https://doi.org/10.3390/ijms22137095

Blanco-Fernandez, B., Castaño, O., Mateos-Timoneda, M. Á., Engel, E., & Pérez-Amodio, S. (2021). Nanotechnology approaches in chronic wound healing. In *Advances in Wound Care* (Vol. 10, Issue 5, pp. 234–256). Mary Ann Liebert Inc. https://doi.org/10.1089/wound.2019.1094

Bombin, A. D. J., Dunne, N. J., & McCarthy, H. O. (2020). Electrospinning of natural polymers for the production of nanofibres for wound healing applications. *Materials Science and Engineering: C*, *114*, 110994.

Cheng, S., Gu, Z., Zhou, L., Hao, M., An, H., Song, K., Wu, X., Zhang, K., Zhao, Z., Dong, Y., & Wen, Y. (2021). Recent progress in intelligent wearable sensors for health monitoring and wound healing based on biofluids. In *Frontiers in Bioengineering and Biotechnology* (Vol. 9). Frontiers Media S.A. https://doi.org/10.3389/fbioe.2021.765987

Chopra, H., Mohanta, Y. K., Mahanta, S., Mohanta, T. K., Singh, I., Avula, S. K., Mallick, S. P., Rabaan, A. A., Alsaihati, H., Alsayyah, A., Alissa, M., Alturaifi, H. R., Alalwan, B., Attia, M. S., Chakraborty, S., & Dhama, K. (2023). Recent updates in nanotechnological advances for wound healing: A narrative review. *Nanotechnology Reviews*, *12*(1). https://doi.org/10.1515/ntrev-2023-0129

Coentro, J. Q., Pugliese, E., Hanley, G., Raghunath, M., & Zeugolis, D. I. (2019). Current and upcoming therapies to modulate skin scarring and fibrosis. In *Advanced Drug Delivery Reviews* (Vol. 146, pp. 37–59). Elsevier B.V. https://doi.org/10.1016/j.addr.2018.08.009

Duan, G., Chen, L., Jing, Z., De Luna, P., Wen, L., Zhang, L., Zhao, L., Xu, J., Li, Z., Yang, Z., & Zhou, R. (2019). Robust antibacterial activity of tungsten Oxide (WO 3-x) nanodots. *Chemical Research in Toxicology*, *32*(7), 1357–1366. https://doi.org/10.1021/acs.chemrestox.8b00399

Fan, Z., Liu, B., Wang, J., Zhang, S., Lin, Q., Gong, P., . . . Yang, S. (2014). A novel wound dressing based on Ag/graphene polymer hydrogel: Effectively kill bacteria and accelerate wound healing. *Advanced Functional Materials*, *24*(25), 3933–3943.

Girija AR, B. S. C. A. J. (2022a). Nanomaterials-based drug delivery approaches for wound healing. *Current Pharmaceutical Design*, *28*(9). https://doi.org/10.2174/1381612828666220328121211

Gupta, K. C., Haider, A., Choi, Y. R., & Kang, I. K. (2014). Nanofibrous scaffolds in biomedical applications. *Biomaterials Research*, *18*(1), 5. https://doi.org/10.1186/2055-7124-18-5

Han, B., Fang, W. H., Zhao, S., Yang, Z., & Hoang, B. X. (2020). Zinc sulfide nanoparticles improve skin regeneration. *Nanomedicine: Nanotechnology, Biology, and Medicine*, *29*. https://doi.org/10.1016/j.nano.2020.102263

Hayat, M., Nawaz, A., Chinnam, S., Muzammal, M., Latif, M. S., Yasin, M., Ashique, S., Zengin, G., & Farid, A. (2023, July 1). Formulation development and optimization of herbo synthetic gel: In vitro biological evaluation and in vivo wound healing studies. *Process Biochemistry*, *130*, 116–126.

Jain, A. S., Pawar, P. S., Sarkar, A., Junnuthula, V., & Dyawanapelly, S. (2021). Bionanofactories for green synthesis of silver nanoparticles: Toward antimicrobial applications. *International Journal of Molecular Sciences*, *22*(21), 11993. https://doi.org/10.3390/ijms222111993

Jiang, Z., Zheng, Z., Yu, S., Gao, Y., Ma, J., Huang, L., & Yang, L. (2023). Nanofiber scaffolds as drug delivery systems promoting wound healing. *Pharmaceutics*, *15*(7). Multidisciplinary Digital Publishing Institute (MDPI). https://doi.org/10.3390/pharmaceutics15071829

Joorabloo, A., & Liu, T. (2022). Recent advances in nanomedicines for regulation of macrophages in wound healing. *Journal of Nanobiotechnology*, *20*(1), 407.

Juncos Bombin, A. D., Dunne, N. J., & McCarthy, H. O. (2020a). Electrospinning of natural polymers for the production of nanofibres for wound healing applications. In *Materials Science and Engineering C* (Vol. 114). Elsevier Ltd. https://doi.org/10.1016/j.msec.2020.110994

Kim, H. S., Sun, X., Lee, J.-H., Kim, H.-W., Fu, X., & Leong, K. W. (2019). Advanced drug delivery systems and artificial skin grafts for skin wound healing. *Advanced Drug Delivery Reviews*, *146*, 209–239. https://doi.org/10.1016/j.addr.2018.12.014

Kushwaha, A., Goswami, L., & Kim, B. S. (2022). Nanomaterial-based therapy for wound healing. *Nanomaterials, 12*(4). https://doi.org/10.3390/nano12040618

Maťátková, O., Michailidu, J., Miškovská, A., Kolouchová, I., Masák, J., & Čejková, A. (2022). Antimicrobial properties and applications of metal nanoparticles biosynthesized by green methods. *Biotechnology Advances, 58*, 107905. https://doi.org/10.1016/j.biotechadv.2022.107905

Motsoene, F., Abrahamse, H., & Dhilip Kumar, S. S. (2023). Multifunctional lipid-based nanoparticles for wound healing and antibacterial applications: A review. *Advances in Colloid and Interface Science, 321*, 103002. https://doi.org/10.1016/j.cis.2023.103002

Nandhini, J., Karthikeyan, E., & Rajeshkumar, S. (2024b). Nanomaterials for wound healing: Current status and futuristic frontier. *Biomedical Technology, 6*(October 2023), 26–45. https://doi.org/10.1016/j.bmt.2023.10.001

Naskar, A., & Kim, K. S. (2020). Recent advances in nanomaterial-based wound-healing therapeutics. *Pharmaceutics, 12*(6), 499. https://doi.org/10.3390/pharmaceutics12060499

Niculescu, A. G., & Grumezescu, A. M. (2022). An up-to-date review of biomaterials application in wound management. *Polymers, 14*(3). https://doi.org/10.3390/polym14030421

Nqakala, Z. B., Sibuyi, N. R. S., Fadaka, A. O., Meyer, M., Onani, M. O., & Madiehe, A. M. (2021). Advances in nanotechnology towards development of silver nanoparticle-based wound-healing agents. *International Journal of Molecular Sciences, 22*(20), 11272. https://doi.org/10.3390/ijms222011272

Osanloo, M., Noori, F., Varaa, N., Tavassoli, A., Goodarzi, A., Moghaddam, M. T., Ebrahimi, L., Abpeikar, Z., Farmani, A. R., Safaei, M., Fereydouni, N., & Goodarzi, A. (2024). The wound healing effect of polycaprolactone-chitosan scaffold coated with a gel containing Zataria multiflora Boiss. volatile oil nanoemulsions. *BMC Complementary Medicine and Therapies, 24*(1). https://doi.org/10.1186/s12906-024-04352-1

Ovais, M., Ahmad, I., Khalil, A. T., Mukherjee, S., Javed, R., Ayaz, M., . . . Shinwari, Z. K. (2018). Wound healing applications of biogenic colloidal silver and gold nanoparticles: recent trends and future prospects. *Applied Microbiology and Biotechnology, 102*, 4305–4318.

Pormohammad, A., Monych, N. K., Ghosh, S., Turner, D. L., & Turner, R. J. (2021). Nanomaterials in wound healing and infection control. *Antibiotics, 10*(5). https://doi.org/10.3390/antibiotics10050473

Shershakova, N. N., Andreev, S. M., Tomchuk, A. A., Makarova, E. A., Nikonova, A. A., Turetskiy, E. A., Petukhova, O. A., Kamyshnikov, O. Y., Ivankov, O. I., Kyzyma, O. A., Tomchuk, O. V., Avdeev, M. V., Dvornikov, A. S., Kudlay, D. A., & Khaitov, M. R. (2023). Wound healing activity of aqueous dispersion of fullerene C60 produced by "green technology." *Nanomedicine: Nanotechnology, Biology and Medicine, 47*, 102619. https://doi.org/10.1016/j.nano.2022.102619

Singh, A. P., Biswas, A., Shukla, A., & Maiti, P. (2019). Targeted therapy in chronic diseases using nanomaterial-based drug delivery vehicles. *Signal Transduction and Targeted Therapy, 4*(1). https://doi.org/10.1038/s41392-019-0068-3

Singh, M., Thakur, V., Kumar, V., Raj, M., Gupta, S., Devi, N., Upadhyay, S. K., Macho, M., Banerjee, A., Ewe, D., & Saurav, K. (2022). Silver nanoparticles and its mechanistic insight for chronic wound healing: review on recent progress. *Molecules, 27*(17). MDPI. https://doi.org/10.3390/molecules27175587

Singh, R., Sharma, A., Saji, J., Umapathi, A., Kumar, S., & Daima, H. K. (2022). Smart nanomaterials for cancer diagnosis and treatment. In *Nano Convergence* (Vol. 9, Issue 1). Springer Nature Singapore. https://doi.org/10.1186/s40580-022-00313-x

Singh, S., Gupta, A., & Gupta, B. (2018). Scar free healing mediated by the release of aloe vera and manuka honey from dextran bionanocomposite wound dressings. *International Journal of Biological Macromolecules, 120*, 1581–1590. https://doi.org/10.1016/j.ijbiomac.2018.09.124

Soni, S., Singla, R., & S. Padwad, Y. (2023). Nanotechnological interventions and mechanistic insights into wound-healing events. In *Wound Healing—Recent Advances and Future Opportunities*. IntechOpen. https://doi.org/10.5772/intechopen.106481

Stoica, A. E., Chircov, C., & Grumezescu, A. M. (2020a). Nanomaterials for wound dressings: An up-to-date overview. *Molecules, 25*(11), 2699.

Stoica, A. E., Grumezescu, A. M., Hermenean, A. O., Andronescu, E., & Vasile, B. S. (2020b). Scar-free healing: Current concepts and future perspectives. *Nanomaterials, 10*(11), 1–18. MDPI AG. https://doi.org/10.3390/nano10112179

Tavakoli, S., & Klar, A. S. (2020a). Advanced hydrogels as wound dressings. *Biomolecules, 10*(8), 1–20. MDPI AG. https://doi.org/10.3390/biom10081169

Tian, J., Wong, K. K., Ho, C. M., Lok, C. N., Yu, W. Y., Che, C. M., . . . Tam, P. K. (2007). Topical delivery of silver nanoparticles promotes wound healing. *ChemMedChem: Chemistry Enabling Drug Discovery, 2*(1), 129–136.

Vijayan, A., C. K., N., & Vinod Kumar, G. S. (2021a). ECM-mimicking nanofibrous scaffold enriched with dual growth factor carrying nanoparticles for diabetic wound healing. *Nanoscale Advances*, *3*(11), 3085–3092. https://doi.org/10.1039/d0na00926a

Wan, A. C. A., & Ying, J. Y. (2010). Nanomaterials for in situ cell delivery and tissue regeneration☆. *Advanced Drug Delivery Reviews*, *62*(7–8), 731–740. https://doi.org/10.1016/j.addr.2010.02.002

Wang, M., Huang, X., Zheng, H., Tang, Y., Zeng, K., Shao, L., & Li, L. (2021a). Nanomaterials applied in wound healing: Mechanisms, limitations and perspectives. *Journal of Controlled Release*, *337*, 236–247. Elsevier B.V. https://doi.org/10.1016/j.jconrel.2021.07.017

Wang, N., Xu, H., Sun, S., Guo, P., Wang, Y., Qian, C., Zhong, Y., & Yang, D. (2020). Wound therapy via a photo-responsively antibacterial nano-graphene quantum dots conjugate. *Journal of Photochemistry and Photobiology B: Biology*, *210*, 111978. https://doi.org/10.1016/j.jphotobiol.2020.111978

Wang, W., Lu, K. J., Yu, C. H., Huang, Q. L., & Du, Y. Z. (2019). Nano-drug delivery systems in wound treatment and skin regeneration. *Journal of Nanobiotechnology*, *17*(1), 82. https://doi.org/10.1186/s12951-019-0514-y

Xu, S., Chang, L., Hu, Y., Zhao, X., Huang, S., Chen, Z., Ren, X., & Mei, X. (2021). Tea polyphenol modified, photothermal responsive and ROS generative black phosphorus quantum dots as nanoplatforms for promoting MRSA infected wounds healing in diabetic rats. *Journal of Nanobiotechnology*, *19*(1), 362. https://doi.org/10.1186/s12951-021-01106-w

Xue, Y., Yang, F., Wu, L., Xia, D., & Liu, Y. (2024). CeO 2 Nanoparticles to promote wound healing: A systematic review. *Advanced Healthcare Materials*, *13*(6). https://doi.org/10.1002/adhm.202302858

Yayehrad, A. T., Siraj, E. A., Matsabisa, M., & Birhanu, G. (2023). 3D printed drug loaded nanomaterials for wound healing applications. *Regenerative Therapy*, *24*, 361–376. https://doi.org/10.1016/j.reth.2023.08.007

Yu, J., Wang, M. Y., Tai, H. C., & Cheng, N. C. (2018). Cell sheet composed of adipose-derived stem cells demonstrates enhanced skin wound healing with reduced scar formation. *Acta Biomaterialia*, *77*, 191–200. https://doi.org/10.1016/j.actbio.2018.07.022

Zhang, Q., Shi, L., He, H., Liu, X., Huang, Y., Xu, D., Yao, M., Zhang, N., Guo, Y., Lu, Y., Li, H., Zhou, J., Tan, J., Xing, M., & Luo, G. (2022). Down-regulating scar formation by microneedles directly *via* a mechanical communication pathway. *ACS Nano*, *16*(7), 10163–10178. https://doi.org/10.1021/acsnano.1c11016

Zhao, M., Shi, J., Cai, W., Liu, K., Shen, K., Li, Z., Wang, Y., & Hu, D. (2021a). Advances on graphene-based nanomaterials and mesenchymal stem cell-derived exosomes applied in cutaneous wound healing. *International Journal of Nanomedicine*, *16*, 2647–2665. Dove Medical Press Ltd. https://doi.org/10.2147/IJN.S300326

Zhou, Y., Dong, Q., Yang, H., Liu, X., Yin, X., Tao, Y., Bai, Z., & Xu, W. (2017). Photocrosslinked maleilated chitosan/methacrylated poly (vinyl alcohol) bicomponent nanofibrous scaffolds for use as potential wound dressings. *Carbohydrate Polymers*, *168*, 220–226. https://doi.org/10.1016/j.carbpol.2017.03.044

Zielinska, A., Carreiró, F., Oliveira, A. M., Neves, A., Pires, B., Nagasamy Venkatesh, D., Durazzo, A., Lucarini, M., Eder, P., Silva, A. M., Santini, A., & Souto, E. B. (2020). Polymeric nanoparticles: Production, characterization, toxicology and ecotoxicology. *Molecules*, *25*(16). MDPI AG. https://doi.org/10.3390/molecules25163731

9 Toxicity Study of Nanomaterials in Wound Healing Purposes

Muhammad Umair,[1] Asima Saeed,[1] Muhammad Imran Khan,[1] and Shahzeb Khan[2*]*

[1] Department of Biomedical Sciences, Pak Austria Fachhochschule: Institute of Applied Sciences and Technology, Haripur, Khyber Pakhtunkhwa, Pakistan

[2] School of Pharmacy and Medical Sciences, University of Bradford, UK.

* Corresponding authors: Dr. Muhammad Imran Khan, Department of Biomedical Sciences, Pak Austria Fachhochschule: IAST.

Dr. Shahzeb Khan, School of Pharmacy and Medical Sciences, University of Bradford, UK.

9.1 INTRODUCTION TO NANOMATERIALS

Nanomaterials are modern-day materials that promise therapeutic roles in various fields, such as biomedical engineering, by providing solutions related to healthcare challenges. Nanotechnology has been built on the building blocks of nanomaterials. The Greek word "nano" is derived from the word "dwarf," meaning reduction in size or small. There is no specific definition of nanomaterials; usually, any material whose size is in the range of 1-100 nm, at least in one of its dimensions, is normally considered a nanomaterial. The development of nanomaterials in biomedical engineering dates back to the 1980s, when researchers first began exploring the potential of nanoparticles for medical applications. Since then, significant advancements have been made in the synthesis, characterization, and application of nanomaterials in the biomedical field. These materials possess specific properties, such as a high surface-to-volume ratio, small size, and size-dependent reactivity, which help them become ideal candidates for various types of applications related to biomedical technology. Related to one of these aspects, nanomaterials have had a significant impact in the delivery of various drugs to their target sites [1]. These nanocarriers can either be liposomes, nanoparticles, or dendrimers that either encapsulate the drug or are bonded to the drug to transport it to the specific site of action, performing the desired effect. This not only improves the efficacy of the drug but also its effect time. In the case of cancer treatment, these carriers have been used to selectively deliver chemotherapeutic drugs to the tumor site, hence performing their action without affecting healthy tissue [2].

Another example of nanomaterials is quantum dots, which are used as highly sensitive and specific agents for imaging related to various disease detection. This type of nanomaterial possesses fluorescent ability; hence, they emit fluorescent light when triggered by specific stimuli, making them ideal candidates for bioimaging technology. However, this technology is still in the production phase, as these have not yet made it to clinical trials. There are some challenges related to their applications. Issues related to nanomaterials, such as toxicity, regulatory approval, manufacturing costs, and biodegradability, are crucial aspects for the safe and effective use of nanomaterials in

 DOI: 10.1201/9781003605966-9

medicine. Nonetheless, biomaterials have bridged the gap between materials science, engineering, and medicine by providing therapeutic and diagnostic solutions [3, 4].

Wound healing is a complex biological process that involves a coordinated sequence of events to restore damaged tissue integrity. Despite advances in medical technology, challenges such as infection, excessive inflammation, and impaired healing persist, resulting in significant healthcare burdens. In recent years, nanomaterials have emerged as promising candidates for enhancing wound healing due to their unique properties and potential to modulate cellular and molecular processes involved in tissue repair.

9.1.1 Background

Nanomaterials with dimensions in the nanometer range (1-100 nm) exhibit distinct physical, chemical, and biological properties compared to their bulk counterparts, primarily due to their high surface-to-volume ratio and quantum effects. These unique properties have sparked tremendous interest in various fields, including medicine, where nanomaterials have shown potential for drug delivery, imaging, and tissue engineering applications [5].

The field of nanomedicine has made significant strides in exploring the use of nanomaterials for wound healing purposes. Nanomaterials can be engineered to mimic the natural extracellular matrix, providing a favorable environment for cell adhesion, proliferation, and differentiation. Additionally, their high surface area and tunable surface chemistry allow for efficient loading and controlled release of therapeutic agents, such as antimicrobials, growth factors, and biomolecules, directly at the wound site.

Despite their promising potential, the introduction of nanomaterials into biological systems raises concerns regarding their potential toxicity. The unique physicochemical properties that make nanomaterials attractive for biomedical applications may also contribute to their toxicity. Therefore, a comprehensive understanding of the toxicological effects of nanomaterials is crucial for ensuring their safe and effective implementation in wound healing applications (**Figure 9.1**).

9.1.2 Importance of Nanomaterials in Wound Healing

Nanomaterials offer several advantages in wound healing applications, including:

1. **Enhanced cell proliferation and migration:** Nanomaterials can interact with cells at the nanoscale, modulating cellular responses and promoting processes crucial for wound healing, such as cell proliferation and migration.
2. **Improved angiogenesis:** The incorporation of nanomaterials into wound dressings or scaffolds can stimulate the formation of new blood vessels, facilitating nutrient and oxygen delivery to the wound site.
3. **Controlled drug delivery:** Nanomaterials can be engineered to encapsulate and release therapeutic agents in a controlled and sustained manner, ensuring localized and prolonged delivery at the wound site.
4. **Antimicrobial properties:** Certain nanomaterials possess inherent antimicrobial properties or can be functionalized with antimicrobial agents, reducing the risk of wound infection and promoting healing.
5. **Biocompatibility and biodegradability:** Nanomaterials can be designed to be biocompatible and biodegradable, minimizing adverse reactions and facilitating integration with the surrounding tissue.

9.2 UNIQUE PROPERTIES OF NANOMATERIALS

Nanomaterials possess distinctive properties that make them attractive for wound healing applications. These properties arise from their unique size, high surface-to-volume ratio, and

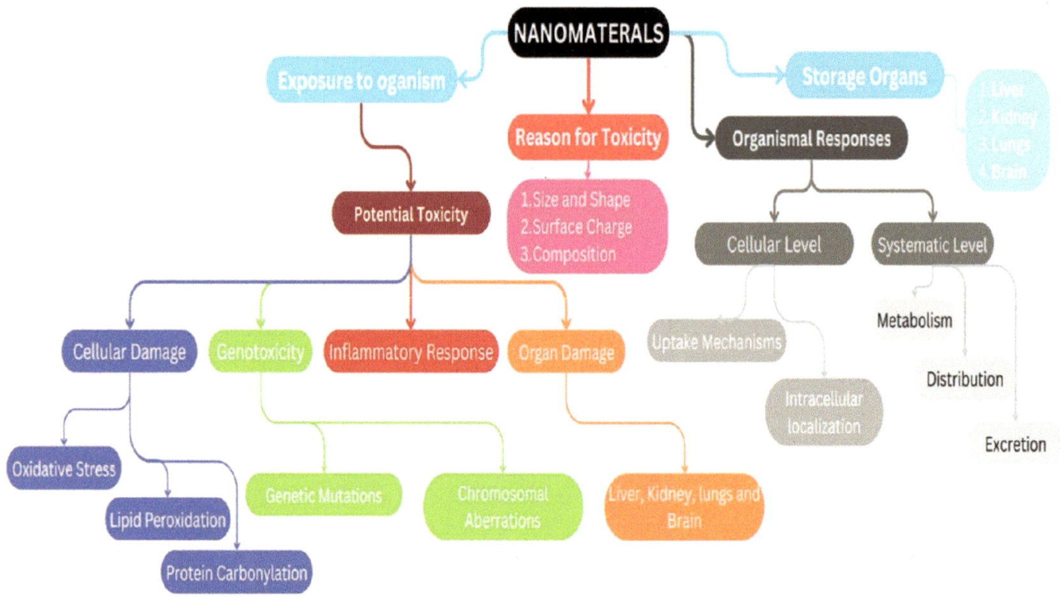

FIGURE 9.1 Pathways of nanomaterial toxicity—an overview. This diagram illustrates the complex interactions between nanomaterials and biological systems, highlighting the potential toxicity based on size, shape, and composition of nanomaterials. It further details the organismal and cellular responses, including genotoxicity, inflammatory response, oxidative stress, lipid peroxidation, and protein carbonylation. The diagram also shows the systemic involvement of storage organs such as the liver, kidney, lungs, and brain in the metabolism, distribution, and excretion of nanomaterials.

quantum effects, which endow them with exceptional characteristics not observed in their bulk counterparts.

9.2.1 Enhanced Cell Proliferation

Cell proliferation is a crucial process in wound healing, as it facilitates the replacement of damaged or lost cells and promotes tissue regeneration. Nanomaterials have demonstrated the ability to enhance cell proliferation through various mechanisms:

a. **Surface topography:** The nanoscale surface features of nanomaterials can influence cell behavior and promote cell adhesion, spreading, and proliferation. Nanostructured surfaces mimic the natural extracellular matrix, providing physical cues that stimulate cellular responses.
b. **Interaction with signaling pathways:** Nanomaterials can interact with specific signaling pathways involved in cell proliferation, such as the MAPK, PI3K/AKT, and Wnt pathways. These interactions can lead to the activation of transcription factors and the upregulation of genes associated with cell cycle progression and proliferation.
c. **Reactive oxygen species (ROS) modulation:** Nanomaterials can generate or scavenge ROS, which play a crucial role in regulating cellular processes, including proliferation. Controlled levels of ROS can trigger signaling cascades that promote cell proliferation, while excessive ROS can induce oxidative stress and impair cell function.
d. **Release of bioactive molecules:** Nanomaterials can be functionalized or loaded with growth factors, cytokines, or other bioactive molecules that stimulate cell proliferation.

The controlled release of these molecules at the wound site can create a favorable microenvironment for cell growth and tissue regeneration.

9.2.2 Migration and Angiogenesis

Cell migration and angiogenesis (the formation of new blood vessels) are essential processes in wound healing, facilitating the recruitment of cells to the wound site and ensuring an adequate supply of nutrients and oxygen for tissue repair.

a. **Cell migration:** Nanomaterials play a crucial role in influencing cell migration through several mechanisms. One significant aspect is surface chemistry; the surface properties of nanomaterials, such as charge and hydrophobicity, can significantly affect cell adhesion and migration. By manipulating these surface characteristics, nanomaterials can enhance or inhibit the attachment and movement of cells, thereby influencing the overall migration process. Additionally, nanomaterials can be engineered to release chemotactic factors, creating chemotactic gradients that guide cells toward the wound site. This targeted release of chemotactic agents ensures that cells move in a directed manner, enhancing the efficiency of the wound healing process. Moreover, the mechanical cues provided by nanostructured surfaces, including their topography and stiffness, can serve as physical signals that direct cell migration. These mechanical properties can mimic the natural extracellular matrix, offering a conducive environment for cells to move and proliferate. By integrating these mechanisms, nanomaterials can effectively modulate cell migration, playing a pivotal role in tissue regeneration and wound healing.
b. **Angiogenesis:** Nanomaterials can promote angiogenesis through several mechanisms. Firstly, they can be loaded with pro-angiogenic factors such as vascular endothelial growth factor (VEGF) or fibroblast growth factor (FGF). These growth factors stimulate the migration and proliferation of endothelial cells, which are crucial for the formation of new blood vessels. Additionally, nanomaterials can modulate signaling pathways involved in angiogenesis, including the VEGF, Notch, and Tie pathways [6]. This interaction enhances the survival, proliferation, and sprouting of endothelial cells. Moreover, nanostructured scaffolds or matrices can mimic the natural extracellular environment, offering physical and biochemical cues that support and enhance angiogenesis. Through these mechanisms, nanomaterials significantly contribute to the development of new vascular networks.

9.2.3 Surface Properties and Functionalization

The surface properties of nanomaterials play a crucial role in their interactions with biological systems and can be tailored for specific wound healing applications through functionalization techniques.

1. **Surface charge:** The surface charge of nanomaterials can influence their interactions with cells, proteins, and biomolecules. Positively charged nanomaterials tend to have higher cellular uptake and can interact with negatively charged cell membranes, while negatively charged nanomaterials may exhibit reduced protein adsorption and improved stability in biological fluids.
2. **Surface chemistry and functionalization:** Surface chemistry and functionalization are critical aspects of nanomaterials, allowing them to be tailored for enhanced biocompatibility, targeted delivery, and therapeutic efficacy. Nanomaterials can be functionalized with various chemical groups, biomolecules, or polymers. One common strategy is the use of polymer coatings; for example, biocompatible polymers like polyethylene glycol (PEG) can be applied to nanomaterials to improve their stability, minimize protein adsorption,

and extend their circulation time in the body. Another approach involves the conjugation of nanomaterials with biomolecules such as antibodies, peptides, or aptamers, facilitating targeted delivery and specific interactions with cells or other biomolecules. Additionally, nanomaterials can be employed as carriers for therapeutic agents, including drugs, growth factors, or genetic materials, which allows for targeted and controlled release directly at the wound site. These functionalization techniques significantly enhance the utility and effectiveness of nanomaterials in biomedical applications, particularly in the context of wound healing.

3. **Surface topography:** The surface topography of nanomaterials, including roughness, patterns, and nanostructures, can influence cell behavior, protein adsorption, and biomolecular interactions. Nanostructured surfaces can mimic the natural extracellular matrix, providing physical cues that promote cell adhesion, proliferation, and differentiation.
4. **Stimuli-responsive properties:** Nanomaterials can be engineered to respond to specific stimuli, such as pH, temperature, or light, allowing for the controlled release of therapeutic agents or modulation of surface properties. For example, pH-responsive nanomaterials can release drugs or growth factors in response to the acidic environment of a wound, while light-responsive nanomaterials can be activated by specific wavelengths of light for targeted therapeutic effects.

9.2.4 Nanomaterial Entry to the Cell

There are different pathways through which nanomaterials can enter and be processed within cells. These pathways are crucial for assessing the potential toxicity, therapeutic effects, and cellular interactions of nanomaterials (see **Figure 9.2**). Here is a comprehensive explanation of each pathway:

1. **Direct Translocation:** This pathway involves the direct transfer of nanomaterials across the cell membrane without the formation of membrane-bound vesicles. This process is typically observed for smaller nanomaterials (<5 nm) or those with specific surface properties that allow them to penetrate the cell membrane directly. Direct translocation can lead to the rapid distribution of nanomaterials within the cytoplasm, potentially interacting with various cellular components and organelles.
2. **Macropinocytosis:** Macropinocytosis is a form of endocytosis in which cells engulf extracellular fluid and suspended particles, including nanomaterials, by forming large vesicles called macropinosomes. This process is often induced by specific signaling pathways or stimuli, such as growth factors or cytokines. Macropinocytosis can facilitate the internalization of relatively large nanomaterials (up to 1 μm) and is commonly observed in certain cell types, such as macrophages and dendritic cells.
3. **Clathrin-mediated endocytosis:** Clathrin-mediated endocytosis is a well-characterized pathway for the internalization of nanomaterials. It involves the formation of clathrin-coated vesicles at the cell membrane, which engulf the nanomaterials and transport them into the cell. This pathway is highly regulated and depends on specific receptor-ligand interactions and cellular signaling events. Once internalized, the clathrin-coated vesicles fuse with early endosomes, where the nanomaterials can be further sorted and trafficked to different cellular compartments.
4. **Caveolae-mediated endocytosis:** Caveolae are small, flask-shaped invaginations of the cell membrane that are involved in the internalization of nanomaterials. This pathway is mediated by caveolin proteins and is particularly relevant for the uptake of nanomaterials with specific surface properties or targeting moieties. Caveolae-mediated endocytosis can lead to the formation of caveosomes, which are distinct from classical endosomes and may have different intracellular trafficking routes.

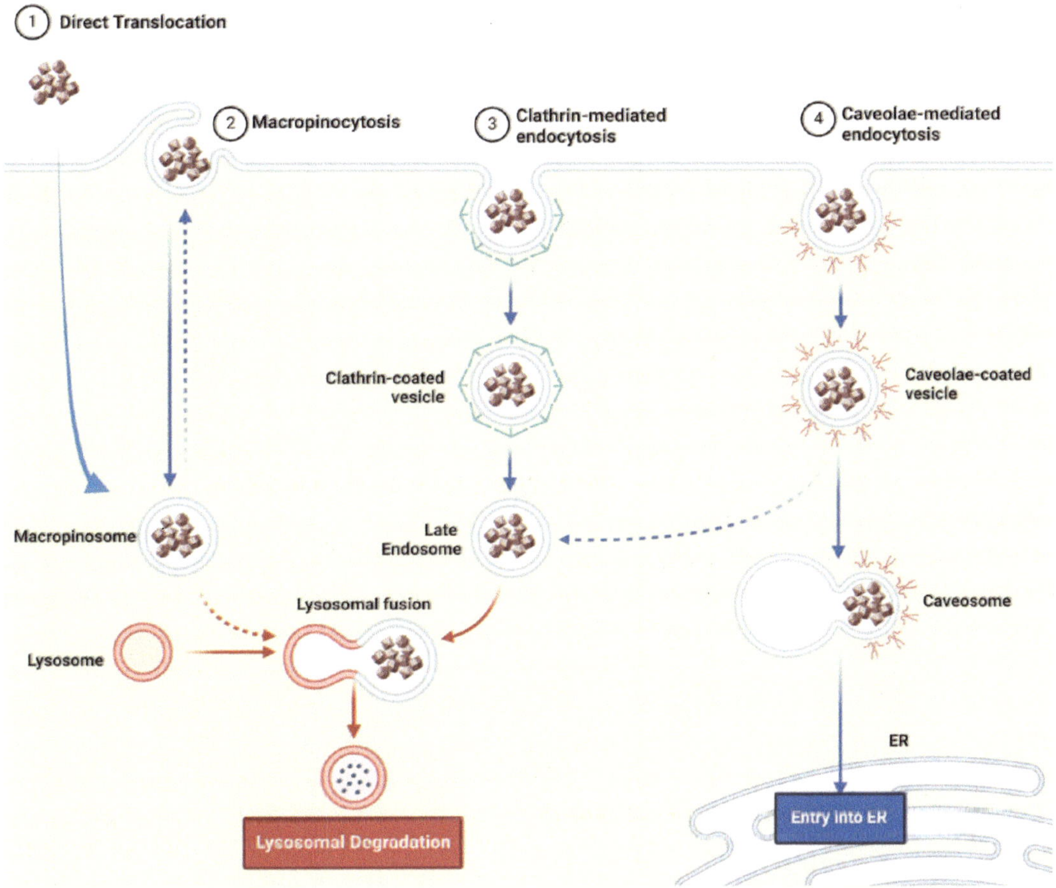

FIGURE 9.2 Comparative Analysis of Cellular Internalization Pathways. This diagram provides a comparative illustration of four different cellular internalization pathways: direct translocation, macropinocytosis, clathrin-mediated endocytosis, and caveolae-mediated endocytosis. Each pathway is depicted with its unique process of substance uptake from the extracellular environment into the cell, leading to various intracellular destinations such as lysosomes or the endoplasmic reticulum (ER).

After internalization through these pathways, nanomaterials can undergo various intracellular processing steps:

Lysosomal fusion and degradation: Nanomaterials internalized via endocytic pathways (micropinocytosis, clathrin-mediated, or caveolae-mediated) are typically trafficked to late endosomes, which eventually fuse with lysosomes. Lysosomes are cellular compartments containing various hydrolytic enzymes and an acidic environment, which can lead to the degradation or dissolution of nanomaterials. This process can release ions, degradation products, or smaller nanoparticles that may interact with cellular components or be excreted from the cell.

Entry into the endoplasmic reticulum (ER): In some cases, nanomaterials internalized through endocytic pathways can escape from the endosomal/lysosomal system and enter the endoplasmic reticulum (ER). The ER is involved in protein synthesis, folding, and trafficking, and exposure to nanomaterials in this compartment can potentially disrupt these processes and induce endoplasmic reticulum stress, which can trigger various cellular responses, including apoptosis or inflammation.

It is important to note that the cellular uptake and processing of nanomaterials can be influenced by various factors, such as the physicochemical properties of the nanomaterials (size, shape, surface charge, and surface modifications), the cell type, and the biological environment. Additionally, different cell types may preferentially utilize specific endocytic pathways, leading to variations in the intracellular fate and potential toxicity of nanomaterials.

9.3 METHODOLOGIES FOR TOXICITY ASSESSMENT

Assessing the toxicity of nanomaterials is a critical step in evaluating their safety and biocompatibility for wound healing applications. Various methodologies have been developed to investigate the potential toxic effects of nanomaterials, including both in vitro and in vivo techniques. In this section, we will focus on in vitro assessment techniques.

9.3.1 In Vitro Assessment Techniques

In vitro assessments provide a controlled environment for evaluating the toxicity of nanomaterials before proceeding to more complex in vivo studies. These techniques involve culturing cells or tissues and exposing them to nanomaterials under controlled conditions. In vitro assessments offer several advantages, including cost-effectiveness, reproducibility, and the ability to study specific mechanisms and cellular responses.

9.3.1.1 Cytotoxicity

Cytotoxicity assessments are essential for evaluating the potential of nanomaterials to cause cellular damage or death. These assessments are a critical first step in toxicity screening, providing valuable insights into the biocompatibility of nanomaterials. Several methods are employed to assess cytotoxicity, each offering specific information about the interactions between nanomaterials and cells.

9.3.1.1.1 Cell Viability Assays

Cell viability assays measure the number of living cells after exposure to nanomaterials. They are widely used to determine the cytotoxic effects of these materials. Common methods include:

MTT Assay (3-(4,5-dimethylthiazol-2-yl)-2,5-diphenyltetrazolium bromide): This colorimetric assay measures the metabolic activity of cells by detecting the reduction of MTT dye to formazan by mitochondrial dehydrogenases in living cells. The amount of formazan produced is proportional to the number of viable cells [7].

WST Assay (Water-Soluble Tetrazolium Salt): Similar to the MTT assay, the WST assay utilizes a water-soluble tetrazolium salt, which is reduced by cellular enzymes to form a water-soluble formazan dye. This method eliminates the need for a solubilization step, making it more convenient [8].

Trypan Blue Exclusion Assay: This dye-based method distinguishes between viable and nonviable cells based on membrane integrity. Viable cells exclude the trypan blue dye, while nonviable cells take it up, allowing for the quantification of cell viability under a microscope [9].

9.3.1.1.2 Membrane Integrity Assays

Membrane integrity assays measure the leakage of intracellular components, such as enzymes or dyes, as an indicator of membrane damage caused by nanomaterials. Examples include:

Lactate Dehydrogenase (LDH) Assay: LDH is an enzyme released from damaged cells. The LDH assay measures the amount of LDH in the culture medium, indicating the extent of cell membrane damage [10].

Neutral Red Uptake Assay: This assay assesses the ability of viable cells to incorporate and bind the neutral red dye within lysosomes. A reduction in dye uptake indicates compromised cell membrane integrity and viability [11].

9.3.1.1.3 Apoptosis and Necrosis Assays

These assays differentiate between programmed cell death (apoptosis) and uncontrolled cell death (necrosis) induced by nanomaterials. Common methods include:

9.3.1.1.4 Annexin V/Propidium Iodide Staining

Annexin V binds to phosphatidylserine, which is externalized on the surface of apoptotic cells, while propidium iodide (PI) penetrates and stains necrotic cells. This dual staining allows for the identification and quantification of apoptotic and necrotic cells using flow cytometry or fluorescence microscopy [12].

9.3.1.1.5 Caspase Activity Assays

Caspases are proteases that play a crucial role in the execution of apoptosis. Caspase activity assays measure the activity of specific caspases (e.g., caspase-3) to determine the level of apoptosis induced by nanomaterials.

9.3.1.1.6 Oxidative Stress Assays

Nanomaterials can generate reactive oxygen species (ROS), leading to oxidative stress and cellular damage. Assays like dichlorofluorescein diacetate (DCFH-DA) and glutathione (GSH) measurements are used to assess oxidative stress levels:

1. **DCFH-DA Assay:** DCFH-DA is a fluorescent probe that is converted to highly fluorescent dichlorofluorescein (DCF) in the presence of ROS. The fluorescence intensity correlates with the amount of ROS produced, providing a measure of oxidative stress [12].
2. **Glutathione (GSH) Measurement:** GSH is a critical antioxidant in cells. The measurement of GSH levels helps assess the cellular redox state and the extent of oxidative stress. A decrease in GSH levels indicates increased oxidative stress and potential cytotoxicity.

By employing these diverse cytotoxicity assays, researchers can comprehensively evaluate the potential adverse effects of nanomaterials on cellular health and viability. This information is crucial for developing safe and effective nanomaterial-based applications in medicine and other fields.

9.3.1.2 Genotoxicity

Genotoxicity assessments are essential for evaluating the potential of nanomaterials to induce genetic damage, which can result in mutations, chromosomal aberrations, or changes in gene expression. Understanding these effects is crucial for determining the long-term safety and potential carcinogenic risks associated with nanomaterials. Several common assays are used to assess genotoxicity, each providing unique insights into different aspects of genetic damage.

1. **Comet assay (single-cell gel electrophoresis):** The comet assay is a sensitive method for detecting DNA strand breaks at the single-cell level. Cells are embedded in agarose gel on a microscope slide, lysed to remove membranes and proteins, and subjected to electrophoresis. Damaged DNA fragments migrate further than intact DNA, forming a comet-like tail when stained and viewed under a fluorescence microscope. The extent of DNA migration indicates the degree of damage [8].
2. **Micronucleus assay:** This assay is used to detect chromosomal damage by identifying micronuclei in cells. Micronuclei are small, extranuclear bodies that form from chromosomal fragments or whole chromosomes that are not incorporated into daughter

nuclei during cell division. Cells are stained and examined microscopically to count the number of micronuclei, providing an indication of chromosomal instability and genotoxic stress [13].

3. **Chromosomal aberration assay:** Chromosomal aberrations are structural changes in chromosomes, such as breaks, gaps, translocations, and aneuploidy. This assay involves culturing cells, arresting them in metaphase, and then staining and examining chromosomes under a microscope. Techniques like metaphase chromosome analysis or fluorescence in situ hybridization (FISH) can identify and quantify these aberrations, offering insights into the structural integrity of chromosomes following nanomaterial exposure [8].
4. **Ames test:** The Ames test is a bacterial reverse mutation assay that evaluates the mutagenic potential of nanomaterials. Specific strains of Salmonella typhimurium or Escherichia coli, which have mutations in genes required for histidine or tryptophan synthesis, are exposed to the test material. If the nanomaterial induces mutations that revert these genes to functional states, the bacteria regain the ability to grow in the absence of the required amino acid, indicating mutagenicity [14].
5. **Gene expression analysis:** Changes in gene expression patterns can reveal the genotoxic effects of nanomaterials. Techniques such as microarrays, RNA-sequencing, and quantitative real-time PCR (qRT-PCR) are employed to measure alterations in the expression levels of genes associated with DNA damage, repair, cell cycle regulation, and apoptosis. By comparing gene expression profiles between treated and control samples, researchers can identify specific genetic pathways affected by nanomaterial exposure [15].

Genotoxicity assessments provide comprehensive insights into the potential genetic risks posed by nanomaterials. These assays are crucial for ensuring the safety of nanomaterials in biomedical applications and for preventing potential long-term health effects such as cancer. As the use of nanomaterials continues to expand, rigorous genotoxicity testing will remain a vital component of their safety evaluation.

9.3.1.3 Immunogenicity

Immunogenicity assessments are essential for evaluating the potential of nanomaterials to trigger an immune response, which could lead to adverse reactions or compromise their therapeutic efficacy. These evaluations are particularly crucial for nanomaterials intended for systemic administration or prolonged exposure. Various assays are employed to assess immunogenicity, including complement activation assays, cytokine release assays, antibody binding assays, lymphocyte proliferation assays, and macrophage uptake and activation assays.

1. **Complement activation assays:** These assays measure the activation of the complement system, an integral part of the innate immune response. The complement system, when activated, leads to the production of complement proteins and the formation of membrane attack complexes. These complexes can disrupt cellular membranes, leading to cell lysis and inflammation. Complement activation assays typically detect the generation of key complement proteins such as C3a, C5a, and the membrane attack complex (C5b-9), providing insight into the potential of nanomaterials to induce an innate immune response [16].
2. **Cytokine release assays:** Cytokine release assays are used to measure the secretion of pro-inflammatory cytokines from immune cells or tissue cultures exposed to nanomaterials. Pro-inflammatory cytokines, such as interleukins (IL-1, IL-6) and tumor necrosis factor-alpha (TNF-α), are critical indicators of an inflammatory response. These assays typically involve exposing immune cells, like peripheral blood mononuclear cells (PBMCs), to nanomaterials and subsequently measuring cytokine levels using techniques such as enzyme-linked immunosorbent assay (ELISA) or multiplex bead-based assays. Elevated levels of these cytokines indicate the immunogenic potential of the nanomaterials [17].

3. **Antibody binding assays:** Antibody binding assays detect the presence of antibodies against nanomaterials in serum or plasma samples, signaling an adaptive immune response. Common techniques for these assays include enzyme-linked immunosorbent assays (ELISAs) and Western blotting. In ELISAs, nanomaterials are immobilized on a surface, and serum samples are added to detect any binding antibodies. Detection is achieved using secondary antibodies conjugated with enzymes that produce a measurable signal. Western blotting involves the separation of proteins by gel electrophoresis, transfer to a membrane, and detection with specific antibodies. The presence of antibodies against nanomaterials indicates sensitization and potential immune memory formation [18].
4. **Lymphocyte proliferation assays:** These assays evaluate the ability of nanomaterials to stimulate lymphocyte proliferation, which is a crucial indicator of an adaptive immune response. Lymphocytes, particularly T cells, play a pivotal role in immune surveillance and response. In lymphocyte proliferation assays, immune cells are exposed to nanomaterials, and cell proliferation is measured using techniques such as [3H]-thymidine incorporation, carboxyfluorescein succinimidyl ester (CFSE) staining, or flow cytometry. An increase in lymphocyte proliferation suggests that the nanomaterial can activate adaptive immune responses, potentially leading to hypersensitivity or other immune-related adverse effects [19].
5. **Macrophage uptake and activation assays:** Macrophages are key players in the innate immune response, responsible for phagocytosing pathogens and presenting antigens. Macrophage uptake and activation assays assess the interaction of nanomaterials with macrophages. Techniques such as flow cytometry, microscopy, and cytokine release measurements are employed to evaluate these interactions. Flow cytometry can quantify the uptake of fluorescently labeled nanomaterials by macrophages, while microscopy provides visual confirmation of phagocytosis. Cytokine release measurements, often performed using ELISA or multiplex assays, assess the activation state of macrophages by measuring the production of pro-inflammatory cytokines. These assays provide valuable information on how nanomaterials may influence innate immune responses and the potential for inflammation and tissue damage [20].

Overall, thorough immunogenicity assessments are crucial for ensuring the safety and efficacy of nanomaterials, particularly those intended for medical applications. By employing a range of assays, researchers can comprehensively evaluate the immune response elicited by nanomaterials, guiding the design of safer and more effective therapeutic agents.

9.3.2 Cytotoxicity, Genotoxicity, and Immunogenicity Studies

The exploration of nanomaterials for biomedical applications, particularly in wound healing, necessitates a comprehensive understanding of their potential cytotoxicity, genotoxicity, and immunogenicity. Recent literature provides valuable insights into the mechanisms and outcomes of these studies, highlighting both the benefits and risks associated with the use of nanomaterials in medical treatments.

9.3.2.1 Cytotoxicity Studies

Recent studies have highlighted the cytotoxic effects of various nanomaterials, such as silver nanoparticles, titanium dioxide nanoparticles, and carbon-based nanomaterials. Silver nanoparticles, for instance, are well-known for their antimicrobial properties, which make them highly suitable for wound healing applications. However, their cytotoxic potential remains a significant concern [21]. A study demonstrated that silver nanoparticles could induce cytotoxicity in fibroblasts and keratinocytes at higher concentrations [22]. The researchers observed that at elevated doses, silver nanoparticles led to increased levels of reactive oxygen species (ROS), causing oxidative stress and subsequent cell damage [23]. These findings suggest the necessity for careful dose optimization to achieve a balance between antimicrobial efficacy and cellular safety. Similarly, titanium dioxide nanoparticles, frequently used in sunscreen and cosmetic products, have been the focus of numerous cytotoxicity studies [24]. Brassolatti P et al. (2022) investigated the effects of titanium dioxide nanoparticles on human dermal fibroblasts and found that these nanoparticles could cause significant oxidative stress and mitochondrial dysfunction [25]. The study highlighted that exposure to titanium dioxide nanoparticles resulted in decreased mitochondrial membrane potential and increased ROS production, leading to apoptosis. These results underscore the critical need for assessing the concentration and exposure duration of nanomaterials to mitigate their cytotoxic effects.

Carbon-based nanomaterials, such as graphene oxide and carbon nanotubes, have also been extensively studied for their cytotoxic potential. Graphene oxide, due to its unique properties, has been explored for various biomedical applications, including drug delivery and biosensing. However, a study by Zhang et al. (2021) revealed that graphene oxide could induce dose-dependent cytotoxicity in human lung epithelial cells [26]. The researchers noted that higher concentrations of graphene oxide led to increased ROS generation, resulting in oxidative stress and cell death. Similarly, carbon nanotubes have been shown to cause cytotoxic effects, primarily due to their ability to penetrate cellular membranes and disrupt normal cellular functions [27].

In addition to in vitro studies, in vivo studies are crucial for understanding the cytotoxicity of nanomaterials in a more complex biological environment. For example, a study by Shehata et al. (2022) evaluated the in vivo toxicity of gold nanoparticles in a mouse model. The researchers found that high doses of gold nanoparticles led to significant liver and kidney damage, as evidenced by elevated levels of liver enzymes and histopathological changes [28]. This study highlights the importance of comprehensive toxicity assessments, including both in vitro and in vivo studies, to ensure the safe application of nanomaterials in medical and cosmetic products.

Recent advancements in nanotoxicology have also focused on developing safer nanomaterials through surface modifications and functionalization. For instance, coating nanoparticles with biocompatible polymers, such as polyethylene glycol (PEG), has been shown to reduce their cytotoxicity. A study by Mitri et al. (2022) demonstrated that PEGylated gold nanoparticles exhibited significantly lower cytotoxicity compared to their non-PEGylated counterparts. The PEG coating reduced protein adsorption and immune recognition, thereby enhancing the biocompatibility of the nanoparticles [28].

Moreover, the development of high-throughput screening methods has enabled more efficient evaluation of nanomaterial cytotoxicity. Techniques such as automated microscopy and flow cytometry allow for rapid and accurate assessment of cellular responses to nanomaterials, facilitating the identification of safer nanomaterial formulations.

9.3.2.2 Genotoxicity Studies

The genotoxic potential of nanomaterials is a critical concern, particularly for applications involving prolonged exposure or systemic administration. Recent advancements in genotoxicity studies have provided valuable insights into the mechanisms and risks associated with nanomaterials.

One prominent study by Amadi et al. (2019) explored the genotoxic effects of graphene oxide nanoparticles. The researchers found that these nanoparticles could cause DNA strand breaks and chromosomal aberrations in human lymphocytes. Their findings highlighted the need for rigorous genotoxicity screening before the clinical use of such materials. The study also emphasized that the size, shape, and surface chemistry of graphene oxide nanoparticles significantly influenced their interaction with cellular DNA. Smaller nanoparticles, with their higher surface area-to-volume ratio, were more likely to induce genotoxic effects due to increased cellular uptake and interaction with intracellular components [29]. Carbon nanotubes have also been extensively studied for their potential genotoxic effects. A researcher in 2024 demonstrated that multiwalled carbon nanotubes (MWCNTs) could induce DNA damage and oxidative stress in lung cells [30]. The study underscored the importance of understanding the physical and chemical properties of nanomaterials, such as size, shape, and surface functionalization, which can influence their genotoxic potential. The findings suggested that the high aspect ratio and durability of MWCNTs contributed to their genotoxicity, potentially leading to chronic inflammatory responses and cellular transformation.

Another recent study focused on titanium dioxide nanoparticles, commonly used in consumer products and medical applications. Researchers found that these nanoparticles could penetrate cell membranes and localize within the nucleus, where they interacted directly with DNA [31]. This interaction led to the formation of DNA adducts and mutations, raising concerns about long-term exposure to titanium dioxide nanoparticles. The study highlighted that surface modifications, such as coating with biocompatible polymers, could mitigate some of these genotoxic effects by reducing direct contact with DNA.

Silver nanoparticles, widely used for their antimicrobial properties, have also been scrutinized for their genotoxic potential. Studies have shown that silver nanoparticles can generate reactive oxygen species (ROS), leading to oxidative stress and DNA damage in exposed cells [23]. The size and surface charge of silver nanoparticles were found to play crucial roles in their genotoxicity, with smaller, positively charged particles causing more significant genetic damage. These findings suggest that careful design and functionalization of silver nanoparticles are necessary to minimize their genotoxic risks.

In addition to the direct genotoxic effects, recent literature has explored the indirect mechanisms through which nanomaterials can cause genetic damage. For instance, exposure to certain nanomaterials has been shown to disrupt cellular homeostasis and interfere with the normal functioning of cellular repair mechanisms [32]. This disruption can lead to an accumulation of genetic damage over time, increasing the risk of mutations and cancer. Studies have also indicated that nanomaterials can affect the epigenetic regulation of gene expression, potentially leading to long-lasting effects on cellular behavior and health.

Factors such as size, shape, surface chemistry, and functionalization significantly influence the genotoxic potential of nanomaterials [33]. Rigorous screening and a comprehensive understanding of these factors are essential to ensure the safe and effective application of nanomaterials in biomedical and consumer products. Ongoing research aims to develop safer nanomaterials through innovative design and functionalization strategies, minimizing their genotoxic risks while harnessing their beneficial properties.

9.3.2.3 Immunogenicity Studies

Understanding how the body's immune system recognizes and responds to nanomaterials is essential to prevent adverse reactions such as inflammation, allergic responses, or immunosuppression. Recent research has provided valuable insights into the immunogenicity of various nanomaterials, particularly those used in wound healing and other medical applications (**Table 9.1**).

A recent study by He et al. (2023) investigated the immunogenicity of gold nanoparticles (AuNPs) [34]. Gold nanoparticles are widely used in biomedical applications due to their unique optical properties, ease of synthesis, and functionalization capabilities. However, their interaction with the immune system is complex and highly dependent on their surface chemistry. He et al. found that AuNPs were generally well-tolerated in vivo, but their surface functionalization significantly influenced their immunogenicity. Specifically, AuNPs functionalized with polyethylene glycol (PEG) showed a reduced immune response compared to unmodified AuNPs. PEGylation, a process where PEG molecules are attached to nanoparticles, helps to shield the nanoparticles from immune recognition, thus enhancing their biocompatibility and prolonging their circulation time in the bloodstream [35]. This study underscores the importance of surface modifications in minimizing the immunogenicity of nanomaterials, making them more suitable for therapeutic applications. Similarly, chitosan-based nanoparticles have been extensively studied for their immunogenic properties. Chitosan, a natural polysaccharide derived from chitin, is known for its excellent biocompatibility and biodegradability, making it an attractive material for drug delivery and wound healing applications. Khalaf et al. (2023) conducted a comprehensive study on the immunogenicity

TABLE 9.1
Various Immunogenicity Assessment Methods for Nanoparticles Used in Wound Healing

Assessment Method	Type	Description	Key Measurements	Advantages	Limitations
Cytokine release assays	*In vitro*	Measures cytokine release from immune cells exposed to nanoparticles	Levels of cytokines (IL-6, TNF-α, IL-1β)	Rapid, detailed information on inflammatory response	Limited to isolated cell responses
Cell viability and proliferation assays	*In vitro*	Assesses the impact of nanoparticles on immune cell viability and proliferation	Cell viability (MTT, XTT, CCK-8), proliferation rates	Easy to perform, high throughput	Does not capture full immune system complexity
Flow cytometry	*In vitro*	Analyzes surface markers and activation status of immune cells exposed to nanoparticles	Expression of surface markers, phagocytic activity	Detailed cellular information	Requires specialized equipment and expertise
Animal models	*In vivo*	Evaluates systemic immune responses to nanoparticles in animal models	Antibody production, immune cell recruitment, cytokine levels	Comprehensive systemic assessment	Ethical concerns, high cost, time-consuming
Histopathological analysis	*In vivo*	Examines tissue samples from treated animals for signs of inflammation and immune cell infiltration	Histological changes, immune cell infiltration	Detailed tissue-level information	Invasive, labor-intensive
Serum analysis	*In vivo*	Analyzes blood samples for antibodies against nanoparticles	Antibody titers (ELISA, SPR)	Quantitative measurement of humoral response	Requires animal models or clinical samples
Molecular dynamics simulations	Computational	Simulates interactions between nanoparticles and biological molecules	Interaction energies, binding affinities	Predictive, noninvasive	Computationally intensive, requires accurate models
QSAR models	Computational	Correlates physicochemical properties of nanoparticles with immunogenic potential	Predicted immunogenicity based on structural features	Predictive, guides design of safer nanoparticles	Requires large datasets for accuracy
High-throughput screening	*In vitro*	Rapidly assesses immunogenicity of large libraries of nanoparticles	Multiple immune parameters simultaneously	Efficient, handles large sample sizes	May miss specific immune interactions
Organ-on-a-chip technology	*In vitro*	Replicates human tissue microarchitecture to study nanoparticle interactions	Immune responses in microfluidic environments	Physiologically relevant, predictive	Complex setup, requires specialized equipment
Omics approaches	*In vitro/in vivo*	Comprehensive analysis of changes in gene expression, protein levels, and metabolic pathways	Transcriptomics, proteomics, metabolomics data	Holistic view of immune response	Data-intensive, requires advanced analysis tools

of chitosan nanoparticles. Their findings indicated that chitosan nanoparticles did not elicit significant immune responses when administered in vivo [36]. This lack of immunogenicity is attributed to the natural origin of chitosan and its ability to degrade into nontoxic, biocompatible byproducts. Additionally, the study highlighted the potential of chitosan nanoparticles as safe carriers for therapeutic agents, capable of delivering drugs and growth factors directly to the wound site without provoking adverse immune reactions.

Another area of interest in immunogenicity studies is the use of carbon-based nanomaterials, such as graphene and its derivatives. These materials have shown promise in various biomedical applications due to their unique mechanical, electrical, and thermal properties. However, their immunogenicity remains a topic of investigation. A study by Yang et al. (2020) explored the immune responses to graphene oxide (GO) nanoparticles [37]. The researchers observed that while GO nanoparticles could induce a mild inflammatory response, functionalization with biocompatible molecules such as amino acids significantly reduced this response. The study concluded that surface modifications are crucial for mitigating the immunogenicity of carbon-based nanomaterials, thereby enhancing their potential for clinical use.

Moreover, the immunogenicity of polymeric nanoparticles has also been a focus of recent research. Poly(lactic-co-glycolic acid) (PLGA) nanoparticles are among the most studied due to their biodegradability and approval by the FDA for various medical applications [38]. A study by Horvath et al. (2023) assessed the immune responses to PLGA nanoparticles with different surface modifications [39]. The results indicated that PLGA nanoparticles coated with poly(vinyl alcohol) (PVA) exhibited lower immunogenicity compared to those with other surface coatings. This finding emphasizes the role of surface chemistry in determining the immunogenic profile of polymeric nanoparticles and highlights the need for careful design in their development for therapeutic purposes.

In conclusion, immunogenicity studies are vital for understanding and mitigating the immune responses elicited by nanomaterials. Recent research has shown that the immunogenicity of nanomaterials is highly dependent on their surface chemistry and functionalization. Modifications such as PEGylation, biocompatible coatings, and the use of natural polymers like chitosan can significantly reduce the immune response, enhancing the safety and efficacy of nanomaterials in biomedical applications. As the field advances, ongoing research will continue to provide insights into the complex interactions between nanomaterials and the immune system, guiding the development of safer and more effective therapeutic nanomaterials. Below is the table for assessing the immunogenicity of various nanoparticles used in wound healing.

9.3.3 Intracellular Pathways and Cellular Responses to Nanomaterials

When nanomaterials enter a cell, they can interact with various cellular components and initiate a cascade of events that may ultimately lead to cellular damage and toxicity (see **Figure 9.3**). The systematic progression of events can be outlined as follows:

1. **Cellular Entry and Localization:** Nanomaterials can gain entry into cells through different pathways, such as direct translocation, endocytosis (including macropinocytosis, clathrin-mediated, and caveolae-mediated endocytosis), or passive diffusion, depending on their size, surface properties, and the cell type (see Figure 9.2). Once internalized, nanomaterials can localize in different cellular compartments, such as endosomes, lysosomes, cytoplasm, or organelles like mitochondria or the endoplasmic reticulum (ER).
2. Initiation of Signaling Cascades: The presence of nanomaterials within cells can trigger various signaling pathways, leading to cellular responses and potential toxicity. These pathways include:

   a. **MAPK Pathway:** Nanomaterials can activate the mitogen-activated protein kinase (MAPK) pathway, which regulates cellular processes such as proliferation,

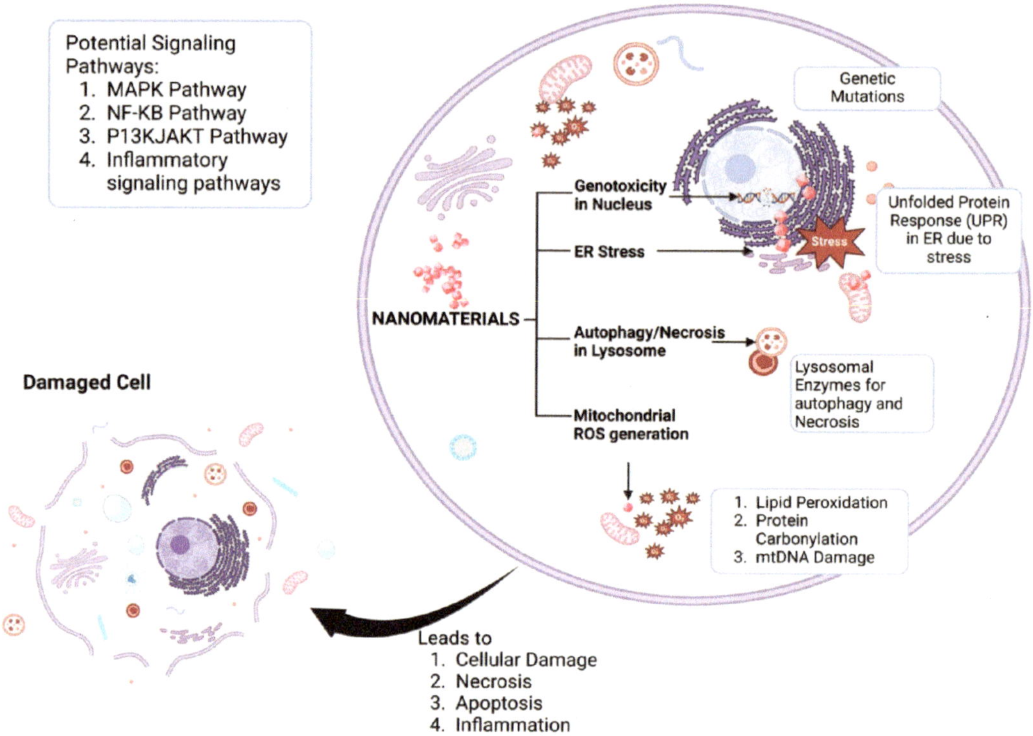

FIGURE 9.3 Intracellular pathways and cellular responses to nanomaterials. This diagram illustrates the potential signaling pathways within a damaged cell due to nanomaterials, including the MAPK Pathway, NF-kB Pathway, and PI3K/AKT Pathway. It also highlights various cellular responses such as necrosis, apoptosis, autophagy/necroptosis, mitochondrial ROS generation, and genetic mutations leading to unfolded protein response (UPR) or ER stress.

differentiation, and stress responses. This activation can lead to the production of inflammatory cytokines, oxidative stress, and cellular damage.

b. **NF-κB Pathway:** Nanomaterials can induce the activation of the nuclear factor kappa B (NF-κB) pathway, a key regulator of immune and inflammatory responses. This pathway leads to the expression of pro-inflammatory genes, cytokine production, and potential tissue damage.

c. **PI3K/AKT Pathway:** The phosphoinositide 3-kinase (PI3K)/protein kinase B (AKT) pathway can be modulated by nanomaterials, resulting in altered cell cycle regulation, apoptosis resistance, or oxidative stress responses.

d. **Inflammatory Signaling Pathways**: Nanomaterials can trigger inflammatory signaling cascades, such as the activation of toll-like receptors (TLRs) or the inflammasome pathway, leading to the production of pro-inflammatory cytokines and the recruitment of immune cells.

3. **Oxidative Stress and Mitochondrial Dysfunction:** Many nanomaterials can induce oxidative stress within cells by generating reactive oxygen species (ROS). This can occur through various mechanisms, such as direct interactions with cellular components or disruption of mitochondrial function. Mitochondria are the primary sites of ROS production, and nanomaterials can disrupt their function, leading to increased ROS generation and oxidative stress.

4. **Cellular Damage and Responses:** The oxidative stress and signaling cascades triggered by nanomaterials can lead to various cellular responses and damage, including:

   a. **Lipid Peroxidation:** ROS can cause the peroxidation of lipids in cell membranes, compromising membrane integrity and cellular function.
   b. **Protein Carbonylation:** Oxidative stress can also lead to the carbonylation of proteins, altering their structure and function and contributing to cellular dysfunction.
   c. **mtDNA Damage:** ROS can damage mitochondrial DNA (mtDNA), leading to mitochondrial dysfunction and further exacerbating cellular stress.
   d. **Necrosis:** Necrosis is an uncontrolled form of cell death characterized by the rupture of the cell membrane and the release of cellular contents, leading to inflammation and tissue damage. Nanomaterials can induce necrosis through various mechanisms, such as direct membrane disruption, oxidative stress, or the depletion of cellular energy reserves.
   e. **Apoptosis:** Apoptosis, or programmed cell death, is a tightly regulated process that plays a crucial role in maintaining cellular homeostasis. Nanomaterials can trigger apoptosis through various pathways, including the induction of oxidative stress, DNA damage, or the activation of specific signaling cascades (e.g., p53, caspase activation).
   f. **Autophagy/Necroptosis:** Autophagy is a cellular process that involves the degradation and recycling of damaged organelles and proteins. Nanomaterials can induce autophagy as a defense mechanism against cellular stress. However, excessive or dysregulated autophagy can lead to cell death, known as autophagic cell death or necroptosis.

      a. Mitochondrial ROS generation: Mitochondria are the primary sites of reactive oxygen species (ROS) production within cells. Nanomaterials can disrupt mitochondrial function, leading to increased ROS generation and oxidative stress, which can contribute to cellular damage, inflammation, and apoptosis.
      b. Genetic mutations: Nanomaterials can induce genotoxicity, resulting in DNA damage and potential genetic mutations. These mutations can lead to altered gene expression, chromosomal aberrations, or the development of cancer.
      c. Unfolded Protein Response (UPR) and ER stress: The endoplasmic reticulum (ER) is responsible for protein folding and quality control. Nanomaterials can disrupt ER function, leading to the accumulation of misfolded or unfolded proteins, triggering the unfolded protein response (UPR) and ER stress. Prolonged ER stress can result in apoptosis or inflammation.

5. **Cellular Consequences and Toxicity:** The cellular responses triggered by nanomaterials can ultimately lead to various forms of cellular damage, including necrosis, apoptosis, and inflammation. These processes can have detrimental effects on tissue function and contribute to the overall toxicity of nanomaterials in wound healing applications.

9.4 CONTEXTUAL RESPONSES TO NANOMATERIAL EXPOSURES DURING WOUND HEALING

Contextual responses to nanomaterial exposures encompass the various biological, physiological, and pathological reactions that occur in living organisms upon contact with nanomaterials. These responses can be influenced by several factors, including the type and composition of the nanomaterial, the route of exposure, and the overall health of the patient. Understanding these responses is crucial for predicting nanomaterial toxicity and ensuring their safe application in wound healing. Nanomaterials interact with a diverse range of cell types and organisms, eliciting varied biological responses influenced by intrinsic factors, extracellular environments, and nanoparticle characteristics. This section explores the nuances of cellular and organismal responses to nanomaterial exposure across different biological systems.

9.4.1 Cellular Responses to Nanomaterial Exposure

9.4.1.1 Cell Type Specificity

Different cell types exhibit distinct sensitivities and responses to nanomaterial exposure due to variations in receptor expression, signaling pathways, and metabolic activities. Immune cells, such as macrophages and dendritic cells, are particularly sensitive to nanomaterials, often mounting inflammatory responses that can exacerbate inflammation and delay wound healing. For example, macrophages exposed to certain nanomaterials may shift towards a pro-inflammatory phenotype, releasing cytokines like TNF-α and IL-1β, which can influence the wound healing milieu [40]. This pro-inflammatory response can be beneficial in some contexts, such as pathogen clearance, but detrimental in chronic wound scenarios where excessive inflammation hinders healing. In contrast, epithelial cells and fibroblasts might primarily undergo cytotoxic or proliferative responses upon nanomaterial exposure. Epithelial cells are crucial for re-epithelialization, while fibroblasts play a significant role in extracellular matrix formation and tissue repair. Understanding these cell-specific responses is essential for designing nanomaterials that promote healing without causing undue harm.

9.4.1.2 Intracellular Localization

Nanomaterials can localize within specific cellular compartments, including endosomes, lysosomes, and mitochondria, influencing their biological fate and toxicity. The intracellular trafficking pathways and nanoparticle dissolution kinetics dictate the duration and intensity of cellular exposure, impacting downstream cellular responses and signaling cascades. For instance, silver nanoparticles localized in lysosomes can generate reactive oxygen species (ROS), leading to oxidative stress and potential cell damage [41]. This oxidative stress can trigger cell death pathways, inflammation, or adaptive responses, depending on the context and extent of exposure. Conversely, nanomaterials designed to release therapeutic agents within specific cellular compartments can enhance targeted treatment efficacy while minimizing off-target effects. For example, nanoparticles engineered to release drugs in the cytoplasm or nucleus can directly interfere with cellular processes involved in disease, providing more effective treatments with fewer side effects.

9.4.1.3 Biological Barriers

Biological barriers, such as the blood-brain barrier (BBB) and the gastrointestinal epithelium, play crucial roles in regulating the entry of nanomaterials into systemic circulation and target tissues. The ability of nanoparticles to traverse these barriers depends on factors like size, surface charge, and surface modifications. For instance, nanoparticles with a hydrophilic surface coating can penetrate the BBB more effectively, facilitating the delivery of therapeutic agents to the brain [42]. This property is particularly valuable in treating neurological conditions where drug delivery to the brain is often challenging. Similarly, the gastrointestinal epithelium can be a barrier to oral nanoparticle delivery, requiring careful design to ensure that therapeutic nanoparticles can be absorbed into the bloodstream and reach their targets. Understanding how nanomaterials interact with these biological barriers is essential for designing nanoparticles that can selectively target wound sites while avoiding unwanted systemic distribution. By optimizing these properties, researchers can develop more effective and safer nanomedicine therapies for wound healing and other medical applications.

9.4.2 Organismal Responses to Nanomaterial Exposure

9.4.2.1 Inflammatory and Immune Responses

Exposure to nanomaterials can elicit a spectrum of inflammatory and immune responses, largely influenced by the physicochemical properties of the nanoparticles. Inflammatory responses typically involve the recruitment of immune cells, the secretion of pro-inflammatory cytokines, and the activation of signaling pathways, which can either facilitate or impede the wound healing process.

For example, a study by Sengul et al. (2023) demonstrated that chronic exposure to certain metal oxide nanoparticles led to sustained inflammatory responses [43]. This prolonged inflammation can hinder healing by maintaining a persistent state of immune activation, thereby disrupting the normal progression of wound healing.

9.4.2.2 Oxidative Stress

Nanomaterials can induce oxidative stress through the generation of reactive oxygen species (ROS), which can damage cellular components, including lipids, proteins, and DNA. Oxidative stress plays a dual role in wound healing: while low levels of ROS are crucial for cell signaling and pathogen defense, excessive ROS can cause cellular damage and delay the healing process. Recent studies have highlighted the necessity of maintaining a balance in ROS levels to optimize wound healing [44]. For instance, Zhang et al. (2021) developed antioxidant-loaded nanoparticles designed to scavenge excessive ROS, thereby creating a more conducive environment for wound healing [45]. These nanoparticles not only reduce oxidative damage but also enhance cell proliferation and tissue regeneration.

9.4.2.3 Systemic Toxicity

Systemic toxicity is a critical concern when evaluating the safety of nanomaterials for wound healing applications. Nanomaterials can enter systemic circulation through various routes, potentially leading to organ-specific toxicity. For example, inhaled nanoparticles can translocate to the liver and spleen, causing hepatic and splenic damage. Similarly, nanoparticles that bypass the gastrointestinal epithelium can accumulate in the kidneys, leading to nephrotoxicity. Recent advances in nanotechnology aim to mitigate these risks by designing biocompatible and biodegradable nanoparticles that minimize systemic exposure and focus their action at the wound site [46].

Recent literature has provided deeper insights into the contextual responses to nanomaterial exposures during wound healing. A study by Sanità et al. (2020) highlighted the role of nanoparticle surface modifications in reducing cytotoxicity and enhancing biocompatibility [47]. By grafting polyethylene glycol (PEG) onto the surface of gold nanoparticles, the researchers were able to reduce protein adsorption and immune recognition, thereby minimizing inflammatory responses and promoting better integration into the wound healing process. Another study by Rao et al. (2020) demonstrated the use of biodegradable polymeric nanoparticles for the controlled release of growth factors in diabetic wound models [48]. These nanoparticles showed improved wound closure rates and enhanced tissue regeneration by providing a sustained release of therapeutic agents, highlighting the potential of nanotechnology in addressing the challenges of chronic wound healing.

9.4.3 Organismal Responses to Nanomaterial Exposure in Wound Healing

9.4.3.1 Species Variability

The response to nanomaterial exposure can vary significantly among different species due to inherent differences in physiology, metabolism, and immune function. Rodent models, such as mice and rats, are extensively used for evaluating the toxicity and efficacy of nanomaterials. However, translating these findings to humans poses significant challenges. Differences in nanoparticle clearance rates, organ distribution patterns, and susceptibility to adverse effects are crucial factors to consider when extrapolating data from animal models to human applications. For instance, the immune system of rodents may react differently to nanoparticles compared to humans, leading to variations in the observed toxicological outcomes. Understanding these species-specific differences is vital for developing safe and effective nanomaterial-based therapies for wound healing in humans. Recent studies have shown that while rodent models provide valuable initial data, the immune response in humans can be more complex, necessitating the use of additional models to better predict human reactions.

9.4.3.2 Tissue Specificity

Following exposure, nanomaterials tend to accumulate in specific tissues and organs, influenced by factors such as blood flow, tissue perfusion, and metabolic activity. The liver, kidneys, and spleen are common target organs for nanomaterial accumulation due to their roles in filtration and metabolism. Each of these organs can exhibit unique responses to nanomaterials, which may include inflammation, fibrosis, or organ dysfunction. For example, the liver, being a primary site for detoxification, is particularly susceptible to nanoparticle-induced oxidative stress and inflammatory responses. Similarly, the kidneys, involved in filtering blood, may experience nephrotoxicity due to nanoparticle accumulation. Studies have indicated that the liver often shows signs of oxidative damage and altered enzyme levels upon exposure to certain nanomaterials, emphasizing the need for targeted safety assessments for each organ. Understanding these organ-specific responses is critical for assessing the safety profile of nanomaterials used in wound healing applications.

9.4.3.3 Developmental Stage

The susceptibility to nanomaterial toxicity can vary significantly depending on the developmental stage of the organism. From embryonic development to adulthood, organisms undergo various phases of tissue growth, organ maturation, and cellular differentiation, all of which can influence their response to nanomaterials. During embryonic development, cells are rapidly dividing and differentiating, making them particularly vulnerable to disruptions caused by nanomaterial exposure. Developmental toxicity studies aim to evaluate the effects of nanomaterials on prenatal and postnatal development, providing crucial insights into potential long-term health effects and transgenerational impacts. For example, exposure to certain nanoparticles during pregnancy has been linked to developmental abnormalities and altered growth patterns in offspring. Research has demonstrated that nanomaterials such as carbon nanotubes and silver nanoparticles can cross the placental barrier, potentially affecting fetal development [49]. These findings underscore the importance of considering developmental stage when assessing the safety and efficacy of nanomaterials in wound healing therapies.

Recent literature has expanded our understanding of the organismal responses to nanomaterial exposure, emphasizing the need for a comprehensive evaluation across different species, tissues, and developmental stages. Advances in nanotoxicology have highlighted the importance of using multiple animal models to capture a broader spectrum of potential human responses. For example, studies involving zebrafish and nonhuman primates are increasingly employed alongside traditional rodent models to provide a more holistic view of nanomaterial toxicity and biocompatibility [50]. Furthermore, cutting-edge imaging techniques and molecular biology tools have enabled more detailed investigations into tissue-specific accumulation and the molecular mechanisms underlying nanomaterial-induced toxicity. Techniques such as single-cell RNA sequencing and high-resolution imaging have provided new insights into how nanomaterials interact with cellular components and disrupt normal physiological processes. These advancements are crucial for designing safer nanomaterials with minimized adverse effects for clinical applications in wound healing [51].

By understanding species variability, tissue specificity, and developmental stage impacts, researchers can better predict human responses and tailor nanomaterials to minimize adverse effects while maximizing therapeutic benefits.

9.5 ASSESSING THE SAFETY OF NANOMATERIALS IN WOUND HEALING

The application of nanomaterials in wound healing is an exciting and rapidly advancing field, offering significant potential for improved therapeutic outcomes. However, this potential comes with the need for rigorous safety assessments to ensure that these materials are safe for use. The unique properties that make nanomaterials advantageous—such as their high surface-to-volume ratio, reactivity, and the ability to be functionalized—can also pose significant risks that must be thoroughly evaluated. Here is a detail of the critical aspects of assessing the safety of nanomaterials in wound healing (**Table 9.2**).

TABLE 9.2
Merits and Demerits of Types of Nanomaterials Used in Wound Healing

Nanomaterial Type	Merits	Demerits
Silver nanoparticles	Potent antibacterial and antifungal properties, reduce inflammation, promote faster healing	Potential cytotoxicity, development of resistance by microorganisms, possible accumulation in tissues
Gold nanoparticles	Antibacterial and anti-inflammatory properties, modulate immune response, promote tissue regeneration	High cost, potential cytotoxicity at high concentrations, possible allergic reactions
Zinc oxide nanoparticles	Antibacterial and anti-inflammatory effects, promote re-epithelialization and collagen synthesis	Potential cytotoxicity, environmental concerns, may cause oxidative stress
Carbon quantum dots	Injectable self-healing hydrogels, antibiofilm properties, cutaneous wound healing capabilities	Limited understanding of long-term effects, potential toxicity at high concentrations
Carbon nanotubes	Enhance cell proliferation and migration, provide structural support, improve mechanical properties of wound dressings	Potential cytotoxicity, difficult to functionalize uniformly, may cause inflammation
Graphene	Antimicrobial and anti-inflammatory properties, high mechanical strength and flexibility	Potential cytotoxicity, challenges in large-scale production, possible environmental impact
Liposomes	Targeted delivery of growth factors, genes, or stem cells, protect encapsulated substances from degradation, enhance stability and efficacy	Potential immunogenicity, stability issues, challenges in large-scale production
Nanoemulsions	Create novel wound dressings with antimicrobial and anti-inflammatory properties, enhance penetration of therapeutic agents	Potential instability, possible skin irritation, challenges in formulation
Nanofibers	Mimic extracellular matrix provides scaffolds for cell growth, promotes a moist environment, facilitates tissue regeneration	Potential cytotoxicity, challenges in uniform drug loading, possible degradation in the body
Nano-dressings	Provide a moist environment, protect against infections, promote tissue regeneration, functionalized with therapeutic agents	Potential for skin irritation, high cost, challenges in maintaining moisture balance

9.5.1 Biocompatibility and Biodegradability

Ensuring the biocompatibility of nanomaterials is paramount for their application in wound healing. Biocompatibility refers to the ability of a material to perform its desired function without causing undesirable local or systemic effects in the host organism. Nanomaterials utilized in wound dressings, scaffolds, or drug delivery systems must not provoke adverse immune responses, inflammation, or toxicity. Recent studies have highlighted the importance of surface modification techniques to enhance the biocompatibility of nanomaterials. For instance, nanoparticles coated with biocompatible polymers or proteins, such as polyethylene glycol (PEG) or albumin, exhibit reduced immunogenicity and improved interaction with biological tissues, leading to better integration and functionality in medical applications.

Biodegradability is another critical factor in the safe application of nanomaterials. Biodegradable nanomaterials can be broken down into nontoxic byproducts that are easily eliminated from the body, preventing the accumulation of these materials in tissues, which could lead to long-term toxicity. Advances in the development of biodegradable nanomaterials, particularly those based on natural polymers like chitosan and collagen, have shown great promise in wound healing applications. These materials provide temporary structural support to the wound site, gradually degrading as the tissue heals, thereby reducing the risk of chronic inflammation or toxicity.

9.5.1.1 Toxicological Evaluation

A comprehensive toxicological evaluation of nanomaterials is essential to understand their potential impacts on human health. This evaluation typically involves both in vitro and in vivo studies to assess various aspects of toxicity, including cytotoxicity, genotoxicity, and systemic toxicity. In vitro studies using cell cultures can provide initial insights into the cytotoxic effects of nanomaterials. These studies can identify potential mechanisms of toxicity, such as oxidative stress, apoptosis, or necrosis. For example, recent research has shown that silver nanoparticles, widely used for their antimicrobial properties, can induce oxidative stress in human keratinocytes at high concentrations, highlighting the importance of dose optimization to mitigate adverse effects. In vivo studies in animal models are crucial for assessing the systemic effects of nanomaterials. These studies can provide valuable information on the distribution, metabolism, and excretion of nanomaterials, as well as their potential for organ-specific toxicity. For instance, a study on the systemic toxicity of zinc oxide nanoparticles in mice revealed that chronic exposure could lead to liver and kidney damage, underscoring the need for long-term safety assessments. These findings emphasize the importance of evaluating not just the immediate effects of nanomaterials, but also their long-term safety and potential for bioaccumulation.

9.5.1.2 Regulatory Guidelines and Standards

Adherence to regulatory guidelines and standards is essential for the safe application of nanomaterials in wound healing. Regulatory bodies such as the U.S. Food and Drug Administration (FDA) and the European Medicines Agency (EMA) have established comprehensive guidelines for evaluating nanomaterials in medical products. These guidelines emphasize the need for extensive preclinical and clinical testing to assess both safety and efficacy. For example, the FDA's guidance on the use of nanotechnology in drug products outlines specific considerations for the characterization, manufacturing, and quality control of nanomaterials. This includes detailed requirements for evaluating the physical and chemical properties of nanomaterials, their stability, and their interactions with biological systems. The EMA similarly provides guidelines that focus on the safety assessment of nanomaterials, including considerations for their biocompatibility, biodegradability, and potential toxicological effects. The importance of harmonizing regulatory frameworks for nanomaterials across different jurisdictions is of high significance. Standardized testing protocols and international collaboration are crucial for ensuring consistent and reliable assessments of nanomaterial safety. The development of comprehensive databases and registries for nanomaterials can facilitate the tracking of their use and potential health impacts, providing valuable data for regulatory decision-making. These efforts are essential for building public trust in the safety of nanomaterial-based medical products and ensuring their responsible and effective use in wound healing and other medical applications.

While the application of nanomaterials in wound healing holds great promise, it is accompanied by significant safety considerations. Ensuring biocompatibility and biodegradability, conducting comprehensive toxicological evaluations, and adhering to stringent regulatory guidelines are all critical steps in the safe development and application of these innovative materials. Continued research and international collaboration will be key to unlocking the full potential of nanomaterials in medical applications, ensuring they are both safe and effective for patient use.

9.6 TYPES OF NANOMATERIALS FOR WOUND HEALING

Materials such as copper oxide, silver oxide, gold, and zinc exhibit strong antibacterial properties, making them valuable in preventing infections and promoting recovery. By integrating these materials with nanotechnology, their effectiveness can be further amplified, leading to improved wound care outcomes [52]. The self-assembling and stimuli-specific nanofibers, as well as nanocarriers, have also shown promise in wound healing by enabling the specific delivery of drugs to the site of action, as well as imparting antibacterial properties. These materials can also export the drug, hence protecting it from various degradative materials, leading to efficient and cost-effective wound healing [53].

9.6.1 Metal Nanoparticles

Metal nanoparticles are among the most widely studied and utilized nanomaterials for wound healing. They possess unique properties that make them effective in combating infections and promoting tissue repair.

9.6.1.1 Silver Nanoparticles

Silver nanoparticles are renowned for their potent antibacterial and antifungal properties. They are extensively used in wound dressings to prevent infections and accelerate the healing process. Silver nanoparticles release silver ions, which interact with bacterial cell membranes, causing structural damage and inhibiting cellular functions. This action not only prevents infection but also reduces inflammation, thereby promoting faster healing.

9.6.1.2 Gold Nanoparticles

Gold nanoparticles have shown promise in wound healing due to their antibacterial and anti-inflammatory properties. These nanoparticles can modulate the immune response, reducing inflammation and promoting tissue regeneration. Additionally, gold nanoparticles can be functionalized with various therapeutic agents, enhancing their efficacy in wound healing applications.

9.6.1.3 Zinc Oxide Nanoparticles

Zinc oxide nanoparticles exhibit both antibacterial and anti-inflammatory effects, making them suitable for wound healing. They can promote re-epithelialization and collagen synthesis, essential steps in the wound healing process. Zinc oxide nanoparticles also protect the wound from UV radiation and other environmental factors, further aiding in recovery.

9.6.2 Carbon-Based Nanomaterials

Carbon-based nanomaterials have unique properties that make them effective in promoting wound healing. These materials are versatile and can be engineered to meet specific healing requirements.

a. **Carbon Quantum Dots:** Carbon quantum dots are used to create injectable self-healing hydrogels with antibiofilm properties and cutaneous wound healing capabilities. These hydrogels can provide a moist environment, essential for wound healing, and prevent bacterial colonization.
b. **Carbon Nanotubes:** Carbon nanotubes, when complexed with chitosan, enhance wound healing by promoting cell proliferation and migration. They provide structural support and improve the mechanical properties of wound dressings, facilitating better tissue regeneration.
c. **Graphene:** Graphene has garnered interest due to its antimicrobial and anti-inflammatory properties. It can be incorporated into wound dressings to provide a barrier against infections while promoting tissue regeneration. Graphene-based materials also have high mechanical strength and flexibility, making them suitable for various wound healing applications.

9.6.3 Liposomes

Liposomes are spherical vesicles that can encapsulate growth factors, genes, or stem cells, offering new treatment options for wound healing. They provide targeted delivery of therapeutic agents, ensuring that they reach the wound site in optimal concentrations. Liposomes protect encapsulated substances from degradation, improving their stability and efficacy. This targeted approach enhances the healing process by delivering essential molecules directly to the site of injury.

9.6.4 Nanoemulsions

Nanoemulsions are fine oil-in-water or water-in-oil dispersions stabilized by surfactants. They are used to create novel wound dressings with antimicrobial and anti-inflammatory properties. Nanoemulsions can enhance the penetration of therapeutic agents into the wound site, providing sustained release and prolonged action. This results in improved infection control and accelerated healing.

9.6.5 Nanofibers

Nanofibers are a game changer in wound healing due to their ability to mimic the extracellular matrix, providing a scaffold for cell growth and proliferation. Electrospun nanofibers can be engineered to have specific mechanical and biological properties, making them ideal for creating wound dressings that promote a moist environment and facilitate tissue regeneration. Nanofibers can also be loaded with therapeutic agents, such as antibiotics and growth factors, for sustained release and enhanced healing.

9.6.6 Nano-Dressings

Nano-dressings are modern and effective wound care solutions that include hydrogels, nanofibers, and films. These materials are extensively used as novel nanomaterial-based bandages. Nano-dressings can provide a moist environment, protect against infections, and promote tissue regeneration. Hydrogels, for example, can absorb exudate while maintaining a moist environment, essential for healing. Nanofiber-based dressings offer structural support and can be functionalized with various therapeutic agents to enhance healing.

9.7 ROLE OF NANOMATERIALS IN TISSUE REGENERATION

Tissue regeneration involves the replacement or repair of damaged tissues by stimulating the body's natural healing processes. This process is critical in wound healing, as it restores the structural and functional integrity of injured tissues. Recent advancements in nanotechnology have provided innovative approaches to enhance tissue regeneration, leveraging the unique properties of nanomaterials to improve healing outcomes [54].

9.7.1 Nanoparticle-Based Therapeutics

Recent advancements in nanoparticle-based therapeutics have shown remarkable potential in enhancing wound healing processes, particularly through the delivery of angiogenic growth factors. Nanoparticles functionalized with factors such as vascular endothelial growth factor (VEGF), fibroblast growth factor (FGF), and endothelial cell proliferation factors have demonstrated significant efficacy in promoting cell migration and proliferation. These nanoparticles can be engineered to deliver these growth factors directly to the wound site, facilitating localized angiogenesis and tissue regeneration. The encapsulation of growth factors within nanoparticles not only enhances their stability but also ensures a controlled and sustained release, thereby optimizing their therapeutic effects. Various studies have provided compelling evidence that nanoparticle-based delivery systems can significantly accelerate wound healing [55]. For example, research published demonstrated that VEGF-loaded nanoparticles significantly improved angiogenesis in a diabetic wound model, leading to faster wound closure and better tissue regeneration [56]. This was attributed to the sustained release of VEGF, which maintained its bioactivity over an extended period and promoted the formation of new blood vessels, crucial for delivering nutrients and oxygen to regenerating tissues.

Another study showcased the use of FGF-loaded nanoparticles in enhancing wound healing and tissue regeneration [57]. The study found that these nanoparticles effectively stimulated endothelial cell proliferation and migration, key processes in angiogenesis. By ensuring a steady supply of FGF at the wound site, the nanoparticles facilitated the formation of a robust vascular network, thereby supporting tissue regeneration and repair.

9.7.2 Nanofiber Scaffolds

Nanofiber scaffolds have emerged as a highly effective approach for wound healing due to their ability to closely mimic the native extracellular matrix (ECM), providing a three-dimensional structural framework that supports cellular functions. These scaffolds are typically created using electrospinning techniques, which produce nanofibrous matrices conducive to cell adhesion, migration, and proliferation. The incorporation of nanoparticles within these nanofiber scaffolds further enhances their therapeutic potential by promoting angiogenesis, a critical factor in tissue regeneration. Recent research has focused on the integration of angiogenic nanoparticles within nanofiber scaffolds to create a synergistic effect that enhances wound healing [58]. For instance, a study published demonstrated that nanofiber scaffolds embedded with VEGF-loaded nanoparticles significantly improved the healing of chronic wounds [56]. The study found that the scaffolds provided a conducive environment for cell growth while the nanoparticles facilitated sustained VEGF release, resulting in enhanced angiogenesis and faster tissue regeneration.

Another innovative approach discussed in Tissue Engineering Part A involved the use of nanofiber scaffolds loaded with both VEGF and FGF nanoparticles. This dual delivery system was shown to effectively support the formation of a well-organized vascular network within the engineered tissues. The presence of both growth factors promoted endothelial cell proliferation and migration, leading to the development of functional blood vessels that were essential for nutrient delivery and waste removal in the regenerating tissue. Furthermore, nanofiber scaffolds embedded with nanoparticles have been shown to significantly improve cellular activities, as evidenced by studies focusing on chronic wound healing [59]. These scaffolds provide not only structural support but also biochemical cues that stimulate cellular processes essential for tissue repair. For example, a study reported that nanofiber scaffolds with embedded nanoparticles facilitated the differentiation of stem cells into endothelial cells, further enhancing the angiogenic response and promoting the regeneration of functional tissues [60].

9.7.3 Nanocomposite Hydrogels

Nanocomposite hydrogels represent a novel approach in wound healing, incorporating biodegradable polymers that carry nanoparticles. These hydrogels are designed to deliver therapeutic agents in a controlled manner, responding to specific stimuli such as changes in pH, temperature, or mechanical stress. The nanoparticles within the hydrogels can be loaded with drugs, growth factors, or other bioactive molecules, which are released upon encountering the appropriate stimuli. This targeted and responsive release mechanism ensures that therapeutic agents are delivered directly to the wound site at the right time, maximizing their effectiveness while minimizing potential side effects. Nanocomposite hydrogels have shown great potential in enhancing wound healing, particularly in the treatment of chronic wounds that are difficult to heal with conventional therapies (**Table 9.3**).

TABLE 9.3
Nanomaterial Toxicity in Wound Healing

Nanomaterial	Size	Shape	Surface Chemistry	Dose-Dependent Toxicity	Wound Healing Application
Silver nanoparticles	10–100 nm	Spherical, rod	Positively charged, functionalized	Low toxicity at 10 µg/mL; significant toxicity at >50 µg/mL	Antibacterial wound dressings, promotes re-epithelialization
Zinc Oxide Nanoparticles	20–80 nm	Spherical	Functionalized, Neutral Charge	Effective at low doses; higher doses can induce cytotoxicity	Enhances collagen deposition and re-epithelialization
Gold Nanoparticles	10–50 nm	Spherical, Rod	Functionalized with PEG, Negatively Charged	Low doses promote angiogenesis; high doses may cause oxidative stress	Promotes angiogenesis, reduces inflammation
Titanium dioxide nanoparticles	30–100 nm	Spherical	Surface modified for biocompatibility	Low toxicity at therapeutic doses; potential toxicity at high concentrations	Antimicrobial, supports tissue regeneration
Silicon nanoparticles	5–50 nm	Porous, spherical	Functionalized for drug delivery	Dose-dependent cytotoxicity observed; optimal at lower doses	Drug delivery systems for wound healing
Copper oxide nanoparticles	15–40 nm	Spherical	Positively charged, antimicrobial	Effective antimicrobial at low doses; high doses toxic to cells	Antimicrobial coatings for wound dressings
Iron oxide nanoparticles	10–100 nm	Spherical	Surface coated with biocompatible polymers	Generally low toxicity; dependent on surface coating and dose	Used in magnetic field-assisted wound healing
Chitosan nanoparticles	20–60 nm	Spherical	Positively charged, biodegradable	Low toxicity; high biocompatibility	Promotes wound healing and antimicrobial activity
Cerium oxide nanoparticles	5–30 nm	Spherical	Surface functionalized with PEG	Antioxidant properties at low doses; toxicity at high doses	Reduces oxidative stress, promotes healing in chronic wounds

9.8 INFLUENCE OF NANOMATERIAL CHARACTERISTICS

The unique properties of nanomaterials, which make them attractive for wound healing applications, are heavily influenced by their physicochemical characteristics. These characteristics play a crucial role in determining the interactions of nanomaterials with biological systems, their biodistribution, cellular uptake, and ultimately, their toxicity profiles. Understanding the influence of these characteristics is essential for designing safe and effective nanomaterials for wound healing purposes.

9.8.1 Size and Shape

Size and shape are among the most fundamental characteristics that govern the behavior and toxicity of nanomaterials. These factors can significantly influence cellular uptake, biodistribution, and interactions with biomolecules.

Size:

a. **Cellular uptake:** Smaller nanoparticles (generally <100 nm) tend to have higher cellular uptake compared to larger particles, as they can more easily cross cell membranes and enter cells through endocytic pathways.

b. **Biodistribution:** The size of nanoparticles can affect their biodistribution and clearance from the body. Smaller particles may be more readily distributed throughout the body but may also be cleared more rapidly by the renal or hepatobiliary systems.
c. **Reactive surface area:** Smaller nanoparticles have a higher surface area-to-volume ratio, which can lead to increased reactivity and potential for generating reactive oxygen species (ROS), causing oxidative stress and cellular damage.

Shape:

a. **Cellular interactions:** The shape of nanomaterials can influence their interactions with cells and biomolecules. For example, rod-shaped or fiber-like nanomaterials may have different cellular uptake and trafficking mechanisms compared to spherical nanoparticles.
b. **Biokinetics:** The shape of nanoparticles can affect their biodistribution, clearance, and accumulation in specific organs or tissues. Elongated or fiber-like nanoparticles may have longer circulation times and higher accumulation in certain organs, such as the lungs or liver.
c. **Toxicity:** The shape of nanomaterials can influence their potential for inducing toxicity. For instance, fiber-like or needle-shaped nanomaterials have been associated with increased cytotoxicity and inflammation, potentially due to their ability to pierce cell membranes or induce frustrated phagocytosis.

9.8.2 Surface Charge

The surface charge of nanomaterials plays a crucial role in their interactions with biological systems, including cellular uptake, protein adsorption, and stability in physiological environments.

a. **Cellular uptake:** Positively charged nanoparticles tend to have higher cellular uptake compared to neutral or negatively charged particles, as they can interact more readily with the negatively charged cell membranes through electrostatic interactions.
b. **Protein adsorption:** The surface charge of nanomaterials can influence the adsorption of proteins, forming a "protein corona" that can alter the particle's surface properties and interactions with cells and biomolecules.
c. **Stability and aggregation:** Surface charge can affect the stability and aggregation behavior of nanomaterials in physiological environments. Particles with similar surface charges tend to repel each other, reducing aggregation, while those with neutral or opposite charges may experience increased aggregation.
d. **Inflammatory responses:** Positively charged nanoparticles have been associated with higher inflammatory responses, potentially due to their increased cellular uptake and interactions with negatively charged cell membranes or proteins involved in inflammatory pathways.

9.8.3 Composition and Coatings

The chemical composition and surface coatings of nanomaterials can significantly influence their toxicity profiles, biodistribution, and biocompatibility.

9.8.3.1 Composition

a. **Inorganic nanomaterials** (e.g., metals, metal oxides, and carbon-based materials) may exhibit varying degrees of toxicity depending on their chemical composition, reactivity, and potential for releasing ions or generating ROS.
b. **Organic nanomaterials** (e.g., polymers, liposomes, and dendrimers) may be more biocompatible but can still induce toxicity depending on their chemical structure, degradation products, and interactions with biological systems.

9.8.3.2 Surface Coatings and Functionalization

a. **Biocompatible coatings**, such as polyethylene glycol (PEG), can improve the stability and circulation time of nanomaterials, reduce protein adsorption, and minimize nonspecific interactions with cells and biomolecules.
b. **Functional groups or biomolecules** (e.g., antibodies, peptides, or aptamers) can be conjugated to the surface of nanomaterials to enhance their targeting specificity, cellular uptake, or therapeutic efficacy.
c. **Surface modifications** can also influence the biodegradability and clearance of nanomaterials, which are crucial for minimizing long-term accumulation and potential toxicity.

9.8.4 Aggregation

The aggregation behavior and solubility of nanomaterials in physiological environments can significantly impact their toxicity profiles and biological interactions. Aggregation can alter the effective size, shape, and surface properties of nanomaterials, potentially leading to changes in cellular uptake and biodistribution.

a. **Highly aggregated nanomaterials** may have reduced bioavailability and cellular interactions, potentially reducing their therapeutic efficacy but also mitigating potential toxicity.
b. **Disaggregation of nanomaterials** in biological environments can lead to the release of individual particles or ions, potentially increasing their reactivity and toxicity.

9.8.5 Solubility

a. **Soluble nanomaterials** may have higher bioavailability and potential for systemic distribution, increasing the risk of off-target effects and toxicity.
b. **Insoluble nanomaterials** may be less bioavailable but can accumulate in specific organs or tissues, potentially leading to localized toxicity or clearance issues.
c. **The solubility of nanomaterials** can be influenced by factors such as pH, ionic strength, and the presence of biomolecules in physiological environments, which can affect their behavior and toxicity profiles.

9.9 COMPOSITION, BIOCOMPATIBILITY, AND TOXICITY OF NANOMATERIALS IN WOUND HEALING

The composition and biocompatibility of nanomaterials are fundamental considerations in their application for wound healing. This section explores the significance of nanomaterial composition and its impact on biocompatibility, along with relevant references to support the toxicity of nanomaterials. Nanomaterials used in wound healing applications encompass a diverse range of materials, including metals, metal oxides, polymers, lipids, and carbon-based materials. Each material possesses unique physicochemical properties that influence its interaction with biological systems and its suitability for wound healing. The composition and biocompatibility of nanomaterials are crucial factors in their application for wound healing. Nanomaterials used in wound healing encompass a diverse range of materials, each with unique physicochemical properties that influence their interaction with biological systems and their effectiveness in promoting wound healing. The primary categories of nanomaterials include metals, metal oxides, polymers, lipids, and carbon-based materials.

9.10 METAL-BASED NANOMATERIALS

Silver nanoparticles (AgNPs) are among the most widely studied nanomaterials for wound healing due to their potent antimicrobial properties. AgNPs can effectively reduce bacterial load and inflammation, which are critical factors in preventing wound infections and promoting healing. For instance, studies have demonstrated that silver nanoparticles facilitate wound healing by reducing microbial colonization and minimizing inflammatory responses. However, the potential cytotoxicity of AgNPs poses a challenge. Research indicates that while AgNPs are effective at antimicrobial action, their biocompatibility must be carefully balanced to avoid adverse effects on human cells. Therefore, optimizing the concentration and exposure time of AgNPs is essential to harness their benefits while minimizing potential risks.

9.10.0.1 Toxicity of AgNPs in Wound Healing

The toxicity of AgNPs is a critical consideration in their application for wound healing. While AgNPs are known for their antimicrobial properties, their interaction with human cells can lead to cytotoxic effects. Research indicates that high concentrations of AgNPs can induce oxidative stress, DNA damage, and apoptosis in human cells. These effects are primarily attributed to the release of silver ions, which can interact with cellular components such as proteins, lipids, and nucleic acids. Oxidative stress is a major concern, as it can lead to cellular damage and hinder the wound healing process. AgNPs can generate reactive oxygen species (ROS) that can damage cellular membranes, proteins, and DNA. This oxidative damage can impair cell proliferation and migration, essential processes in wound healing. Therefore, it is crucial to carefully regulate the dosage of AgNPs to avoid excessive ROS production and oxidative stress.

Additionally, the size and shape of AgNPs play a significant role in their toxicity. Smaller nanoparticles have a larger surface area-to-volume ratio, which can increase their reactivity and potential for toxicity. Studies have shown that smaller AgNPs are more likely to penetrate cells and tissues, leading to higher levels of cytotoxicity. Therefore, selecting the appropriate size and shape of AgNPs is essential to minimize their toxic effects while maintaining their antimicrobial efficacy. The route of administration and exposure duration also influence the toxicity of AgNPs. Prolonged exposure to AgNPs can lead to the accumulation of silver ions in tissues, potentially causing long-term adverse effects. To mitigate these risks, it is important to optimize the duration of AgNP application and consider using coatings or carriers that can control the release of silver ions, reducing the likelihood of toxicity. Recent literature highlights various strategies to mitigate the toxicity of AgNPs while maintaining their therapeutic benefits. One approach involves the use of biocompatible coatings, such as polyethylene glycol (PEG) or chitosan, which can reduce the release of silver ions and improve the stability of AgNPs. These coatings can enhance the biocompatibility of AgNPs, reducing their cytotoxic effects and promoting safer use in wound healing. Another strategy is the development of hybrid nanomaterials that combine AgNPs with other therapeutic agents. For example, combining AgNPs with anti-inflammatory drugs or growth factors can enhance their therapeutic efficacy while minimizing their cytotoxicity. These hybrid materials can provide a synergistic effect, promoting wound healing through multiple mechanisms.

9.10.1 Metal Oxide Nanoparticles

Zinc oxide nanoparticles (ZnO NPs) have garnered attention for their multifaceted role in wound healing. ZnO NPs exhibit antibacterial and anti-inflammatory properties, which are beneficial for wound management. Additionally, they promote wound closure and tissue regeneration, making them valuable in accelerating the healing process. Studies indicate that ZnO NPs exhibit low cytotoxicity toward mammalian cells at appropriate concentrations, which enhances their suitability for wound healing applications. By promoting cell proliferation and differentiation, ZnO NPs aid in the restoration of tissue integrity and function.

9.10.1.1 Toxicity of Zinc Oxide Nanoparticles in Wound Healing

Despite the promising benefits of ZnO NPs in wound healing, it is crucial to address their potential toxicity to ensure their safe application. The cytotoxic effects of ZnO NPs are dose-dependent, and excessive concentrations can lead to adverse effects such as oxidative stress, inflammation, and cell death. Recent literature has highlighted that while ZnO NPs are generally safe at low concentrations, higher doses can induce the production of reactive oxygen species (ROS) in cells. These ROS can cause oxidative damage to cellular components, including lipids, proteins, and DNA, leading to cell apoptosis or necrosis. Therefore, it is essential to optimize the concentration of ZnO NPs in wound healing formulations to maximize their therapeutic benefits while minimizing potential risks. Another concern is the long-term exposure to ZnO NPs, which can lead to the accumulation of nanoparticles in tissues, potentially causing chronic inflammation and toxicity. Studies have shown that repeated application of ZnO NPs can result in their penetration through the skin barrier, especially in cases where skin integrity is compromised due to wounds. This penetration can lead to systemic distribution of nanoparticles, raising concerns about their long-term biocompatibility and safety.

To mitigate these risks, researchers are exploring various strategies, such as surface modification and functionalization of ZnO NPs, to enhance their safety profile. Coating ZnO NPs with biocompatible materials can reduce their potential toxicity by preventing excessive ROS generation and improving their interaction with biological tissues. Additionally, incorporating ZnO NPs into controlled-release systems can help in maintaining their therapeutic concentrations at safe levels, thereby minimizing the risk of adverse effects.

9.10.2 Polymeric Nanoparticles

Polymeric nanoparticles, such as those composed of chitosan, gelatin, and poly(lactic-co-glycolic acid) (PLGA), have garnered significant attention for their potential in enhancing wound healing. These nanoparticles offer several advantages, including tunable degradation kinetics and sustained drug release, which are particularly beneficial for long-term wound care. The versatility and functionality of polymeric nanoparticles make them ideal candidates for various wound healing applications.

Chitosan Nanoparticles: Chitosan, a natural polysaccharide derived from chitin, has been extensively studied for its wound healing properties. Chitosan nanoparticles have demonstrated the ability to enhance wound healing through multiple mechanisms. They promote cell proliferation, which is essential for the formation of new tissue. Additionally, chitosan nanoparticles facilitate angiogenesis, the process of new blood vessel formation, which is crucial for supplying nutrients and oxygen to the wound site. Their inherent antimicrobial activity helps in preventing infections, a common complication in wound healing. The biocompatibility and biodegradability of chitosan make it a suitable material for developing wound healing therapies that support tissue regeneration without adverse immune responses.

Gelatin Nanoparticles: Gelatin, a protein derived from collagen, is another material used to create polymeric nanoparticles for wound healing. Gelatin nanoparticles are biocompatible and biodegradable, providing a supportive matrix for cell attachment and growth. These nanoparticles can be engineered to deliver growth factors and other therapeutic agents directly to the wound site, enhancing the healing process. The controlled release of these agents from gelatin nanoparticles ensures a sustained therapeutic effect, promoting efficient tissue repair and regeneration.

PLGA Nanoparticles: Poly(lactic-co-glycolic acid) (PLGA) is a synthetic polymer widely used in biomedical applications due to its excellent biocompatibility and tunable degradation rates. PLGA nanoparticles can encapsulate a variety of drugs and bioactive molecules, providing controlled and sustained release profiles. This controlled release mechanism is particularly advantageous in wound healing, where a steady supply of therapeutic agents is needed to support ongoing tissue repair. PLGA nanoparticles have been shown to enhance cell migration, proliferation, and angiogenesis, all of which are critical for effective wound healing.

9.10.2.1 Toxicity of Polymeric Nanoparticles in Wound Healing

While polymeric nanoparticles offer numerous benefits for wound healing, their potential toxicity must be carefully evaluated to ensure safety and efficacy. The unique properties of these nanoparticles, such as their small size and large surface area, can lead to unexpected interactions with biological systems, raising concerns about their biocompatibility and potential adverse effects.

Despite their biocompatibility, chitosan nanoparticles can induce toxicity under certain conditions. High concentrations of chitosan nanoparticles have been associated with cytotoxic effects, including reduced cell viability and increased oxidative stress. Additionally, the degree of deacetylation and molecular weight of chitosan can influence its toxicity profile. Therefore, optimizing these parameters is crucial to minimizing adverse effects while maintaining therapeutic efficacy. Similarly, gelatin nanoparticles are generally considered safe due to their biodegradability and biocompatibility. However, their potential to induce immunogenic responses cannot be overlooked. The source of gelatin and the method of nanoparticle preparation can affect their immunogenicity. Careful selection of gelatin sources and purification processes is necessary to reduce the risk of immune reactions. PLGA nanoparticle degradation products, lactic acid and glycolic acid, can accumulate and lead to localized acidosis, which may cause inflammation and cytotoxicity. The size, surface charge, and composition of PLGA nanoparticles also play critical roles in determining their toxicity. Tailoring these properties can help mitigate potential adverse effects.

9.10.3 Lipid-Based Nanoparticles

Lipid-based nanoparticles, including liposomes and solid lipid nanoparticles, are utilized for their ability to encapsulate and deliver therapeutic agents directly to wound sites. These nanoparticles can carry growth factors, cytokines, and antimicrobial agents, providing a controlled release of these therapeutic molecules to enhance wound healing. Lipid-based nanocarriers offer several benefits, such as biocompatibility, improved drug stability, and enhanced bioavailability. Their ability to protect encapsulated agents from degradation while providing a sustained release at the wound site makes them particularly effective in promoting tissue regeneration and reducing infection.

9.10.4 Toxicity of Lipid-Based Nanoparticles in Wound Healing

Lipid-based nanoparticles (LNPs), including liposomes and solid lipid nanoparticles (SLNs), are another class of nanomaterials extensively studied for their potential in wound healing. These nanoparticles are biocompatible and can encapsulate various therapeutic agents, providing targeted and sustained release at the wound site.

9.10.4.1 Potential Toxicity Concerns

Despite their advantages, the potential toxicity of lipid-based nanoparticles needs careful consideration. Studies have shown that certain formulations of LNPs can induce cytotoxicity, inflammation, and immune responses. The size, charge, and composition of these nanoparticles can influence their interaction with biological tissues and cells, potentially leading to adverse effects. For example, cationic liposomes, which are often used to deliver genetic material, have been associated with higher cytotoxicity compared to neutral or anionic liposomes. This cytotoxicity is primarily due to the interactions between the positively charged liposomes and the negatively charged cell membranes, leading to membrane disruption and cell death. Additionally, the degradation products of lipid nanoparticles can induce oxidative stress and inflammatory responses, further complicating the healing process.

9.10.5 Carbon-Based Nanomaterials

Carbon-based nanomaterials, such as carbon nanotubes and graphene oxide, have shown promise in wound healing due to their unique structural and electrical properties. These materials can enhance cell adhesion and proliferation, which are essential for tissue regeneration. Additionally, carbon-based nanomaterials can be functionalized with various bioactive molecules to improve their therapeutic efficacy. However, their biocompatibility and potential cytotoxic effects require thorough investigation to ensure safe application in wound healing.

9.10.5.1 Functionalization with Bioactive Molecules

One of the significant advantages of carbon-based nanomaterials is their ability to be functionalized with various bioactive molecules. This functionalization can enhance their therapeutic efficacy by providing targeted delivery of growth factors, antimicrobial agents, and other therapeutic compounds directly to the wound site. For example, functionalizing CNTs with growth factors such as epidermal growth factor (EGF) can accelerate wound healing by promoting cell proliferation and migration. Graphene oxide can be conjugated with antimicrobial peptides to prevent infections, a common complication in wound healing.

9.11 BIOCOMPATIBILITY CONCERNS AND MITIGATION STRATEGIES FOR NANOMATERIALS IN WOUND HEALING

The integration of nanomaterials in wound healing has demonstrated significant potential due to their unique properties, such as high surface-to-volume ratio, antibacterial capabilities, and ability to promote cellular growth and tissue regeneration. However, these benefits come with critical biocompatibility concerns that need to be addressed to ensure safe and effective application in medical settings. This section explores the biocompatibility issues associated with nanomaterials in wound healing and discusses various strategies to mitigate these concerns, incorporating insights from recent literature.

9.11.1 Biocompatibility Concerns

One of the primary concerns with the use of nanomaterials in wound healing is their potential to induce cytotoxicity. The high reactivity of nanoparticles, attributed to their large surface area, can lead to the generation of reactive oxygen species (ROS). Excessive ROS production can cause oxidative stress, damaging cellular components such as lipids, proteins, and DNA. This oxidative damage can impair cell function and viability, hindering the wound healing process. Another significant concern is the inflammatory response triggered by nanomaterials. While inflammation is a crucial part of the natural healing process, excessive or prolonged inflammation can be detrimental. Some nanomaterials can exacerbate inflammatory responses, leading to chronic inflammation, which can delay wound healing and contribute to fibrosis and scarring.

Furthermore, the long-term persistence of nanomaterials in the body raises concerns about their potential to cause chronic toxicity. Unlike traditional biomaterials that degrade over time, some nanomaterials may accumulate in tissues, leading to prolonged exposure and potential adverse effects. This persistence can result in sustained inflammation and other deleterious effects.

9.11.2 Mitigation Strategies

To address the biocompatibility concerns associated with nanomaterials in wound healing, several strategies have been developed and refined through recent research. These strategies aim to enhance the safety and efficacy of nanomaterials while maintaining their beneficial properties.

Surface Modification: Modifying the surface of nanomaterials can significantly improve their biocompatibility. Coating nanoparticles with biocompatible polymers, such as polyethylene glycol (PEG), can reduce their reactivity and prevent the formation of ROS. Surface modification can also minimize protein adsorption and immune recognition, thereby reducing inflammatory responses.

Recent studies have shown that PEGylation of silver nanoparticles not only enhances their stability and biocompatibility but also retains their antibacterial properties.

Controlled Release Systems: Developing controlled release systems for nanomaterials can mitigate their potential toxicity by ensuring a sustained and localized release of therapeutic agents. Encapsulating nanoparticles within biodegradable polymers, such as polylactic-co-glycolic acid (PLGA), can provide a controlled release profile, minimizing the risk of burst release and associated cytotoxicity. These systems can also reduce the frequency of application, improving patient compliance and overall treatment outcomes.

Biodegradable Nanomaterials: Utilizing biodegradable nanomaterials can address concerns about long-term persistence and chronic toxicity. Materials such as chitosan, collagen, and gelatin are naturally degradable and have been shown to support wound healing while minimizing adverse effects. Recent advancements in the synthesis of biodegradable nanomaterials have led to the development of novel composites that combine the mechanical strength and antibacterial properties of synthetic materials with the biocompatibility of natural polymers.

Targeted Delivery Systems: Engineering nanomaterials for targeted delivery can enhance their therapeutic efficacy and reduce off-target effects. Functionalizing nanoparticles with targeting ligands, such as antibodies or peptides, can direct them to specific cell types or tissues, ensuring that the therapeutic agents are delivered precisely where needed. This targeted approach minimizes systemic exposure and reduces the risk of adverse effects. For example, nanoparticles functionalized with RGD peptides have shown enhanced targeting to wound sites, promoting efficient healing with reduced systemic toxicity.

Comprehensive In Vitro and In Vivo Testing: Rigorous preclinical testing is essential to evaluate the biocompatibility of nanomaterials. Comprehensive in vitro studies, including cytotoxicity assays and oxidative stress measurements, provide initial insights into the potential hazards. Subsequent in vivo studies in relevant animal models help assess the systemic and local effects of nanomaterials over extended periods. Recent advancements in 3D tissue models and organ-on-a-chip technologies offer promising alternatives for more accurate and predictive biocompatibility assessments.

9.11.3 Research Insight into Biocompatibility Concerns and Mitigation Strategies

Recent studies have provided valuable insights into the biocompatibility of nanomaterials in wound healing. A study by Batool et al. (2021) investigated the effects of gold nanoparticles on wound healing and found that surface modification with biocompatible polymers significantly reduced inflammatory responses and enhanced tissue regeneration [61]. Similarly, research by Lee et al. (2021) demonstrated that biodegradable chitosan nanoparticles loaded with growth factors promoted faster wound closure and reduced scar formation in diabetic wound models [62].

Moreover, advancements in nanotechnology have led to the development of smart nanomaterials that respond to specific stimuli, such as pH or temperature changes, to release therapeutic agents. These stimuli-responsive nanomaterials offer precise control over drug delivery, minimizing potential side effects and enhancing healing outcomes. For instance, a study highlighted the effectiveness of pH-responsive nanogels in delivering antibiotic and anti-inflammatory agents to intracellular bacterial infections and wound sites, resulting in accelerated healing and reduced inflammation [63].

9.12 FUTURE DIRECTION IN NANOMATERIAL RESEARCH

The future of nanomaterial research in wound healing holds immense promise, driven by continuous advancements in nanotechnology and an ever-growing understanding of wound healing mechanisms. As the field progresses, several key areas are poised to revolutionize how wounds are treated, enhancing both the efficacy and safety of nanomaterial applications.

9.12.1 Multifunctional Nanomaterials

One of the most exciting directions in nanomaterial research is the development of multifunctional nanomaterials. These advanced materials can simultaneously perform multiple roles, such as promoting cell proliferation, providing antimicrobial effects, and delivering therapeutic agents. Recent studies highlight the potential of combining various functionalities into a single nanomaterial platform. For example, a study by Guo et al. (2022) demonstrated a nanocomposite hydrogel that not only exhibited excellent antibacterial properties but also promoted angiogenesis and accelerated diabetic wound healing [64]. Such multifunctional approaches are particularly beneficial for chronic wound management, where multiple challenges need to be addressed concurrently.

9.12.2 Stimuli-Responsive Nanomaterials

Stimuli-responsive nanomaterials, which change their behavior in response to specific environmental triggers, represent a significant leap forward in targeted wound healing. These materials can respond to stimuli such as pH, temperature, or enzymatic activity, providing controlled and localized delivery of therapeutics. Recent innovations include pH-responsive nanoparticles that release drugs in the acidic environment of infected wounds, thereby minimizing systemic side effects. Researchers showcased the use of enzyme-responsive nanoparticles that released anti-inflammatory agents in the presence of matrix metalloproteinases, enzymes typically upregulated in chronic wounds [65]. This targeted approach ensures that therapeutic effects are maximized where they are needed most, enhancing healing outcomes.

9.12.3 Bioactive Nanomaterials

The integration of bioactive molecules with nanomaterials is another promising area of research. Bioactive nanomaterials can be engineered to release growth factors, cytokines, or other biologically active substances that promote tissue regeneration and modulate immune responses. Recent developments have seen the incorporation of growth factors such as EGF and VEGF into nanofiber scaffolds, resulting in enhanced epithelialization and angiogenesis. A notable study by Kim et al. (2022) developed a bioactive nanoparticle system that released transforming growth factor-beta (TGF-β) to stimulate fibroblast proliferation and collagen synthesis, significantly improving the healing of full-thickness skin wounds in animal models [66, 67].

9.12.4 Nanotechnology-Integrated Smart Dressings

Smart dressings that incorporate nanotechnology offer a futuristic approach to wound care. These dressings can monitor wound conditions in real-time and deliver therapeutics as needed, providing a dynamic and responsive treatment method. Innovations in this area include the development of dressings embedded with nanosensors that detect changes in pH, temperature, and moisture levels, which are indicative of infection or other complications. Upon detecting such changes, these smart dressings can release antimicrobial agents or other therapeutics to address the issue. Research by Pang et al. (2023) introduced a smart wound dressing that combined nanosensors with a drug delivery system, allowing for the automatic release of antibiotics in response to bacterial infection markers, significantly improving infection control and healing rates [68].

9.12.5 Personalized Nanomedicine

The future of nanomaterial research also lies in personalized nanomedicine, where treatments are tailored to individual patient needs based on their genetic makeup and specific wound characteristics. Advances in genomics and personalized medicine are paving the way for nanomaterials that can be customized to deliver patient-specific therapies. This approach ensures that the therapeutic agents are optimally suited to the patient's unique biological environment, enhancing efficacy and minimizing adverse effects. Personalized nanomedicine could involve the use of nanoparticle-based gene therapy to correct genetic deficiencies that impair wound healing or the use of patient-derived stem cells encapsulated in nanomaterials for targeted tissue regeneration.

9.12.6 Safety and Biocompatibility

As the use of nanomaterials in wound healing advances, ensuring their safety and biocompatibility remains a critical area of research. Long-term studies are needed to assess the potential cytotoxicity, immunogenicity, and environmental impact of nanomaterials. Recent efforts have focused on developing biodegradable nanomaterials that can safely break down in the body, reducing the risk of long-term accumulation and adverse effects. Research by Wang et al. (2023) emphasized the importance of comprehensive biocompatibility testing, demonstrating that their newly developed biodegradable nanoparticle system showed no signs of toxicity or adverse immune responses in prolonged animal studies [69].

9.13 CONCLUSION: ADVANCING WOUND HEALING THROUGH NANOMATERIALS

The advent of nanomaterials has opened new horizons in the field of wound healing, offering unprecedented opportunities to enhance and expedite the healing process. The unique properties of nanomaterials, such as their high surface-to-volume ratio, biocompatibility, and the ability to functionalize their surfaces with therapeutic agents, make them ideal candidates for innovative wound care solutions. Recent literature underscores the transformative impact nanotechnology can have on wound management, highlighting both the scientific advancements and practical applications of these materials. Nanomaterials have shown exceptional promise in creating wound dressings that not only protect the wound but also actively contribute to healing. Their antibacterial properties are particularly beneficial in preventing infections, a common and serious complication in wound care. For example, silver nanoparticles have been extensively studied for their potent antimicrobial activity, which can significantly reduce bacterial load and promote a sterile healing environment. Beyond their antimicrobial capabilities, nanomaterials can be engineered to deliver therapeutic agents directly to the wound site. This targeted delivery system ensures a localized and sustained release of drugs, growth factors, or other bioactive molecules, optimizing their efficacy and reducing systemic side effects. For instance, the use of nanofibrous scaffolds loaded with growth factors has been shown to enhance tissue regeneration by providing a conducive environment for cell proliferation and differentiation. These scaffolds mimic the natural extracellular matrix, facilitating cell attachment and promoting the formation of new tissue. Chronic wounds, such as diabetic ulcers, pose significant challenges due to their resistance to conventional treatments. Nanotechnology offers a promising solution to this problem. Research has highlighted the effectiveness of nanomaterials in managing chronic wounds by enhancing the body's natural healing mechanisms. Nanoparticles designed to respond to specific stimuli, such as pH changes or enzymatic activity, can release their therapeutic payload precisely when and where it is needed, thus addressing the unique pathological environment of chronic wounds. Moreover, the integration of nanotechnology with advanced materials science has led to the development of smart wound dressings. These dressings can monitor the wound environment in real-time and respond accordingly, such as by releasing antimicrobial agents or changing their physical properties to enhance healing. For example, hydrogel dressings incorporating nanomaterials can maintain optimal moisture levels in the wound bed, absorb exudates, and provide a barrier against pathogens while releasing therapeutic agents in a controlled manner. The potential of nanomaterials extends beyond traditional wound care into regenerative medicine. Techniques such as 3D printing of nanomaterial-based scaffolds are being explored to create customized wound care solutions tailored to the specific needs of patients. These scaffolds can be designed to fit the exact contours of a wound, ensuring optimal contact and promoting faster and more effective healing. Additionally, incorporating stem cells into these nanoscaffolds can further enhance their regenerative capabilities, paving the way for new treatments in tissue engineering and regenerative medicine.

In conclusion, the integration of nanomaterials into wound healing represents a significant leap forward in medical science. By leveraging the unique properties of these materials, researchers and clinicians can develop advanced wound care solutions that offer enhanced antibacterial protection, targeted drug delivery, and improved tissue regeneration. As the field continues to evolve, ongoing research and development will undoubtedly lead to even more innovative applications, ultimately improving patient outcomes and setting new standards in wound care. The future of wound healing lies in the continued exploration and application of nanotechnology, promising a new era of more effective and efficient treatments.

REFERENCES

1. Ji, J., et al., Research trends of piezoelectric nanomaterials in biomedical engineering. *Advanced NanoBiomed Research*, 2023.3(1): p. 2200088.
2. Hayat M, Nawaz A, Chinnam S, Muzammal M, Latif MS, Yasin M, Ashique S, Zengin G, Farid A. Formulation development and optimization of herbo synthetic gel: In vitro biological evaluation and in vivo wound healing studies. *Process Biochemistry*, 2023.130 Jul 1: p. 116–126.
3. Mohammadi, R., et al., Fluorescence sensing and imaging with carbon-based quantum dots for early diagnosis of cancer: A review. *Journal of Pharmaceutical and Biomedical Analysis*, 2022.212: p. 114628.
4. Kargozar, S., et al., Quantum dots: A review from concept to clinic. *Biotechnology Journal*, 2020.15(12): p. 2000117.
5. Tiwari, A. and A. Tiwari, *Nanomaterials in Drug Delivery, Imaging, and Tissue Engineering*. 2013, John Wiley & Sons.
6. Matutule, L. J., *Effects of Gold Nanoparticles on Vascular Endothelial Growth Factor-A-Induced Melanoma Cell Growth and Angiogenesis*. 2020, University of Pretoria (South Africa).
7. de la Parra, S., et al., Comparative toxicological analysis of two pristine carbon nanomaterials (graphene oxide and aminated graphene oxide) and their corresponding degraded forms using human in vitro models. *Toxicology*, 2024.504: p. 153783.
8. Fontaine, M., et al., Nanomaterial genotoxicity evaluation using the high-throughput p53-binding protein 1 (53BP1) assay. *PLoS One*, 2023.18(9): p. e0288737.
9. Fortis, S.P., et al., Effect of silica-based mesoporous nanomaterials on human blood cells. *Chemico-Biological Interactions*, 2024.387: p. 110784.
10. Liu, S., et al., Toxicity of energetic nanomaterials assessed on L929 fibroblasts. *Chemical Papers*, 2024: p. 1–8.
11. Mourya, D., et al., In vitro effects of zirconia nanoparticles: uptake, genotoxicity, and mutagenicity in V-79 cells. *Biological Trace Element Research*, 2024.202(3): p. 927–940.
12. Tunç, T., C. Hepokur, and A. Kari̇per, Synthesis and characterization of paclitaxel-loaded silver nanoparticles: evaluation of cytotoxic effects and antimicrobial activity. *Bioinorganic Chemistry and Applications*, 2024.2024(1): p. 9916187.
13. Farabaugh, C. S., et al., In vitro micronucleus assay: method for assessment of nanomaterials using cytochalasin B. *Frontiers in Toxicology*, 2023.5: p. 1171960.
14. da Cunha Goldstein, A., et al., In vitro genotoxicity assessment of graphene quantum dots nanoparticles: A metabolism-dependent response. *Mutation Research/Genetic Toxicology and Environmental Mutagenesis*, 2023.885: p. 503563.
15. Šíma, M., et al., Gene expression profiles and protein-protein interaction networks in THP-1 cells exposed to metal-based nanomaterials. *Environmental Toxicology and Pharmacology*, 2024.108: p. 104469.
16. Neun, B.W., A.N. Ilinskaya, and M.A. Dobrovolskaia, Analysis of complement activation by nanoparticles. *Characterization of Nanoparticles Intended for Drug Delivery*, 2018: p. 149–160.
17. Ruijter, N., et al., The state of the art and challenges of in vitro methods for human hazard assessment of nanomaterials in the context of safe-by-design. *Nanomaterials*, 2023.13(3): p. 472.

18. Chen, H., et al., Colorimetric biosensing assays based on gold nanoparticles functionalized/combined with non-antibody recognition elements. *TrAC Trends in Analytical Chemistry*, 2024: p. 117654.
19. Devanabanda, M., S. A. Latheef, and R. Madduri, Immunotoxic effects of gold and silver nanoparticles: Inhibition of mitogen-induced proliferative responses and viability of human and murine lymphocytes in vitro. *Journal of Immunotoxicology*, 2016.13(6): p. 897–902.
20. Bancos, S., D. L. Stevens, and K. M. Tyner, Effect of silica and gold nanoparticles on macrophage proliferation, activation markers, cytokine production, and phagocytosis in vitro. *International Journal of Nanomedicine*, 2015: p. 183–206.
21. Ahn, E.-Y., H. Jin, and Y. Park, Assessing the antioxidant, cytotoxic, apoptotic and wound healing properties of silver nanoparticles green-synthesized by plant extracts. *Materials Science and Engineering: C*, 2019.101: p. 204–216.
22. Galandáková, A., et al., Effects of silver nanoparticles on human dermal fibroblasts and epidermal keratinocytes. *Human & Experimental Toxicology*, 2016.35(9): p. 946–957.
23. Flores-López, L.Z., H. Espinoza-Gómez, and R. Somanathan, Silver nanoparticles: Electron transfer, reactive oxygen species, oxidative stress, beneficial and toxicological effects. Mini review. *Journal of Applied Toxicology*, 2019.39(1): p. 16–26.
24. Dreno, B., et al., Safety of titanium dioxide nanoparticles in cosmetics. *Journal of the European Academy of Dermatology and Venereology*, 2019.33: p. 34–46.
25. Brassolatti, P., et al., Functionalized titanium nanoparticles induce oxidative stress and cell death in human skin cells. *International Journal of Nanomedicine*, 2022: p. 1495–1509.
26. Zhang, L., et al., Graphene oxide induces dose-dependent lung injury in rats by regulating autophagy. *Experimental and Therapeutic Medicine*, 2021.21(5): p. 1–11.
27. Cui, H.-F., et al., Interfacing carbon nanotubes with living mammalian cells and cytotoxicity issues. *Chemical Research in Toxicology*, 2010.23(7): p. 1131–1147.
28. Shehata, A.M., et al., Evaluation of the ameliorative effect of zinc nanoparticles against silver nanoparticle–induced toxicity in liver and kidney of rats. *Biological Trace Element Research*, 2022: p. 1–11.
29. Amadi, E.E., *Effects of Graphene Oxide in vitro on DNA Damage in Human Whole Blood and Peripheral Blood Lymphocytes from Healthy individuals and Pulmonary Disease Patients: Asthma, COPD, and Lung Cancer*. 2019, University of Bradford.
30. Solorio-Rodriguez, S.A., et al., Single-walled vs. multi-walled carbon nanotubes: Influence of physicochemical properties on toxicogenomics responses in mouse lungs. *Nanomaterials*, 2023.13(6): p. 1059.
31. Tomankova, K., et al., Cytotoxicity, cell uptake and microscopic analysis of titanium dioxide and silver nanoparticles in vitro. *Food and Chemical Toxicology*, 2015.82: p. 106–115.
32. Xuan, L., et al., Nanoparticles-induced potential toxicity on human health: Applications, toxicity mechanisms, and evaluation models. *MedComm*, 2023.4(4): p. e327.
33. Niroumand, U., et al., The effect of size, morphology and surface properties of mesoporous silica nanoparticles on pharmacokinetic aspects and potential toxicity concerns. *Frontiers in Materials*, 2023.10: p. 1189463.
34. He, C., et al., Gold nanoparticles enhance the ability of radiotherapy to induce immunogenic cell death in glioblastoma. *International Journal of Nanomedicine*, 2023: p. 5701–5712.
35. Krishnan, J., et al., A concise review on effect of PEGylation on the properties of lipid-based nanoparticles. *ASSAY and Drug Development Technologies*, 2024.
36. Khalaf, E.M., et al., Recent progressions in biomedical and pharmaceutical applications of chitosan nanoparticles: A comprehensive review. *International Journal of Biological Macromolecules*, 2023.231: p. 123354.
37. Yang, Z., et al., Cytotoxicity and immune dysfunction of dendritic cells caused by graphene oxide. *Frontiers in Pharmacology*, 2020.11: p. 1206.
38. Chavan, Y., et al., Redefining the importance of polylactide-co-glycolide acid (PLGA) in drug delivery. In *Annales Pharmaceutiques Françaises*. 2022, Elsevier.
39. Horvath, D. and M. Basler, PLGA particles in immunotherapy. *Pharmaceutics*, 2023.15(2): p. 615.
40. Sim, S.L., et al., Macrophages in skin wounds: functions and therapeutic potential. *Biomolecules*, 2022.12(11): p. 1659.
41. Shang, M., et al., Silver nanoparticle-induced impaired autophagic flux and lysosomal dysfunction contribute to the microglia inflammation polarization. *Food and Chemical Toxicology*, 2022.170: p. 113469.
42. Ding, S., et al., Overcoming blood–brain barrier transport: Advances in nanoparticle-based drug delivery strategies. *Materials Today*, 2020.37: p. 112–125.
43. Sengul, A.B. and E. Asmatulu, Toxicity of metal and metal oxide nanoparticles: a review. *Environmental Chemistry Letters*, 2020.18(5): p. 1659–1683.

44. Dunnill, C., et al., Reactive oxygen species (ROS) and wound healing: the functional role of ROS and emerging ROS-modulating technologies for augmentation of the healing process. *International Wound Journal*, 2017.14(1): p. 89–96.
45. Zhang, W., et al., Antioxidant therapy and antioxidant-related bionanomaterials in diabetic wound healing. *Frontiers in Bioengineering and Biotechnology*, 2021.9: p. 707479.
46. Korrapati, P.S., et al., Recent advancements in nanotechnological strategies in selection, design and delivery of biomolecules for skin regeneration. *Materials Science and Engineering: C*, 2016.67: p. 747–765.
47. Sanità, G., B. Carrese, and A. Lamberti, Nanoparticle surface functionalization: how to improve biocompatibility and cellular internalization. *Frontiers in Molecular Biosciences*, 2020.7: p. 587012.
48. Rao, S.S., et al., Natural polymeric biomaterials in growth factor delivery for treating diabetic foot ulcers. *Journal of Drug Delivery Science and Technology*, 2020.55: p. 101385.
49. Wang, Z. and Z. Wang, Nanoparticles induced embryo–fetal toxicity. *Toxicology and Industrial Health*, 2020.36(3): p. 181–213.
50. Chakraborty, C., et al., Zebrafish: A complete animal model to enumerate the nanoparticle toxicity. *Journal of Nanobiotechnology*, 2016.14: p. 1–13.
51. Arias-Hidalgo, C., et al., Single-cell proteomics: The critical role of nanotechnology. *International Journal of Molecular Sciences*, 2022.23(12): p. 6707.
52. Nethi, S.K., et al., Recent advances in inorganic nanomaterials for wound-healing applications. *Biomaterials Science*, 2019.7(7): p. 2652–2674.
53. Kushwaha, A., L. Goswami, and B.S. Kim, Nanomaterial-based therapy for wound healing. *Nanomaterials*, 2022.12(4): p. 618.
54. Yang, Y., et al., Applications of nanotechnology for regenerative medicine; healing tissues at the nanoscale. In *Principles of regenerative medicine*. 2019, Elsevier. p. 485–504.
55. Nqakala, Z.B., et al., Advances in nanotechnology towards development of silver nanoparticle-based wound-healing agents. *International Journal of Molecular Sciences*, 2021.22(20): p. 11272.
56. Garcia-Orue, I., et al., Nanotechnology-based delivery systems to release growth factors and other endogenous molecules for chronic wound healing. *Journal of Drug Delivery Science and Technology*, 2017.42: p. 2–17.
57. Firoozi, N. and Y. Kang, Immobilization of FGF on poly (xylitol dodecanedioic acid) polymer for tissue regeneration. *Scientific Reports*, 2020.10(1): p. 10419.
58. Lai, H.-J., et al., Tailored design of electrospun composite nanofibers with staged release of multiple angiogenic growth factors for chronic wound healing. *Acta biomaterialia*, 2014.10(10): p. 4156–4166.
59. Memic, A., et al., Latest progress in electrospun nanofibers for wound healing applications. *ACS Applied Bio Materials*, 2019.2(3): p. 952–969.
60. Mathapati, S., et al., Nanofibers coated on acellular tissue-engineered bovine pericardium supports differentiation of mesenchymal stem cells into endothelial cells for tissue engineering. *Nanomedicine*, 2014.9(5): p. 623–634.
61. Batool, F., et al., Modulation of immune-inflammatory responses through surface modifications of biomaterials to promote bone healing and regeneration. *Journal of Tissue Engineering*, 2021.12: p. 20417314211041428.
62. Lee, Y.-H., Y.-L. Hong, and T.-L. Wu, Novel silver and nanoparticle-encapsulated growth factor co-loaded chitosan composite hydrogel with sustained antimicrobility and promoted biological properties for diabetic wound healing. *Materials Science and Engineering: C*, 2021.118: p. 111385.
63. Obuobi, S., et al., Liposomal delivery of antibiotic loaded nucleic acid nanogels with enhanced drug loading and synergistic anti-inflammatory activity against S. aureus intracellular infections. *Journal of Controlled Release*, 2020.324: p. 620–632.
64. Guo, C., et al., Development of a microenvironment-responsive hydrogel promoting chronically infected diabetic wound healing through sequential hemostatic, antibacterial, and angiogenic activities. *ACS Applied Materials & Interfaces*, 2022.14(27): p. 30480–30492.
65. Sobczak, M., Enzyme-responsive hydrogels as potential drug delivery systems—state of knowledge and future prospects. *International Journal of Molecular Sciences*, 2022.23(8): p. 4421.
66. Bjelić, D. and M. Finšgar, The role of growth factors in bioactive coatings. *Pharmaceutics*, 2021.13(7): p. 1083.
67. Ren, L.-L., et al., Transforming growth factor-β signaling: from tissue fibrosis to therapeutic opportunities. *Chemico-biological Interactions*, 2023.369: p. 110289.
68. Pang, Q., et al., Smart wound dressing for advanced wound management: Real-time monitoring and on-demand treatment. *Materials & Design*, 2023: p. 111917.
69. Wang, E.Y., et al., Recent advances in micro-and nano-scale carrier systems for controlled delivery of vaccines. *Biomaterials*, 2023: p. 122345.

10 The Emergence of High-Tech Tools in Topical Dermal Drug Delivery

3D and 4D Printing

*Jyotsna Vitore,[1] Chantibabu Katta,[1] Srushti Mahajan,[1] Pankaj Kumar Singh,[1] Neelesh Kumar Mehra,[1] and Jitender Madan[1]**

[1] Department of Pharmaceutics, National Institute of Pharmaceutical Education and Research, Hyderabad, Telangana, India

* Corresponding author: Dr. Jitender Madan,
Associate Professor, Department of Pharmaceutics,
National Institute of Pharmaceutical Education and
Research, Hyderabad, Telangana, India

10.1 INTRODUCTION

Presently, new advancements in the medical field have enriched the science with ground-breaking technologies. In the pharmaceutical industry, the exploitation of novel technologies that are advanced is widely increasing. Targeted drug delivery therapies and customized manufacturing techniques have triggered research in this area and have spread their footprint in the marketplace. Furthermore, there is a need for the emergence of transformable techniques for robust manufacturing of personalized medicines. As there is an increase in demand for personalized medicines for the treatment of complex disorders, the need for novel technologies is growing to resolve the issue [1]. Nowadays, there is a continuing trend for customized healthcare employing technologies such as (three-dimensional) and 4D (four-dimensional) printing. The technology has found its multidisciplinary application in tissue engineering, medical devices, drug delivery, dentistry, aerospace, construction, and automotive. Like every technology, it has its own pros and cons over conventional manufacturing processes, as these processes are often unfit for the production of tailored drug delivery, being innately labor-intensive, dose-inflexible, and time-consuming procedures [2]. This generates a necessity for the healthcare system to acclimatize and incorporate innovative stages for personalized treatment [3, 4].

3D printing is also denoted as the additive process of producing a compact object from a digital model. It is accomplished using an additive approach in which a printing process positions successive layers of materials in various shapes [5]. Using computer-aided design (CAD) software, imaging techniques such as magnetic resonance imaging (MRI) or computed tomography (CT) scans, desired structures can be produced from a digital 3D file to easily generate structures that are individualized for each patient [6]. In pharmaceutical formulation growth, 3DP has gained rising attention as an effective strategy to resolve some challenges of the traditional pharmaceutical process. The advances in technology can make it possible for every patient to customize care [7–9].

DOI: 10.1201/9781003605966-10

The 4D printing idea was recently emerged to describe the 3D printing of objects that have the capability of changing their configuration (shape, property, and functionalities) over time due to either a clever involvement of numerous materials in a single print or additive manufacturing of smart, stimuli-responsive ingredients. 4D printing can generate products that individualistically alter their morphology in response to certain outside stimuli [10]. Pharmaceutical formulation with smart materials can therefore perform as predictors of probable future applications for 4D printing by combining 3D printing capable of making personalized medicines and smart materials which are able to modify their individual morphology [11, 12]. The various research subjects in 4D printing lie within the following categories: development of equipment, deformation mechanisms, and mathematical modeling. The American Society for Testing and Materials (ASTM) classified these technologies into seven categories of machines based on the additive process involved; namely binder jetting, photopolymerization, material extrusion, material jetting, powder bed fusion, sheet lamination, and directed energy deposition. Among all available technologies, the one that is commonly used is FDM and SLA [13].

The productivity, time savings, bulk manufacturing, and restricted labor power of 3D and 4D printing have increased their use. 3D printing has so far been used to prepare complex formulas that would be difficult to manufacture using traditional manufacturing methodologies. This technique allows for a wide range of controlled release products, rapid dissolving pills, and multiple drug combinations, as well as customized release profiles and designs [14]. The printlet geometry, infill percentage, and polymer inclusion are the three most important constraints that can be changed to monitor drug release.

The present review emphasizes more on the advancements in topical dermal drug delivery through the emergence of high-tech tools such as 3D and 4D technologies. Numerous examples of 3D and 4D printing use in drug delivery are reported in the literature, including drug-eluting implants, medical devices, drug-loaded patches, and hydrogels. The application of 3D and 4D printing in different dosage forms and dermal drug deliveries, along with their mechanisms and present limitations, will provide great scope for readers. Moreover, the regulatory and future aspects of high-tech printing are also overviewed in this article.

10.2 IS CONVENTIONAL TOPICAL DRUG DELIVERY SYSTEM (TDDS) BECOMING FUTILE?

Topical drug delivery is becoming favorable in treating skin diseases due to its suitability and affordability. Among all existing routes of administration for skin diseases, topical dermal drug delivery has been found to be more beneficial in scientific exercises for treatment directed to the diseased location in the body. Nowadays, there are a number of treatment options available in dermatological science due to the unique characteristics of dermatological diseases and the external nature of the skin. Numerous products applied to the skin might penetrate the skin layers to a limited extent, where their properties are anticipated. Some of the topical formulations for skin conditions like acne and cutaneous inflammations such as dermatitis, erythematous lupus, and psoriasis are in this state. The challenge for the dermatologist during treatment is the selection of an appropriate topical therapy to treat the presented disease [15]. The selection of the active drug and its suitable vehicle for formulation is crucial for a formulation scientist. A number of dosage forms are also available to treat topical diseases, having specific mechanisms and improved patient compliance, such as gels, ointments, lotions, and creams. However, patient compliance with topical dermal drug delivery for the conventional manufacturing of dermal dosage forms is limited because of the generalized content of all active ingredients and excipients in the formulation. In addition, further limitations regarding the manufacturing of topical formulations need various apparatus handling and involve a tedious process for transferring the formulation into the desired container. Therefore, the manufacturer may incur excessive material loss due to increased processing steps, which leads to a less robust manufacturing method with material loss, thus affecting final yield. All these factors contribute to increased manufacturing costs for the company. The aforementioned limitations of conventional topical drug delivery can be enhanced by means of new manufacturing technologies like 3D and 4D printing. These revolutionary technologies have been widely utilized in medical and tissue engineering. Moreover, these technologies have wider applications in the pharmaceutical industry, particularly for dosage forms such as the manufacturing of bilayer and multilayer tablets, capsules, gels, and patches.

10.3 HIGHLIGHTS OF 3D PRINTING

In the early 70s of the twentieth century, the application of 3D printing evolved. Pierre A. L. Ciraud defined a process for applying powdered material and then solidifying each layer with the help of a high-energy laser [7]. Meltable materials like plastics and metals can be utilized for the preparation of objects. In the 1980s, Ross Housholder identified an idea of sand binding by means of different materials, and Carl Deckard developed a process of solidification of powdered bed with the help of a laser beam known as selective laser sintering (SLS). Chuck Hull's SLA was the first commercially available technique. The technique relied on ultraviolet light to photopolymerize liquid resin. Scott Crump filed a patent by the end of the 1980s for a method which prepares objects using thermoplastic material known as FDM. The first commercially available technology created by Chuck Hull was SLA [16]. In the 1990s, MIT scientist Emanuel Sachs and his colleagues invented the three-dimensional printing technique focused on binding material connecting selective regions of powder.

10.4 HOW 3D AND 4D PRINTING SOFTWARE WORKS

We live in an era of technology where each coming day is described as the next industrial revolution, and things became cloud-based computing, cashless expenses, artificial intelligence, and three-dimensional (3D) printing were considered a delusion. 3DP is important to appreciate not only for its potential benefits toward healthcare but also for the several innovations that have contributed to the development of such revolutionary technology in all other fields. Consequently, technologies such as medical imaging techniques, image processing, and reconstruction have been used to develop rapid prototyping technology. As with every technology, there is a method of usage; similarly, 3DP has the following steps. At first, a computer-assisted virtual design (CAD) file is created using software such as Onshape, SolidWorks, Creo Parametric, AutoCAD, Autodesk, and others, and then converted to a "standard tessellation language or stereolithography" (STL). The knowledge about the surface of a three-dimensional model is given by triangulated facets. The .stl file can be transformed into a G file by cutting the template with slicer software, which must first be mounted on the 3D printer. Finally, the print head is shifted to the x-y axis to configure the base of the 3D object and then back to the z-axis to finish the object [17]. Additionally, some steps may involve the removal of extra material and the cleaning of the environment. Essentially, the object construction process is labor-free and fully computer-regulated. This has significant advantages in relation to creation and expense, but any desired improvement in the final object is accomplished by changing the previous CAD file.

10.5 WHY 4D PRINTING OVER 3D PRINTING?

Scientists were determined to find other 3DP-compatible techniques after the invention of additive manufacturing. Since traditional 3D printing techniques concentrate on constructing static assemblies from a single material, they are unable to meet all of the demands for dynamic roles that are needed in industries such as pharmaceuticals, biomedical devices, tissue engineering, and automobiles. 4D bioprinting allows for precise spatial control of different machinery that has the capability of self-folding or self-spreading to encapsulate and release capsules or cells in a regulated manner. Tiny oil droplets, known as "multisomes," can be revealed in water. The similar technique can be

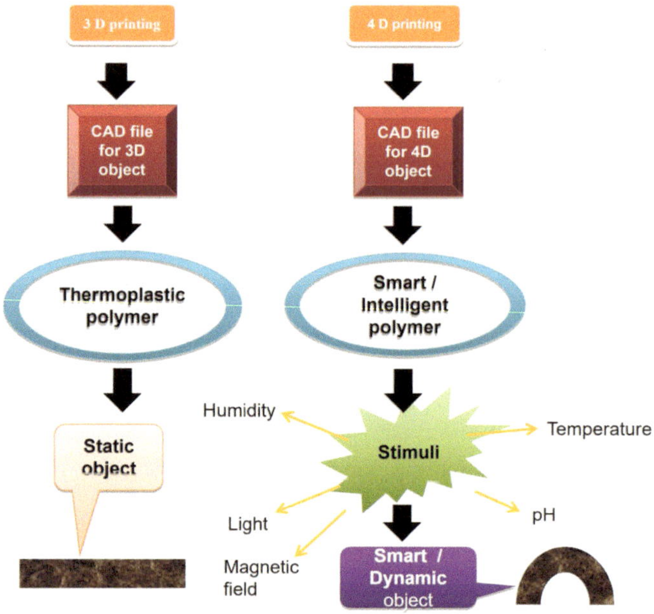

FIGURE 10.1 Variance between 3D and 4D printing4D printing is pioneering.

utilized to extend various nano as well as micro particulate systems that react to various micro-environmental changes such as temperature, humidity, osmolarity, mild or electrically driven, and magnetic stimuli, etc. The creation of four-dimensional print technology has sparked a lot of interest among researchers in intelligent or smart materials. **Figure 10.1** depicts the distinction between both printing techniques.

The main variation between 3D and 4D printing is the addition of one more dimension and making conventional additive techniques smarter by using smart materials having specific triggers or stimuli. 4D printing was first acquainted with in 2013 by the Self-Assembly Lab at MIT. The main mechanism of this technique is based on the use of smart materials in 3D printing. In smart materials, a certain kind of stimuli-responsive polymers have the ability to transform with the onset of stimuli such as to modify 4D printing, which can generate products that individually vary their morphology in response to certain external stimuli in a predefined manner. This means that the 3D printed object can transform its geometry as well as its function when given stimuli. To obtain the specific shape and time-associated parameter, the material needs to be programmable and receptive to numerous external stimuli like pH, light, heat, or magnetic field. Materials having the capacity to alter their nature, morphology, or shape with time are commonly known as smart materials [18]. Two major necessities for 4D printing are responsive raw materials to create programmable objects and the presence of external stimuli like heat, light, pH, or water that can lead to the change occurring. Only if these two conditions are met can the four-dimensional printed objects be powered to attain their maximum efficiency.

10.6 INDUSTRIALIZED PRINTING TECHNIQUES FOR TOPICAL PRODUCTS

10.6.1 Stereolithographic 3D Printing (SLA)

SLA, also known as SLA three-dimensional printing, is one of the most popularly used techniques in the world of additive manufacturing. Chuck Hull, the founder of a three-dimensional printing company, invented the term as well as the technology in 1986. It works by hardening liquid resin

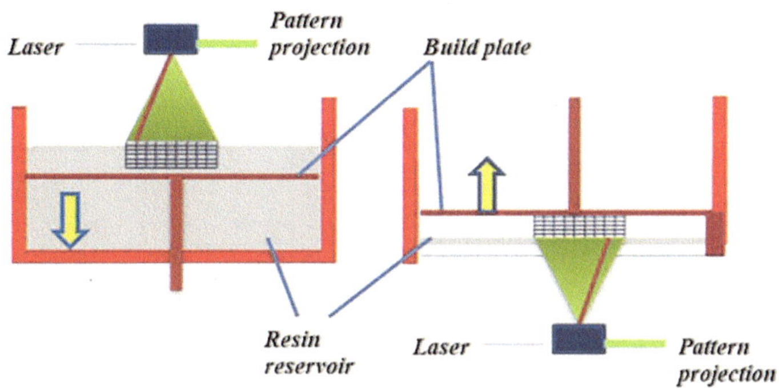

FIGURE 10.2 Schematic representation of SLA printing.

in a reservoir with a high-power laser to generate the desired 3D form [19]. In short, this method utilizes a low-power laser and photo-polymerization to transform light-sensitive liquid into 3D solid plastics layer by layer. The main ingredient for SLA 3D printing is photosensitive polymer [20, 21].

The first layer of the print is "drawn" onto the light-sensitive resin by a laser at the start of the process. The liquid solidifies wherever the laser beam reaches. A computer-aided mirror directs the laser to the correct coordinates. It's worth noting that most SLA printers operate backward; that is, the laser is aimed upward toward the build base. The build base is initially low and is eventually elevated. Following the first layer, the surface is elevated to the thickness of the layer, which is typically 0.1 mm. The additional resin is then permitted to flow under the previously printed section. The laser then solidifies the next cross-section and repeats the procedure until the entire component is finished. The resin that does not come into contact with the laser rests in the vat and can be reused [22]. The platform is removed from the tank, and the extra resin is drained after the material polymerization. The model is removed at the end of the operation from the platform. The excess resin is washed off and put into a UV oven for final curing. Post-print curing helps to achieve the maximum strength and stability possible for objects [23]. **Figure 10.2** depicts a schematic description of the SLA technique.

The biggest disadvantage of this approach is that the resins used have carcinogenic potential and may pose a health risk. A further drawback is that the resins are photosensitive, so long-term stability is an issue. The slow speed of SLA printing, at about 1–3 cmh-1, can also consume a lot of time [24]. As we have already stated, digital light treatment is a descendant of the SLA; DLP has a DLP screen to flash the entire platform with a single image of each sheet. As the projection screen is a digital one, it consists of square pixels for every layer. Therefore, the resolution of a DLP printer is equivalent to the pixel dimension, and the laser spot is equivalent to the SLA [25].

10.6.2 Fused Deposition Modeling (FDM)

Due to the extremely lower cost of printers; printing precision, which is critical to assuring drug quality constraints; and hot-melt extrusion (HME), a technological process previously combined in the pharmaceutical industry. Fused deposition modeling is a form of 3D printing technique that deals with the manufacturing of drug delivery systems [26, 27]. FDM is an additive manufacturing process that involves fusing layers of materials together in a pattern to produce an object. The material is generally melted past the temperature of glass transition. It is then extruded to or on the front of previous extrusions, creating a layer of the object [27–29]. FDM and fused filament fabrication (FFF) are similar, but the term FDM was trademarked in 1991, which created the need for the second name [29]. Many layers of molten thermoplastic material combine to create a three-dimensional object.

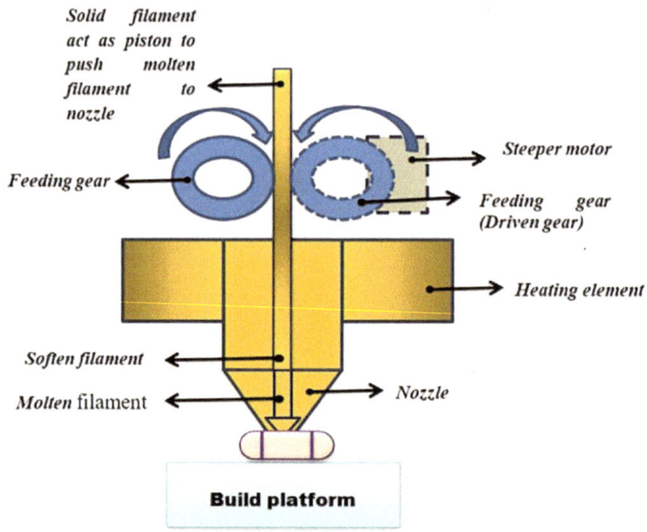

FIGURE 10.3 FDM printing.

In simpler terms, FDM 3D printing typically takes a plastic filament, squeezing it through a hot end. Then it is melted and deposited in layers on the print bed. The layers are then fused and built up throughout the print, which forms the finished portion [30, 31]. **Figure 10.3** represents the FDM technique diagrammatically.

MakerBot from the United States is one of the world's renowned FDM 3D printer producers [24]. It has manufactured an FDM dual extruder, the MakerBot Replicator 2X. In comparison with other 3D printers, FDM offers a number of advantages but also has some disadvantages. 3D printing, therefore, makes the production of some good quality products more economical [32]. Thermoplastic polymers like polyvinyl alcohol (PVA) have recently also been used in the medical industry as medicinal carriers [2]. The FDM 3D printer is portable, easy to use, and equipped for use in compounding drugstores.

One of the major advantages of FDM 3D printing is scalability. It can be easily scaled to any size. FDM printers are continually being made superior and less expensive due to low part costs and the simple designs involved. In addition to that, its material flexibility is also consumer-friendly. On any FDM printer, a wide variety of thermoplastic materials and exotic filaments can be printed with relatively few upgrades and modifications, something that cannot be said of other styles where a material must be a resin or fine powder [33]. FDM technologies have also been used for dermal adhesives intended for cutaneous drug delivery. Scalability is one of the greatest benefits of FDM 3D printing. It can be scaled to any size effortlessly. Due to the low costs of components and the simple designs involved, FDM printers are continuously becoming less costly. Besides that, it also offers a good level of material flexibility. On any FDM printer, there is relatively little possible upgrade and modification of a wide range of thermoplastic materials and exotic filaments, which is not possible in other types where a resin or fine pulver is required for this material [33]. For dermal adhesives intended for the delivery of skin medicine, FDM technology is also used. The major drawback of FDM printing is that the thermoplastic material needs to be melted at high temperatures. A large number of drugs and active medicines can easily be destroyed by high temperatures. The part quality or detail is one of the most frequently mentioned downsides of FDM 3D printing. As the substance must be extruded into layers and its predefined thickness is certain, high-definition prints are difficult to obtain and often require a great deal of after-working to acquire a professional, finished look [34].

10.7 INSIGHT INTO 3DP FOR TOPICAL DRUG DELIVERY

The 3D printing (3DP) process was patented in 1986. Spritam (levetiracetam) is the first FDA-approved 3D printing technology product, which caused an exceptional change in pharmaceutical manufacturing processes [35]. It created a lot of attraction for pharmaceutical scientists because of its flexible manufacturing process and its capability of developing personalized medicines. Recently, researchers are focusing more on various dosage forms like oral, mucosal, injectable hydrogels, and topical dosage forms. Only a handful of literature is available on three-dimensional printing of topical dosage forms, and a lot of research is ongoing to commercialize 3D printing topical products in the market. In this review, we have discussed case studies of research papers in which 3D printing technology has been used for developing topical dosage forms.

10.7.1 Case Study 1

Alvaro Goyanes et al. have utilized two different 3D printing techniques, namely, FDM and SLA, for making 3D printed personalized anti-acne medication-laden masks/patches for topical drug delivery. The authors have used polycaprolactone (PCL), a biocompatible polymer that is approved by the FDA. They also selected widely used commercially available flexible filaments NinjaFlex® (NF) and Flex EcoPLA™ (FPLA). Both polymers are biocompatible, have good mechanical properties, and low toxicity.

10.7.1.1 FDM 3D Printing

The FDM 3D printing technique is a two-step process, i.e., HME followed by FDM 3D printing. However, in FDM 3D printing, the use of HME for FPLA-salicylic acid as well as PCL-salicylic acid has produced good results in terms of uniformity, shape, and size, but the NF-salicylic acid thread was found to be delicate for FDM 3D printing, and the theoretical drug loading was 2% w/w, but practically it was 0.4% w/w and 1.2% w/w for the FPLA-salicylic acid and PCL-salicylic acid respectively, showing thermal decomposition of salicylic acid during HME and 3D printing because of the higher temperature used in HME. Even though drug loading was higher in PCL-salicylic acid by HME, it did not have the properties to print the required shape (nose-shaped anti-acne 3D patch), but the printing characteristics and resolution were good enough for the 3D printing technique using HME threads of FPLA-salicylic acid. Franz diffusion cells were used for drug diffusion tests of FDM 3D printed FPLA-salicylic acid and PCL-salicylic acid, which found a lower diffusion rate, i.e., 53 and 187 µg/cm2, respectively, within 3 hours.

10.7.1.2 SLA Printing

SLA printing is a single-step process. Various SLA 3DP polymer discs PEGDA/PED (4:6) and PEGDA/PED (8:2) were used and achieved higher drug loading, up to 1.9% w/w, than in the PCL-salicylic acid filament of FDM 3DP. It achieved greater resolution than in FDM FPLA-salicylic acid and exhibited higher drug diffusion rates with various SLA 3DP polymer discs, i.e., 291 µg/cm2 in 3 h for PEGDA/PED (4:6)-salicylic acid and 229 µg/cm2 for the PEGDA/PED (8:2)-salicylic acid under the same conditions applied for FDM 3DP masks. Therefore, it was concluded that the SLA printing technique is a more suitable and promising 3D printing technique to formulate anti-acne devices in a nose shape with salicylic acid [24].

10.7.2 Case Study 2

In another study, Junhui Fu et al. prepared 3D-printed vaginal rings with different shapes for progesterone controlled release using the FDM technique. Here, in this study, the authors screened the formulations based on the printability of filaments because, in most FDM 3D printing, printability is challenging after the HME process. The authors selected the polymers PVA, PLA, and PCL, which are commonly used in FDM technology; they are biocompatible and nontoxic polymers, and screened out PVA because it is a water-soluble polymer and did not provide a long-term sustained release pattern. For better results in 3D printing through FDM, they experimented with blending PLA (a glassy polymer) with a high melting point and PCL (an elastomer) with thermoplastic characteristics in different ratios. The results showed that the higher the amount of PLA in the blend, the harder the filament formed, and conversely, when PCL exceeded 30% in the mixture, printing became difficult. Pure PLA and the ratios with PLA and PCL (9:1, 8:2 w/w) effortlessly formed rings with decent thermoplastic properties and elastic nature. Since progesterone, PLA, and PCL are hydrophobic in nature, they made a solid dispersion of progesterone with PEG (2:1) and used 2% Tween 80 to improve drug release from the rings and to impart hydrophilicity to the PCL/PLA particles, respectively. They observed that progesterone made a difference in the properties of the filaments because they found gaps in drug-loaded filaments when compared to blank filaments, which they considered useful for drug release. The optimized formulation selected for printing was progesterone 5%; PEG 10%; Tween 80 2%; PLA 66%; and PCL 17%. With the above optimal formulation filaments, the FDM printer was utilized to print "O," "Y," and "M" shaped rings with a suitable size for insertion into female adult vaginas. Thermal analysis showed that the drug is in amorphous form and did not degrade in the vaginal rings. Among the differently shaped rings, the O-shaped ring showed higher dissolution because of its larger surface area compared to the Y and M-shaped rings. The vaginal rings demonstrated sustained release of the drug for up to 7 days [36].

10.7.3 Case Study 3

Hydrogels have a wide range of drug delivery applications. They can be used in oral delivery, intravenous (injectable hydrogels) delivery, and in topical (ophthalmic, rectal, vaginal, and skin) delivery systems. In 2017, Pamela RoblesMartinez and her group prepared drug-loaded hydrogels using 3D printing. It was found that among many types of 3D printing, SLA printing is favorable for making hydrogels by cross-linking polymers because it contains a laser, which is the reason for the photopolymerization of resins and gives a cross-linked polymeric matrix. Then it can be fabricated layer by layer by SLA 3D printing. PEGDA and 300 PEG as cross-linking polymers, riboflavin and triethanolamine as photoinitiators are shown to be feasible to print drug-loaded hydrogels successfully. In the case of DPPO as a photoinitiator, the printed hydrogel showed poor results in terms of shape and size. Varied amounts of water (0–30%) in the hydrogels have shown consistency in shape and size but have shown faster drug release with higher amounts of water content in the hydrogel [37].

10.7.4 Case Study 4

In the study performed by Hassan et al, they formulated wound dressing using antimicrobial metals. They used zinc, copper, and silver for incorporation into the FDA-approved polymer polycaprolactone (PCL). These metals enhanced wound healing because of their broad-spectrum antimicrobial properties. 3D models of the nose and ears were made using 3D scanning, which provides the chance to tailor the shape and size of the dressing. HME was used to make pellets of polymer and metals. Wound dressings were produced with various shapes and concentrations using different filaments. Metals showed differential release from the dressing; for example, the first rapid release occurred up to 24 hours, followed by a slow release up to 72 hours. They tested efficacy using a thermal activity monitor, which revealed that silver and copper wound dressings showed more potent antibacterial activity. This study demonstrated that the 3D technique is a simple way of potentially and affordably manufacturing personalized wound dressings.

10.8 POTENTIAL BENEFITS OF 3D AND 4D PRINTED PRODUCTS

The conventional manufacturing processes of pharmaceutical products were introduced over 200 years ago and are still in use due to cost-effectiveness and large-scale manufacturing [38]. However,

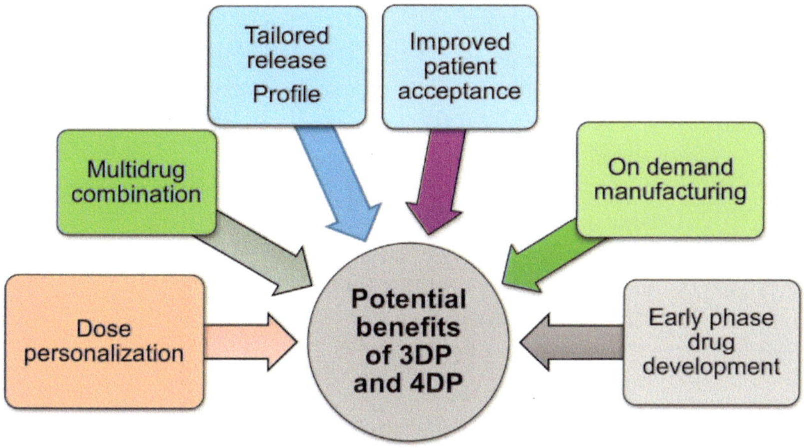

FIGURE 10.4 Potential benefits of 3D and 4D printing.

these processes are time-consuming and labor-intensive as well as dose inflexible. The pharmaceutical industry has never explored transformative technology in anticipation of 3D and 4D printing. This technique changes the face of healthcare industries and could best serve its potential benefits toward patient compliance and the manufacturing of personalized medicine, as discussed below and shown in **Figure 10.4**.

10.9 DOSE PERSONALIZATION

With less strength available for treating diseases, there is a need for different and specific dosing of active drugs for the elderly and children. Proper dosing is not possible due to the unavailability of personalized dosing in the current system. Now, it's peak time to change from the conventional system to the newer 3D and 4D techniques so that we can overcome all the issues [39, 40].

10.10 MULTIDRUG COMBINATION

There are many clinical conditions that cannot be treated with single-drug dosing, and they need to be cured with multiple drug combinations in different proportions. This medication is also known as polypharmacy. The conventional manufacturing process does not support personalized multidrug manufacturing, as it can only produce fixed-dose products in bulk. Thus, patient requirements in terms of medication are not taken into consideration [12, 41].

3D and 4D printing are flexible technologies by which multiple drugs can be incorporated into printed medicine. In the literature, it is also reported to have different release profiles. Complex controlled drug delivery can also be one of the major benefits of 3D and 4D printing [42].

10.11 TAILORED RELEASE PROFILE

In the pharmaceutical sector, different rate-controlling polymers are being used for manufacturing purposes. With the proper selection of polymer, compatible printer, and appropriate processing parameters, we can achieve the desired release profile. The novel strategy of 4D printing has promising potential to manufacture controlled drug delivery products that will have predetermined release patterns influenced by exterior stimuli (pH, temperature, light), etc. [35].

10.12 IMPROVED PATIENT ACCEPTABILITY

Patient acceptance can increase because of 3D and 4D printed products. The conventional mass manufacturing process does not account for individualized prescriptions and compliance with formulations. Patient-friendly formulations can be fabricated with the involvement of the patient or caretaker in terms of digitalized prescriptions.

10.13 ON-DEMAND PRINTING

This printing technology empowers its use in immediate manufacturing. It is the best suitable technique to manufacture less stable products or those with a shorter shelf life. In this case, the desired dosage form can be prepared prior to its administration.

10.14 EARLY-PHASE DRUG DEVELOPMENT

In this stage, the transition from the preclinical stage to phase 1 clinical trials is studied. During this study, different doses are administered to gain information regarding safety, toxicity, quality, and efficacy. Depending on the data from early development, further studies of specific APIs are accepted or rejected. In drug development, at every stage, formulations need to be prepared for study. Cost-effectiveness, higher bioavailability, dose flexibility, and rapid progression should be present in an ideal formulation. Formulations created by conventional systems do not support such requirements. 3D printing could be advantageous as it enables the production of small-scale batches at a lower cost and efficiently. These types of techniques could help expedite clinical trial entrance. They also facilitate speedy clinical go or no-go conclusions in the early development of drugs.

10.15 REGULATORY ASPECTS OF 3D AND 4 D PRINTING

3D and 4D printing are being termed the next big thing in revolutionizing the pharmaceutical industry. However, the approval process for medicinal products employing these technologies faces many challenges. These include technological and regulatory aspects. Pharmaceutical products manufactured by novel technologies like 3D and 4D printing must meet the proposed regulatory guidelines from bodies like the FDA and EMA, etc., for market approval [43]. Agencies such as the FDA classify medical devices into three different classes, each requiring specific technical data for clinical approval. Similarly, drug products manufactured using 3D printing must undergo a similar regulatory approval process as conventional products. Most medical devices manufactured using 3D printing technologies were approved via 510 (k) pathways. Class III devices, which are in the high-risk zone, should be subjected to stringent clinical evaluation for approval. However, there has been no 3D printed device with pre-market approval (PMA). Similarly, the foremost 3D printed tablet formulation was permitted by the FDA through 505 (b) 2 filing [44]. Since the manufacturing variables for pharmaceutical products made by 3D printing are different, the regulations regarding these technologies need to be altered to address the safety and efficacy aspects of these formulations. Considering the significance of the discrepancies that may arise in the approval process, the FDA has released draft guidance on "Technical Considerations for Additive Manufactured Devices" in May 2016. The guideline emphasizes various aspects of regulatory requirements for a 3D printed medical device or product [45].

These aspects can be categorized into various sub-sections, such as software, printing device, raw material, product validation, process validation, revalidation, personnel validation, data integrity, cybersecurity, and aspects related to the liability of personnel involved in the manufacture of on-demand personalized products.

1. **Software:** These are used in 3D printing to design the digital format of the intended device structure. All the software registrations and standards should conform to the individual country guidelines and norms. Similarly, proper file formatting is essential to ensure a

smooth approval process. The FDA suggested that the manufacturer follow the guidelines as specified in the ISO/ASTM 52915 Standard specification aimed at additive manufacturing file format (AMF) [39].

2. **Printing device:** It is an integral part of the manufacturing process. Generally, the FDA doesn't regulate the design of 3D printers. However, the body has provided a certain framework to familiarize manufacturers with printer design. The machine setup and maintenance shall follow the existing regulatory guidelines to ensure the robustness of the manufacturing process. Similarly, any support structure used to aid any specific layering procedure should also be included in the documentation so that the specified standards will be maintained.
3. **Raw material:** It forms a critical aspect of the 3D printed formulations. The information regarding the incoming material shall be documented properly to aid in the validation and subsequent filing. The raw material details that shall be stored include CAS identifiers, certificates of analysis (COA), and standard testing procedures in compliance with ASTM/ISO. The state of the material and relevant physiochemical properties should be clearly documented and provided to the regulator.
4. **Product characterization:** of 3D printed products needs to be evaluated for various parameters to ensure the continuity of quality standards of each batch. The finished product will be characterized for mechanical testing, dimensional testing, and material testing. The final product shall adhere to the specified dimensional and mechanical attributes to ensure consistency in quality management. Material testing has the highest importance among the above as it mainly deals with the safety aspects of the final product. All the raw materials, such as polymers, shall be tested for chemical and physical parameters. The materials, such as polymers, are usually subjected to various melting and re-solidification processes where there is a scope for the formation of new chemical moieties which, if not properly evaluated, may pose severe health risks to patients. The material details shall be properly stated by the manufacturer, including the Chemical Abstract Service (CAS) number and device master file (MAF) wherever applicable. Similarly, the physical properties of materials used, such as adhesive/cohesive properties, microstructure, material homogeneity, microstructural voids, percentage cross-linking, degree of curing, and crystal morphology, should be determined and documented [46]. All the data should be properly documented for regulatory filings.
5. **Process validation and revalidation:** These play a major role in setting the quality standards. Variability of various parameters during manufacturing may affect the quality of the final product. Hence, the process validation needs to be done along with the in-process monitoring to confirm the quality of the produce. Similarly, any change in the process parameter (e.g., software, raw material, and machine orientation) needs to be subjected to revalidation procedures. All the validation data should be documented using existing guidelines.
6. **Cleaning and sterility:** Since the materials or devices manufactured by 3D printing have varied geometry and porosity, the sterilization and cleaning processes in such cases pose a challenge to the manufacturer. Increased surface area as a result of the varied geometry of the final device may make it difficult to ensure the removal of residues. In such cases, the manufacturer needs to provide necessary validation documents for sterilization and cleaning. It is advised that the manufacturer follow guidelines such as "Submission and Review of Sterility Information in Premarket Notification (510(k)) Submissions for Devices Labelled as Sterile."
7. **Data integrity and cyber security:** Since 3D printing can be a remotely operated and fully automated process, proper data integrity measures should be followed as per individual regulatory bodies. These remotely operated processes pose special cybersecurity concerns as they may be liable to hacking, process tampering, and formatted file tampering, which may compromise product quality and thus adversely affect patients. Such activities should be strictly controlled, and manufacturers should follow the procedures to ensure there are no discrepancies in these areas.

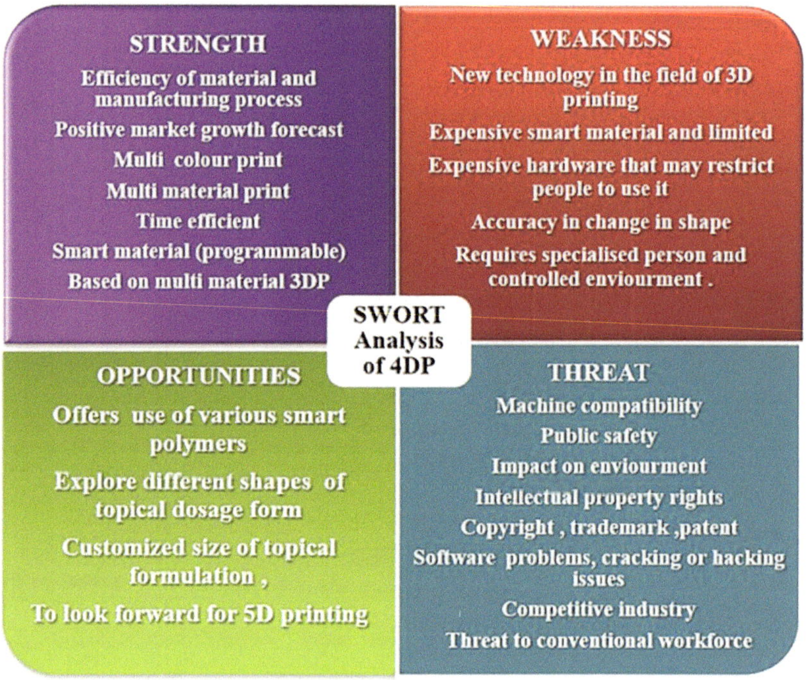

FIGURE 10.5 SWORT analysis of 4D printing.

Until now, drug formulations employing 3D printing techniques were approved using the data that would be required for conventional formulations, as the types of excipients used for the tablets were the same. However, with changes in manufacturing techniques, the above-mentioned parameters need to be considered to make safe and effective 3D-printed pharmaceutical formulations.

The strengths, weaknesses, opportunities, and threats of 3D and 4D printing are depicted in **Figures 10.5 and 10.6**.

10.16 FUTURE CONSIDERATIONS

The 3D printing and 4D printing are currently rapidly growing in the pharmaceutical industry, where they are providing superior product features, easy manufacturing, and the ability to personalize medicines. Besides their extensive usage in medical devices and tissue engineering, in recent years, 3D and 4D printing have been widely used in the pharmaceutical industry for manufacturing solid oral dosage forms and in the construction of transdermal drug delivery systems. However, limited research has still been done in the area of topical dosage forms. In the near future, we can expect more research on topical dosage forms and more personalized products by addressing existing problems of 3D and 4D printing related to topical drug delivery. Scientists may even come up with new types of dosage forms, as a limited variety of topical dosage forms can currently be printed using 3D and 4D printing. Pharmaceutical scientists can take inspiration from Spritam for their innovative projects.

Although the conventional manufacturing of topical dosage forms can be altered by 3D and 4D printing in terms of greater efficiency and customized drug delivery, this will lead to improved patient compliance. Topical formulations such as gels, creams, pastes, lotions, and some patches or masks can be prepared with the help of this rapidly growing technology. Skin disorders such as psoriasis, hyperpigmentation, allergic reactions, or inflammations need customized products to cure them, depending on the type of clinical condition. Future applications of 3DP could involve printing doses based on

FIGURE 10.6 SWORT analysis of 3D printing.

the genetic profile, metabolic trajectory, and biological markers of the patient with the development of genetic testing. Their individual medication and dosage could then be directly dispensed at the point of treatment, which would pose a challenge for centralized production processes.

10.17 CONCLUSION

The 3D printing and 4D printing technologies are rapidly growing in the pharmaceutical field. 3D printing applied in topical drug delivery for making nose-shaped masks, vaginal rings, and drug-loaded hydrogels has shown promising results, advantageous features, and greater chances of personalizing the dosage forms easily. Among the various technologies used in 3D printing, FDM and SLA technologies have been used for topical drug delivery, and they are feasible. A 4D printed product is the product of 3D printing with smartness by stimuli-responsive materials. 4D printed products are more beneficial in terms of treating diseases because of their smartness. There are different challenges compared to other pharmaceutical products for these 3D printed and 4D printed products from regulatory bodies for their approval. The principal concerns for these items include material controls, parameter analysis during designing and printing, evaluation of the final appliance's physical characteristics, and safety considerations in relation to cleaning, sterility, and biocompatibility of the final appliance. With all the issues addressed, 3D and 4D printing can dominate formulation technology in topical drug delivery.

REFERENCES

1. M. Camović *et al.*, "Coated 3D printed PLA microneedles as transdermal drug delivery systems antimicrobial local applications view project lipid nanoparticles as drug delivery system for micro RNA view

project coated 3D printed PLA microneedles as transdermal drug delivery systems," *Springer*, 2020, vol. 73, pp. 735–742.
2. J. J. Schwartz and A. J. Boydston, "Multimaterial actinic spatial control 3D and 4D printing," *Nature Communications*, 2019, vol. 10, no. 1, pp. 1–10.
3. A. Alhnan and T. Chijioke, "Emergence of 3D printed dosage forms: Opportunities and challenges," *Pharmaceutical Research*, 2016, vol. 33, no. 8, p. 33.
4. M. Palo, J. Holländer, J. Suominen, J. Yliruusi, and N. Sandler, "3D printed drug delivery devices: Perspectives and technical challenges," *Expert Review of Medical Devices*, 2017, vol. 14, no. 9. Taylor and Francis Ltd, pp. 685–696, Sep. 02.
5. D. Yu, L. Zhu, C. Branford-White, and X. L. Yang, "Three-dimensional printing in pharmaceutics: Promises and problems," *Journal of Pharmaceutical Sciences*, 2008, vol. 97, no. 9, pp. 3666–3690.
6. N. Elkasabgy, A. Mahmoud, A., and A. Maged, "3D printing: An appealing route for customized drug delivery systems," *International Journal of Ph*, 2020, vol. 5, p. 119732.
7. L. K. Prasad and H. Smyth, "3D Printing technologies for drug delivery: A review," *Drug Development and Industrial Pharmacy*, 2016, vol. 42, no. 7. Taylor and Francis Ltd., pp. 1019–1031.
8. J. Goole and K. Amighi, "3D printing in pharmaceutics: A new tool for designing customized drug delivery systems," *International Journal of Pharmaceutics*, 2016, vol. 499, no. 1–2, pp. 376–394.
9. Y. Sun, S. Soh, Y. Sun, and S. Soh, "Printing tablets with fully customizable release profi les for personalized medicine," *Advanced Research*, 2015, vol. 27, no. 47, pp. 7847–7853.
10. J. N. Farhang, M. M. Seyed, and L. Xun, "A review of 4D printing," *Mater Des*, 2017, vol. 112, pp. 42–79.
11. J. Firth, S. Gaisford, and A. W. Basit, "A new dimension: 4D printing opportunities in pharmaceutics," in *AAPS Advances in the Pharmaceutical Sciences Series*, vol. 31, Springer Verlag, 2018, pp. 153–162.
12. M. Alomari, F. Mohamed, A. W. Basit, and S. Gaisford, "Personalised dosing: Printing a dose of one's own medicine," *International Journal of Pharmaceutics*, 2015, vol. 494, no. 2, pp. 568–577.
13. A. Sydney Gladman, E. A. Matsumoto, R. G. Nuzzo, L. Mahadevan, and J. A. Lewis, "Biomimetic 4D printing," *Nature Materials*, 2016, vol. 15, no. 4, pp. 413–418.
14. Y. Zhou *et al.*, "From 3D to 4D printing: Approaches and typical applications," *Journal of Mechanical Science and Technology*, 2015, vol. 29, no. 10, pp. 4281–4288.
15. L. L. Jorge, C. C. Feres, and V. E. P. Teles, "Topical preparations for pain relief: Efficacy and patient adherence," *Journal of Pain Research*, 2011, vol. 4. pp. 11–24.
16. U. Jammalamadaka and K. Tappa, "Functional biomaterials review recent advances in biomaterials for 3D printing and tissue engineering," *Mdpi*, 2018, vol. 9, no. 1, p. 22.
17. I. Ursan, L. Chiu, and A. Pierce, "Three-dimensional drug printing: A structured review," *Journal of the American Pharmacists Association*, 2013, vol. 53, no. 2, pp. 136–144.
18. W. Jamróz, J. Szafraniec, M. Kurek, and R. Jachowicz, "3D Printing in pharmaceutical and medical applications—recent achievements and challenges," *Pharmaceutical Research*, 2018, vol. 35, no. 9. Springer New York LLC, pp. 1–22.
19. T. Ngo, A. Kashani, G. Imbalzano, K. T. Nguyen, and D. Hui, "Additive manufacturing (3D printing): A review of materials, methods, applications and challenges," *Composites Part B: Engineering*, 2018, vol. 143, pp. 172–196.
20. K. Arcaute, B. Mann, and R. Wicker, "Stereolithography of spatially controlled multi-material bioactive poly(ethylene glycol) scaffolds," *Acta Biomaterialia*, 2009, vol. 6, no. 3, pp. 1047–1054.
21. K. Arcaute, "Stereolithography of three-dimensional bioactive poly(ethylene glycol) constructs with encapsulated cells," *Annals of Biomedical Engineering*, 2016, vol. 34, no. 9, pp. 1429–1441.
22. F. Melchels, J. Feijen, and D. Grijpma, "A review on stereolithography and its applications in biomedical engineering," *Biomaterials*, 2010, vol. 31, no. 24, pp. 6121–6130.
23. A. Awad, S. Trenfield, A. Goyanes, S. G.-D. discovery Today, and U. 2018, "Reshaping drug development using 3D printing," *Drug Discovery Today*, 2018, vol. 23, no. 8, pp. 1547–1555, Mar. 26.
24. A. Goyanes, U. Det-Amornrat, J. Wang, . . . A. B.-J. of controlled, and U. 2016, "3D scanning and 3D printing as innovative technologies for fabricating personalized topical drug delivery systems," *Elsevier*, 2016, vol. 234, pp. 41–48.
25. E. Mathew, G. Pitzanti, E. Larrañeta, and D. A. Lamprou, "Pharmaceutics 3D printing of pharmaceuticals and drug delivery devices," 2020, p. 266.
26. M. Cunha-Filho, M. R. Araújo, G. M. Gelfuso, and T. Gratieri, "FDM 3D printing of modified drug-delivery systems using hot melt extrusion: A new approach for individualized therapy," *Therapeutic Delivery*, 2017, vol. 8, no. 11, pp. 957–966.

27. D. Khong Tan, M. Maniruzzaman, and A. Nokhodchi, "Pharmaceutics advanced pharmaceutical applications of hot-melt extrusion coupled with fused deposition modelling (FDM) 3D printing for personalised drug delivery," *Pharmaceutics*, 2018, vol. 10, no. 4, p. 203.
28. G. Kollamaram, D. Croker, G. M. Walker, A. Goyanes, A. W. Basit, and S. Gaisford, "Low temperature fused deposition modeling (FDM) 3D printing of thermolabile drugs," *International Journal of Pharmaceutics*, 2018, vol. 545, no. 1–2, pp. 144–152.
29. J. Long, H. Gholizadeh, J. Lu, and A. Seyfoddin, "Review: Application of fused deposition modelling (FDM) method of 3D printing in drug delivery immobilisation of drugs in aerogels view project stimuli responsive drug delivery: Personalized therapy view project," *Current Pharmaceutical Design*, 2017, vol. 23, no. 3, pp. 433–439.
30. V. Surange, P. G. E. and T. (IRJET), U. 3, and U. 2016, "3D printing process using fused deposition modelling (FDM)," *International Research Journal of Engineering and Technology*, 2016, vol. 3, no. 3.
31. M. Alhijjaj, P. Belton, and S. Qi, "An investigation into the use of polymer blends to improve the printability of and regulate drug release from pharmaceutical solid dispersions prepared via," *European Journal of Pharmaceutics and Biopharmaceutics*, 2016, vol. 108, pp. 111–125.
32. T. C. Okwuosa, D. Stefaniak, B. Arafat, A. Isreb, K.-W. Wan, and M. A. Alhnan, "A lower temperature FDM 3D printing for the manufacture of patient-specific immediate release tablets," 2016.
33. M. Hayat, A. Nawaz, S. Chinnam, M. Muzammal, M. S. Latif, M. Yasin, S. Ashique, G. Zengin, A. Farid, "Formulation development and optimization of herbo synthetic gel: In vitro biological evaluation and in vivo wound healing studies," *Process Biochemistry*, 2023, vol. 130, pp. 116–126, Jul. 1.
34. T. Feuerbach, S. Kock, and M. Thommes, "Characterisation of fused deposition modeling 3D printers for pharmaceutical and medical applications," *Pharmaceutical Development and Technology*, 2018, vol. 23, no. 10, pp. 1136–1145.
35. M. Bakhatwar, S. Chikkala, and V. Krishna, "Three-dimensional printing in pharmaceutical technology–An overview of innovations," 2019.
36. J. Fu, X. Yu, and Y. Jin, "3D printing of vaginal rings with personalized shapes for controlled release of progesterone," *International Journal of Pharmaceutics*, 2018, vol. 539, no. 1–2, pp. 75–82.
37. P. R. Martinez, A. Goyanes, A. W. Basit, and S. Gaisford, "Fabrication of drug-loaded hydrogels with stereolithographic 3D printing," *International Journal of Pharmaceutics*, 2017, vol. 532, no. 1, pp. 313–317.
38. G. Chen, X. Yihua, P. Kwok, and L. Kang, "Pharmaceutical applications of 3D printing," *Additive Manufacturing*, 2020, p. 101209.
39. M. Preis and H. Öblom, "3D-printed drugs for children—are we ready yet?" *AAPS PharmSciTech*, 2017, vol. 18, no. 2, pp. 303–308.
40. K. Pietrzak, A. Isreb, and M. Alhnan, "A flexible-dose dispenser for immediate and extended release 3D printed tablets," *European Journal of Pharmaceutics and Biopharmaceutics*, 2015, vol. 96, pp. 380 387.
41. N. Masnoon, S. Shakib, L. Kalisch-Ellett, and G. E. Caughey, "What is polypharmacy? A systematic review of definitions," *BMC Geriatrics*, 2017, vol. 17, no. 1. BioMed Central Ltd., pp. 1–10.
42. R. L. Maher, J. Hanlon, and E. R. Hajjar, "Clinical consequences of polypharmacy in elderly," *Expert Opinion on Drug Safety*, 2014, vol. 13, no. 1. pp. 57–65.
43. A. Khairuzzaman, "Regulatory perspectives on 3D printing in pharmaceuticals," in *Advances in the Pharmaceutical Sciences Series*, vol. 31, Springer Verlag, 2018, pp. 215–236.
44. A. Bandyopadhyay and B. Heer, "Additive manufacturing of multi-material structures," *Materials Science and Engineering R: Reports*, 2018, pp. 1–16, Jul. 01.
45. C. M. Madla, S. J. Trenfield, A. Goyanes, S. Gaisford, and A. W. Basit, "3D printing technologies, implementation and regulation: An overview," in *Advances in the Pharmaceutical Sciences Series*, vol. 31, Springer Verlag, 2018, pp. 21–40.
46. A. M. Nissan, "Regulating the three-dimensional future: How the FDA should structure a regulatory mechanism for additive manufacturing (3D printing)," *Boston University Journal of Science & Technology Law*, 2016, vol. 22, p. 267.

11 Polymeric Nanomaterials in Wound Healing and Tissue Remodeling

Blessy Baby Mathew,[*1,2] Parinika Ajoy Raj,[3] G. Shivashirin,[4] Juwariah Fazal,[5] Shreya Prakash,[6] and Ankitha H. K.[7]*

[1] School of Basic & Applied Sciences, Dayananda Sagar University, Bengaluru-11, Karnataka, India

[2,3,4,5,6,7] Department of Biotechnology, Dayananda Sagar College of Engineering, Bengaluru-11, Karnataka, India

11.1 INTRODUCTION

Any formulation that could introduce the useful components inside the host organism in the active state where it can prove to be efficient is called a drug-delivery system. The safety parameters can be wisely selected, and the concentration of the drug and its release can be modified based on certain factors. The drug must be able to act at its full potential at the site of action, passing through the biological membranes to deliver its therapeutic properties. These routes might have various impacts, relying upon how they are applied (1, 2). The route of oral administration restricts the utilization of pH-safe or exceptionally hydrophilic drugs to guarantee the expected assimilation and uptake by the digestive epithelium cells. Similarly, the obtrusive nature of infusions has been related to undesirable outcomes and a poor prognosis of disease. This emphasizes the need for a favorable process and a better-controlled drug release system, maintaining therapeutic viability. The dynamic focus is to target the site of action for the desired function and activity of the active component (3, 4). Nanoparticles are such materials that can be used to enhance the bioavailability of the drug by providing a greater surface area where a desirable drug molecule can be linked or attached. This can significantly enhance absorption and facilitate easy penetration across the biological membranes to reach the target site. These particles can be combined with a suitable polymer or a polymer matrix by incorporating specific cross-linking agents to reach an optimum composition and behavior where the functionality of the drug molecule can be made suitable for a required application (5–8). When nanoparticles are used as efficient drug carriers, they can display unique features such as a high surface area-to-volume ratio, better drug release, and site-target specificity. Over the course of time, nanoparticles have become very appealing for their varied applications in science and medicine. These nanometric-sized particles can possibly balance biopharmaceutical highlights, pharmacokinetic properties, and the restorative viability of captured drugs (9, 10).

Nanoparticles that are combined with a polymer consisting of active components but still range from 1 to 1000 nm in size are called polymeric nanoparticles. The incorporation of any suitable polymer can result in interior entrapment or just a surface phenomenon, as seen in the case of adsorption, and lead to the development of different morphological assemblies (11). The polymeric nanoparticles can respond well to internal and external stimuli and can take up and hold within themselves an assorted range of therapeutic freight, competently releasing the encapsulated drug load at the target location. This is mainly due to their wide range of surface functionality, varied size, and shapes (12). The morphology and charge of these polymeric nanoparticles can be easily

 DOI: 10.1201/9781003605966-11

changed as per the requirements of the application. Various polymers such as polylactic acid, gelatin, polycaprolactone, chitosan, polyvinyl alcohol, alginic acid, and polylactide-co-glycolide have been explored for building efficient drug delivery systems by incorporating nanomaterials into them. By adding sulfide bonds to the available polymeric nanoparticles, the ability to control drug release can be handled well. Anionic polymers have been found to be a good choice, but cationic polymers are usually toxic and are not degradable or biocompatible, due to which they are not preferred for drug delivery and tissue remodeling applications (13). Not only do these particles showcase huge potential in sustained and targeted drug delivery, but they can also be used as latent drug carriers for drug protection, controlled release, and reduction of overactivity in a short period to avoid drug accumulation, thus promoting better metabolism. This shields the specific biological activity from the external environment and enhances the bioavailability of the drug, thereby increasing its therapeutic potential as well. The continuous polymeric network that underpins nanospheres allows for the retention of drugs inside or adsorbed onto their surface, whereas nanocapsules consist of a core containing oil in which the drug is found in dissolved form and is protected by a polymer coating that regulates drug release (11). Numerous uses for polymer nanoparticles have been discovered in the biomedical industry. Biological molecules have been concentrated, purified, and confined using polymer nanoparticles. When combined with a complicated biological mixture, a responsive polymer nanoparticle can bind specifically to one target molecule, separate it from the medium, and then release the desired molecule. Both in vitro and in vivo biosensing have been achieved with polymer nanoparticles. An artificial extracellular matrix made of polymer nanoparticles can be effectively used in the culture of different biological tissues. Nonetheless, polymer nanoparticles are mostly employed in the biomedical field as medication delivery systems (14, 15).

When compared to other drug delivery platforms like liposomes or inorganic nanoparticles, polymer nanoparticles provide a number of advantages. Polymer nanoparticles, for instance, have exceptional stability even in intricate settings or at low concentrations. They can be reasonably easily functionalized with a range of active targeting ligands, and their size and shape can be modified for passive targeting (16). When compared to alternative drug delivery methods, polymer nanoparticles often exhibit higher loading efficiency, improving the mass ratio of therapeutic medicines to carrier materials. Regardless of the hydrophilicity and molecular weight of the therapeutic agents, the flexibility in the synthesis also permits the codelivery of various therapeutic agents in a single system.

Polymer nanoparticles can be used to release the encapsulated cargo in a controlled and selective manner. The polymer nanoparticles that are utilized to deliver drugs have multiple purposes. The medicine can be shielded by the polymer nanoparticle from the severe conditions found in biological systems, such as the digestive system's acidic pH or the presence of enzymes that can break down a wide range of compounds, including therapeutic drugs (17, 18). To do this, the active chemical can be encapsulated inside polymer particles that will mitigate the effects of the surrounding environment. For the intended uses, the drug's encapsulating vehicle must also maximize the active molecules' circulation time. The bioavailability of the medication may be increased by functionalizing the polymer nanoparticles with "stealth" properties (19, 20). Polymeric nanoparticles are used to create nanostructured polymers, which have better physical qualities like flexibility, sensitivity to temperature and moisture, durability, and resistance to gas or water intrusion. The antibacterial capabilities of certain polymeric nanoparticles, including those derived from chitosan, are well-established. These kinds of polymeric nanoparticle composite materials can be made using a variety of techniques, including electrospinning from nanofibers, layer-by-layer deposition, and direct emulsification of monomers into form (21–23). It has been demonstrated that a polymeric nanoparticle's physicochemical properties, such as its surface charge and functional groups, can influence how well the phagocytic system's cells absorb it. Prior research has demonstrated that polystyrene microparticles featuring a primary amine on their surface experienced a notably higher rate of phagocytosis in contrast to microparticles with carboxyl, hydroxyl, and sulfate groups. Therefore, it is widely acknowledged that formulations with a positive charge have a higher rate of cell absorption than those with a neutral or negative charge. Apart from their roles in opsonization and improved hepatic absorption, the precise function of each protein adsorbed on the surface of the nanoparticles in the clearance and biodistribution remains unclear. It is widely believed that receptor-mediated processes of absorbed blood proteins (opsonins) onto their surface and complement activation cause the fast uptake of administered nanoparticles. A high rate of phagocytic uptake may be caused by a particular type of protein, a mixture of proteins, or even a change in the structure of the protein (24).

11.2 DESIGN AND DEVELOPMENT OF POLYMERIC NANOPARTICLES

Colloidal systems, which usually have sizes ranging from 5–10 nm to an upper limit of 1000 nm, can be defined as polymer nanoparticles. They act as transporters for molecules that are physiologically active and can encapsulate the particles or be adsorbed on the exterior of the sphere. Medications, genes, nucleic acids, and other functional materials are considered physiologically active materials. Both natural polymers and synthetic polymers are being used in the preparation process. Most natural polymers are biodegradable, whereas synthetic polymers are not. Therefore, natural polymers are preferred for preparing polymeric nanoparticles. Natural polymers such as starch, albumin, sodium alginate, chitin, gelatin, and cellulose are commonly used, whereas common synthetic polymers like polyvinyl alcohol (PVA), polyethylene glycol (PEG), polyvinyl pyrrolidone (PVP), and polyethylene (PE) are used in the making of polymeric nanoparticles. Metal ions, tiny molecules, surfactants, or polymers can be used on the surface of polymer nanoparticles to improve targeting, bind with ligands, and prevent certain immunological responses (25).

The development of polymeric nanoparticles consists of two parts: (i) micro-ionization, which converts a material into nanoparticles, and (ii) the stabilization of the resulting nanoparticles. For the micro-ionization part, bulk polymers or small monomers are used. Emulsion polymerization is a traditional technique that can help make polymeric particles in the range of 102-103 nm. A large amount of surfactant or cosurfactant is added to make smaller nanoparticles. Adding surfactants to the nanoparticles disrupts the polymer solid content in the dispersion and also limits its applications. Removing a surfactant from a dispersion can affect stability and is extremely difficult. The preparation of a concentrated, uniform polymeric nanoparticle without any surfactants that is also stable in water has been a challenge until now. Natural polyelectrolytes and proteins are being used as surfactants, as they help in the stabilization of nanoparticles due to their low molar mass. They are also stabilized by surface modification, copolymerization, and by grafting the polymer chains on the particle surface (26).

Polymeric nanoparticles have started to receive more interest in recent times due to their versatile nature. Their structures can be modified into packages that deliver the desired substance to the target site and bring about the specific reaction required by the substance. Incorporating certain functionalities into the nanoparticles can help improve responsiveness under different conditions like temperature, pH, and different enzymatic environments. Variation in these conditions should typically not affect the nanoparticles, and they should carry out their functions effectively. Therapeutics based on polymeric nanoparticles are under development to improve the diagnosis of viral, pulmonary, and urinary infections and also help with cancer treatment. Improvements in traditional methods of controlled polymerization have helped engineer several multifunctional polymeric nanoparticles, where there is very good control over the shape, size, and architecture of the nanoparticles. Control over these properties ensures that the nanoparticles have good versatility in different applications. There are some barriers that the nanoparticles have to go through to reach their target. Some of these are mentioned below:

11.2.1 Biological Barriers

Drugs have to reach their site of action to obtain any therapeutic benefit. These drugs can either be free or loaded into nanoparticles, and their target is usually on the molecular or cellular level.

The barriers that these nanoparticles encounter on their way to the target site can be classified into external barriers, which include skin and mucosa. Others are *en-route* barriers, which include the extracellular matrix and the blood. Cellular barriers consist of degradation by endosomes or lysosomes and can also be limited by cellular uptake.

11.2.2 External Barriers

The entire surface of our body is protected by the skin and mucus. They hinder the uptake of polymeric nanoparticles by the target sites that usually lie under the tissues or are present in the blood. Skin and mucus are different structures, and they prevent polymeric nanoparticle uptake through different mechanisms. They can also alter some of the characteristics of the nanoparticles and affect their stability before they reach their target.

11.2.2.1 En-Route Barriers

Injecting the nanoparticles helps them cross barriers like skin and mucus easily. However, injecting these nanoparticles can be unfavorable, as it is invasive and increases the costs of administering these substances. The nanoparticles can also get distributed into tissues and organs during circulation. When blood is filtered in the kidneys, some of the blood is reabsorbed back and sent into circulation. A part of the blood is then further processed, which secretes some components into the urine. This can help with the excretion of the nanoparticles.

11.2.3 Rational Design of Polymeric Nanoparticles

Efforts are being made to design and develop polymeric nanoparticles with specific biological, chemical, and physical properties. Some characteristics, like size, surface, and structure of the nanoparticles, determine their stability in blood and their intracellular distribution, as well as the bioavailability of the prepared nanoparticles. Some significant key points and segments to keep in mind while designing these particles are as follows:

11.2.3.1 Stealth Nanocarriers

Coating the nanoparticles with a dense layer of neutral hydrophilic polymers helps avoid interaction with biological fluids and also prevents adsorption into plasma proteins. It also helps in avoiding immunological recognition of the nanoparticles.

11.2.3.2 Smart Ingredients

Incorporating certain functions onto the surface or in the shell of the nanoparticle can help with their diffusion into biological barriers. It aids in the biodistribution of the nanoparticles into specific cells or organelles. They also help control and modulate the responsiveness of the nanoparticles to certain stimuli.

11.2.3.3 Accumulation at a Specific Site

Nanoparticles that have ligands present on their surface are specifically designed to enhance their binding to receptors that are expressed in excess in the target cells. This is beneficial for the nanoparticles, as it helps decrease the total time that normal cells are exposed to the drug. It also aids in accumulation at the target site.

11.2.3.4 Route of Administration

Nanoparticles can be administered through oral, nasal, skin, vaginal, pulmonary, rectal, and ocular routes, even though injection is the most common. The route of administration that is selected is just as important as the nanoparticle design. There are challenges and barriers in every route, and being aware of them is very important for designing a good nanocarrier (27).

11.3 PREPARATORY METHODS OF POLYMERIC NANOPARTICLES

11.3.1 Solvent Evaporation Method

Organic solvents are typically used to dissolve a polymer and make a solution. The drug is then dissolved into the polymer solution. These two phases are then emulsified using a surfactant and stirred continuously. After a certain period of time, a stable emulsion is obtained, and the organic solvent that remains can either be evaporated through continuous heating. The polymer nanoparticles are obtained through centrifugation and washed with distilled water to remove any impurities that are present. Then the obtained product is lyophilized.

11.3.2 Emulsification-Solvent Diffusion Method

This technique is specifically developed for preparing a polymer nanoparticle. The polymer, which will encapsulate the drug, is first dissolved in a water-soluble solvent. The polymer is precipitated, and the nanoparticles obtained as a result are diluted with water or a different organic solvent. The polymer, which is saturated with water, is emulsified in an aqueous solution that has a stabilizer present in it. The solvent is then removed through evaporation.

11.3.3 Monomer Polymerization

This technique is used for monomers that have a low solubility in water. The monomer, initiator, and surfactant are mixed with water at a certain temperature for a specific amount of time to aid in the polymerization. Polymeric nanoparticles are obtained after this time period and have some amount of surfactant trapped in them. Removal of these surfactants is very difficult and it is cost-effective (28). Many other methods of polymerization are used for making the polymeric nanoparticles. Some of them are solvent displacement, salting out, dialysis, and interfacial polymerization (29).

11.3.4 Mini-Emulsion Polymerization

Research on mini-emulsion-based polymerization has gained popularity in recent years, and it has contributed to the development of a wide variety of polymeric nanoparticles and their improvement. Monomer blends, stabilizers, and surfactants are the components that are commonly used in the process of mini-emulsion polymerization. Using molecules with lower molecular masses as co-stabilizers and utilizing a high-shear device is the main distinction between mini-emulsion polymerization and emulsion polymerization. The mini-emulsion is stabilized using great shear, and it attains a steady state condition (30).

11.3.5 Microemulsion Polymerization

This is another polymerization method that has proven to be an effective method for producing polymeric nanoparticles. Both emulsion and microemulsion polymerization produce colloidal particles due to the emulsion of particles with large molecular mass, but they differ kinetically. The number of strings and particle size are comparatively smaller in this method. Initiators that are soluble in water are added to a thermodynamic emulsion in an aqueous state. The steady state generated from this aqueous mixture is where the polymerization begins. Adding a larger quantity of surfactant completely coats the particles. This causes changes in the osmotic impact of the particles, helping the nanoparticles grow and assisting in secondary nucleation and the creation of empty micelles (30).

11.3.6 Reverse Salting-Out

When a solvent that is miscible in water is separated from an aqueous solution, the salting-out process is said to take place. The salting-out method is based on the separation of a hydro-miscible solvent from an aqueous solution through a salting-out agent with a gel and a colloidal stabilizer. Some salting-out agents are calcium chloride, magnesium acetate, and magnesium chloride. The water and acetone are saturated with the aqueous phase, and their miscibility is reduced. The emulsion is then prepared by stirring continuously at room temperature. Deionized water is added to the emulsion to dilute it and precipitate out the polymer. The remainder of the salting-out agent and solvent are removed by cross-flow filtration (31).

11.3.7 Nanoprecipitation

The process of nanoprecipitation uses two solvents that are miscible. The polymer is dissolved using an organic solvent such as acetonitrile or acetone. These organic solvents can be removed with the help of evaporation, as they are miscible in water. The polymer is first dissolved in a solvent with moderate solubility. This polymeric solution is then added drop by drop into an aqueous solution with continuous stirring. Since the polymer spontaneously diffuses into the aqueous solution, nanoparticles are formed as they try to avoid the water molecules present in the solution. Further, the solvent diffuses, and the polymer precipitates in the form of nanoparticles (31).

11.3.8 Supercritical-Fluid Technology

Solvents are used at a pressure and temperature that are well above their critical point in this method. This is known as supercritical-fluid technology. It is one of the most researched methods for the production of nanoparticles. It has gained a lot of attention more recently, as it typically uses solvents that are environmentally safe, and the contamination of these solvents in the final product is found to be very low.

11.3.9 Rapid Expansion of Supercritical Solution (RESS)

This method falls under supercritical-fluid technology. In this particular process, compounds are solvated in a supercritical fluid and expanded using a nozzle. Some of the supercritical solvents that are usually used are carbon dioxide and water. However, it is a rather impractical method, as the polymers that have been used for this method have very little solubility.

11.3.10 Small-Angle Scattering (SAS)

This method provides a solution to the limited solubility observed in the RESS method. The drug and the polymers are first dissolved in an organic solvent. Then the supercritical solvent is used as an anti-solvent, which helps precipitate the polymers from the organic solvent. The anti-solvent is made to flow through another vessel to remove any remnant organic solvent. The product is collected from the vessel after being depressurized. The only drawback of this method is that there is no control over the size of the nanoparticle and its properties (32).

11.4 EXAMPLES OF VARIOUS DRUG-LOADED POLYMERS AND NANOPARTICLES UTILIZED IN THE PREPARATION OF POLYMERIC NANOPARTICLES

11.4.1 Metal Oxide-Loaded Polymeric Nanoparticles

Metal oxide and metal nanoparticles can be effective against antibiotic-resistant bacteria. Polymeric carriers can encapsulate metal or metal oxide nanoparticles to improve antimicrobial activity. The advantage of this encapsulation is that it reduces the toxic effects of the metal/metal oxide nanoparticles, such as lipid peroxidation, proteins/nucleic acid damage, and cytoskeleton degradation (33).

Polymeric nanoparticles are colloidal and possess properties such as a higher surface-to-mass ratio, biodegradability, reduced toxicity, and the ability to absorb and carry other molecules. These have now been extensively used in drug delivery (34). Based on the synthesis, polymeric nanoparticles form two types of structures that are involved in drug delivery: Nanocapsules (drug is embedded in a cavity which is then surrounded by a polymer membrane) and Nanospheres (drug is uniformly dispersed in a matrix). The major component of the polymeric nanoparticle is the polymer, and this includes the drug/polymer conjugates, polymer micelles, and nanoparticles. The two important features taken into account when producing polymeric nanoparticles are: (1) the action of the active ingredients should not be compromised by the chemical characteristics of the polymer, and (2) the physical properties must be reproducible and consistent (34). The optimal size range is approximately__ to 200nm. They prevent drugs from being degraded, increase the halfx0life, improve bioavailability, and provide sustained release. Their surface area could be modified to improve effects such as targeted delivery, reduction of drug toxicity due to systemic spread, and other side effects. Polymeric nanoparticles are widely chosen for drug delivery due to their high stability, solubility in water, biodegradability, biocompatibility, and nonimmunogenicity. They also have the ability to encapsulate hydrophilic and hydrophobic drugs. By adjusting their composition or surface charge, their loading capacity, drug release kinetics, and biological performance can be regulated (35).

11.4.2 Mupirocin-Loaded Polymeric Nanoparticles

Nanoparticles can incorporate antibiotics and/or anti-inflammatory drugs to accelerate wound healing. One such antibiotic that is effective against bacterial skin infections and promotes wound healing is mupirocin. Mupirocin loaded in chitosan microspheres boosts the wound healing activity of a collagen scaffold (36). Based on this article, the extent of blood vessel formation was semi-quantitatively evaluated by immunohistochemical (IHC) labeling for alpha-smooth muscle actin. It was found that the number of vessels increased over time. The peak vascularization in the control, nanoparticle, and nanoparticle with mupirocin was recorded on the 21st day, whereas the peak vascularization for dressing with fish oil, mupirocin, and PLGA-PVA NPs with entrapped fish oil was on the 14th day (36). Mupirocin promotes angiogenesis during wound healing and increases the key mediator of angiogenesis, Vascular Endothelial Growth Factor (36). A hydrogel formulation loaded with mupirocin nanoparticles was successfully developed. Mupirocin can be released from hydrogels through various mechanisms, including diffusion, swelling, and chemically controlled methods. Mupirocin released from nanoparticles and hydrogel systems has the ability to permeate the cell wall and bind to isoleucyl-tRNA synthetase (37).

11.4.3 Chlorhexidine-Loaded Calcium Phosphate Polymeric Nanoparticles

Chlorhexidine and its derived products have a rapid kill rate against broad-spectrum bacteria and fungus and are hence used widely in medicine. Chlorhexidine is positively charged and easily binds to negatively charged cell surfaces. It causes the cell surface rupture at the interface through osmosis and leads to cell death through the attack on the cytoplasmic membrane. It has both bacteriostatic and bactericidal actions based on its concentration. It also has the ability to kill 100% of bacteria within 20 minutes, based on in vitro studies. Calcium phosphates are able to imitate bone mineral with perfect biocompatibility and are hence used extensively as bone substitutes in bone tissue engineering. They aid in delivering growth factors and drugs in tissue engineering. In the study [6], a calcium phosphate nanoparticle mixed hybrid mixture containing chlorhexidine was developed. Calcium phosphates have the ability to easily disintegrate in acidic pH, which results in the release of calcium ions onto the wound surface. The dressing created maintains a slightly acidic condition of the skin along with chitosan, gelatin, and chlorhexidine, which effectively inhibits bacteria. This acidic environment helps in the release of calcium ions from the calcium phosphate nanoparticles (38).

11.4.4 Alginate Polymer-Loaded Nanoparticles

A polysaccharide widely recognized as biocompatible and biodegradable, applied in various areas such as food manufacturing, cosmetics, tissue engineering, drug delivery, and wound management, is alginate. Dressings made from alginate can be used to treat superficial wounds and lacerations. Alginate is widely recommended in wound management of moderate to high amounts of exudate wounds, as it is a strongly absorbent material (39).

In recent years, metal and metal oxide nanoparticles have been used as antimicrobial agents as well as in the prevention of wound infections. Nanoparticles of silver, copper, zinc, gold, and their oxides are seen as potential tools against the alarming context of antimicrobial resistance. They have been applied in many biopolymeric hydrogels, especially alginate materials, in the field of wound healing (39). The interesting biopharmaceutical properties exhibited by alginate that make it suitable for modified release of drugs include biocompatibility, biodegradability, pH sensitivity, mucoadhesiveness, nontoxicity, and no immunogenicity. It has usually been orally administered for the treatment of reflux esophagitis, making it a popular pharmaceutical excipient (40). Various cross-linking techniques can be used to prepare alginate hydrogels, which have gained popularity in wound healing, drug delivery, tissue engineering, and the delivery of bioactive agents, as they retain structural similarity to extracellular matrices and can be modified to aid in many crucial roles. Wound dressings made from alginate have the ability to create a moist microenvironment, reduce infection by bacteria at the wound site, and promote wound healing. Depending on the types and cross-linking techniques used, alginate gels can release small chemical drugs to macromolecular proteins in a controlled manner. These gels can be injected or orally administered into the body in a minimally invasive way, providing a wide range of applications in the pharmaceutical industry (40).

11.4.5 Nanozyme-Reinforced Copper Coordination Polymeric Nanoparticles

Because of their low cost, wide surface area, and stability in challenging settings, nanomaterials with enzyme-mimicking activity show promise for controlling cellular redox equilibrium and decreasing oxidative damage. Several nanomaterials have been developed as effective nanozymes that mimic antioxidants. These materials include ceria, manganese dioxide, polydopamine nanoparticles, metal-organic frameworks, and Prussian blue. Compared to endogenous antioxidants, these materials have higher stability and availability, and they may all be able to reduce oxidative damage and inflammatory reactions in vivo (41). Metalloenzymes are a class of oxidoreductases that use proton translocation-related lone pair electrons to initiate cellular redox reactions with biochemicals. Because numerous reactive oxygen species (ROS) are present in diabetes, oxidative damage may not be efficiently alleviated by endogenous antioxidants and nanozymes that imitate single-component oxidoreductases alone. As a result, creating nanomaterials for biomedical applications that have a straightforward structure and exceptional ROS scavenging capabilities is difficult. Enzymes like tyrosinase, laccase, and Cu/Zn-SODs require copper ions as redox cofactors to function properly in cellular redox processes. In these processes, copper ions alternate between Cu(II) and Cu(I) states (41).

Keeping cellular copper in the Cu(I) valence state is essential for optimal biological utilization. The Cu(I) form is the key component that travels and senses intracellularly. Therefore, it is crucial to create copper-based nanomaterials that can scavenge endogenous ROS and have a Cu(I) state component. Furthermore, it was noted that copper ions significantly influence wound healing by supporting collagen deposition and angiogenesis (41).

11.4.6 Chitosan-Loaded Polymeric Nanoparticles

Numerous biopolymers have been utilized for the preparation of nanoparticles. Chitosan and PVA have both been used mostly for their biocompatibility, low cost, and nontoxic effects on cells (42).

Chitosan is a polysaccharide biopolymer that is the deacetylated form of chitin. It consists of amino groups, which provide cationic and hydroxyl functional groups. These features are responsible for the chelation and polycationic characteristics that allow it to form films easily. It is an immensely beneficial wound healing agent, as it has proven to be helpful in the stimulation of hemostasis, enhancement of cell proliferation, as well as the acceleration of wound healing. It is also highly biocompatible and does not cause any adverse effects after contact with human skin. Certain enzymes present in the human body also possess the ability to degrade chitosan without releasing any toxic residues. Chitosan can aid in the enhancement of blood clotting by binding to red blood cells and also helps in the remaining phases of wound healing by stimulating fibroblasts, macrophages, and other inflammatory cells, hence fast-tracking the inflammatory phase (42). However, chitosan on its own exhibits weak mechanical properties and low solubility in water as well as in organic solvents. To combat these disadvantages, chitosan is often blended or cross-linked with other polymers like polyvinyl alcohol (PVA). Polyvinyl alcohol is a hydrophilic synthetic polymer synthesized from the hydrolysis of polyvinyl acetate. PVA is biodegradable in nature and can be degraded by biological organisms and even water. It is widely used in the biomedical, commercial, and food industries as it possesses nontoxic, biocompatible, and ideal solubility characteristics. It is also widely used by blending it with other polymers of hydrophilic properties to enhance the mechanical characteristics of films and hydrogels (43). Additionally, chitosan is also used to increase the stability of certain drugs by preparing chitosan nanoparticles. Chitosan-PVA combinations are ideal candidates to produce wound dressings as they can offer quick healing of both chronic and acute wounds in a composition that has high mechanical strength. Silver and zinc oxide nanoparticles are also used in combination with chitosan and PVA to form an ideal wound dressing system (44).

11.4.7 PLGA-Loaded Polymeric Nanoparticles

Poly(lactic-co-glycolic) acid (PLGA) is a linear copolymer. PLGA can be prepared at different ratios between lactic acid and glycolic acid. PLGA is a biodegradable and biocompatible polymer and is commonly used in tissue engineering and drug delivery. PLGA has also garnered a lot of attention for its biomedical applications because of the tailored biodegradation rate, ability to alter the surface properties to improve interaction with various biological materials, and its approval for clinical use (45). A simple and reasonable solution to improve the wound healing process is the application of an external lactate source, which can aid in the acceleration of angiogenesis, engagement of endothelial progenitor cells, and the activation of procollagen factors. Hence, by releasing exogenous lactate continuously, PLGA can be employed to encourage wound healing (46). PLGA is a biocompatible and biodegradable polymer and has also been approved by both the FDA (Food and Drug Administration) and the European Medicines Agency as an excipient in parenteral drugs. Hence, PLGA plays a dual role in drug delivery systems as a wound healing agent and in the sustainable release of the drug to the wound (47). The physical properties of poly(lactic-co-glycolic) acid (PLGA) can be altered based on the ratio of lactide to glycolide. The PLGA copolymers that are rich in glycolide are hydrophilic when the drug-free PLGA is diluted with a different compound. PLGA can be blended with substances like vildagliptin, which is linked to hydrophilicity, to alter its degradation and releasing properties (48). PLGA-based conjugates can also be utilized to form nanoparticles. Nanotechnology can be used in the development of topical treatments with enhanced aqueous solubility and bioavailability. These allow long-term and continuous administration of the drugs to the target site. PLGA facilitates a higher drug carrying capacity in comparison to other polymers, provides better stability, and helps in lower dose frequency (49).

11.4.8 Gentamicin-Loaded Polymeric Nanoparticles

Gentamicin is an aminoglycoside antibiotic that has emerged as a popular agent to treat topical infections. The various desirable characteristics, such as low cost, broad spectrum, and favorable pharmacokinetics and pharmacodynamics, make gentamicin a beneficial antimicrobial agent. Gentamicin provides high efficiency against Gram-negative bacterial infections. The aminoglycosides can bind to the 16s RNA and damage the ribosomal subunit, causing the termination of the translation process during protein biosynthesis. The topical application of gentamicin can cause a significant decrease in the duration of wound healing for local infections. Gentamicin exhibits antimicrobial activity to protect the wound from harmful microorganisms (50). Gentamicin is also used to treat infections caused by multiresistant bacteria. It is effective against multiple Gram-positive bacterial strains. It has been majorly utilized for the topical treatment of infections on the skin, as it is effective against many aerobic Gram-negative and a few aerobic Gram-positive bacteria. However, its major limitation is its toxicity (51). The intravenous or intramuscular administration of gentamicin can cause the accumulation of this compound in the renal cortex, as well as the perilymph and endolymph of the inner ear. This accumulation can, in turn, lead to injury in the kidney and hearing loss, respectively. These disadvantages can be avoided through the local administration of gentamicin to prevent systemic toxicity (52). Additionally, studies have demonstrated that gentamicin has anti-inflammatory qualities through the inhibition of NADPH oxidase in neutrophils. It has also been proven that the intraperitoneal administration of gentamicin into the system can lead to enhancement in angiogenesis in mice (53).

11.4.9 Simvastatin-Loaded Polymeric Nanoparticles

Simvastatin is a statin that is utilized for the effective treatment of hyperlipidemia through the competitive inhibition of 3-hydroxyl-3-methylglutaryl coenzyme A (HMG-CoA) reductase. Hence, this statin is highly beneficial in the reduction of cholesterol biosynthesis. Apart from this anti-hyperlipidemic property, Simvastatin also exhibits anti-inflammatory, immunomodulatory, and angiogenic abilities that can aid in the enhancement of wound healing. It has been proven to increase the production and release of vascular endothelial growth factors near the wound region, which is a crucial process for the production of new blood vessels. Simvastatin has also exhibited its ability to enhance the epithelialization of the skin epidermal barrier through the reduction of isoprenylation of certain downstream targets like mevalonate and farnesyl pyrophosphate. The decrease in farnesyl pyrophosphate levels can help in the activation of the migration of keratinocytes as well. Simvastatin can also cause various cells to secrete vascular endothelial growth factor (VEGF), which leads to an increase in angiogenesis (54). However, Simvastatin also has a few disadvantages, such as low solubility in water, which results in its poor absorption. This drawback can be mitigated by the administration of Simvastatin in a nanoparticle for topical delivery. This modification can lead to an increase in the efficiency of wound healing. Simvastatin has also displayed suitable antimicrobial activity due to its ability to suppress Gram-positive bacteria (55). Simvastatin is a lipid-lowering agent and has been shown to decrease oxidative stress, improve microvascular functions, and improve endothelial function. In addition, it has also displayed anti-inflammatory activity in both chronic and acute wounds through the regulation of pro-inflammatory cytokines like TNF-α, IL-1, IL-6, and IL-8. It can also inhibit the trans-endothelial migration of leukocytes to reduce inflammation and provide an overall enhancement in the wound healing process (56).

11.4.10 Insulin-Loaded Polymeric Nanoparticles

Patients with diabetes mellitus (DM)-related wounds frequently have delayed healing and other impairments due to variables such as vascular damage and neuropathy. Insulin insufficiency in diabetic individuals causes delayed metabolic processes during wound healing, increasing the potential for infections, chronic wounds, and, if left untreated, amputations. Insulin is quickly degraded when applied topically; thus, putting it into polymeric nanoparticles is a novel technique for improving diabetic wound healing (57). Previously conducted studies investigated the application of such nanoparticles using a diabetic animal model, showing their potential to promote healing by delivering insulin in a continuous and regulated manner directly to the wound site. The nanoparticles are made of biodegradable and biocompatible polymers, often with an emphasis on chitosan and PLGA. These types of polymers were chosen for their capacity to efficiently encapsulate insulin, preserve it from deterioration, and provide progressive release.

Insulin-loaded chitosan nanoparticles perform better because of their mucoadhesive qualities, a result of their high zeta potential. Because of their size, they have a large cell density and surface area, making them ideal for granular tissue reformation and re-epithelialization. The study found that these loaded polymeric particles caused more polymorph nuclear penetration than control groups. However, zeta potential, drug diffusion, and nanoparticle size can all impact their stability. Increased zeta potential and other variables can reduce insulin release, lowering bioavailability while perhaps boosting wound healing uniformity (58). Insulin-loaded nanoparticles aid in lowering inflammation at the wound site, which is critical for optimal healing in diabetes patients who often suffer from chronic inflammation. It also revealed greater collagen deposition, which is crucial for the structural integrity and strength of the treated tissue. The main benefits of utilizing polymeric nanoparticles loaded with insulin for diabetic wound healing are as follows: they ensure localized and sustained insulin release at the wound site, maximizing therapeutic effects while reducing negative systemic effects. Additionally, they may reduce the frequency of wound treatment applications, making the process less burdensome for patients (59).

11.4.11 Clindamycin-Loaded Polymeric Nanoparticles

Bacterial infections can seriously affect the wound healing process. The amount of disruption is determined by the bacterial load in the tissue, the kind of bacteria present, the host's immunological response, and the release of bacterial products such as metalloproteinases and endotoxins. MRSA infections in cutaneous wounds have grown dramatically during the last decade. MRSA has become a widespread community-acquired illness, resulting in worldwide hospitalizations. MRSA wound infections are a major source of worry because they can spread to other organs, creating deep, invasive infections that can lead to systemic bacteremia, resulting in severe morbidity and death globally (60). Similarly, the growing prevalence of species resistant to antibiotics necessitates the creation of novel forms of nanomedicine that may be applied to treat wounds produced by bacteria, including diabetic foot ulcers. The process of wound healing requires longer to epithelialize because people with diabetes have an ineffective defensive system against reactive oxygen species (ROS), which are formed in our bodies as a byproduct of oxygen reduction (61).

Clindamycin, a semisynthetic derivative of lincomycin, is an FDA-approved antibiotic that is frequently used due to its lesser chance of adverse effects and safety measures. Other antibiotics frequently have severe drug-drug interactions and associated hazards, such as myelosuppression or neuropathy, with extended treatment. It inhibits ribosomal translocation and protein synthesis in bacteria. However, its traditional methods of delivery have significant disadvantages, including unpleasant side effects, uneven antibiotic dispersion, and the promotion of bacterial resistance. Furthermore, the necessity for daily treatment for a week or longer might lead to low patient compliance (62). To address these concerns, studies analyze the effect of surface charge on the adherence of Clindamycin-loaded nanoparticles, in addition to their antibacterial activity and potential for wound healing. They created nanoparticles made of biocompatible and biodegradable polymers like poly(lactic-co-glycolic acid) (PLGA), PVA, and PEI. Such polymers are chosen for their capacity to efficiently encapsulate clindamycin, preventing it from degradation and enabling regulated release while also allowing for the fine-tuning of its surface features. It has been discovered that nanoparticles with a positive surface charge had a greater adherence to negatively charged MRSA cell surfaces. This enhanced adhesion raises the local concentration of clindamycin at the infection site, which improves antibacterial effectiveness (63, 64). Studies conducted for their effectiveness in diabetic wound treatments and acne vulgaris treatments found faster wound healing rates and reduced inflammation (64, 65).

11.4.12 Asiaticoside Polymeric Nanoparticles

Chinese herbal medicine has traditionally utilized the plant Centella asiatica to manufacture the triterpenoid compound known as asiaticoside (AC). Several studies have demonstrated the many therapeutic benefits of AC, including anti-ulcer, anti-inflammatory, antidepression, anti-dementia, antibacterial, and antioxidant benefits. AC promotes wound healing in a number of ways, including lowering oxidative stress, boosting collagen synthesis, encouraging cell proliferation, improving angiogenesis, modifying collagen fiber secretion and organization, and suppressing scar formation. The medicinal and pharmacological benefits of AC in wound healing are impacted by its low solubility and lipophilicity. Thus, an efficient delivery method is required to carry AC to the site of skin damage (66). Studies assess these nanoparticles' effectiveness in vivo and in vitro, providing insights on how they may be used to treat diabetes patients' slow healing of wounds. Tests using in vitro cell cultures showed that the inclusion of AC in the formulation significantly increased fibroblast proliferation and migration. Administering AC-loaded polymeric nanoparticles to diabetic rats improved wound healing efficacy by increasing collagen biosynthesis, upregulating COL-1 protein levels, and enhancing α-SMA expression compared to control groups (67). AC-loaded nanoparticles for topical administration encapsulate both dry (D) and glycolic (G) extracts in NLC. Two distinct formulations, NLC-D and NLC-G, for optimal skin delivery were developed. The rapid release of AC from NLC-G improved skin permeation, and the increased SC fluidity caused by NLC-G appeared to facilitate AC penetration into deeper skin layers (68, 69).

11.4.13 Curcumin-Loaded Polymeric Nanoparticles

Curcumin, a natural molecule, has a variety of biological actions, including antibacterial, antifungal, anti-inflammatory, and antioxidant properties. Furthermore, it promotes wound healing, epithelial regeneration, and fibroblast proliferation, making it a great option for the creation of wound dressings. Curcumin's limited bioavailability and instability in neutral and alkaline aqueous solutions, as well as in hydrophilic topical formulations, present challenges (70). Polymers like chitosan, polyvinyl alcohol, and carboxymethyl cellulose encourage wound healing by boosting cell activity. Combining curcumin nanoparticles with these polymers improves stability and accelerates healing, overcoming curcumin's limitations. A distinct peak of curcumin in all polymeric nanoparticles indicates proper drug incorporation in the polymeric matrix. The *in vivo* studies found that curcumin-loaded chitosan nanoparticles promote wound healing as a result of the synergistic effect of curcumin and chitosan (71).

11.5 PROPERTIES AND FEATURES OF POLYMERIC NANOPARTICLES

11.5.1 Size

Manipulation of a nanoparticle's size can have a major effect on how it behaves in the body, influencing things like tissue distribution, circulation time, clearance rates, and cellular uptake. Larger nanoparticles (about 41 nm) have a tendency to opsonize and gather in organs like the spleen and liver, which can cause small blood vessels to aggregate and become blocked. Conversely, smaller nanoparticles (about 5 nm), especially those that are fewer than 10 nm in size, are rapidly eliminated from the bloodstream by mechanisms including extravasation or renal clearance. Smaller nanoparticles can have benefits in some situations, such as when used as contrast agents or to help bridge the blood-brain barrier, but they might not stay in the bloodstream long enough to cause a noticeable buildup in the brain. Moreover, lymphatic outflow makes it easier for smaller nanoparticles to reach lymph nodes. For *in vivo* applications, polymeric nanoparticles ranging in size from 20 to 100 nm show great a promise (72).

11.5.2 Morphology

The most prevalent form of polymeric nanoparticles is spherical; however, different morphologies can be achieved based on the composition of the nanoparticles and the polymer structure. Spherical entities that elongate into cylindrical or vesicular architectures may exhibit a variety of properties, including cellular uptake, solubilization capacity, and *in vivo* circulation time. It is typically found that spherical nanoparticles penetrate cells more deeply than long, cylindrical ones (72, 73).

11.5.3 Charge

The studies revealed that the fate of nanoparticles at several stages, such as physical stability, penetration through skin and mucous membranes, interactions with the extracellular matrix, blood stability, cellular absorption, and endocytosis, can be significantly impacted by their surface charge. Excess positive charge in blood causes opsonization, particle aggregation, and clearance, which may obstruct blood arteries and capillaries. The charge of the nanoparticles affects renal clearance; larger, negatively charged particles have a lower excretion rate. The negatively charged proteoglycans on the cell surface cause the surface charge to be directly correlated with cellular binding and interaction at the cellular level. In general, when zeta potential values rise, so does the absorption of nanoparticles. On the other hand, an overabundance of positive charge can cause toxicity and trigger immune responses (74).

11.6 TECHNIQUES TO ASSESS THE PROPERTIES AND CHARACTERISTICS OF POLYMERIC NANOPARTICLES

11.6.1 Transmission Electron Microscopy (TEM)

It is an effective method for examining the size and form of nanoparticles. TEM offers spectroscopic, diffraction, and imaging data. Despite operating on distinct principles, TEM and scanning electron microscopy (SEM) provide comparable types of information (75). The preparation of samples for TEM is a complex and time-consuming process, as the sample must be extremely thin for electron penetration. During TEM characterization, nanoparticle samples are precipitated into films and subsequently immobilized using a staining agent. The sample can survive the vacuum environment thanks to this procedure. An ultra-thin sample interacts with the electron beam as it is transmitted through it, displaying the surface properties (76).

11.6.2 Scanning Electron Microscopy (SEM)

It works on the principle of scanning the surface of a sample with high-energy electrons that focus on a little portion. Specifically, it is based on creating images by reflecting electrons from the sample. SEM provides various benefits for morphological and dimensional analysis; however, it provides little information about size distribution (75, 77). An electron cannon produces the first electrons in the process, which are then accelerated by an anode. Magnetic lenses are used to align these electrons in a parallel fashion. The condenser electromagnetic lens directs, and the objective lens concentrates the electron beam, which is ready for measurement. An electromagnetic deflector is used to deflect the beam over the sample surface during the scanning process, enabling detailed imaging of the morphology and structure of the sample. SEM is an effective instrument for analyzing sample surface features at high magnifications, but it is not a perfect means of obtaining detailed statistical information (78).

11.6.3 Atomic Force Microscopy (AFM)

Based on the surface forces between a sample and a sharp tip affixed to a cantilever, it produces detailed images. This tip, which is usually less than 10 nm in diameter and 10–20 microns long, interacts with the sample surface to produce positional variations that are quantified to create topographic pictures. There are three main modes of operation with AFM. When in contact mode, the cantilever's deflections are recorded to map the topography once the tip makes gentle physical contact with the surface. Vibrating the cantilever in the noncontact mode reduces the possibility of sample damage by vibrating the cantilever near the surface, where variations in vibration frequency brought about by attractive Van der Waals forces offer surface information. To minimize lateral forces that can damage fragile samples and enable high-resolution imaging, the tapping mode oscillates the cantilever so that the tip occasionally taps the surface. AFM is a flexible and effective method for high-resolution surface analysis of different materials because of these modes (75).

11.6.4 X-Ray Diffraction (XRD)

An essential method for figuring out a nanoparticle's crystal structure is X-ray diffraction, or XRD. This method allows identification of the types of atoms at lattice points, crystal planes, and plane distances without destroying the sample. The crystal's shape and the wavelength of the light interacting with it produce different diffraction patterns. When reflected light from several layers of the material interferes with one another within its periodic structure, diffraction takes place. Bragg's law, which connects the wavelength and the spacing between crystal planes to the angle of incidence and reflection of X-rays, defines the prerequisite for diffraction to occur. This makes it possible to analyze the crystal structure in detail using the patterns of diffraction that are recorded (75, 79).

11.6.5 Fourier Transform Infrared Spectroscopy (FTIR)

It is a surface phenomenon for chemical functional group analysis. FTIR works by directing infrared radiation onto a sample, allowing a small amount of energy to pass through while the remainder is absorbed by the sample. The molecules of the sample produce absorbance and transmittance spectra as a result, which are distinctive to the material. The absorption and transmission peaks in these spectra are indicative of the vibrational frequencies of the atom-to-atom bonds in the material. The exact wavelengths at which the peaks emerge determine the links between the atoms, and the intensity of these peaks tells us how much material there is. As a result, FTIR spectroscopy is a helpful technique for describing the molecular makeup and structure of materials (75, 80).

11.6.6 Dynamic Light Scattering (DLS)

Particle size distribution and characterization can be achieved by the commonly used method of dynamic light scattering (DLS), also known as photon correlation spectroscopy. Its working principle is the analysis of how a particle's size affects its Brownian motion in a solvent. The Stokes-Einstein equation is used to find the hydrodynamic dimension of the diffusion coefficient in a homogeneous solution. DLS is a flexible technique for a range of sample types because it may be used with both monodisperse and polydisperse materials. It is limited, nonetheless, by the fact that large particles or samples with high dimensions might make precise measurement difficult. Large samples like these need to be thoroughly dispersed in a solvent to prevent agglomeration and guarantee accurate results (77). One of DLS's primary benefits is its efficiency; measurements can be completed quickly, and the necessary equipment is reasonably priced (81).

11.6.7 Electrophoretic Light Scattering (ELS)

The method known as electrophoretic light scattering (ELS) is essential for precisely determining the zeta potential of suspended particles. The electrical potential at the sliding plane of a suspended particle is a measurement that is important for understanding the firmness of colloidal dispersions and has implications for a range of industrial and scientific applications. Electrophoretic mobility can be directly measured with ELS by observing the migration of particles toward electrodes that are oppositely charged with the application of an electric field. Particle velocity is strongly affected by variables like viscosity, medium dielectric constant, and electric field strength. The zeta potential, which directly affects electrophoretic mobility and colloidal stability, can be precisely determined by ELS analysis (75). A low zeta potential denotes insufficient repulsive forces, which could result in precipitation and instability, whereas a high zeta potential indicates significant electrostatic repulsion between particles, assuring stability by preventing aggregation. All things considered, ELS is essential for comprehending and managing the behavior of colloidal systems, which makes it useful in a variety of industries, including materials science and medicine (75).

11.6.8 X-ray Photoelectron Spectroscopy (XPS)

X-ray electron spectroscopy for chemical analysis (ESCA), often known as photoelectron spectroscopy (XPS), is a top surface analysis method. Because it is nondestructive, it is used to know chemical properties like binding constants, oxidation states, and atomic composition of a sample. By irradiating the sample with high-energy X-rays that only penetrate 5 to 20 Å into the sample, the approach allows for surface-specific chemical analysis as opposed to bulk analysis. An electron in the K-shell, which is the atom's lowest energy shell, is expelled when an atom absorbs these X-rays. The kinetic energy (KE) carried by this expelled electron is closely linked to the energy of the incident X-ray beam (hv), the electron binding energy (BE), and the spectrometer's work function (φ). In fact, in XPS analysis, the shielding effect is a major factor in determining the electron binding energy. Lower binding energies result from shielded electrons, which are easier to extract from the atom because they are less electrostatically attracted to the nucleus due to inner-shell electrons. The associated peaks in the XPS spectra move in the direction of lower energy within the predicted range as a result (82).

11.7 CONCLUSION

Polymeric biomaterials are one of the foundations of tissue design. A large number of materials have been utilized. Approaches have shown expanding complexity over recent years, utilizing the specific functions of sustained or targeted drug delivery, microfluidics, micropatterning, and other advancements. Difficulties, for example, creating three-dimensional lattices and delivering them through minimally invasive methods have been accomplished using polymeric nanoparticles. Only a few polymer nanoparticle medication delivery methods are currently authorized for use in clinical settings. Recent developments in the synthesis of polymer nanoparticles, along with the flexibility in the types of polymers and design strategies, indicate that polymer nanoparticles are a perfect platform on which to construct delivery vehicles with more sophisticated architecture and functionality. This can be achieved with accuracy and precision in the generation of synthetic architecture, placement of functionality, and ease of conjugation. As a result, the number of therapies utilizing polymer nanoparticles will only increase over time. A significant recent improvement is the design of biomaterials for tissue engineering scaffolds to achieve specific biological effects on cells, as well as the other way around. Much still needs to be accomplished, especially in integrating other new advances into the field.

REFERENCES

1. Tiwari, G., Tiwari, R., Sriwastawa, B., Bhati, L., Pandey, S., Pandey, P., & Bannerjee, S. K. (2012). Drug delivery systems: An updated review. *International Journal of Pharmaceutical Investigation*, *2*(1), 2.
2. Adepu, S., & Ramakrishna, S. (2021). Controlled drug delivery systems: Current status and future directions. *Molecules*, *26*(19), 5905.
3. Abuhelwa, A. Y., Williams, D. B., Upton, R. N., & Foster, D. J. (2017). Food, gastrointestinal pH, and models of oral drug absorption. *European Journal of Pharmaceutics and Biopharmaceutics*, *112*, 234–248.

4. Gavhane, Y. N., & Yadav, A. V. (2012). Loss of orally administered drugs in GI tract. *Saudi Pharmaceutical Journal*, *20*(4), 331–344.
5. Jia, L. (2005). Nanoparticle formulation increases oral bioavailability of poorly soluble drugs: Approaches, experimental evidences and theory. *Current Nanoscience*, *1*(3), 237–243.
6. Patra, J. K., Das, G., Fraceto, L. F., Campos, E. V. R., Rodriguez-Torres, M. D. P., Acosta-Torres, L. S., . . . Shin, H. S. (2018). Nano based drug delivery systems: Recent developments and future prospects. *Journal of Nanobiotechnology*, *16*, 1–33.
7. Chenthamara, D., Subramaniam, S., Ramakrishnan, S. G., Krishnaswamy, S., Essa, M. M., Lin, F. H., & Qoronfleh, M. W. (2019). Therapeutic efficacy of nanoparticles and routes of administration. *Biomaterials Research*, *23*(1), 20.
8. Begines, B., Ortiz, T., Pérez-Aranda, M., Martínez, G., Merinero, M., Argüelles-Arias, F., & Alcudia, A. (2020). Polymeric nanoparticles for drug delivery: Recent developments and future prospects. *Nanomaterials*, *10*(7), 1403.
9. Mogoşanu, G. D., Grumezescu, A. M., Bejenaru, C., & Bejenaru, L. E. (2016). Polymeric protective agents for nanoparticles in drug delivery and targeting. *International Journal of Pharmaceutics*, *510*(2), 419–429.
10. Gagliardi, A., Giuliano, E., Venkateswararao, E., Fresta, M., Bulotta, S., Awasthi, V., & Cosco, D. (2021). Biodegradable polymeric nanoparticles for drug delivery to solid tumors. *Frontiers in Pharmacology*, *12*, 601626.
11. Carreiró, F., Oliveira, A. M., Neves, A., Pires, B., Nagasamy Venkatesh, D., Durazzo, A., . . . Souto, E. B. (2020). Polymeric nanoparticles: Production, characterization, toxicology and ecotoxicology. *Molecules*, *25*(3731).
12. Beach, M. A., Nayanathara, U., Gao, Y., Zhang, C., Xiong, Y., Wang, Y., & Such, G. K. (2024). Polymeric nanoparticles for drug delivery. *Chemical Reviews*, *124*, 5505–5616.
13. Dhanjal, D. S., Mehta, M., Chopra, C., Singh, R., Sharma, P., Chellappan, D. K., . . . Satija, S. (2021). Novel controlled release pulmonary drug delivery systems: Current updates and challenges. In *Modeling and control of drug delivery systems* (pp. 253–272). Academic Press.
14. Ulbrich, K., Hola, K., Subr, V., Bakandritsos, A., Tucek, J., & Zboril, R. (2016). Targeted drug delivery with polymers and magnetic nanoparticles: Covalent and noncovalent approaches, release control, and clinical studies. *Chemical Reviews*, *116*(9), 5338–5431.
15. Kim, B. S., Park, I. K., Hoshiba, T., Jiang, H. L., Choi, Y. J., Akaike, T., & Cho, C. S. (2011). Design of artificial extracellular matrices for tissue engineering. *Progress in Polymer Science*, *36*(2), 238–268.
16. Elsabahy, M., & Wooley, K. L. (2012). Design of polymeric nanoparticles for biomedical delivery applications. *Chemical Society Reviews*, *41*(7), 2545–2561.
17. Khalid, M., & El-Sawy, H. S. (2017). Polymeric nanoparticles: Promising platform for drug delivery. *International Journal of Pharmaceutics*, *528*(1–2), 675–691.
18. Sur, S., Rathore, A., Dave, V., Reddy, K. R., Chouhan, R. S., & Sadhu, V. (2019). Recent developments in functionalized polymer nanoparticles for efficient drug delivery system. *Nano-Structures & Nano-Objects*, *20*, 100397.
19. Mishra, D., Hubenak, J. R., & Mathur, A. B. (2013). Nanoparticle systems as tools to improve drug delivery and therapeutic efficacy. *Journal of Biomedical Materials Research Part A: An Official Journal of the Society for Biomaterials, the Japanese Society for Biomaterials, and the Australian Society for Biomaterials and the Korean Society for Biomaterials*, *101*(12), 3646–3660.
20. Klinkova, A., & Thérien-Aubin, H. (2023). *Nanochemistry: Chemistry of nanoparticle formation and interactions*. Elsevier.
21. Vauthier, C. (2019). A journey through the emergence of nanomedicines with poly (alkylcyanoacrylate) based nanoparticles. *Journal of Drug Targeting*, *27*(5–6), 502–524.
22. Zielińska, A., Carreiró, F., Oliveira, A. M., Neves, A., Pires, B., Venkatesh, D. N., Durazzo, A., Lucarini, M., Eder, P., Silva, A. M., Santini, A., & Souto, E. B. (2020). Polymeric nanoparticles: Production, characterization, toxicology and ecotoxicology. *Molecules*, *25*(16), 3731.
23. Vasile, C. (2018). Polymeric nanocomposites and nanocoatings for food packaging: A review. *Materials*, *11*(10), 1834.
24. Alexis, F., Pridgen, E., Molnar, L. K., & Farokhzad, O. C. (2008). Factors affecting the clearance and biodistribution of polymeric nanoparticles. *Molecular Pharmaceutics*, *5*(4), 505–515.
25. Lu, X. Y., Wu, D. C., Li, Z. J., & Chen, G. Q. (2011). Polymer nanoparticles. *Progress in Molecular Biology and Translational Science*, *104*, 299–323.
26. Zhang, G., Niu, A., Peng, S., Jiang, M., Tu, Y., Li, M., & Wu, C. (2001). Formation of novel polymeric nanoparticles. *Accounts of Chemical Research*, *34*(3), 249–256.

27. Elsabahy, M., & Wooley, K. L. (2012). Design of polymeric nanoparticles for biomedical delivery applications. *Chemical Society Reviews*, *41*(7), 2545–2561.
28. Lu, X. Y., Wu, D. C., Li, Z. J., & Chen, G. Q. (2011). Polymer nanoparticles. *Progress in Molecular Biology and Translational Science*, *104*, 299–323.
29. Banik, B. L., Fattahi, P., & Brown, J. L. (2016). Polymeric nanoparticles: The future of nanomedicine. *Wiley Interdisciplinary Reviews. Nanomedicine and Nanobiotechnology*, *8*(2), 271–299.
30. Ganeshan, P., Sahoo, S., Gopalan, A., Ramesh, S., Nirmala, P., Ramkumar, G., Agnes Shifani, S., Subbiah, R., & Isaac JoshuaRamesh Lalvani, J. (2021). Preparation of polymeric nanomaterials using emulsion polymerization. *Advances in Materials Science and Engineering*, *2021*, 1–9.
31. Zielińska, A., Carreiró, F., Oliveira, A. M., Neves, A., Pires, B., Venkatesh, D. N., Durazzo, A., Lucarini, M., Eder, P., Silva, A. M., Santini, A., & Souto, E. B. (2020). Polymeric nanoparticles: Production, characterization, toxicology and ecotoxicology. *Molecules (Basel, Switzerland)*, *25*(16), 3731.
32. Szymusiak, M. (2016). *Design, formation, and characterization of polymeric nanoparticles for drug delivery (version 1)*. University of Illinois at Chicago.
33. Alavi, M., & Rai, M. (2020). Topical delivery of growth factors and metal/metal oxide nanoparticles to infected wounds by polymeric nanoparticles: An overview. *Expert Review of Anti-Infective Therapy*, *18*(10), 1021–1032.
34. Wu, S., Helal-Neto, E., Matos, A. P. D. S., Jafari, A., Kozempel, J., Silva, Y. J. A., Serrano-Larrea, C., Alves Junior, S., Ricci-Junior, E., Alexis, F., & Santos-Oliveira, R. (2020). Radioactive polymeric nanoparticles for biomedical application. *Drug Delivery*, *27*(1), 1544–1561.
35. Nunes, D., Andrade, S., Ramalho, M. J., Loureiro, J. A., & Pereira, M. C. (2022). Polymeric nanoparticles-loaded hydrogels for biomedical applications: A systematic review on in vivo findings. *Polymers*, *14*(5), 1010.
36. Komprda, T., Sládek, Z., Vícenová, M., Simonová, J., Franke, G., Lipový, B., Matejovičová, M., Kacvinská, K., Sabliov, C., Astete, C. E., Levá, L., Popelková, V., Bátik, A., & Vojtová, L. (2022). Effect of polymeric nanoparticles with entrapped fish oil or mupirocin on skin wound healing using a porcine model. *International Journal of Molecular Sciences*, *23*(14), 7663.
37. Kamlungmak, S., Nakpheng, T., Kaewpaiboon, S., Mudhar Bintang, M. A. K., Prom-In, S., Chunhachaichana, C., Suwandecha, T., & Srichana, T. (2021). Safety and biocompatibility of mupirocin nanoparticle-loaded hydrogel on burn wound in rat model. *Biological & Pharmaceutical Bulletin*, *44*(11), 1707–1716.
38. Viswanathan, K., Monisha, P., Srinivasan, M., Swathi, D., Raman, M., & Dhinakar Raj, G. (2016). Chlorhexidine-calcium phosphate nanoparticles—Polymer mixer based wound healing cream and their applications. *Materials Science & Engineering. C, Materials for Biological Applications*, *67*, 516–521.
39. Froelich, A., Jakubowska, E., Wojtyłko, M., Jadach, B., Gackowski, M., Gadziński, P., Napierała, O., Ravliv, Y., & Osmałek, T. (2023). Alginate-based materials loaded with nanoparticles in wound healing. *Pharmaceutics*, *15*(4), 1142.
40. Severino, P., da Silva, C. F., Andrade, L. N., de Lima Oliveira, D., Campos, J., & Souto, E. B. (2019). Alginate nanoparticles for drug delivery and targeting. *Current Pharmaceutical Design*, *25*(11), 1312–1334.
41. Zhao, J., Xu, T., Sun, J., Yuan, H., Hou, M., Li, Z., Wang, J., & Liang, Z. (2023). Multifunctional nanozyme-reinforced copper-coordination polymer nanoparticles for drug-resistance bacteria extinction and diabetic wound healing. *Biomaterials Research*, *27*(1), 88.
42. Gaaz, T. S., Sulong, A. B., Akhtar, M. N., Kadhum, A. A., Mohamad, A. B., & Al-Amiery, A. A. (2015). Properties and applications of polyvinyl alcohol, halloysite nanotubes and their nanocomposites. *Molecules (Basel, Switzerland)*, *20*(12), 22833–22847.
43. Sathiyaseelan, A., Saravanakumar, K., Mariadoss, A. V. A., & Wang, M. H. (2021). Antimicrobial and wound healing properties of FeO fabricated chitosan/PVA nanocomposite sponge. *Antibiotics (Basel, Switzerland)*, *10*(5), 524.
44. Kalantari, K., Mostafavi, E., Saleh, B., Soltantabar, P., & Webster, T. J. (2020). Chitosan/PVA hydrogels incorporated with green synthesized cerium oxide nanoparticles for wound healing applications. *European Polymer Journal*, *5*, 109853.
45. Gentile, P., Chiono, V., Carmagnola, I., & Hatton, P. V. (2014). An overview of poly(lactic-co-glycolic) acid (PLGA)-based biomaterials for bone tissue engineering. *International Journal of Molecular Sciences*, *15*(3), 3640–3659.
46. Chereddy, K. K., Her, C. H., Comune, M., Moia, C., Lopes, A., Porporato, P. E., Vanacker, J., Lam, M. C., Steinstraesser, L., Sonveaux, P., Zhu, H., Ferreira, L. S., Vandermeulen, G., & Préat, V. (2014). PLGA

nanoparticles loaded with host defense peptide LL37 promote wound healing. *Journal of Controlled Release*, *28*(194), 138–147.

47. Lee, C. H., Huang, C. H., Hung, K. C., Huang, S. C., Kuo, C. C., & Liu, S. J. (2022). Nanofibrous Vildagliptin/PLGA membranes accelerate diabetic wound healing by angiogenesis. *Pharmaceuticals (Basel, Switzerland)*, *15*(11), 1358.
48. Gourishetti, K., Keni, R., Nayak, P. G., Jitta, S. R., Bhaskaran, N. A., Kumar, L., Kumar, N., Krishnadas, N., & Shenoy, R. R. (2020). Sesamol-loaded PLGA nanosuspension for accelerating wound healing in diabetic foot ulcer in rats. *International Journal of Nanomedicine*, *15*, 9265–9282.
49. Hsieh, C. F., Chen, C. H., Kao, H. H., Govindaraju, D. T., Dash, B. S., & Chen, J. P. (2022). PLGA/Gelatin/Hyaluronic acid fibrous membrane scaffold for therapeutic delivery of adipose-derived stem cells to promote wound healing. *Biomedicines*, *10*(11), 2902.
50. Hayat, M., Nawaz, A., Chinnam, S., Muzammal, M., Latif, M. S., Yasin, M., Ashique, S., Zengin, G., & Farid, A. (2023, July 1). Formulation development and optimization of herbo synthetic gel: In vitro biological evaluation and in vivo wound healing studies. *Process Biochemistry*, *130*, 116–126.
51. Wang, P., Long, Z., Yu, Z., Liu, P., Wei, D., Fang, Q., Ma, D., & Wang, J. (2019). The efficacy of topical gentamycin application on prophylaxis and treatment of wound infection: A systematic review and meta-analysis. *International Journal of Clinical Practice*, *73*(5), e13334.
52. Budiatin, A. S., Gani, M. A., Samirah, Ardianto, C., Raharjanti, A. M., Septiani, I., Putri, N. P. K. P., & Khotib, J. (2021). Bovine hydroxyapatite-based bone scaffold with gentamicin accelerates vascularization and remodeling of bone defect. *International Journal of Biomaterials*, *2021*, 5560891.
53. Hwang, M. R., Kim, J. O., Lee, J. H., Kim, Y. I., Kim, J. H., Chang, S. W., Jin, S. G., Kim, J. A., Lyoo, W. S., Han, S. S., Ku, S. K., Yong, C. S., & Choi, H. G. (2010). Gentamicin-loaded wound dressing with polyvinyl alcohol/dextran hydrogel: Gel characterization and in vivo healing evaluation. *AAPS PharmSciTech*, *11*(3), 1092–1103.
54. Ahmed, L. M., Hassanein, K. M. A., Mohamed, F. A., & Elfaham, T. H. (2023). Formulation and evaluation of simvastatin cubosomal nanoparticles for assessing its wound healing effect. *Scientific Reports*, *13*(1), 17941.
55. Ahmed, L. M., Hassanein, K. M. A., Mohamed, F. A., & Elfaham, T. H. (2023). Formulation and evaluation of simvastatin cubosomal nanoparticles for assessing its wound healing effect. *Scientific Reports*, *13*(1), 17941.
56. Yasasvini, S., Anusa, R. S., VedhaHari, B. N., Prabhu, P. C., & RamyaDevi, D. (2017). Topical hydrogel matrix loaded with Simvastatin microparticles for enhanced wound healing activity. *Materials Science & Engineering. C, Materials for Biological Applications*, *72*, 160–167.
57. Quitério, M., Simões, S., Ascenso, A., Carvalheiro, M., Leandro, A. P., Correia, I., . . . Reis, C. P. (2021). Development of a topical insulin polymeric nanoformulation for skin burn regeneration: An experimental approach. *International Journal of Molecular Sciences*, *22*(8), 4087.
58. Ribeiro, M. C., Correa, V. L. R., Da Silva, F. K. L., Casas, A. A., De Lima Das Chagas, A., De Oliveira, L. P., Miguel, M. P., Diniz, D. G. A., Amaral, A. C., & De Menezes, L. B. (2020). Wound healing treatment using insulin within polymeric nanoparticles in the diabetes animal model. *European Journal of Pharmaceutical Sciences*, *150*, 105330.
59. Macedo, A. S., Mendes, F., Filipe, P., Reis, S., & Fonte, P. (2021b). Nanocarrier-mediated topical insulin delivery for wound healing. *Materials*, *14*(15), 4257.
60. Hasan, N., Cao, J., Lee, J., Hlaing, S. P., Oshi, M. A., Naeem, M., Ki, M. H., Lee, B. L., Jung, Y., & Yoo, J. W. (2019). Bacteria-targeted clindamycin loaded polymeric nanoparticles: Effect of surface charge on nanoparticle adhesion to MRSA, antibacterial activity, and wound healing. *Pharmaceutics*, *11*(5), 236.
61. Saha, K., Ghosh, A., Bhattacharya, T., Ghosh, S., Dey, S., & Chattopadhyay, D. (2023). Ameliorative effects of clindamycin—nanoceria conjugate: A ROS responsive smart drug delivery system for diabetic wound healing study. *Journal of Trace Elements in Medicine and Biology*, *75*, 127107.
62. Kim, J. O., Choi, J. Y., Park, J. K., Kim, J. H., Jin, S. G., Chang, S. W., Li, D. X., Hwang, M. R., Woo, J. S., Kim, J. A., Lyoo, W. S., Yong, C. S., & Choi, H. G. (2008b). Development of clindamycin-loaded wound dressing with polyvinyl alcohol and sodium alginate. *Biological & Pharmaceutical Bulletin*, *31*(12), 2277–2282. https://doi.org/10.1248/bpb.31.2277
63. Balderrama-González, A. S., Piñón-Castillo, H. A., Ramírez-Valdespino, C. A., Landeros-Martínez, L. L., Orrantia-Borunda, E., & Esparza-Ponce, H. E. (2021). Antimicrobial resistance and inorganic nanoparticles. *International Journal of Molecular Sciences*, *22*(23), 12890.
64. Madawi, E. A., Al Jayoush, A. R., Rawas-Qalaji, M., Thu, H. E., Khan, S., Sohail, M., . . . Hussain, Z. (2023). Polymeric nanoparticles as tunable nanocarriers for targeted delivery of drugs to skin tissues for treatment of topical skin diseases. *Pharmaceutics*, *15*(2), 657.

65. Mandegari, M., Ghasemi-Mobarakeh, L., & Varshosaz, J. (2022). Fabrication and characterization of a novel wound dressing with clindamycin loaded PVA nanoparticles for acne treatment. *Fibers and Polymers/Fibers and Polymers*, *23*(12), 3369–3379.
66. Hasan, N., Cao, J., Lee, J., Hlaing, S. P., Oshi, M. A., Naeem, M., Ki, M. H., Lee, B. L., Jung, Y., & Yoo, J. W. (2019b). Bacteria-targeted clindamycin loaded polymeric nanoparticles: Effect of surface charge on nanoparticle adhesion to MRSA, antibacterial activity, and wound healing. *Pharmaceutics*, *11*(5), 236.
67. Narisepalli, S., Salunkhe, S. A., Chitkara, D., & Mittal, A. (2023). Asiaticoside polymeric nanoparticles for effective diabetic wound healing through increased collagen biosynthesis: In-vitro and in-vivo evaluation. *International Journal of Pharmaceutics*, *631*, 122508.
68. Da Rocha, P. B. R., Souza, B. D. S., Andrade, L. M., Anjos, J. L. V. D., Mendanha, S. A., Alonso, A., Marreto, R. N., & Taveira, S. F. (2019). Enhanced asiaticoside skin permeation by Centella asiatica-loaded lipid nanoparticles: Effects of extract type and study of stratum corneum lipid dynamics. *Journal of Drug Delivery Science and Technology*, *50*, 305–312.
69. Kumar, M., Kumar, D., Mahmood, S., Singh, V., Chopra, S., Hilles, A. R., & Bhatia, A. (2024b). Nanotechnology-driven wound healing potential of asiaticoside: A comprehensive review. *RSC Pharmaceutics*, *1*(1), 9–36.
70. Guadarrama-Acevedo, M. C., Mendoza-Flores, R. A., Del Prado-Audelo, M. L., Urbán-Morlán, Z., Giraldo-Gomez, D. M., Magaña, J. J., González-Torres, M., Reyes-Hernández, O. D., Figueroa-González, G., Caballero-Florán, I. H., Florán-Hernández, C. D., Florán, B., Cortés, H., & Leyva-Gómez, G. (2019). Development and evaluation of alginate membranes with curcumin-loaded nanoparticles for potential wound-healing applications. *Pharmaceutics*, *11*(8), 389.
71. Shende, P., & Gupta, H. (2020). Formulation and comparative characterization of nanoparticles of curcumin using natural, synthetic and semi-synthetic polymers for wound healing. *Life Sciences*, *253*, 117588.
72. Elsabahy, M., & Wooley, K. L. (2012). Design of polymeric nanoparticles for biomedical delivery applications. *Chemical Society Reviews*, *41*(7), 2545–2561.
73. Geng, Y. A. N., Dalhaimer, P., Cai, S., Tsai, R., Tewari, M., Minko, T., & Discher, D. E. (2007). Shape effects of filaments versus spherical particles in flow and drug delivery. *Nature Nanotechnology*, *2*(4), 249–255.
74. Adler, A. F., & Leong, K. W. (2010). Emerging links between surface nanotechnology and endocytosis: Impact on nonviral gene delivery. *Nano Today*, *5*(6), 553–569.
75. Haider, A., Ikram, M., Shahzadi, I., & Asif Raza, M. (2023). Properties and characterization of polymeric nanomaterials. In *Polymeric nanoparticles for bovine mastitis treatment* (pp. 33–48). Springer Nature Switzerland.
76. Molpeceres, J., Aberturas, M. R., & Guzman, M. (2000). Biodegradable nanoparticles as a delivery system for cyclosporine: Preparation and characterization. *Journal of Microencapsulation*, *17*(5), 599–614.
77. Zhou, W., Apkarian, R., Wang, Z. L., & Joy, D. (2007). Fundamentals of scanning electron microscopy (SEM). In *Scanning microscopy for nanotechnology: Techniques and applications* (pp. 1–40). Springer.
78. Hamdy, A., Mostafa, M. K., & Nasr, M. (2018). Zero-valent iron nanoparticles for methylene blue removal from aqueous solutions and textile wastewater treatment, with cost estimation. *Water Science and Technology*, *78*(2), 367–378.
79. Gregory, N. W. (1957). Elements of X-ray diffraction. *Journal of the American Chemical Society*, *79*(7), 1773–1774.
80. Skoog, D. A., Holler, F. J., & Crouch, S. R. (2019). Textbook "principles of instrumental analysis". *Cengage Learning*, *7*, 695–815.
81. Bootz, A., Vogel, V., Schubert, D., & Kreuter, J. (2004). Comparison of scanning electron microscopy, dynamic light scattering and analytical ultracentrifugation for the sizing of poly (butyl cyanoacrylate) nanoparticles. *European Journal of Pharmaceutics and Biopharmaceutics*, *57*(2), 369–375.
82. Raja, P. M., & Barron, A. R. (2019). X-ray photoelectron spectroscopy. *Chemistry LibreTexts*, *1*, 1–91.

12 Application of Green Synthesized Nanomaterials for Wound Healing and Cell Growth Stimulation

Sharath Kumar N.[1] *and Aamir Hussain Dar*[2]*

[1] ICAR-Central Institute of Temperate Horticulture, Srinagar, J&K, India

[2] Department of Food Technology, Islamic University of Science and Technology, Kashmir, India

* Corresponding author: Dr. Aamir Hussain Dar

12.1 INTRODUCTION

Nanotechnology, a multidisciplinary cutting-edge technology with the process of converting materials to the nanoscale, has shown its omnipresence and importance and is undergoing a significant change in its perspective and application. It can be regarded as an emerging field of the 21st century. But the origin of nanomaterials dates back to the historical perspective of the 'Lycurgus cup,' which is a Roman treasure containing glass with gold–silver-alloyed nanoparticles. This unique cup has a color-changing property upon incident light due to the presence of nanoparticles. Close to 4500 years ago, asbestos nanofibers were used as a reinforcing material in ceramic blends (Heiligtag & Niederberger, 2013). The scope and significance of nanotechnology have aroused interest in the scientific community when Nobel laureate and American Physicist Richard Feynman, in 1959, said in a lecture note of the American Physical Society meeting at the California Institute of Technology—"There's Plenty of Room at the Bottom"—which signified that the majority of the work can be done at atomic and molecular levels, provided suitable technology exists. This offers a unique opportunity to explore natural materials, which then sparked the interest of researchers to tap the benefits of nanomaterials (Hornyak et al., 2008). Way back in 1914, the word nanometer was referred to by Richard Adolf Zsigmondy. Later, Norio Taniguchi is credited with coining the term nanotechnology in 1974 and is referred to as the father of nanotechnology (Santamaria, 2012). The word *nano* in Greek refers to dwarf, which can be referred to as one-millionth of a micron (Mansoori, 2005). It refers to the control of size and structures at the atomic and molecular events in the range of 1-100 nm or one of its size dimensions in the range of 1-100 nm. Restructuring the dimensionality of materials provides unique and novel surface properties with distinct physical, chemical, material, and biological properties that are purely influenced by size, shape, and composition. The fabrication, synthesis, and modification of nanomaterials present a significant opportunity for several applications ranging from agriculture, food, medicine, cosmetics, sensor development, biomolecular studies, material science, energy, water purification, the environment, electronics, high-performance computing, information technology, semiconductor mechanics, etc. The uniqueness of the size factor offers nanomaterials a plethora of changes in optical, magnetic, and electric properties with relative advantages over other materials. The unique properties of these nanomaterials are of significant importance in providing beneficial impacts. Often, nanomaterials are confused with several closely

DOI: 10.1201/9781003605966-12

related terminologies, each having specific size dimensions, methods of fabrication, and unique material properties catering to diverse end applications (Krishnaswamy & Orsat, 2017; Baig et al., 2021).

i) Zero-dimensional nanomaterials where all the materials are in the dimensions below 100 nm. Nanoparticles and quantum dots fall in this category.
ii) One-dimensional nanomaterials where at least two of its dimensions are below 100 nm. Nanotubes, nanowires, and nanorods can be classified as one dimensional nanomaterials.
iii) Two-dimensional nanomaterials in which at least one of its dimensions is below 100 nm, encompassing nanofilms and nanoplates.
iv) Three-dimensional nanomaterials with none of the dimensions in the nanolevel and aggregated by the nanosized elementary structures. Nanocrystalline grains and engineered nanomaterials have been defined in this category.

The method of fabrication or synthesis of nanomaterials has much more to alter the chemistry and structural arrangement at atomic, molecular, and submolecular levels. Few of the properties include size, shape, morphology, surface area, diffusivity, magnetic property, electric property, thermal property, optical property, quantum effects, mechanical property, catalytic benefits, and antimicrobial property that make the nano-derived materials unique and useful, offering significant importance. The medicine and pharmacology applications of nanomaterials are of prime importance in wound healing and cell simulation strategies. Wounds are chronic medical pathologies that have affected nearly 1–2% of the population in developed countries (Olsson et al., 2019). Global spending on wound management is expected to reach $18.7 billion with a compounded annual growth rate of 6.6% (Reportlinker, 2020; Nandhini et al., 2024). The increasing cases of wounds and their associated negative health emergencies have drawn the attention of medical professionals and have put a global burden on the public health infrastructure system. Wound care is a multicompounded and multipathological silent epidemic that culminates and is exacerbated by lifestyle disorders such as hypertension, diabetes, and chronic heart diseases, leading to severe morbidity and mortality. The impact of chronic wounds has affected the physical, mental, social, and psychological behavior of humans and continues to exert considerable pressure on the existing healthcare systems. Given the extreme importance of wound management, it becomes essential to design wound healing procedures for restoring the skin pattern both in structure, composition, and function. The objective of healthy wound healing is not only to reduce the secondary associated pathology but also to control the infection rate and prevent the occurrence of traumatic deaths. Employing nanomaterials for wound healing procedures reduces the drudgery associated with conventionally followed wound healing treatments.

Synthesis of nanomaterials occurs from organic, inorganic, and biological systems either by following a top-down approach or a bottom-up approach (Bhushan, 2016). The top-down approach refers to the breakdown of bulky materials to achieve the required size reduction, mainly by physical methods such as etching, mechanical milling or ball milling, laser ablation sputtering, and electro-explosion, whereas the bottom-up approach refers to the self-assembling of atoms and molecules to the nanomaterial level, which can be initiated by organic and inorganic materials, including chemical vapor deposition, solvothermal, hydrothermal methods, and reverse micellar methods. The increased application and proven beneficial impact have made industries opt for larger industrial-level nanomaterial synthesis, commonly referred to as engineered nanomaterials. In light of the relative advantages of nanomaterials, it has become necessary to discuss the potential health and environmental impacts caused by increased nanomaterial synthesis and provide alternative and sustainable solutions with equivalent or improved material properties. The time has come to understand the negative impacts of several conventional nanoparticles; for example, TiO2 nanoparticles have the potential to reach the kidney and liver through the pulmonary pathway, the deposition of which can lead to carcinogenesis via the development of reactive oxygen species and inducing toxicity (Shakeel et al., 2016). Some research studies have also accounted for immune disruption and histopathological aberrations. There is a lack of investigations on the release profile rates in the environment, their potential toxicity, and degradation behavior. To effectively mitigate the human and ecological impacts of surrogating nanomaterials, it is important to deliver alternative sourcing materials for nanomaterials along with an ecofriendly approach for greener synthesis. Underlining the global importance of wound healing, it becomes imperative to allow the use of green synthesized nanomaterials in the subject of wound healing and cell stimulation strategies. Therefore, this chapter deliberately addresses the physiology and types of wounds occurring from a global perspective, with the available greener nanomaterials sourced from natural counterparts in a greener manufacturing pathway. The applications of these green synthesized nanomaterials for several types of wound healing procedures are highlighted in detail.

12.2 PHYSIOLOGY OF WOUND HEALING

Wound is referred to as abruption or disruption in the skin or tissue or the mucosal surface, which then triggers a cascading chemical and biomolecular event in the system, initiating the process of healing (Wolcott et al., 2010). It can become chronic if the regular healing and regulation mechanisms do not facilitate healing or repair. Aging is generally a factor for chronic wounds, where the healing fails to re-attain the regular format. The cause of a wound can be accidental, due to disease effects, or due to age-related chronic effects. The broader classification of wounds relates to pressure scars, moisture-related dermatitis, skin tears, vascular wounds, and neuropathic-related wounds (Whitney, 2005). Wound healing is a highly structured and cascading set of events managed by immune cells and a plethora of activated factors that are secreted in response to the wound. Broadly, the process of wound healing can be divided into four stages—hemostasis, inflammation, proliferation, and maturation (Fitridge & Thompson, 2011); but it can also include coagulation, immune cell initiation, and angiogenesis. Each step cannot be stated to follow a prescribed sequence; they often overlap depending on the type of wound, the health status of the tissues, and associated comorbidities.

12.2.1 Hemostasis

Immediately after the wound is created, it exposes the inner tissue of collagen to the circulating platelets; thereby, vasoconstriction of the tissue starts, leading to adhesion and aggregation at the site of injury or wound (Rumbaut, 2010). The vasoconstriction is made possible by an increase in cytoplasmic calcium levels. This is added upon by changes in the vicinity pH levels and increased hypoxia. Upon succession of wound to the subendothelial layer, specific integrin receptors and cell surface receptors interact with the platelets, which then undergo activation, resulting in the emission of growth factors, cyclic AMP, and adhesion growth factors responsible for aggregation and coagulation. The activated platelets are known to host close to 300 activated substances in their dense intracellular granules (Golebiewska & Poole, 2015). The aggregation is followed by fibrin mesh deposition and clotting, both of which result in the cleaning and removal of foreign materials from the wounded site, followed by further aggregation. Fibrin serves as a scaffold for cells such as neutrophils, monocytes, fibroblasts, and endothelial cells, and it helps to strengthen the thrombus. Research investigations have identified the presence of cytokines, chemokines, and growth factors such as platelet α granules (PAF), platelet-derived growth factor (PDGF), transforming growth factor beta (TGF-β), transforming growth factor alpha (TGF-α), basic fibroblast growth factor (BFGF), insulin-like growth factor-1 (IGF-1), and vascular endothelial growth factor (VEGF), which are responsible for the successful hemostasis initiated by the activated platelets. These growth factors assist in the migration and deployment of inflammatory response cells such as macrophages, neutrophils, and fibroblasts. The release of vascular endothelial growth factor, transforming growth factor beta, and transforming growth factor alpha is responsible for initiating the process of angiogenesis.

These growth factors then interact with the extracellular matrix for the next step of the inflammatory process.

12.2.2 Inflammation

The succession of hemostasis is the inflammation phase, where there is infiltration of various inflammatory responses to the wound site, leading to the release of several biomolecules responsible for making the wound for the healing process. The macrophages (which are derivatives of monocytes), polymorphonuclear neutrophils, fibroblasts, cytokines, and chemokines are major factors involved in the inflammatory response. During inflammation, soluble factors such as pro-inflammatory cytokines (IL-1, IL-6, IL-8, and TNF-α) and growth factors (PDGF, TGF-β, TGF-α, IGF-1, and FGF) that are involved in the recruitment and activation of fibroblasts and epithelial cells are received at the site of the blocked wound and are ready for the healing process. This is supported by the presence of prostaglandins and histamines, which cause vasodilation of the vessels for efficient movement. Neutrophils are involved in the phagocytosis of the infectious agents, supported by the presence of cytokines and chemokines that release antimicrobial peptides. This stage, where coordination between innate and adaptive immunity plays a crucial role, can have beneficial effects; however, overproduction can have deleterious effects on the wounded tissues. The essential function of the inflammatory process is to cleanse the wound and prepare it for healing, starting with a pro-inflammatory response and ending with anti-inflammatory response factors, for which balance is necessary and important (Turabelidze & Dipietro, 2011). An imbalance in the inflammatory response can lead to chronic wound effects. Prolonged inflammation can sometimes lead to keratinocyte activation and dysregulation in the process of cell differentiation. Chronic inflammation can make healthy cells prone to cancerous events. It is also mentioned that the prolonged inflammatory phases can lead to the release of various antimicrobial substances such as reactive oxygen species (ROS), cationic peptides, eicosanoids, proteases, and myeloperoxidase (Landén et al., 2016). The secreted angiogenic factors in the hemostatic phase serve to stimulate the production of newer blood vessels (MacLeod & Kwock, 2018). The inflammatory response usually lasts for 3–5 days but often gets delayed in some wound types due to microbial infection, aging, imbalance in the concentration and release of growth factors, lower degree of inflammatory response activation, and reduced proteolytic activity.

12.2.3 Proliferation

The migration of the wound healing process from the inflammatory to the proliferative phase is highly acknowledged due to the deposition of newer epithelial cells, restoring the lost blood vessel network and granulation tissue forming phase. Re-epithelization is promoted by the secretion of keratinocytes generated due to changes in electrochemical gradients and exposure to factors such as hydrogen peroxide, pathogens, growth factors, and cytokines that have accumulated during the inflammatory phase (Shaw & Martin, 2015). The mode of working of keratinocytes changes the polarity and gradient from a top-down to a front-rear approach to initiate epithelial generation by undergoing epithelial–mesenchymal transition. The cell adhesion structures, such as desmosomes and hemidesmosomes, participate in the reorganization of cytoskeleton development for the smoother migration of these keratinocytes. The fibroblasts' secreted metalloproteinases assist in the smoother epithelial migration by forming the basement with extracellular matrix proteins (ECM). Often, this process is triggered due to nitric oxide response from macrophages (Landén et al., 2016). Restoration of lost blood vessels occurs by angiogenesis (generated by endothelial cells in response to angiogenic factors) and vasculogenesis (generated by bone marrow cells). Here, there is the development of newer capillary molecules being synthesized to replace the damaged vessels. The angiogenic factors include vascular endothelial growth factor (VEGF), platelet-derived growth factor (PDGF), basic fibroblast growth factor (bFGF), and the serine protease thrombin, which play a part in the restoration and creation of newer blood vessels. The fibroblasts perform a multifunctional role of secreting matrix metalloproteinases (MMP) that serve to degrade the initially developed fibrin matrix and deposit granulation tissue containing collagen, fibroblasts, fibronectins, granulocytes, macrophages, blood vessels, and other extracellular matrix components such as proteoglycans, hyaluronic acid, glycosaminoglycans, and fibronectin.

12.2.4 Remodeling

Remodeling is the final phase of wound healing. As the name suggests, this is the stage of maturation of the deployed granular matrix into scar components, thereby increasing tissue strength and remodeling the extracellular matrix deposited during the proliferative phase. This is followed by the release of extracellular matrix that is similar to skin tissue in morphology and homogeneity, which can replace this scar component. During the process, there is a significant reduction in capillary movements owing to aggregation. The tissue strengthening is achieved by the replacement of type III collagen (embryo-associated collagen III) with type I collagen, with more pronounced cross fibers helping in the strengthening of the matrix in conjunction with enzymes such as lysyl oxidase, collagenases, and metalloproteinases that assist in covalent cross-linking (Darby et al., 2014). In addition, tropoelastin releases elastin that helps provide flexibility to the developed extracellular matrix skin. Glycosaminoglycans, proteins, proteoglycans, and secreted protein acidic rich in cysteine get synthesized, which aid in successful maturation and matrix remodeling. The process can take anywhere from 6 months to a maximum of 2 years, with corresponding changes in tissue, matrix, and color development (Singh et al., 2017).

12.3 ROLE OF NANOMATERIALS IN THE WOUND HEALING PROCESS

Currently practiced wound management practices revolve around the use of wound dressings that act as a passive barrier, thus preventing exposure to the environment and microbes. The area in the proliferating and remodeling state must be kept sterile, made hypoxic, and allowed to maintain sufficient moisture to support proper division and creation of cell structures. The material of wound dressings also depends on the type of wound and its intensity. Ideal selective wound dressings should be biocompatible, cost-effective, able to regulate the movement of moisture and air particles, and possess enough strength and rigidity (Nguyen et al., 2023). Additionally, the selected material should be nontoxic, nonallergenic, and provide a biodegradable option to the environment. Traditionally, wound dressings were made of cotton and wool; however, innovative approaches in material design and chemistry have allowed for the development of foams, hydrogels, alginates, films, and hydrocolloids as the current generation of wound dressing materials. The oral intake of pharmaceutical drugs, such as broad-spectrum antibiotics, nonsteroidal anti-inflammatory drugs, and glucocorticoid steroids, has been recommended in certain wound healing cases to create a sterile area and allow for cell growth. This approach of using antibiotic drugs is more recommended for surgical chronic wounds as a prophylactic measure (El-Ela et al., 2019). It is also recommended to avoid possible infections that can arise due to bacterial or any microbial growth on the wound site. The antibiotics of choice include penicillins, cephalosporins, macrolides, clindamycins, tetracyclines, and fluoroquinolones, which mainly aim to alter the physiological state of the bacteria, eventually leading to their death. Again, the choice of antibiotics depends on the etiology and the severity of the infection. The limitation of oral antibiotics lies in their possible nonsupply to the indicated vascular chains upon entering the digestive tract. Inappropriate and nonregulated use of these antibiotics can lead to toxicity and resistance development in microbes. Topical antibiotics and antiseptics are used for wound healing to address the limitation of inadequate vascular supply from oral antibiotics and to maintain the wounded area sterile, free of microbes, and cell remnants or debris, which may predispose the healing site to infection (Atiyeh et al., 2009). Here, the antibiotic is topically concentrated in the area of the wound, allowing for faster healing and potentially avoiding the undesirable side effects of oral antibiotics that can cause digestive disorders (Heal et al., 2016). Commonly used topical antibiotics and antiseptics include silver sulfadiazine, fusidic acid, metronidazole, sodium chloride, chlorhexidine, and povidone-iodine, respectively. Some traditional methods of dressing include the addition of essential oils (Mori et al., 2016), aloe vera (Garcia-Orue et al., 2017), astragalus (Wong et al., 2001), burdock (Pomari et al., 2013), honey (Negut et al., 2018), maggots (Sherman, 2014), and propolis (Oryan et al., 2018) to promote wound healing. Modern approaches that have emerged in the medical field include hyperbaric oxygen treatments, negative pressure wound therapies, bioengineered cell constructs, and ultimately resorting to vascular surgery. Needless to say, there is no single-stop solution for scientific wound healing with the accuracy needed to maintain desirable conditions at the wound site that allow for granulation succession and effective healing.

Nanomaterials in this line can be seen as interesting and opportunistic materials with significant moldable properties to suit various requirements. The dimension reduction to extremely lower scales allows for increased surface active regions that may have more active sites to interact with the wounded region. In addition, the antimicrobial efficacy of certain nanomaterials can promote faster healing due to the creation of aseptic surfaces. The multitasking effect of nanomaterials can influence the enhancement of the vascular supply chain, cell regrowth dynamics promoting scaffold development, and granulation improvement, modulate the release of certain growth factors and immune cells, modulate cellular responses by means of contact or interaction, and provide skeletal support for the growing scaffolds (Lin et al., 2014). Nanomaterials, by means of their unique size and composition, can act for wound healing due to their ability to penetrate and provide antimicrobial action or can act as entrapping materials for novel drugs to carry to the site of the wound. The choice of nanomaterials can lead to dramatic improvements in different stages of wound healing. Biodegradability, nontoxicity, biocompatibility, ease of synthesis, reduced side effects, increased permeability, the ability to aid in angiogenesis, the ability to stimulate immune cells, and antimicrobial activity are unique advantages that have increased the scope of utilizing nanomaterials in wound healing.

12.4 GREENER SOURCES OF NANOMATERIALS

The commonly synthesized nanomaterials include organic and inorganic derived sources. The magnetic, metal, metal oxide, and ceramic-based nanomaterials are included in the inorganic source, whereas the carbon-based nanomaterials, dendrimers, ferritins, polymeric micelles, and liposomes are classifications under the organic source. These organic-derived nanoparticles are limited by their shorter shelf life and reduced stability (Naseri et al., 2015). There is increased use and demand for the inorganic source due to its hydrophilic properties, stability, and nontoxicity behavior (Alshammari et al., 2023). They are believed to possess excellent material, optical, mechanical, corrosion resistance, and thermal properties and can act as vehicles for drug and bioactive delivery to the site of action (Bajpai et al., 2016). The conventional methods followed for nanomaterial synthesis include laser ablation, electrochemical, photochemical, irradiation, emulsion, and microwave techniques, all of which lead to the consumption of more energy with simultaneous release of toxic gases. Additionally, during fabrication, various hazardous chemicals and solvents are utilized as reducing and stabilizing agents, which may have a significant impact on environmental emissions. In this regard, a shift has been made toward natural sources of nanomaterials as a sustainable and green concept with a view to considerably reducing the use of solvents.

The commonly followed top-down approach involves physical forces of crushing, milling, and breaking to synthesize nanoparticles, while the bottom-up approach involves the use of chemicals to synthesize nanoparticles with the required characteristics. While the top-down approach for nanoparticle synthesis is backlogged with the requirement of huge space and consumption of higher energy, the bottom-up approach involves considerable volumes of solvents and chemicals, thus adding to environmental residues and emission impacts. On the other hand, following the

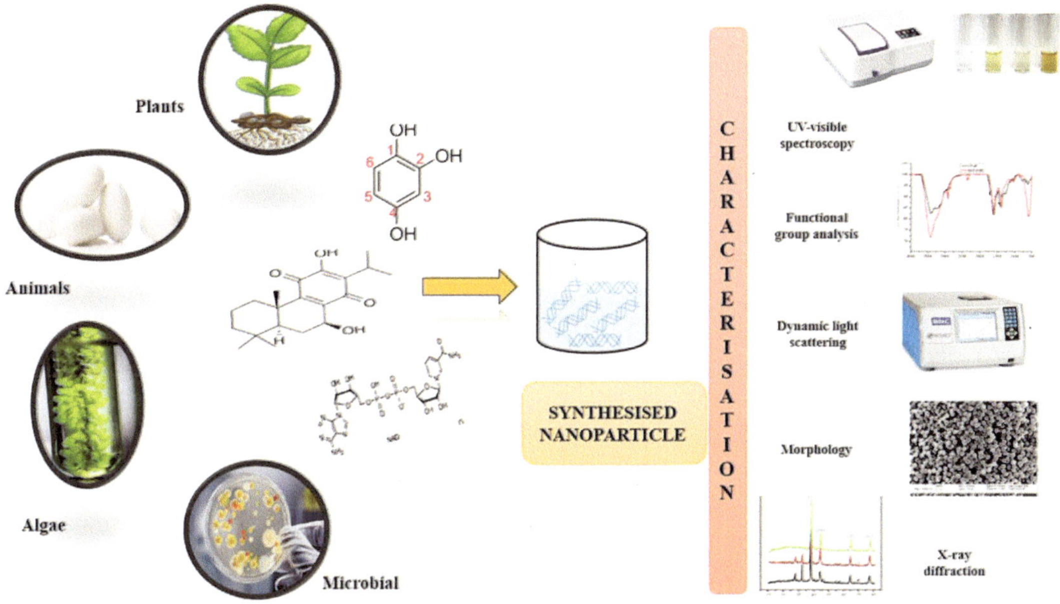

FIGURE 12.1 Green synthesis and characterization of nanomaterials.

greener approach to nanoparticle synthesis revolves around using biological agents such as plants, animals, microbes, and algae, either whole or in extracts, as sources of nanoparticles (**Figure 12.1**). It can be viewed as reducing toxicity and offering energy, time, and cost-effectiveness (Reddy et al., 2012). This can be seen as a combination of biology, ecology, biotechnology, and nanotechnology for the creation of nanoparticles.

12.4.1 Plant-Derived Nanomaterials

Plants, their parts, and extracts are increasingly being used for nanomaterial synthesis, with plant extracts presenting an interesting source for this synthesis owing to the presence of numerous bioactive compounds and their ability to act as reducing agents. Aptly called phytonanotechnology, this field explores potential sources for the increased efficiency of nanoparticle synthesis. The nanomaterial synthesis in plants can be a twofold strategy. The first is the biotransformation of inorganic minerals in the soil to nanoparticle format in the plant itself, while the second, which is more commonly followed, involves the use of plant extracts to reduce commercially available minerals to create nanoparticles. Gold, silver, copper, zinc, and their corresponding oxide-based nanoparticles can be synthesized using plant extracts. Here, the plants are phytomined for increased reducing ability and are utilized for the respective synthesis. It has been shown that the reducing and stabilizing effects of natural plant extracts are far superior to their synthetic counterparts (Iravani, 2011). Additionally, the process is simple, easy, and reduces toxicity emissions compared to the synthetic route for nanoparticle production. The basic mechanism of nanoparticle synthesis includes the reduction of metal ions by the phytochemicals present in plant extracts, followed by nucleation, after which the nanoparticles start to synthesize, form clusters, and coalesce, leading to larger nanoparticle synthesis, and finally, the termination phase, which defines the final size and shape of the nanoparticles. The reduction of metal ions is crucial, and its synthesis rate is found to be influenced by pH, temperature, and length of reaction time.

The preparation of extracts for the reaction follows the sequence of collecting the intended plants, washing, drying, homogenizing, followed by the extract preparation using Soxhlet apparatus or the decoction methods (Chatterjee et al., 2016). Several reports suggest that the presence of various plant-derived metabolites such as phenolics, sugars, terpenoids, saponins, alkaloids, tannins, vitamins, polyphenols, and proteins, polysaccharides are known to act as reducing as well as stabilizing agents (Savoia, 2012; Qamar & Ahmad, 2021). The presence of hydroxyl groups is known to provide hydrogen ions for the reduction reactions. These metabolites are known to participate in tuning the final size, shape, and dimensions of the nanomaterials. Furthermore, the size, shape, and composition of nanomaterials vary as per the source of the plant and plant parts, the type of metal used, the method used for synthesis, its operating conditions, and its interaction with the metal ions (Shankar et al., 2004). Different parts of plants are known to produce nanoparticles with altered effectiveness. In an experiment evaluating the nanoparticles synthesized from leaves, flowers, and bark of *Hibiscus rosa-sinensis*, Periasamy et al. (2022) observed a deviation from the regular spherical shape of nanoparticles in bark-derived ones, and this nanoparticle exhibited the highest antimicrobial efficacy toward tested microbes *Escherichia coli*, *Pseudomonas aeruginosa*, *Staphylococcus aureus*, and *Bacillus subtilis*. *Ananas comosus* extract was able to reduce silver nitrate ($AgNO_3$), leading to the generation of silver nanoparticles, which was confirmed by sharp diffraction peaks at 38.45°, 44.48°, 64.69°, and 77.62° corresponding to silver, and a broader peak base indicated a smaller size ratio. Though the extract was a rich source of bioactive compounds, including hydrocinnamic acids, caffeic acids, rosmarinic acid, and chlorogenic acids, it is believed that the presence of ferulic acids was responsible for the reduction reactions due to its resonance-stabilized phenoxyl radicals (Ahmad & Sharma, 2012). An interesting insight into proteins playing the role of a reducing agent is observed in the studies of Li et al. (2007), in which capsicum extracts were used for silver nanoparticle synthesis. The obtained peaks at 210 and 280 nm confirmed the presence of peptide bonds and the aromatic amino acids (tyrosine, tryptophan, phenylalanine). The perfectly similar nanoparticle shape was possible due to the isotropic growth of silver nanoparticles, and it was found to follow the recognition reduction—limited nucleation and growth model (Burda et al., 2005). Since silver nanoparticles are well-recognized antimicrobial agents, Yassin et al. (2022) approached their synthesis from *Origanum majorana* extracts and found that they exhibited antimicrobial activity tested using disk diffusion assays against *Staphylococcus aureus*, *Klebsiella pneumoniae*, and *Aspergillus baumannii*. Zinc oxide nanoparticles (ZnO-P) are shown to have photo protective properties, which were assayed using photocatalytic degradation of methyl orange (azo-based cationic dye), which proved that exposing the synthesized *Salvinia officinalis* leaf-based nanoparticles was able to exert 92.47% degradation after 120 minutes of exposure, resulting in a color change from orange to colorless (Abomuti et al., 2021). Silver nanoparticles synthesized from *Eugenia roxburghii* were able to reduce biofilm production by *Staphylococcus aureus*, as confirmed by colorless indication on Congo Red assay media (CRA); the control media was able to change color from red to black due to biofilm developed due to the secretion of exopolysaccharides (Giri et al., 2022).

To determine the effect of plasma on the Dzimitrowicz et al. (2019) prepared the gold nanoparticles from aqueous extracts of Lamiaceae groups of plants with tetrachloroauric (III) acid hydrate ($HAuCl_4$ X $4H_2O$). The basic step of gold nanoparticle synthesis is the reduction of Au (III) ions to their metallic forms (Au^0) to the final concentrations of 50, 100, and 200 mg L^{-1}, respectively, for which the reduction was confirmed by the color shift studies of UV-visible spectroscopy. Further, the developed gold nanoparticles were subjected to direct current atmospheric pressure glow microdischarge (dc-APGD) and observed that the plasma treatment was able to induce aggregation leading to size enlargement of the nanoparticles. The shifts in peaks subjected to FTIR confirm the participation of phenols in the bioreduction reaction, which was analyzed by the Folin-Ciocalteu assay that underwent reduction owing to the participation of phenols in reducing the Au(III) ions. Similar participation of phenols is noticed in the gold nanoparticle synthesis from *Nepenthes khasiana* extracts (Dhamecha et al., 2016). In an interesting study to determine the effects of light on silver nanoparticles from almond extract, Srikar et al. (2016) used four approaches—darker conditions, diffused sunlight conditions, and direct sunlight conditions—and found that the sunlight condition was able to exert increased conversion reactions between silver ions to silver nanoparticles.

The increase was close to 60 times that of the control. It was also found that the amide and alcohol groups were responsible for stabilizing and reducing reactions.

12.4.2 Animal-Derived Nanomaterials

The byproducts from animal husbandry and processing can be successfully converted to nanomaterials using a bottom-up approach. Both polysaccharides and proteins can be used as polymer base materials for nanoparticle synthesis. Prominent feedstocks from animal sources include chondroitin sulfate, collagen, chitosan, and silk fibroin. They are biodegradable, biocompatible, nontoxic, and possess excellent mechanical properties (Patil et al., 2020). Silk fibroins are produced from silkworms, lacewings, spiders, and wasp beetles. They are protein polymers with two chains bonded by disulfide bonds. As a naturally occurring protein biopolymer, fibroin has been approved as a biomaterial by the US-FDA (Zhao et al., 2015). The strong intra- and intermolecular hydrogen bonds and van der Waals forces contribute to the polymer's strength, rigidity, aqueous resistance, and degradation factors (Keten et al., 2010). The repetitive sequence of hydrophilic and hydrophobic amino acid side chains yields antiparallel crystalline sheets. These properties make fibroin protein an excellent biopolymer with higher mechanical strength. Moreover, the amino acid chains include a glycine-alanine-glycine-alanine-glycine-serine repeat sequence, with the dominant presence of tyrosine and methanolic serine sidechains. It is these tyrosine and methanolic serine sidechains that play an important role in nanoparticle synthesis by acting as reducing agents (Si et al., 2007). The nanoparticles can be used as excellent biomaterials for drug delivery, wound dressings, and various biomedical devices.

Das and Dhar (2014) fabricated gold, silver, palladium, and platinum nanoparticles from thiol-modified fibroin by in situ redox method. The thiol modification was necessary to act as a stabilizing agent for the developed nanoparticles. Alkali conditions were maintained (pH 10) during the reaction to prevent cross-linking due to precipitation. The catalytic reduction property was observed by the conversion of *p*-nitrophenol to *p*-aminophenol in the presence of sodium borohydride (NaBH4). Collagen represents the accumulated structural protein and offers close to one-third of the body weight of the animal (Nimni, 2018). The amino acids bind themselves into a triple helical model by fibrils, and these three polypeptide chains form a unique triple helical structure that is intervened in a helical pattern with repetitive amino acid repetition of glycine, proline—hydroxyproline. Generally, it is extracted by acid methods or enzyme (collagenase, papain, pepsin)-based methods. In addition, collagen mimics the exact scaffold required for wound healing. El-Sawah et al. (2024) biosynthesized collagen nanoparticles in the actinomycetes strain *Streptomyces xinghaiensis* NEAA-1, with yield optimized at 8.92 mg/mL, and the collagen nanoparticles exhibited antihemolytic, antioxidant, and cytotoxic activity. Chitosan is an abundantly available polysaccharide with D-*glucosamine* and *N*-acetyl-D-glucosamine linked by β (1–4) obtained by deacetylation of the chitin molecule. It is commonly found in the exoskeletons of crustaceans, cuticles in insects, and fungal cell walls. The polysaccharide is characterized by its cationic charge, molecular weight, and degree of acetylation, which are known to influence its antimicrobial, mucoadhesive, gelling, and stabilizing properties. Therefore, chitosan presents an excellent opportunity for nanoparticle fabrication due to its role as a reducing, capping, and stabilizing agent, along with its role as a size-controlling agent (Venkatesham et al., 2014; Phan et al., 2019). The free hydroxyl group undergoes oxidation, releasing electrons that can participate in nucleation, reducing the metal ions to nanoforms.

Wongpreecha et al. (2018) developed silver nanoparticles using a one-pot green approach with chitosan as a reducing and stabilizing agent. They used an autoclave at 105, 110, 115, and 120 °C for 50 min. The pressure generated during the autoclave enhanced the reduction reaction, which was confirmed by X-ray diffraction (XRD) and transmission electron microscopy (TEM), and the nanoparticles were found to have lower cytotoxicity against mouse fibroblast cells. On the other hand, some research investigations find that higher temperatures during autoclaving can promote aggregation and transform the developed nanoparticles into microparticles (Pourali et al.,

2013). Low-temperature reaction conditions were successfully tested and proved effective for gold nanoparticle synthesis. The selected temperature conditions were ambient room temperature for 6 hours and an hour at 35 °C; the synthesized nanoparticles were stable when analyzed using X-ray photoelectron spectroscopy (XPS), Fourier transform infrared spectroscopy (FTIR), and UV-visible spectroscopy (Wu et al., 2014). An electrochemical oxidation approach with a potential of 2V, followed by UV irradiation (254 nm), was used for reducing chitosan-based silver nanoparticles as a greener approach. The tested conditions included electrochemical oxidation for 6, 12, 18, and 24 h, followed by UV exposure for 1, 3, and 6h. This technique presents an opportunity to reduce the significant vacuum requirement for current adjustment while obtaining higher purity nanoparticles (Reicha et al., 2012). Recently, animal blood plasma, due to its constituent enzymes, proteins, and bioactive peptides, has been used in nanomaterial biosynthesis (Kakakhel et al., 2021).

12.4.3 Algae

Algae contribute excellent biomachinery for the source and extraction of nanoparticles. They are aquatic photosynthetic and filamentous prokaryotes mainly performing the role of photosynthesis. Microalgae and macroalgae are the types of algae differing in type, nature, and source of availability (Leaf et al., 2020). Due to a wider cultivation environment (open ponds or photobioreactors), algae accumulate significant concentrations of polysaccharides, antioxidants, phenolics, flavonoids, alkaloids, organic acids, reductases, carotenoids, polyunsaturated fatty acids, and several smaller amounts of minor bioactives (Santhoshkumar et al., 2021). These constituents are known to aid in nanoparticle synthesis. The process includes the mixing of selected algal cells and nanoparticle precursors for a predefined time and temperature, leading to the synthesis of nanoparticles. Sometimes the algal cell, such as cyanobacteria, itself hosts several nanoparticles due to photosynthesis and respiration activities in conjunction with enzymes such as nitrogenase that act as a reducing agent, leading to nanoparticle synthesis (Mukherjee et al., 2021).

Deepak et al. (2018) synthesized silver nanoparticles using brown algae *Sargassum wightii*, and the developed nanoparticles were found to exhibit enzymatic inhibition against α-amylase and α-glucosidase and antioxidant properties by 1,1-diphenyl-2-picryl-hydrazyl (DPPH) and ferric ion reducing power (FRAP) assays, which can be used for diabetes management in reducing blood sugar spikes. Spirulina is used for the biosynthesis of several nanoparticles, including gold and silver nanomaterials (Govindaraju et al., 2008). Viera et al. (2015) developed a green method of silver nanomaterial synthesis using aqueous extracts of *Laurencia aldingensis* and *Laurenciella* spp., which exhibited excellent cytotoxicity against sarcoma tumor cells. This suggests the potential use of several algae-derived nanoparticles in various pharmacological and medical applications. Higher pH among the examined conditions of 6.2, 6.8, 7.8, and 8.4 during the synthesis phase was found to correlate with an increased yield of silver nanomaterials in the nanomaterial synthesis from brown seaweed *Sargassum longifolium* (Rajeshkumar et al., 2014). Several investigations through a green approach have examined the antifungal and antibacterial efficiency of synthesized nanomaterials.

12.4.4 Microbes

Microbial nanoparticles present an excellent opportunity for the biosynthesis of nanoparticles that can occur by means of templating, biomineralization, or enzyme-induced reduction approaches. Bacteria, fungi, yeast, viruses, and actinomycetes can be successfully utilized for nanomaterial synthesis (Salem et al., 2021). These microbes offer a cost-friendly and time-saving approach for greener synthesis compared with plant systems. Nanoparticle synthesis can be intracellular or extracellular. While the intracellular method essentially involves culturing microbial biomass with selected aqueous nanoparticle precursors, which will undergo further bioreduction (Li et al., 2011), the extracellular approach refers to the use of microbial extracts and supernatant cultures (Das et al., 2014) in synthesizing nanomaterials with corresponding nanomaterial precursors such as gold, silver, zinc, copper, etc. The exact mechanism involves the solubilization of inorganic minerals using enzymes or the precipitation of soluble to insoluble states, which is dependent on the culture conditions and the nature of the microbe selected. The conversion of metallic precursors to nanoparticles can be via enzymatic or nonenzymatic approaches. The extracts and supernatants are believed to host numerous reducing enzymes capable of facilitating nanoparticle synthesis. The presence of NADH enzyme machinery (NADH nitrate reductase, NADH sulfite reductase), cysteine desulfhydrase, and nitrate reductase is observed in actinomycetes when used for nanoparticle synthesis (Shah et al., 2015). Meanwhile, the presence of oxidoreductases and β-glucans is responsible for yeast-derived nanoparticles (Luo et al., 2015). The nonenzymatic method includes the occurrence of polysaccharides, peptides, and their constituent functional groups like aldehydes, carboxyl, ketones, and amides, which participate in the reduction reactions (Maneerung et al., 2008).

Rajivgandhi et al. (2022) synthesized zinc oxide nanoparticles with soil-derived actinomycetes and observed antibacterial efficacy against *Klebsiella pneumoniae*. The zone of microbial inhibition increased in a dose-dependent manner with concentration. They carried out microwave treatment for mixed cultures of actinomycetes and zinc nanomaterial precursors, which accelerated the reduction reaction by generating intense and higher heat to the flask and is responsible for the more crystalline nature of the obtained nanoparticles. Ali et al. (2018) isolated bacteria from soil samples, and isolate T10 extract was selected for extracellular silver nanomaterial synthesis, which was confirmed by the color change effects in UV-visible spectroscopy. The silver nanoparticle was identified by distinctive peaks in XRD. Extracellular electron transfer (EET) was observed in *Shewanellaoneidensis* and *Geobactersulfurreducens* in response to external environmental conditions. This presents an exciting opportunity for intracellular nanoparticle synthesis. They are also referred to as electroactive microorganisms and provide an opportunity for bacteria to control the dimension of growing nanoparticles without the need to transport metal ions into the cell matrix (Zou et al., 2021).

12.5 APPLICATIONS OF GREEN SYNTHESIZED NANOMATERIALS IN WOUND HEALING AND CELL STIMULATION

Wound healing is an extremely important procedure, as the delay in normalized healing can lead to chronic conditions and complications, especially in the case of diabetes mellitus and heart diseases, sometimes leading to mortality and morbidity. Compared with conventionally derived nanomaterials, greener-sourced nanomaterials offer an edge with respect to the reduction in energy, solvents, and residues for their synthesis, along with an ecofriendly approach for their use. Green synthesized nanomaterials can be altered to meet desired needs, providing a larger surface area-to-volume ratio that can be biocompatible with other natural drugs of choice without any residual or side effects. The green route for nanomaterials biosynthesis presents an opportunity to efficiently utilize natural resources (plants, animals, microbes, algae, byproducts) to derive comparatively greater beneficial nanomaterials. Furthermore, the synthesized nanomaterials can themselves be used as antimicrobial agents due to their improved material and therapeutic properties. Wound healing strategies can be designed as incision-based or excision-based. Incision-based wounds are created by surgical procedures without the loss of tissue, while excision-based wounds involve loss of tissue. An essential wound healing agent should be biocompatible with human-derived tissues, provide requirements for regeneration, and be suitable for various critical wound healing needs across different tissue types. It should also be nontoxic and provide in vivo efficiency for improved wound healing. Additionally, it should participate in the release of immune cells and growth factors (Barroso et al., 2020). The depth of penetration, colloidal stability, size and shape, dose, and format of nanomaterials in which they are applied play a critical role in effective wound healing, i.e., re-epithelization, revascularization, angiogenesis, and cell proliferation.

Ahn et al. (2019) examined the silver nanoparticles developed from seven plant extracts for antioxidant, cytotoxicity, and wound healing properties. *In vitro* wound healing activity was assayed by cell scratch test on NIH3T3 cells and found that silver nanoparticles synthesized with *Lindera strychnifolia* had higher wound healing efficacy. Nanomaterials, when formulated as nanofibers, provide better porosity for aeration and adequate moisture control. The silver nanoparticles, among other related metal counterparts, are widely used in the wound healing process on account of their excellent antimicrobial action. The cumulative efforts of silver nanoparticles lie in their antimicrobial effect coupled with modulation of the fibrocyte microclimate (Tyavambiza et al., 2021). The antimicrobial effect derives from the blockage of the respiratory pathway of bacterial cells, thereby inducing microbial death. It is also shown that silver nanomaterials have the unique ability to interact with sulfur- and phosphorus-rich side chains in the bacterial cell membrane, eventually leading to apoptosis and cell death. This potential allows silver nanoparticles to conjugate with plant extracts, leading to higher bioactivity, as observed in the studies of Ashour et al. (2015), in which cranberry extract (a potent antioxidant) was used for silver nanomaterial synthesis, which possessed excellent wound healing properties. The immunomodulatory performance of silver nanoparticles was confirmed by Dhapte et al. (2014) in *Bryonia laciniosa*-derived nanoparticles, where the wound healing properties were compared with commercially available topical silver sulfadiazine cream. A similar effect was observed in the studies of Tian et al. (2007), in which the selected silver nanoparticles were comparatively more beneficial than silver sulfadiazine. The synthetic cream was able to induce considerable scar formation and increased inflammatory response, completing only 78% wound healing on the 14th day of application, whereas the biosynthesized nanoparticles were able to achieve close to 92% healing. There was a reduced concentration of IL-6, which is a proinflammatory cytokine responsible for the activation of monocytes and fibroblast proliferation. A similar increase in wound recovery of 98% compared to the control was observed in *Catharanthus roseus*-derived nanomaterials (Al-Shmgani et al., 2017). Zangeneh et al. (2019) investigated the synthesis of copper-based nanoparticles from the aqueous extract of the *Falcaria vulgaris* medicinal plant, and the wound healing properties were assessed against synthetic treatments. It was observed that there was a substantial reduction in the inflammatory count, including neutrophils and lymphocytes, with a corresponding increase in the concentration of hydroxylproline and hexuronic acid. There was also an observed increment in the fibrocyte-to-fibroblast ratio. Fibrocytes are proinflammatory agents secreted in response to wounds that aid in angiogenesis, tissue regeneration, collagen deposition, and wound contraction. It is also believed they participate in phagocytosis by releasing lysozyme complexes (de Oliveira et al., 2020). The use of medicinal plants and extracts is known to maintain a stable concentration of free radicals due to their ability to act as antioxidants, which regulate enzymes such as superoxide dismutase, catalase, and peroxidase. The presence of connective tissue markers, such as hydroxyproline, hexosamine, and hexuronic acid, signifies the deposition of collagen matrix and acts as a foundation for the development of the extracellular matrix. In fact, the presence of proline and its conversion product, hydroxyproline, is recognized as a marker for the initiation of collagen synthesis. The hexosamine and hexuronic acids contain glycosaminoglycans as the major component (Murthy et al., 2019). There is an increasing trend in using natural medicinal and aromatic plants for their bioactive potential, which can be used as effective wound healers and natural cell stimulators. They are known to possess antimicrobial, antioxidant, and anti-inflammatory properties, which can significantly intensify the wound repair process. The wound retraction method (as a percentage of wound closure) for the patients was considered in the investigation of Chai et al. (2018), in which the biosynthesized silver nanoparticles from *Delonix elata* leaves were incorporated into wounded areas covered with dressing. The silver nanoparticle-dressed wounds had a higher rate of wound decrement, measured by dimensions of 0.02 mm in length and 0 mm in width on the 17th day of observation, whereas the control had 6.68 mm in length and 2.22 mm in width of wound exposure. In contrast, the relative migration rate was analyzed with a wound scratch test on L929 fibroblast cells in the studies of Veeraraghavan et al. (2021), in which *Scutellaria barbata* was used as a plant-derived bioreducing agent. The obtained results showed that there was a concentration-dependent reduction in the migration of cells when incubated with silver nanoparticles. The synergistic ability of biosynthesized nanoparticles from the aqueous leaf extract of *Urtica dioica* with synthetic antibiotics was observed in the studies of

Jyoti et al. (2016), in which the selected antibiotics (amikacin, kanamycin, vancomycin, tetracycline, cefetaxime, amoxicillin) showed excellent synergy in inhibiting microbial growth. The synergy combinations are beneficial for reducing the spread and intensity of antimicrobial infections, as the microbes may find it difficult to synthesize resistant machinery for the combined effects.

Pourali and Yahyaei (2016) isolated the bacterial colonies and characterized them by genotyping and phenotyping studies in which a selective representative of Gram-positive and Gram-negative bacteria was taken from the Bacillaceae and Enterobacteriaceae families, respectively. The selected microbes were used for extracellular silver nanoparticles preparation. The wound healing property was analyzed by in vivo wound healing experiments on a mouse model and found that the treated groups had comparatively higher wound healing progress on the 7th day and subsequently on the 14th and 21st days. Among the selected microbial nanoparticles, the one with *Escherichia fergusonii* had a higher re-epithelialization percentage than the control with synthetic silver cream. Application of spirulina-based pectin nanoparticles was developed by Rajapaksha et al. (2020) using the sonication method, in which the developed nanoparticle was used for *in vitro* biological assays, including wound healing in adult zebrafish. The wound closure and % pigment accumulation were used as criteria for wound healing completion and were found to undergo rapid healing and melanin pigment restoration. To quantify the response, wound healing percentage was calculated, which was found to be 48.90% corresponding to the control, and complete wound closure occurred on the 24th day postwounding (dpw). Histological evidence showed the presence of inflammatory cells, denser epithelial cells (with a thickness of 53 μm), and formation of granulation tissue.

There was a comparative reduction observed as against the hydroxylamine hydrochloride in the biofilm of *Pseudomonas aeruginosa* exposed to green silver nanoparticles in the studies of Franceschinis et al. (2023), which was developed using aqueous extract of *Aloe maculata*, but failed to provide a significant reduction in the preformed biofilm even at the higher minimum inhibitory concentration (MIC) range. In a similar line of study, there was a 60% biofilm inhibition in the studies of Veeraraghavan et al. (2021). Due to the relatively lower wound healing effect exhibited by the straight synthesized nanoparticles, they are always accompanied or incorporated in certain forms to increase their effectiveness. Gelatin/Neem extract-based nanocomposite incorporated green silver nanoparticles from *Praecitrullus fistulosus* were studied by Sheokand et al. (2022), which possessed excellent antimicrobial activity and a pH of 5.9, close to skin physiological pH, capable enough for acting as wound dressing material. There was aggregation of synthesized nanoparticles observed due to lower electrostatic repulsion and variation in sizes for green synthesized ones, as the developed nanoparticles over six months of storage resulted in higher values for the polydispersity index than the synthetically produced nanoparticles (Franceschinis et al., 2023).

The biosynthesized nanoparticles can be explored in the form of nanofibers, nanosponges, nanoemulsions, nanoliposomes, hydrogels, and composites, which ultimately are required for increasing the material performance and enhanced delivery of the bioactive compound to the targeted site to promote enhanced wound healing. Interesting research investigation by Chen et al. (2017) wherein they fabricated an innovative wound healing approach by blending green synthesized silver nanomaterials and konjac glucomannan polysaccharides leading to the development of freeze-dried porous sponges. The freeze-drying method was adopted to create more open and porous structures as desired, and the porous structures are required for the oxygen entry to the wounds, which is known to play a role in angiogenesis, re-epithelialization, and induce collagen synthesis. Egg white was used as a green reducing agent for the silver nanomaterial synthesis. The wound healing property was observed by invitro mouse skin fibroblast cells and in vivo rabbit model study. The use of MTT 3–4, 5-dimethylthiazol-2-yl-2, 5-diphenyltetrazolium bromide blue for color change study on cytotoxicity was done. Several researchers also use lactate dehydrogenase assay (LDH), neutral red (NR), and crystal violet (CV) for analyzing the cytotoxic effects. All these dye-based methods involve the study of cell viability by a color-changing pattern through staining biomolecules such as DNA and cell membranes, etc. Inductively coupled plasma mass spectrometry (ICP-MS) analysis performed on developed sponges showed that the synergistic blend had a positive action in providing a slow and sustained release rate. It is highly essential that the incorporated nanomaterial should be released slowly into the wound environment. There was an increased wound closure rate of 99% in 0.3% dosed silver nanoparticles. The selection of a biopolymer matrix in composites and sponges' development is essential as the selected polymer provides solid support and creates a matrix for the incorporation of nanomaterials. In addition, the preference for higher water-absorbing and moisture-retaining agents is welcome. A biocompatible nanocomposite network was created by *Syzygium cumini* leaf extract-derived nanocomposites impregnated into bamboo leaves-derived cellulose nanocrystals by Singla et al. (2016) and performed in vivo wound healing was observed in the mice model study. Their cytotoxicity evaluation was conducted against the growing keratinocytes with SRB assay and observed a comparatively lower cytotoxicity (<40%) against the synthetic vinblastine. Silver nanoparticles is known to affect cell viability, cell proliferation, cell metabolism and the cell death process. Topical application of developed wound healing matrix (CNC and nanoparticles) was seen to provide synergistic effect by the hydrophilic nature of CNC increasing the flow of silver nanoparticles and the area had better tissue regeneration which was confirmed by biochemical and histopathological studies. There was significant reduction in the inflammatory agents such as cytokines in the treated groups due to anti-inflammatory response of synthesized nanomaterials which accounted for increased fibroblasts proliferation and keratinocytes differentiation. The work also recognized the increase in angiogenic effect due to lysozyme induced fragmentation of CNC to derived components thus aiding in the development of more compact collagen bundles. Red cabbage extract derived silver nanoparticles in the form of hydrogels was incorporated in the chitosan/polyvinyl alcohol blends and was recommended for wound dressing applications as a result of its antibacterial activity against _Pseudomonas_ _aeruginosa_ and _Klebsiella pneumonia_ strains. Fabrication in terms of hydrogels provides unique opportunity to tune the material properties and are known to absorb the wound emanated exudates and prevent dehydration. The prerequisite of higher mechanical strength is essential for hydrogel preparation as each of the contributing molecule plays a unique role for addressing as wound healing strategy (Kaliaperumal et al., 2023). In a similar line of study, hydrogels were developed from gelatin/polyvinyl alcohol with silver nanoparticles from the aqueous peel extract of *Musa balbisiana,* and its biocompatibility was tested with human foreskin fibroblast cells with an increased cell viability count of > 67% (Bag et al., 2023). Interesting research investigation by Chen et al. (2017) wherein they fabricated an innovative wound healing approach by blending green synthesized silver nanomaterials and konjac glucomannan polysaccharides leading to the development of freeze-dried porous sponges. Egg white was used as a green reducing agent for the silver nanomaterial synthesis. The wound healing property was observed by invitro mouse skin fibroblast cells and in vivo rabbit model study. The use of MTT 3–4, 5-dimethylthiazol-2-yl-2, 5-diphenyltetrazolium bromide blue for color change study on cytotoxicity was done. Several researchers also use lactate dehydrogenase assay (LDH), neutral red (NR), and crystal violet (CV) for analyzing the cytotoxic effects. All these dye-based methods involve the study of cell viability by a color-changing pattern through staining biomolecules such as DNA and cell membranes, etc. Inductively coupled plasma mass spectrometry (ICP-MS) analysis performed on developed sponges showed that the synergistic blend had a positive action in providing a slow and sustained release rate. It is highly essential that the incorporated nanomaterial should be released slowly into the wound environment. There was an increased wound closure rate of 99% in 0.3% dosed silver nanoparticles. The selection of a biopolymer matrix in composites and sponges' development is essential as the selected polymer provides solid support and creates a matrix for the incorporation of nanomaterials. In addition, the preference for higher water-absorbing and moisture-retaining agents is welcome. The selection of plant or bioreducing agents also assists in dictating the process methodology for the film formation. In one of the studies, silver nanoparticles developed by *Momordica charantia* fruit extract conferred the property of electrospinnability, which led to the synthesis of nanofibers and exhibited limited cytotoxicity and better wound healing (Alippilakkotte et al., 2017). A narrative-based use of various nanomaterials on wound healing property is being depicted in **Table 12.1**.

TABLE 12.1
Green Synthesized Nanomaterial Applications in Wound Healing and Cell Stimulation

Nanomaterial Precursor	Source of Bioreducer	Reaction Conditions	Key Effects on Wound Management	References
Plant-Based Nanomaterial Biosynthesis				
Zinc	*Trianthema portulacastrum*	Incubation—50 minutes Purification—centrifugation at 8000 rpm for 15 minutes at 60 °C	✓ Reduction in wound area at third day and highest wound contraction observed. ✓ Rate of re-epithelialization was more in 2% incorporated zinc nanoparticles ✓ Increase in hydroxyproline content ✓ Increase in levels of antioxidant enzymes (superoxide dismutase, catalase and peroxidase) ✓ Reduced accumulation of C-reactive proteins (marker of inflammation)	(Yadav et al., 2018)
Silver and zinc	Seeds of *Prosopis farcta* for silver nanoparticles; coffee seeds for zinc nanoparticles	Silver nanoparticle (incubation—2 days; room temperature; filtered with Whatman paper) Zinc nanoparticle (incubation—3 hour; drying at 60 °C and calcination at 600 °C)	✓ Bandages impregnated with biosynthesized silver and zinc nanoparticles showed enhanced antibacterial activity. ✓ Silver nanoparticle showed higher antimicrobial effect in terms of minimum inhibitory concentration (MIC) as compared with zinc and blends (silver/zinc)	(Khatami et al., 2018)
Zinc oxide	Leaf extract of *Barleria gibsoni*	Combustion method—at 400 °C Reaction time—< 2 minutes	✓ Gel-formulated zinc oxide nanoparticles had a significant wound healing rate but comparatively lower than the synthetic dermazine drug	(Shao et al., 2018)
Zinc oxide	Leaf extract of *Strychnos nux-vomica*	Incubation—100 °C for 2 hours Centrifugation—1957 × *g* for 10 minutes Annealing—400 °C for 2 hours	✓ Dosage at 200 µg/ml had a higher wound closure rate (28.81%) assayed by scratch wound assay at 12 hours	(Steffy et al., 2018)
Cerium dioxide hydrogel based	Fruit extract of *Abelmoschus esculentus*	Drying—120 °C for hour Calcination—600 °C for 4 hours	✓ Decrease in the swelling rate of the formulated chitosan hydrogels on increasing nanoparticles ✓ Significant increase in the wound closing rate at 1 and 5% incorporation after 11 days of incision ✓ Increase in collagen synthesis and tensile strength of skin noticed	(Ahmed et al., 2021)

(Continued)

TABLE 12.1 (*Continued*)
Green Synthesized Nanomaterial Applications in Wound Healing and Cell Stimulation

Nanomaterial Precursor	Source of Bioreducer	Reaction Conditions	Key Effects on Wound Management	References
Silver	Younger leaf extract of *Curcuma longa*	Incubation—30 minutes Centrifugation—15,000 rpm for 30 minutes	✓ 20 μg loaded *Curcuma longa* nanoparticles had a better cell migration with enhanced fibroblast cell proliferation and migration (L929 cell)	(Maghimaa & Alharbi, 2020)
Titanium oxide	Leaf extracts of *Strychnos spinosa* and *Blighia sapida*	Incubation—70 °C for 4 hours Centrifugation—2000 rpm for 20 minutes Drying—40 °C for 30 minutes	✓ Initiation of wound contraction by the fourth day on topical application of 10 and 20% biosynthesized nanoparticles ✓ 94% wound closure on 14th day of application as against 86% wound closure in control gentamycin sulfate	(Hassan et al., 2020)
Silver	Leaf and aerial extracts of *Cratoxylum formosum*, *Phoebe lanceolata*, *Scurrula parasitica*, *Ceratostigma minus*, Mucuna *birdwoodiana*, *Myrsine africana* and *Lindera strychnifolia*.	Incubation—24 hours	✓ Wound healing was assayed by cell scratch method ✓ Among the tested samples the *Lindera strychnifolia* derived silver nanoparticles was able to induce 64% wound closure	(Ahn et al., 2019)
Silver	Plant extract of *Scutellaria barbata*	Incubation for 30 minutes	✓ 5 and 7.5 μg of biosynthesized nanoparticles were used for wound closure assay on L929 fibroblast cells. ✓ Reduced size of wound by proliferation, differentiation, and migration of the L929 fibroblast cells	(Veeraraghavan et al., 2021)
Animal-Based Nanomaterial Biosynthesis				
Silver	Trimmed wastes of raw cowhide as a source of collagen	Incubation—for 3 hours at 80 °C Centrifugation—50,000 rpm for 1 hour Freeze dried at 4 °C	✓ Topical application of biohybrid-based hydrogel lead to increase in collagen deposition ✓ Reduction in wound size by 91 % on the 12th day of dpw and 98% on the 16th day of wounding ✓ Comparative better tissue regeneration as compared with povidone iodine solution	(Ragothaman et al., 2021)
Gold	Sericin from silk cocoon of *Bombyx mori*	Incubation with UV radiation treatment for 24 hours Centrifugation at 11,000 rpm for 25 minutes	✓ Wound scratch test on HaCaT cells performed with 5, 10, and 25 μg/mL led to enhance migration and better wound closure with 70.96 and 69.76% for 5 and 10 μg/mL, respectively	(Das et al., 2022)
Silver	Silkworm cocoons	Incubation—room temperature for 4 hours	✓ Epidermis thickening occurred at 14 days of healing.	(Yu et al., 2017)

Microbe-Based Nanomaterial Biosynthesis				
Zinc oxide	Endophytic fungal extract of *Xylaria arbuscula*	Incubation—26 °C for 24 hours Centrifuged at 10,000 rpm for 15 minutes Drying at 150 °C	✓ *In vitro* wound scratch assay tested at 25, 50, 75, and 100 μg/ml ✓ 95.37 % wound healing for 100 μg/ml ✓ Enhanced cell migration at the rate of 99.24% after 72 hours	(Nehru et al., 2023)
Silver	*Streptomyces griseorubens* AU2	Incubation—28 °C for 24–48 hours	✓ Dosage at 20 μg/mL Increased cell migration with enhanced wound closure rate in (NIH-3T3 fibroblast cell lines)	(Baygar et al., 2022)
Silver	Cyanobacteria—*Phormidium* spp.	Incubation—aid of illumination with (2000 ± 200 Lux) carried for 72 hours	✓ No signs of inflammation in treated wound as observed against control with inflammation ✓ Accelerated scab development and increase in the accumulation of hydroxyproline indicating of faster wound healing ✓ Reduction in the number of days for re-epithelialization and was found to be dose dependent (17.5, 16 and 15 days for 10, 30 and 50 μg/kg) respectively	(Younis et al., 2021)
Selenium	Soil extracted strain M10A62 actinomycetes	Incubation for 72 hours Centrifugation—20,000 rpm for 1 hour	✓ Wound healing studies were made with 5 and 10 % incorporation in ointment base. ✓ 10% incorporated biosynthesized nanomaterial had wound closure at 18 days compared with antibiotic lyramycin treated 21 days	(Ramya et al., 2015)

12.6 FUTURE PERSPECTIVES

Green synthesized nanoparticles have created limitless opportunities for their synthesis and role delivery in wound healing and related aspects. But there are quite important gaps that have arisen alongside their benefits. The major issue is the toxicity of nanoparticles as a function of their size, shape, and morphology. It is widely believed that synthesized nanoparticles, irrespective of the source or process, are harmful due to their extremely small size, which allows them to enter the human system and leave harmful residues and toxic effects. Especially, free silver ions, when exposed to skin, can cause permanent skin discoloration in humans, leading to argyrosis (bluish discoloration of the eyes). If they enter the human system, they get solubilized and are known to cause damage to the kidneys, liver, and respiratory tract infections. Hence, there is a critical need for safety evaluation studies of these green synthesized nanomaterials for their scale-up and continuation for commercial use. The biosafety requirements for green synthesis and its further application need to be evaluated. Furthermore, the end products of green synthesis methods present subtle variations in their properties, as observed in many investigations, and there is always a challenge of reproducibility. The properties may be related to uniformity issues in yield, purity at a larger scale, and size, shape, morphology, and variable stability at the individual level (Franceschinis et al., 2023). The causal agents can be numerous, including the diverse nature of raw materials, non-standardized protocols for extract derivation, and subsequent nanomaterial synthesis. This causes major quality issues if undertaken on a commercial scale as an alternative to synthetic counterparts. Among the particle properties, the precise management of particle dimensions poses a serious challenge and is found to be responsible for varying effects (Pal et al., 2022). There is still room for advancements in purification and handling patterns to standardize processes. The affordability of raw materials poses a significant challenge if well-documented materials are to be sourced for larger-scale production. Another significant hindrance limiting larger-scale uptake, as observed in many studies, is the lack of scientific investigations on the methods of synthesis, the mechanisms of nanoparticle synthesis, and the correlation of processes followed with the material properties observed. Hence, a scientific approach needs to be standardized and streamlined, starting from sourcing until the end application.

As a reference to the wound healing properties of green synthesized nanomaterials, many of the investigated materials are limited to laboratory-scale investigations, and the retention of bioactive activity poses a challenge in its stability. The behavior of nanoparticles toward different steps of wound healing presents a varied response to different cell lines selected with respect to their cytotoxicity levels, and there is a lack of uniformity in the response observed. As an example IC_{50} values was found to vary in the selected zebra fish animal was 200 μg/mL as opposed to the previously reported literature of 50 μg/mL which may lead to uncertainty and confusion (Rajapaksha et al., 2020). The resistance states and their possible impact on the use of nanomaterials should be considered, as microbial infection prevention plays a crucial role in wound management, and the risk of antimicrobial resistance toward this approach should always be cautioned. There is some reported evidence of bacterial resistance to silver nanoparticles when subjected to long-term exposure at sub-inhibitory levels (Panáček et al., 2018). The possible risks of silver-based nanomaterials impacting genetic machinery (DNA), causing apoptosis, and inducing oxidative stress with damage to mitochondrial machinery should be a matter of concern when animal-based studies are implemented on human populations. Overall, the concepts of safety evaluation, toxicity studies, and microbial resistance physiology are prima facie requirements for the advancement of green synthesized nanomaterials in scientific wound management practice.

12.7 CONCLUSION

Currently followed wound management practices are backed by the problems of allergenicity, antimicrobial resistance, side effects of drugs, and nonbiocompatibility with tissues. Nanotechnology, which is regarded as the technology of the 21st century, is spearheading its importance in diverse areas spanning agriculture, food, bioprocessing, medicine, pharma, energy, etc. The perspective of green synthesis of nanomaterials and its application in wound healing has delivered results of biocompatibility, safety, limited or reduced solvents and residual problems, and nontoxicity, ultimately leading to the concept of sustainability. It presents an opportunity to tap into natural renewable bio-resources (plants, animals, microbes, algae) as bioreducing agents for nanomaterial synthesis and applications. The material properties (size, shape, composition, concentration) of nanoparticles are synergized with the bioactive potential of selected bioreducing agents that may confer antibacterial, anti-inflammatory, and antioxidant effects, accelerating the process of wound healing. The process parameters such as pH, temperature, length of incubation, method of incubation, and downstream processes play a critical role in the successful formulation of nanoparticles. Additionally, the development of various formats of nanoparticles (nanosponges, nanohydrogels, nanoemulsions) is explored to provide overall support for rapid wound healing. The ability of nanomaterials to participate in wound healing via re-epithelialization, collagen synthesis, fibroblast differentiation, keratinocyte generation, and tissue granulation, with the absence of allergens, has added to the wound healing benefits of green synthesized nanomaterials.

REFERENCES

Abomuti, M. A., Danish, E. Y., Firoz, A., Hasan, N., & Malik, M. A. (2021). Green synthesis of zinc oxide nanoparticles using salvia officinalis leaf extract and their photocatalytic and antifungal activities. *Biology*, *10*(11), 1075.

Ahmad, N., & Sharma, S. (2012). Green synthesis of silver nanoparticles using extracts of Ananas comosus. *Green and Sustainable Chemistry*, *2*, 141–147.

Ahmed, H. E., Iqbal, Y., Aziz, M. H., Atif, M., Batool, Z., Hanif, A., . . . Ahmad, H. (2021). Green synthesis of CeO2 nanoparticles from the Abelmoschus esculentus extract: Evaluation of antioxidant, anticancer, antibacterial, and wound healing activities. *Molecules*, *26*(15), 4659.

Ahn, E. Y., Jin, H., & Park, Y. (2019). Assessing the antioxidant, cytotoxic, apoptotic and wound healing properties of silver nanoparticles green-synthesized by plant extracts. *Materials Science and Engineering: C*, *101*, 204–216.

Ali, I., Qiang, T. Y., Ilahi, N., Adnan, M., & Sajjad, W. (2018). Green synthesis of silver nanoparticles by using bacterial extract and its antimicrobial activity against pathogens. *International Journal of Biosciences*, *13*(5), 1–5.

Alippilakkotte, S., Kumar, S., & Sreejith, L. (2017). Fabrication of PLA/Ag nanofibers by green synthesis method using Momordica charantia fruit extract for wound dressing applications. *Colloids and Surfaces A: Physicochemical and Engineering Aspects*, *529*, 771–782.

Alshammari, B. H., Lashin, M. M., Mahmood, M. A., Al-Mubaddel, F. S., Ilyas, N., Rahman, N., . . . Khan, R. (2023). Organic and inorganic nanomaterials: Fabrication, properties and applications. *RSC Advances*, *13*(20), 13735–13785.

Al-Shmgani, H. S., Mohammed, W. H., Sulaiman, G. M., & Saadoon, A. H. (2017). Biosynthesis of silver nanoparticles from Catharanthus roseus leaf extract and assessing their antioxidant, antimicrobial, and woundhealing activities. *Artificial Cells, Nanomedicine, and Biotechnology*, *45*, 1234–1240.

Ashour, A. A., Raafat, D., El-Gowelli, H. M., & El-Kamel, A. H. (2015). Green synthesis of silver nanoparticles using cranberry powder aqueous extract: Characterization and antimicrobial properties. *International Journal of Nanomedicine*, *10*, 7207–7221.

Atiyeh, B. S., Dibo, S. A., & Hayek, S. N. (2009). Wound cleansing, topical antiseptics and wound healing. *International Wound Journal*, *6*(6), 420–430.

Bag, D. S. S., Bora, A., Golder, A., Raina, K., Haridhasapavalan, K. K., & Thummer, R. P. (2023). Gelatin-PVA-AgNPs triad composite as wound healing hydrogel with wounded skin surface protective efficiency. *Available at SSRN 4219683*.

Baig, N., Kammakakam, I., & Falath, W. (2021). Nanomaterials: A review of synthesis methods, properties, recent progress, and challenges. *Materials Advances*, *2*(6), 1821–1871.

Bajpai, A. K., Bajpai, J., Saini, R. K., Agrawal, P., & Tiwari, A. (2016). *Smart biomaterial devices: Polymers in biomedical sciences*. CRC Press.

Barroso, A., Mestre, H., Ascenso, A., Simões, S., & Reis, C. (2020). Nanomaterials in wound healing: From material sciences to wound healing applications. *Nano Select*, *1*(5), 443–460.

Baygar, T., Saraç, N., & Uğur, A. (2022). Cytotoxicity and wound healing capacity of biologically synthesized silver nanoparticles. *Mugla Journal of Science and Technology*, *8*(1), 63–68.

Bhushan, B. (2017). Introduction to nanotechnology. In *Springer handbook of nanotechnology* (pp. 1–19). Springer.

Burda, C., Chen, X., Narayanan, R., & El-Sayed, M. A. (2005). Chemistry and properties of nanocrystals of different shapes. *Chemical Reviews*, *105*(4), 1025–1102.

Chai, S. H., Wang, Y., Qiao, Y., Wang, P., Li, Q., Xia, C., & Ju, M. (2018). Bio fabrication of silver nanoparticles as an effective wound healing agent in the wound care after anorectal surgery. *Journal of Photochemistry and Photobiology B: Biology*, *178*, 457–462.

Chatterjee, A., Archana, L., Niroshinee, V., & Abraham, J. (2016). Biosynthesis of lanthanum nanoparticles using green gram seeds and their effect on microorganisms. *Research Journal of Pharmaceutical Biological and Chemical Sciences*, *7*(2), 1462–1470.

Chen, H., Lan, G., Ran, L., Xiao, Y., Yu, K., Lu, B., . . . Lu, F. (2018). A novel wound dressing based on a Konjac glucomannan/silver nanoparticle composite sponge effectively kills bacteria and accelerates wound healing. *Carbohydrate Polymers*, *183*, 70–80.

Darby, I. A., Laverdet, B., Bonté, F., & Desmoulière, A. (2014). Fibroblasts and myofibroblasts in wound heal ing. *Clinical, Cosmetic and Investigational Dermatology*, 301–311.

Das, G., Seo, S., Yang, I. J., Nguyen, L. T. H., Shin, H. S., & Patra, J. K. (2023). Synthesis of biogenic gold nanoparticles by using sericin protein from Bombyx mori silk cocoon and investigation of its wound healing, antioxidant, and antibacterial potentials. *International Journal of Nanomedicine*, 17–34.

Das, S., & Dhar, B. B. (2014). Green synthesis of noble metal nanoparticles using cysteine-modified silk fibroin: Catalysis and antibacterial activity. *RSC Advances*, *4*(86), 46285–46292.

Das, V. L., Thomas, R., Varghese, R. T., Soniya, E. V., Mathew, J., & Radhakrishnan, E. K. (2014). Extracellular synthesis of silver nanoparticles by the Bacillus strain CS 11 isolated from industrialized area. *3 Biotech*, *4*, 121–126.

de Oliveira, R. C., & Wilson, S. E. (2020). Fibrocytes, wound healing, and corneal fibrosis. *Investigative Ophthalmology & Visual Science*, *61*(2), 28–28.

Deepak, P., Amutha, V., Birundha, R., Sowmiya, R., Kamaraj, C., Balasubramanian, V., . . . Perumal, P. (2018). Facile green synthesis of nanoparticles from brown seaweed Sargassum wightii and its biological application potential. *Advances in Natural Sciences: Nanoscience and Nanotechnology*, *9*(3), 035019.

Dhamecha, D., Jalalpure, S., & Jadhav, K. (2016). Nepenthes khasiana mediated synthesis of stabilized gold nanoparticles: Characterization and biocompatibility studies. *Journal of Photochemistry and Photobiology B: Biology*, *154*, 108–117.

Dhapte, V., Kadam, S., Moghe, A., & Pokharkar, V. (2014). Probing the wound healing potential of biogenic silver nanoparticles. *Journal of Wound Care*, *23*, 431–436.

Dzimitrowicz, A., Jamróz, P., Sergiel, I., Kozlecki, T., & Pohl, P. (2019). Preparation and characterization of gold nanoparticles prepared with aqueous extracts of Lamiaceae plants and the effect of follow-up treatment with atmospheric pressure glow microdischarge. *Arabian Journal of Chemistry*, *12*(8), 4118–4130.

El-Ela, F. I. A., Farghali, A. A., Mahmoud, R. K., Mohamed, N. A., & Moaty, S. A. (2019). New approach in ulcer prevention and wound healing treatment using doxycycline and amoxicillin/LDH nanocomposites. *Scientific Reports*, *9*(1), 6418.

El-Sawah, A. A., El-Naggar, N. E. A., Eldegla, H. E., & Soliman, H. M. (2024). Green synthesis of collagen nanoparticles by Streptomyces xinghaiensis NEAA-1, statistical optimization, characterization, and evaluation of their anticancer potential. *Scientific Reports*, *14*(1), 3283.

Fitridge, R., & Thompson, M. (2011). *Mechanisms of vascular disease: A reference book for vascular specialists* (p. 587). University of Adelaide Press.

Franceschinis, G., Beverina, M., Corleto, M., Sosa, A. M., Lillo, C., Casará, L. A., . . . Calienni, M. N. (2023). Green-synthesized silver nanoparticles using Aloe maculata extract as antibacterial agent for potential topical application. *OpenNano*, *12*, 100148.

Garcia-Orue, I., Gainza, G., Gutierrez, F. B., Aguirre, J. J., Evora, C., Pedraz, J. L., . . . Igartua, M. (2017). Novel nanofibrous dressings containing rhEGF and Aloe vera for wound healing applications. *International Journal of Pharmaceutics*, *523*(2), 556–566.

Giri, A. K., Jena, B., Biswal, B., Pradhan, A. K., Arakha, M., Acharya, S., & Acharya, L. (2022). Green synthesis and characterization of silver nanoparticles using Eugenia roxburghii DC. extract and activity against biofilm-producing bacteria. *Scientific Reports*, *12*(1), 8383.

Golebiewska, E. M., & Poole, A. W. (2015). Platelet secretion: From haemostasis to wound healing and beyond. *Blood Reviews*, *29*(3), 153–162.

Govindaraju, K., Basha, S. K., Kumar, V. G., & Singaravelu, G. (2008). Silver, gold and bimetallic nanoparticles production using single-cell protein (Spirulina platensis) Geitler. *Journal of Materials Science*, *43*(15), 5115–5122.

Hassan, H., Omoniyi, K. I., Okibe, F. G., Nuhu, A. A., & Echioba, E. G. (2020). Assessment of wound healing activity of green synthesized titanium oxide nanoparticles using strychnos spinosa and blighia sapida. *Journal of Applied Sciences and Environmental Management*, *24*(2), 197–206.

Heal, C. F., Banks, J. L., Lepper, P. D., Kontopantelis, E., & van Driel, M. L. (2016). Topical antibiotics for preventing surgical site infection in wounds healing by primary intention. *Cochrane Database of Systematic Reviews*, *11*.

Heiligtag, F. J., & Niederberger, M. (2013). The fascinating world of nanoparticle research. *Materials Today*, *16*(7–8), 262–271.

Hornyak, G. L., Tibbals, H. F., Dutta, J., & Moore, J. J. (2008). Natural nanomaterials. In *Introduction to nanoscience and nanotechnology* (pp. 695–748). CRC Press.

Iravani, S. (2011). Green synthesis of metal nanoparticles using plants. *Green Chemistry*, *13*(10), 2638–2650.

Jyoti, K., Baunthiyal, M., & Singh, A. (2016). Characterization of silver nanoparticles synthesized using Urtica dioica Linn. leaves and their synergistic effects with antibiotics. *Journal of Radiation Research and Applied Sciences*, *9*(3), 217–227.

Kakakhel, M. A., Sajjad, W., Wu, F., Bibi, N., Shah, K., Yali, Z., & Wang, W. (2021). Green synthesis of silver nanoparticles and their shortcomings, animal blood a potential source for silver nanoparticles: A review. *Journal of Hazardous Materials Advances*, *1*, 100005.

Kaliaperumal, K., Subramanian, K., Thirunavukkarasu, R., Varadharajan, R. K., Binsuwaidan, R., Alabdallah, N. M., . . . Karunakaran, R. (2023). Antibacterial wound dressing with hydrogel from chitosan and polyvinyl alcohol from the red cabbage extract loaded with silver nanoparticles. *Green Processing and Synthesis*, *12*(1), 20230035.

Keten, S., Xu, Z., Ihle, B., & Buehler, M. J. (2010). Nanoconfinement controls stiffness, strength and mechanical toughness of β-sheet crystals in silk. *Nature Materials*, *9*, 359–367.

Khatami, M., Varma, R. S., Zafarnia, N., Yaghoobi, H., Sarani, M., & Kumar, V. G. (2018). Applications of green synthesized Ag, ZnO and Ag/ZnO nanoparticles for making clinical antimicrobial wound-healing bandages. *Sustainable Chemistry and Pharmacy*, *10*, 9–15.

Krishnaswamy, K., & Orsat, V. (2017). Sustainable delivery systems through green nanotechnology. In *Nano- and microscale drug delivery systems* (pp. 17–32). Elsevier.

Landén, N. X., Li, D., & Ståhle, M. (2016). Transition from inflammation to proliferation: A critical step during wound healing. *Cellular and Molecular Life Sciences*, *73*, 3861–3885.

Leaf, M. C., Gay, J. S., Newbould, M. J., Hewitt, O. R., & Rogers, S. L. (2020). Calcareous algae and cyanobacteria. *Geology Today*, *36*(2), 75–80.

Li, S., Shen, Y., Xie, A., Yu, X., Qiu, L., Zhang, L., & Zhang, Q. (2007). Green synthesis of silver nanoparticles using Capsicum annuum L. extract. *Green Chemistry*, *9*(8), 852–858.

Li, X., Xu, H., Chen, Z. S., & Chen, G. (2011). Biosynthesis of nanoparticles by microorganisms and their applications. *Journal of Nanomaterials*, *2011*(1), 270974.

Lin, P. C., Lin, S., Wang, P. C., & Sridhar, R. (2014). Techniques for physicochemical characterization of nanomaterials. *Biotechnology Advances*, *32*(4), 711–726.

Luo, C. H., Shanmugam, V., & Yeh, C. S. (2015). Nanoparticle biosynthesis using unicellular and subcellular supports. *NPG Asia Materials*, *7*(8), e209.

MacLeod, A. S., & Kwock, J. T. (2018). Inflammation in wound repair: Role and function of inflammation in wound repair. In *Wound healing: Stem cells repair and restorations, basic and clinical aspects* (pp. 177–194). https://doi.org/10.1002/9781119282518.ch14

Maghimaa, M., & Alharbi, S. A. (2020). Green synthesis of silver nanoparticles from Curcuma longa L. and coating on the cotton fabrics for antimicrobial applications and wound healing activity. *Journal of Photochemistry and Photobiology B: Biology*, *204*, 111806.

Maneerung, T., Tokura, S., & Rujiravanit, R. (2008). Impregnation of silver nanoparticles into bacterial cellulose for antimicrobial wound dressing. *Carbohydrate Polymers*, *72*(1), 43–51.

Mansoori, G. A. (2005). *Principles of nanotechnology: Molecular-based study of condensed matter in small systems*. World Scientific.

Mori, H. M., Kawanami, H., Kawahata, H., & Aoki, M. (2016). Wound healing potential of lavender oil by acceleration of granulation and wound contraction through induction of TGF-β in a rat model. *BMC Complementary and Alternative Medicine*, *16*, 1–11.

Mukherjee, A., Sarkar, D., & Sasmal, S. (2021). A review of green synthesis of metal nanoparticles using algae. *Frontiers in Microbiology*, *12*, 693899.

Murthy, S., Gautam, M. K., Goel, S., Purohit, V., Sharma, H., & Goel, R. K. (2013). Evaluation of in vivo wound healing activity of Bacopa monniera on different wound model in rats. *BioMed Research International, 2013*(1), 972028.

Nandhini, J., Karthikeyan, E., & Rajeshkumar, S. (2024). Nanomaterials for wound healing: Current status and futuristic frontier. *Biomedical Technology, 6*, 26–45.

Naseri, N., Valizadeh, H., & Zakeri-Milani, P. (2015). Solid lipid nanoparticles and nanostructured lipid carriers: Structure, preparation and application. *Advanced Pharmaceutical Bulletin, 5*, 305–313.

Negut, I., Grumezescu, V., & Grumezescu, A. M. (2018). Treatment strategies for infected wounds. *Molecules, 23*(9), 2392.

Nehru, L., Kandasamy, G. D., Sekar, V., Alshehri, M. A., Panneerselvam, C., Alasmari, A., & Kathirvel, P. (2023). Green synthesis of ZnO-NPs using endophytic fungal extract of Xylaria arbuscula from Blumea axillaris and its biological applications. *Artificial Cells, Nanomedicine, and Biotechnology, 51*(1), 318–333.

Nguyen, H. M., Le, T. T. N., Nguyen, A. T., Le, H. N. T., & Pham, T. T. (2023). Biomedical materials for wound dressing: Recent advances and applications. *RSC Advances, 13*(8), 5509–5528.

Nimni, M. E. (Ed.). (2018). *Collagen* (pp. 1–78). CRC Press.

Olsson, M., Järbrink, K., Divakar, U., Bajpai, R., Upton, Z., Schmidtchen, A., & Car, J. (2019). The humanistic and economic burden of chronic wounds: A systematic review. *Wound Repair and Regeneration, 27*(1), 114–125.

Oryan, A., Alemzadeh, E., & Moshiri, A. (2018). Potential role of propolis in wound healing: Biological properties and therapeutic activities. *Biomedicine & Pharmacotherapy, 98*, 469–483.

Pal, K., Chakroborty, S., & Nath, N. (2022). Limitations of nanomaterials insights in green chemistry sustainable route: Review on novel applications. *Green Processing and Synthesis, 11*(1), 951–964.

Panáček, A., Kvítek, L., Smékalová, M., Večeřová, R., Kolář, M., Röderová, M., . . . Zbořil, R. (2018). Bacterial resistance to silver nanoparticles and how to overcome it. *Nature Nanotechnology, 13*(1), 65–71.

Patil, P. P., Reagan, M. R., & Bohara, R. A. (2020). Silk fibroin and silk-based biomaterial derivatives for ideal wound dressings. *International Journal of Biological Macromolecules, 164*, 4613–4627.

Periasamy, S., Jegadeesan, U., Sundaramoorthi, K., Rajeswari, T., Tokala, V. N. B., Bhattacharya, S., . . . Nellore, M. K. (2022). Comparative analysis of synthesis and characterization of silver nanoparticles extracted using leaf, flower, and bark of hibiscus rosasinensis and examine its antimicrobicidal activity. *Journal of Nanomaterials, 2022*(1), 8123854.

Phan, T. T. V., Hoang, G., Nguyen, T. P., Kim, H. H., Mondal, S., Manivasagan, P., . . . Junghwan, O. (2019). Chitosan as a stabilizer and size-control agent for synthesis of porous flower-shaped palladium nanoparticles and their applications on photo-based therapies. *Carbohydrate Polymers, 205*, 340–352.

Pomari, E., Stefanon, B., & Colitti, M. (2013). Effect of Arctium lappa (burdock) extract on canine dermal fibroblasts. *Veterinary Immunology and Immunopathology, 156*(3–4), 159–166.

Pourali, P., Baserisalehi, M., Afsharnezhad, S., Behravan, J., Ganjali, R., Bahador, N., & Arabzadeh, S. (2013). The effect of temperature on antibacterial activity of biosynthesized silver nanoparticles. *Biometals, 26*, 189–196.

Pourali, P., & Yahyaei, B. (2016). Biological production of silver nanoparticles by soil isolated bacteria and preliminary study of their cytotoxicity and cutaneous wound healing efficiency in rat. *Journal of Trace Elements in Medicine and Biology, 34*, 22–31.

Qamar, S. U. R., & Ahmad, J. N. (2021). Nanoparticles: Mechanism of biosynthesis using plant extracts, bacteria, fungi, and their applications. *Journal of Molecular Liquids, 334*, 116040.

Ragothaman, M., Villalan, A. K., Dhanasekaran, A., & Palanisamy, T. (2021). Bio-hybrid hydrogel comprising collagen-capped silver nanoparticles and melatonin for accelerated tissue regeneration in skin defects. *Materials Science and Engineering: C, 128*, 112328.

Rajapaksha, D. C., Edirisinghe, S. L., Nikapitiya, C., Dananjaya, S. H. S., Kwun, H. J., Kim, C. H., . . . De Zoysa, M. (2020). Spirulina maxima derived pectin nanoparticles enhance the immunomodulation, stress tolerance, and wound healing in zebrafish. *Marine Drugs, 18*(11), 556.

Rajeshkumar, S., Malarkodi, C., Paulkumar, K., Vanaja, M., Gnanajobitha, G., & Annadurai, G. (2014). Algae mediated green fabrication of silver nanoparticles and examination of its antifungal activity against clinical pathogens. *International Journal of Metals, 2014*(1), 692643.

Rajivgandhi, G., Gnanamangai, B. M., Prabha, T. H., Poornima, S., Maruthupandy, M., Alharbi, N. S., . . . Li, W. J. (2022). Biosynthesized zinc oxide nanoparticles (ZnO NPs) using actinomycetes enhance the anti-bacterial efficacy against K. Pneumoniae. *Journal of King Saud University-Science, 34*(1), 101731.

Ramya, S., Shanmugasundaram, T., & Balagurunathan, R. (2015). Biomedical potential of actinobacterially synthesized selenium nanoparticles with special reference to anti-biofilm, anti-oxidant, wound healing, cytotoxic and anti-viral activities. *Journal of Trace Elements in Medicine and Biology, 32*, 30–39.

Reddy, G. A. K., Joy, J. M., Mitra, T., Shabnam, S., & Shilpa, T. (2012). Nano silver–a review. *International Journal of Advanced Pharmaceutical*, *2*(1), 09–15.

Reicha, F. M., Sarhan, A., Abdel-Hamid, M. I., & El-Sherbiny, I. M. (2012). Preparation of silver nanoparticles in the presence of chitosan by electrochemical method. *Carbohydrate Polymers*, *89*(1), 236–244.

Reportlinker. Global advanced wound care products industry. 2020. www.reportlinker.com/p05961345/global-advanced-wound-care-products-industry.html?utm_source=gnw (last accessed February4, 2021).

Rumbaut, R. E. (2010). *Platelet-vessel wall interactions in hemostasis and thrombosis*. Morgan & Claypool Life Sciences.

Salem, S. S., & Fouda, A. (2021). Green synthesis of metallic nanoparticles and their prospective biotechnological applications: An overview. *Biological Trace Element Research*, *199*(1), 344–370.

Santamaria, A. (2012). Historical overview of nanotechnology and nanotoxicology. *Nanotoxicity: Methods and Protocols*, 1–12.

Santhoshkumar, J., Rajeshkumar, S., & Kumar, S. V. (2017). Phyto-assisted synthesis, characterization and applications of gold nanoparticles—a review. *Biochemistry and Biophysics Reports*, *11*, 46–57.

Savoia, D. (2012). Plant-derived antimicrobial compounds: Alternatives to antibiotics. *Future Microbiology*, *7*(8), 979–990.

Shah, M., Fawcett, D., Sharma, S., Tripathy, S. K., & Poinern, G. E. J. (2015). Green synthesis of metallic nanoparticles via biological entities. *Materials*, *8*(11), 7278–7308.

Shakeel, M., Jabeen, F., Shabbir, S., Asghar, M. S., Khan, M. S., & Chaudhry, A. S. (2016). Toxicity of nano-titanium dioxide (TiO 2-NP) through various routes of exposure: A review. *Biological Trace Element Research*, *172*, 1–36.

Shankar, S. S., Rai, A., Ahmad, A., & Sastry, M. (2004). Rapid synthesis of Au, Ag, and bimetallic Au core–Ag shell nanoparticles using Neem (Azadirachta indica) leaf broth. *Journal of Colloid and Interface Science*, *275*(2), 496–502.

Shao, F., Yang, A., Yu, D. M., Wang, J., Gong, X., & Tian, H. X. (2018). Bio-synthesis of Barleria gibsoni leaf extract mediated zinc oxide nanoparticles and their formulation gel for wound therapy in nursing care of infants and children. *Journal of Photochemistry and Photobiology B: Biology*, *189*, 267–273.

Shaw, T. J., & Martin, P. (2016). Wound repair: A showcase for cell plasticity and migration. *Current Opinion in Cell Biology*, *42*, 29–37.

Sheokand, B., Pathak, S. R., Srivastava, C. M., Kumar, A., Bahadur, I., & Vats, M. (2022). Green silver nanoparticles functionalised gelatin nanocomposite film for wound healing: Construction and characterization. *Journal of Molecular Liquids*, *368*, 120561.

Sherman, R. A. (2014). Mechanisms of maggot-induced wound healing: What do we know, and where do we go from here?. *Evidence-Based Complementary and Alternative Medicine*, *2014*(1), 592419.

Si, S., Dinda, E., & Mandal, T. K. (2007). In situ synthesis of gold and silver nanoparticles by using redox-active amphiphiles and their phase transfer to organic solvents. *Chemistry–A European Journal*, *13*(35), 9850–9861.

Singh, S., Young, A., & McNaught, C. E. (2017). The physiology of wound healing. *Surgery (Oxford)*, *35*(9), 473–477.

Singla, R., Soni, S., Kulurkar, P. M., Kumari, A., Mahesh, S., Patial, V., . . . Yadav, S. K. (2017). In situ functionalized nanobiocomposites dressings of bamboo cellulose nanocrystals and silver nanoparticles for accelerated wound healing. *Carbohydrate Polymers*, *155*, 152–162.

Srikar, S. K., Giri, D. D., Pal, D. B., Mishra, P. K., & Upadhyay, S. N. (2016). Light induced green synthesis of silver nanoparticles using aqueous extract of Prunus amygdalus. *Green and Sustainable Chemistry*, *6*(1), 26–33.

Steffy, K., Shanthi, G., Maroky, A. S., & Selvakumar, S. (2018). Potential bactericidal activity of S. nux-vomica–ZnO nanocomposite against multidrug-resistant bacterial pathogens and wound-healing properties. *Journal of Trace Elements in Medicine and Biology*, *50*, 229–239.

Tian, J., Wong, K. K., Ho, C. M., Lok, C. N., Yu, W. Y., Che, C. M., . . . Tam, P. K. (2007). Topical delivery of silver nanoparticles promotes wound healing. *ChemMedChem: Chemistry Enabling Drug Discovery*, *2*(1), 129–136.

Turabelidze, A., & Dipietro, L. A. (2011). Inflammation and wound healing. *Endodontic Topics*, *24*(1), 26–38.

Tyavambiza, C., Dube, P., Goboza, M., Meyer, S., Madiehe, A. M., & Meyer, M. (2021). Wound healing activities and potential of selected African medicinal plants and their synthesized biogenic nanoparticles. *Plants*, *10*(12), 2635.

Veeraraghavan, V. P., Periadurai, N. D., Karunakaran, T., Hussain, S., Surapaneni, K. M., & Jiao, X. (2021). Green synthesis of silver nanoparticles from aqueous extract of Scutellaria barbata and coating on the cotton fabric for antimicrobial applications and wound healing activity in fibroblast cells (L929). *Saudi Journal of Biological Sciences*, *28*(7), 3633–3640.

Venkatesham, M., Ayodhya, D., Madhusudhan, A., Veera Babu, N., & Veerabhadram, G. (2014). A novel green one-step synthesis of silver nanoparticles using chitosan: Catalytic activity and antimicrobial studies. *Applied Nanoscience*, *4*, 113–119.

Vieira, A. P., Stein, E. M., Andreguetti, D. X., Colepicolo, P., & da Costa Ferreira, A. M. (2016). Preparation of silver nanoparticles using aqueous extracts of the red algae Laurencia aldingensis and Laurenciella sp. and their cytotoxic activities. *Journal of Applied Phycology*, *28*, 2615–2622.

Whitney, J. D. (2005). Overview: Acute and chronic wounds. *Nursing Clinics*, *40*(2), 191–205.

Wolcott, R. D., Cutting, K. F., Dowd, S. E., & Percival, S. L. (2010). Types of wounds and infections. *Microbiology of Wounds*, 219–32.

Wong, M. W., Leung, P. C., & Wong, W. C. (2001). Limb salvage in extensive diabetic foot ulceration-a preliminary clinical study using simple debridement and herbal drinks. *Hong Kong Medical Journal*, *7*(4), 403–407.

Wongpreecha, J., Polpanich, D., Suteewong, T., Kaewsaneha, C., & Tangboriboonrat, P. (2018). One-pot, large-scale green synthesis of silver nanoparticles-chitosan with enhanced antibacterial activity and low cytotoxicity. *Carbohydrate Polymers*, *199*, 641–648.

Wu, Y., Zuo, F., Lin, Y., Zhou, Y., Zheng, Z., & Ding, X. (2014). Green and facile synthesis of gold nanoparticles stabilized by chitosan. *Journal of Macromolecular Science, Part A*, *51*(5), 441–446.

Yadav, E., Singh, D., Yadav, P., & Verma, A. (2018). Ameliorative effect of biofabricated ZnO nanoparticles of Trianthema portulacastrum Linn. on dermal wounds via removal of oxidative stress and inflammation. *RSC Advances*, *8*(38), 21621–21635.

Yassin, M. T., Mostafa, A. A. F., Al-Askar, A. A., & Al-Otibi, F. O. (2022). Facile green synthesis of silver nanoparticles using aqueous leaf extract of Origanum majorana with potential bioactivity against multidrug resistant bacterial strains. *Crystals*, *12*(5), 603.

Younis, N. S., El Semary, N. A., & Mohamed, M. E. (2021). Silver nanoparticles green synthesis via cyanobacterium Phormidium sp.: Characterization, wound healing, antioxidant, antibacterial, and anti-inflammatory activities. *European Review for Medical & Pharmacological Sciences*, *25*(7).

Yu, K., Lu, F., Li, Q., Chen, H., Lu, B., Liu, J., . . . Lan, G. (2017). In situ assembly of Ag nanoparticles (AgNPs) on porous silkworm cocoon-based wound film: Enhanced antimicrobial and wound healing activity. *Scientific Reports*, *7*(1), 2107.

Zangeneh, M. M., Ghaneialvar, H., Akbaribazm, M., Ghanimatdan, M., Abbasi, N., Goorani, S., . . . Zangeneh, A. (2019). Novel synthesis of Falcaria vulgaris leaf extract conjugated copper nanoparticles with potent cytotoxicity, antioxidant, antifungal, antibacterial, and cutaneous wound healing activities under in vitro and in vivo condition. *Journal of Photochemistry and Photobiology B: Biology*, *197*, 111556.

Zhao, Z., Li, Y., & Xie, M. B. (2015). Silk fibroin-based nanoparticles for drug delivery. *International Journal of Molecular Sciences*, *16*(3), 4880–4903.

Zou, L., Zhu, F., Long, Z. E., & Huang, Y. (2021). Bacterial extracellular electron transfer: A powerful route to the green biosynthesis of inorganic nanomaterials for multifunctional applications. *Journal of Nanobiotechnology*, *19*(1), 12.

Index

K

L

M

N

O

P

R

S

T

U

V

W

Z